Zagros • Hindu Kush • Himalaya Geodynamic Evolution

Geodynamics Series

The Final Reports of the International Geodynamics Program sponsored by the Inter-Union Commission on Geodynamics.

Zagros • Hindu Kush • Himalaya Geodynamic Evolution

Edited by H. K. Gupta
F. M. Delany

Geodynamics Series
Volume 3

American Geophysical Union
Washington, D.C.

Geological Society of America
Boulder, Colorado

1981

Final Report of Working Group 6, Geodynamics of
the Alpine-Himalaya Region, East, coordinated by
Frances M. Delany on behalf of the Bureau of the
Inter-Union Commission on Geodynamics

American Geophysical Union, 2000 Florida Avenue, N.W.
  Washington, DC 20009

Geological Society of America, 3300 Penrose Place; P.O. Box 9140
  Boulder, Colorado 80301

Library of Congress Cataloging in Publication Data

Main entry under title:

Zagros, Hindu Kush, Himalaya, geodynamic evolution.

   (Geodynamics series; v. 4)
   Includes bibliographies.
   1. Geodynamics. 2. Geology--Iran--Zagros
Mountains. 3. Geology--Hindu Kush Mountains
(Afghanistan and Pakistan) 4. Geology--Himalaya
Mountains. I. Delany, (Frances M.) II. Gupta,
(Harsh K.), 1942-   III. Series.
QE501.Z26   551.4'32'0954   81-15014
ISBN 0-87590-507-2   AACR2

Printed in the United States of America

CONTENTS

After a decade of intense and productive scientific cooperation between geologists, geophysicists and geochemists the International Geodynamics Program formally ended on July 31, 1980. The scientific accomplishments of the program are represented in more than seventy scientific reports and in this series of Final Report volumes.

The concept of the Geodynamics Program, as a natural successor to the Upper Mantle Project, developed during 1970 and 1971. The International Union of Geological Sciences (IUGS) and the International Union of Geodesy and Geophysics (IUGG) then sought support for the new program from the International Council of Scientific Unions (ICSU). As a result the Inter-Union Commission on Geodynamics was established by ICSU to manage the International Geodynamics Program.

The governing body of the Inter-Union Commission on Geodynamics was a Bureau of seven members, three appointed by IUGG, three by IUGS and one jointly by the two Unions. The President was appointed by ICSU and a Secretary-General by the Bureau from among its members. The scientific work of the Program was coordinated by the Commission, composed of the Chairmen of the Working Groups and the representatives of the national committees for the International Geodynamics Program. Both the Bureau and the Commission met annually, often in association with the Assembly of one of the Unions, or one of the constituent Associations of the Unions.

Initially the Secretariat of the Commission was in Paris with support from France through BRGM, and later in Vancouver with support from Canada through DEMR and NRC.

The scientific work of the Program was coordinated by ten Working Groups.

WG 1 Geodynamics of the Western Pacific-Indonesian Region

WG 2 Geodynamics of the Eastern Pacific Region, Caribbean and Scotia Arcs

WG 3 Geodynamics of the Alpine-Himalayan Region, West

WG 4 Geodynamics of Continental and Oceanic Rifts

WG 5 Properties and Processes in the Earth's Interior

WG 6 Geodynamics of the Alpine-Himalayan Region, East

WG 7 Geodynamics of Plate Interiors

WG 8 Geodynamics of Seismically Inactive Margins

WG 9 History and Interaction of Tectonic, Metamorphic and Magmatic Processes

WG 10 Global Syntheses and Paleoreconstruction

These Working Groups held discussion meetings and sponsored symposia. The papers given at the symposia were published in a series of Scientific Reports. The scientific studies were all organized and financed at the national level by national committees even when multinational programs were involved. It is to the national committees, and to those who participated in the studies organized by those committees, that the success of the Program must be attributed.

Financial support for the symposia and the meetings of the Commission was provided by subventions from IUGG, IUGS, UNESCO and ICSU.

Information on the activities of the Commission and its Working Groups is available in a series of 17 publications: Geodynamics Reports, 1-8, edited by F. Delany, published by BRGM; Geodynamics Highlights, 1-4, edited by F. Delany, published by BRGM; and Geodynamics International, 13-17, edited by R. D. Russell. Geodynamics International was published by World Data Center A for Solid Earth Geophysics, Boulder, Colorado 80308, USA. Copies of these publications, which contain lists of the Scientific Reports, may be obtained from WDC A. In some cases only microfiche copies are now available.

This volume is one of a series of Final Reports summarizing the work of the Commission. The Final Report volumes, organized by the Working Groups, represent in part a statement of what has been accomplished during the Program and in part an analysis of problems still to be solved. This volume from Working Group 6 (Chairman, Hari Narain) was edited by H. Gupta and F. Delany.

At the end of the Geodynamics Program it is clear that the kinematics of the major plate movements during the past 200 million years is well understood, but there is much less understanding of the dynamics of the processes which cause these movements.

Perhaps the best measure of the success of the Program is the enthusiasm with which the Unions and national committees have joined in the establishment of a successor program to be known as: Dynamics and evolution of the lithosphere: The framework for earth resources and the reduction of the hazards.

To all of those who have contributed their time so generously to the Geodynamics Program we tender our thanks.

C. L. Drake, President ICG, 1971-1975

A. L. Hales, President ICG, 1975-1980

Members of Working Group 6:

H. Narain
S. Abdullah
N.N. Ambraseys
P. Bordet
A.R. Crawford
J. Eftekharnezhad
A. Gansser
H.K. Gupta

A. Marussi
A.A. Nowroozi
D.K. Ray
Ibrahim Shah
J. Stocklin
R. Tahirkheli
M. Takin
J.M. Tater

PREFACE

The International Geodynamics Project focussed attention on processes within the earth responsible for the movement of the lithospheric blocks. At any one time, strong tectonic activity appears limited to a few mobile belts. Most of the present-day seismic activity is confined to the Circum-Pacific belt, the Alpide belt and the mid-oceanic ridges. These belts include oceanic and continental rift systems, the island arcs and young folded mountains. Continent to continent collision of the Eurasian and the Indian plates is generally believed to be responsible for the origin of the Himalaya, the tectonics of this region and the neighbouring south and central Asia. To focus attention on geodynamic problems in this relatively much less known Alpine-Himalayan region bounded by Iran in the West and Burma in the East, the Inter-Union Commission on Geodynamics formed a separate Working Group 3b under the Chairmanship of Hari Narain. Later, in 1975, this Working Group 3b on 'Geodynamics of the Alpine-Himalayan region, East' was given independant status and re-numbered as Working Group 6.

The first meeting of the Working Group was convened at the National Geophysical Research Institute, Hyderabad in March 1973 where a number of general and specific recommendations were made. The general recommendations included early publication of geological reports and maps for wider circulation, invitations to countries which included the Himalayan chain of mountains to participate in the project and the programme of the Working Group 3b, use of ERTS imagery and aerial photographs for mapping of large- and small-scale geological features. It was recommended that data obtained and interpretation carried out in marine research be properly integrated and correlated with the data obtained on the continental areas of the Himalayan belt. It was further recommended that field colloquia be held in different parts of the region to study and compare the critical areas and stimulate exchange of information and ideas. Considering the great geographic extent of the wider Himalayan belt and the variable degree of precision of information on its different sectors, specific recommendations included multidisciplinary studies along critical zones and in all, ten geotraverses, beginning in the Zagros thrust zone in the West, to the Arakan Yoma in the East, were identified. Additionally, investigations using specific methods were recommended.

During the following years, the Working Group 6 met on an average once a year, reviewed progress and defined ways and means of making its work more effective and purposeful. Minutes of these Working Group meetings were published in the ICG Reports and other publications and widely distributed.

During the period 1973-1979, the following symposia were organized:

- Seminar on Geodynamics of the Himalayan Region, held at Hyderabad from March 5 to 8, 1973. The proceedings were published by the National Geophysical Research Institute, Hyderabad (ICG Scientific Report Series No 5, edited by Harsh K. Gupta, 221 pages).

- International Colloquium on the Geotectonics of the Kashmir Himalaya-Karakorum-Hindu Kush-Pamir Orogenic Belts, held in Rome from June 25 to 27, 1974. The colloquium was convened by the Accademia dei Lincei, Rome and the proceedings were published by them (Atti dei Convegni Lincei, 21, 1976, 314 pages).

- Symposium on the Geodynamics of Southeast Asia, held in Tehran from Sept. 8 to 15, 1975. The symposium was convened by the Geological Survey of Iran and the proceedings published in a special publication by them (ICG Scientific Report Series, No 18). The symposium was followed by very successful field trips to Mashshad, Kopet Dagh and Gonabad (NE Iran) and to Tabriz and Rezaiyeh (NW Iran).

- Himalayan Geology Seminar, held in New Delhi from Sept 13 to 17, 1976. The symposium was convened by the Geological Survey of India and was followed by a number of field trips. The proceedings have been published by the Geological Survey of India in seven volumes (G.S.I. Miscellaneous Publication, No 41, 1979).

- Symposium on Geodynamics of the Central Himalaya held at Kathmandu from March 26 to 31, 1978. The symposium was convened by the Geological Survey of Nepal. The proceedings are published in Tectonophysics (ICG Scientific Report Series, No 47), edited by J.M. Tater. (Tectonophysics, 62, No 1-2, 1980, 164 pages). The symposium was followed by a very interesting field trip to the Kali Gandaki valley.

- Symposium on the Geodynamics of Pakistan, held at Peshawar from November 23 to 29, 1979. The symposium was convened by the National Center of Excellence in Geology, University of Peshawar and the proceedings were edited by R.A. Tahirkheli, M.Q. Jan and M. Majid (Special issue of the Geological Bulletin, University of Peshawar, Vol. 13, 1980, 213 pages). An excellent geological excursion was arranged along the newly opened Karakorum highway after the symposium.

This final report of the Working Group includes in all seventeen technical papers. Papers in the first part (Geology) deal with regional and local field geology, give new data extending from Iran (Berberian, Stocklin) to Afghanistan (Abdulla), to the Central Himalaya (Lefort, Valdiya, Sinha) and include a summary report from the Geological Survey of India on their work in the Indian Himalaya (Krishnaswamy). Detailed up-to-date geological maps of Afghanistan and the Central Himalaya (Nepal and Kumaon) are included in the volume. The general tectonics of the entire region have been reviewed by Gansser. In the Geochronology-geochemistry section, Crawford gives up-to-date information on isotopic dating of the Himalayan rocks and Rao discusses the geochemistry of an undisturbed ophiolite sequence from the Kargil-Dras sector of the Kashmir Himalaya. In the third section (Geophysics), Chandra reviews focal mechanism solutions and their tectonic implications for the entire region. Seeber addresses the problems of seismicity and continental subduction along the Chaman and subsidiary faults. Kaila uses deep seismic sounding to interpret the seismotectonics of central Asia. Gupta has reviewed the surface wave dispersion and attentuation studies for the entire region and Singh has estimated the stresses in the northern part of the Indian shield. Klootwijk debates the northern extent of greater India on the basis of the paleomagnetic data.

Both the tectonics and the logistics make this a very difficult area in which to work. The success achieved by our Working Group in its deliberations has been due to untiring efforts of the Geological Surveys of the countries involved, and various research institutes and individual scientists from the countries within the area and outside. Continued support of the Geodynamics Bureau, especially in organizing international conferences and field colloquia has been very helpful. We thank one and all who have directly or indirectly helped us all these years.

Hari Narain
Chairman, W.G. 6

Harsh K. Gupta
Secretary, W.G. 6

TECTONO-PLUTONIC EPISODES IN IRAN

F.Berberian and M.Berberian

Department of Earth Sciences, University of Cambridge,
Cambridge, England

Abstract. Data on the extensive plutonism which
has occurred in the Iranian continental crust has
been compiled and critically reviewed in an
attempt to interpret their geological evolution.
Notwithstanding many assumptions due to scarcity
of data, eight tectono-plutonic phases are
recognized from Precambrian to Pliocene, which are
separated by relatively stable intervals. The
elongated belt of the Precambrian calc-alkaline
intrusives in east Central Iran seems to be
generated in an 'island-arc environment' above an
ocean-ocean subduction zone. The three phases of
the Mesozoic plutonic episodes are mainly found
along and above the active continental margin of
southwestern Central Iran. They represent parts
of an 'Andean-type magmatic-arc' formed by partial
melting of mantle and or subducting Tethyan
oceanic crust underneath southwestern Central
Iran. Intrusion of the late-Cretaceous batholith
above the Makran subduction zone in southeast Iran
indicates that the subduction of the Oman oceanic
crust was well established prior to Cretaceous
time. However, there are three phases of plutonic
activity of Tertiary age, of which the
petrogeneses and evolution need more detailed
study to solve the existing controversy on the
time of the Arabian-Iranian continental collision.
The late Eocene plutonic belt shifted inland, away
from the Mesozoic magmatic-arc belt, and then
shifted back to a new position along the active
continental margin during Oligo-Miocene time.
Detailed isotopic and trace-major element studies
of the Oligo-Miocene plutonic rocks indicated that
the magmas were derived from melting of a mantle
or oceanic crust source. Therefore the Oligo-
Miocene Andean-type plutonic belt suggests that
the subduction was active beneath Central Iran
during Paleogene time and the Arabian-Central
Iranian collision might have taken place during
the Miocene.

Introduction

The intrusive rocks of Iran have been little
studied in detail but the preliminary
reconnaissance level investigations are scattered
among various publications. Trezel [in Jaubert
1821] seems to be the first investigator
mentioning granites in the area southwest of the
Caspian sea. On the geological map of Iran [NIOC
1959] the intrusive rocks were divided into acid,
intermediate, basic and ultrabasic units without
regard to their stratigraphic position. Stocklin
[1972] divided the Iranian granites into five
groups. On the tectonic map of Iran [Stocklin and
Nabavi 1973] the intrusive rocks were divided into
three groups: Precambrian, Mesozoic and Tertiary.
The geological map of Iran [Huber 1978] also gave
limited information on the stratigraphic position
of the intrusive bodies. None of the previous
studies analysed the tectonic setting of the
igneous activity, which was then largely unknown.

In order to deduce the major Iranian tectono-
plutonic episodes and their geological evolution,
the existing scattered data in literature are
critically reviewed and compiled. This review was
supported where necessary by our own observation
and interpretation, and has resulted in maps of
the intrusive rocks for each plutonic episode,
which are presented here for the first time. After
a brief review of the plutonic rocks of each
episode, a short discussion on the possible
mechanisms responsible for their development and
evolution is given. The discussion presented in
this paper is mainly based on geological observations, and indicates a complex history of plutonic
emplacement, compositional variations and possible
mixture of different genetic types. Most plutons
are composite masses, with clear-cut intrusive
contacts between rocks of different composition.
Due to lack of geochemical studies, radiometric
age dating and detailed mapping of the Iranian
plutonic bodies, this study cannot suffice to
unravel the evolution and genesis of the
constituent magmas. However, this attempt which
resulted in establishment of tectono-plutonic
episodes, their siting, distribution and relationship to the major Iranian tectono-sedimentary
units, is a basic and necessary prerequisite for
further studies. Due to the fragmentary nature of
the data, any analysis must be largely speculative
but it is offered in the absence of a detailed
account.

Tectono-Sedimentary Units Of Iran

To simplify the complicated geoology and tectonics of Iran, the country can be divided into two marginal active fold-thrust mountain belts located in NE (Kopeh Dagh) and in SW (Zagros) Iran, resting on the Hercynian Turan and the Precambrian Arabian plates respectively. Between these marginal fold belts, are the Central Iran,

Alborz, Zabol-Baluch and Makran units (Fig. 1). No intrusive body is exposed in the Kopeh Dagh and the Zagros marginal fold-thrust belts in Iran.

Kopeh Dagh Fold-Thrust Belt in NE Iran

The northeastern active fold belt of Iran, the Kopeh Dagh, is formed on the Hercynian

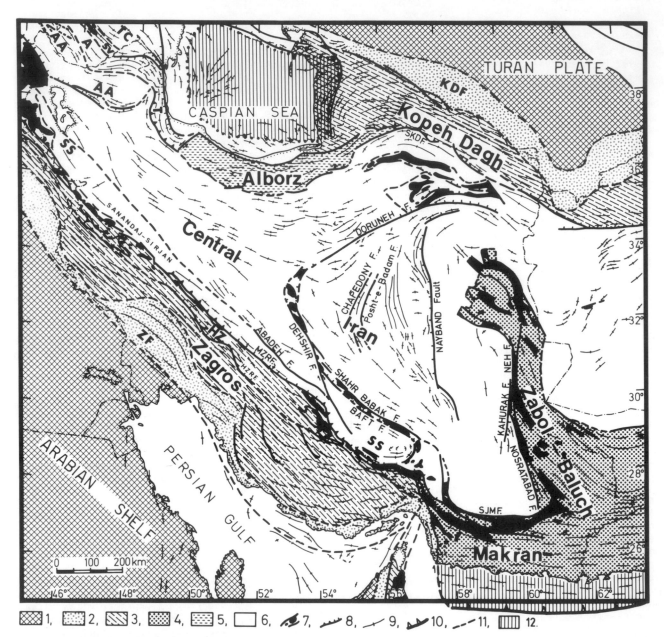

Fig. 1. Iranian major tectono-sedimentary units.
1 - Stable areas: Arabian Precambrian platform in the SW and Turanian Hercynian plate in the NE. The low dipping, relatively flat lying beds south and southwest of the Persian Gulf comprise the Arabian shelf over the buried Precambrian stable shield. 2 and 3 - Marginal mobile and active fold-thrust belts peripheral to the stable areas (Zagros and High Zagros -HZ- in the SW, and Kopeh

Dagh in the NE) indicating Mesozoic–Tertiary intracratonic subsiding sedimentary basins along several deep seated multi-role longitudinal faults. 2 – Neogene-Quaternary foredeeps, transitional from unfolded forelands to marginal fold zones, with strong late Alpine subsidence. ZF: Zagros Foredeep in the SW; KDF: Kopeh Dagh Foredeep in the NE. 3 – Main sector of the marginal active fold-thrust belt. 4 – Zabol-Baluch (east Iran) and Makran (southeast Iran) post-ophiolite flysch troughs. Late Tertiary seaward accretion and landward under-thrusting seem to be responsible for the formation of the present Makran ranges. 5 – Alborz Mountains, bordering southern part of the Caspian Sea. 6 – Central Iranian Plateau (Central Iran) lying between the two marginal active fold belts. In northwestern part of the country, Central Iran joins the Transcaucasian early Hercynian Median Mass (TC), the Sevan-Akera ophiolite belt (SV), and the Little Caucasus (A: Armenian (Miskhan-Zangezurian) Late Hercynian belt with a possible continuation to the Iranian Talesh Mountains (T) along the western part of the Caspian Sea; AA: the Araxian-Azarbaijan zone of the Caledonian consolidation, with the Vedi (V) ophiolite belt). SS: Sanandaj-Sirjan belt, a narrow intracratonic mobile belt (during Paleozoic) and active continental margin (Mesozoic), forming the southern margin of the Central Iran in contact with the Main Zagros reverse fault (MZRF). The belt bears the imprints of several major crustal upheavals (severe tectonism, magmatism and metamorphism). The Central Iranian province joins Central Afghanistan in the east. 7 – Postulated Upper Cretaceous High Zagros-Oman ophiolite-radiolarite (75 Ma) and the Central Iranian ophiolite-melange belts (65 Ma), with outcrops indicated in black. The southeastern parts of the Middle Cretaceous (110 Ma) Sevan-Akera and Vedi ophiolites of the Little Caucasus are shown in the NW of the country. The extensive belts of ophiolites mark the original zone of convergence between different blocks. The position of ophiolites are modified by post emplacement convergent movements. 8 – Major facies dividing basement faults, bordering different tectono- sedimentary units. Contrasting tectono-sedimentary regimes, belts of ophiolites and associated oceanic sediments, together with paleogeographic contrasts along the Main Zagros (MZRF) and the High Zagros (HZRF) reverse faults in southwest, and the South Kopeh Dagh fault (SKDF) in northeast, indicate existence of old geosutures along these lines. The Chapedony and Posht-e-Badam faults delineate the possible Precambrian island-arc in eastern Central Iran. SJMF: South Jaz Murian Fault in SE. 9 – Late Alpine fold axes. 10 – Active subduction zone of Makran in the Gulf of Oman. 11 – Province boundary. 12 – Areas with oceanic crust (south Caspian and Gulf of Oman).
Based on M. Berberian [this volume]. Lambert Conformal Conic Projection.

---

metamorphosed basement at the southwestern margin of the Turan platform (Fig. 1). The belt is composed of about 10 km of Mesozoic-Tertiary sediments (mostly carbonates) and, like the Zagros, was folded into long linear NW-SE trending folds during the last phase of the Alpine orogeny, in Plio-Pleistocene time. No magmatic rocks are exposed in Kopeh Dagh except for some Triassic basic dykes.

Zagros Fold-Thrust Belt in SW Iran

The southwestern marginal active fold belt of Iran, the Zagros, is formed on the northeastern margin of the Arabian continental crust (Fig. 1). The geological history of the belt is simply marked by relatively quiet sedimentation (with some minor disconformities) continuing from late Precambrian to Miocene time. The sedimentation was of platform-cover type in the Paleozic, miogeosynclinal from Middle Triassic to Miocene, and synorogenic with conglomerates in late Miocene-Pleistocene times [James and Wynd 1965, Stocklin 1968a, Berberian 1976]. The belt was folded during Plio-Pleistocene orogenic movements. Its northeastern margin (the High Zagros) had undergone an earlier deformation during late Cretaceous times. Several salt horizons decouple the above structures from those at depth, and the

upper Precambrian salt deposits (the Hormoz Salt) [Stocklin 1968b, 1972] are intruded the folded beds as salt plugs.

Although the suture zone between the Zagros active fold-thrust belt and Central Iran lies along the present Main Zagros reverse fault line [Berberian and King 1981], Alavi [1980] suggested that it lies to the northeast of the Sanandaj-Sirjan belt of southwest Central Iran (see fig.1).

No in situ intrusive body is exposed in the Zagros belt, but some Upper Precambrian granite, granodiorite, tonalite, and gabbro are brought up by salt domes [Kent 1970, 1979].

Central Iran and Alborz

The Central Iran and Alborz areas are situated between the two marginal fold-thrust belts, but unlike them have undergone several major orogenic phases (Fig. 1). The zone is characterized by several episodes of syntectonic regional metamorphism and magmatism especially during the late Paleozoic (Hercynian), Middle Triassic, late Jurassic and late Cretaceous orogenic phases.

Zabol-Baluch and Makran

The Zabol-Baluch in east and the Makran in southeast Iran, are post-Cretaceous flysch-molasse belts which join together in southeast Iran and continue to the Pakistan Baluchestan ranges (Fig. 1). It is believed that the flysch sediments were deposited on the Upper Cretaceous ophiolite-melange basement with a basal conglomerate [Huber 1978].

Tectono-Plutonic Episodes

Based on stratigraphy and the few radiometric ages available, eight plutonic episodes have been recognized and defined in this study. Unfortunately the ages of the Iranian plutonic rocks, as well as their petrochemical characters, are not well established. Detailed petrochemical analysis and mapping of the rocks may assist in deciding a mechanism for the formation of the magmatic episodes through crustal anatexis, deformation of primitive mantle and/or descending ocean slabs.

Precambrian Intrusives

The Precambrian rocks in Iran were folded, metamorphosed, granitized and finally uplifted by the late Precambrian orogenic movements. There is evidence that during the Upper Precambrian, Iran and Arabia were possibly united and formed a continuous continental crust in the northeastern part of Gondwanaland [Stocklin 1974, 1977, Berberian and King 1981]. Therefore there should be similarities between the 'basement consolidation' and cratonization of Iran and Arabia. The Precambrian shield of Arabia (a stable block composed chiefly of Precambrian metamorphic and igneous rocks) which dips at a very low angle

TABLE 1. Orogenic events and plutonic rocks in Arabia with their possible equivalents in Iran (modified after Brown 1972, Schmidt et al. 1973 and 1978, Greenwood et al. 1976, Fitch 1978, Kroner et al. 1979)

HIJAZ (Pan African) TECTONIC CYCLE	Orogenic Events	Episode	Orogeny	Arabia	Iran
				post Najd Granite, Granodiorite and subordinate Syenite	Alkali Granite
	NAJD Faulting — 510–540 Ma				
				Granite–Granodiorite	Granite Diorite, Granite
	BISHAH Orogeny — 550 Ma				
		3rd Episode		Granite, Quartz Monzonite, Migmatite and Folded Granite (550–600 Ma)	Anatectic Granite, Migmatite
			Yafikh Orogeny		?
		2nd Episode		Quartz Monzonite, Diorite, Tonalite (720–785 Ma)	?
			Ranyah Orogeny		
				2nd Dioritic Series (800 Ma)	?
			Aqiq Orogeny		
		1st Episode		1st Dioritic Series (960 Ma)	?
	ASIR Tectonism — 1050 Ma				
	BASEMENT Tectonism				?
					?

northeastward below the sedimentary cover, continues to dip below the Zagros active fold-thrust belt in SW Iran. Apparently, because of the small amount of rubidium in the Precambrian intrusive rocks and the repeated orogenic movements, the few attempts to date the Iranian basement mainly by Rb/Sr total rock techniques have failed [Crawford 1977]. Hence, the identification of various plutonic episodes discovered in Arabia (Table 1) and the establishment of the nature of the crust upon which the Upper Precambrian sediments were deposited, has not been possible so far in Iran.

Several exposed plutonic bodies in Central Iran and Alborz are assigned a Precambrian age by their stratigraphic positions below the late Precambrian and Cambrian sediments. These plutonic bodies are usually intruded into the Upper Precambrian metamorphic rocks. In this study the Precambrian intrusive rocks of Iran are divided into:
- anatectic granite-diorite and migmatite,
- calc-alkaline, and
- alkaline groups.

Precambrian Anatectic Granite-Diorite and Migmatite

The Chapedony metamorphic complex exposed in eastern Central Iran, consists of gneiss, anatectic granite and diorite, migmatite, amphibolite and marble. The complex is assumed to be stratigraphically the lowest of the metamorphic complexes of Central Iran. It is metamorphosed in the highest subfacies of the amphibolite facies and its lower part contains migmatites and anatectites. The anatectic granites are mainly coarse grained and contain oriented alkali feldspar (20-60 %), plagioclase (An 20-27, 30-60 %), quartz and biotite (0-20 %). A little clinopyroxene or hornblende is present, but no orthopyroxene has been found. The granodiorites and diorites contain hornblende and plagioclase (An 60-40) with rare alkali feldspars [Haghipour and Pelissier 1968, Hushmandzadeh 1969, Stocklin 1972, Haghipour 1974, 1978, Haghipour et al. 1977]. Whole-rock Rb/Sr dating of the anatectic granite gives ages of 541 and 505 Ma [Haghipour 1974]. There is some uncertainty about this age and it may possibly reflect a metamorphic imprint.

Precambrian Calc-Alkaline Intrusives

The late Precambrian calc-alkaline intrusives ranging from gabbro to granite, are emplaced in Upper Precambrian rocks (Fig. 2). Only one small exposure of a pyroxenite of this age has been reported.

Pyroxenite. A small body of pyroxenite, mainly composed of augite (70 %) and interstitial olivine (30 %) partly altered to serpentine, is reported from a tectonically disturbed zone of the Posht-e-Badam area (east of Chapedony in Figs. 1 and 2) in Central Iran [Haghipour 1974, 1978, Haghipour et al. 1977]. The pyroxenite and serpentine lenses

are mixed with amphibolites, mica schists and metagraywackes.

Gabbro. Upper Precambrian gabbros and norites forming small intrusive bodies in the Precambrian rocks are exposed in the Chapedony area in Central Iran (Fig. 2). They consist essentially of plagioclase (An 55-60), olivine, augite, orthopyroxene (hypersthene, enstatite), amphibole and biotite [Haghipour 1974, 1978, Haghipour et al. 1977].

Diorite. The diorite in the Chapedony area contains plagioclase (An 35-50, 40-60 %), amphibole (10-35 %), biotite (5-15 %). Quartz diorite with 10-20 % quartz is also present. The diorites are cut by doleritic dykes, lamprophyres, pegmatites and aplites.

Granite, Granodiorite. Upper Precambrian calc-alkaline granites and granodiorites are reported from two localities in Central Iran at Kuh-e-Sefid (Chapedony) and Muteh (Fig. 2):

Sefid Calc-Alkaline Granite-Granodiorite. The Sefid calc-alkaline granodiorites of the Chapedony area (Fig. 2) contain plagioclase (oligoclase-andesine, 40-50 %), hornblende (5-25 %), biotite (5-25 %), quartz (10-25 %) and alkali feldspars (5-15 %, mostly orthoclase but some carry microcline) [Haghipour 1974, 1978, Haghipour et al. 1977]. Similar granodiorite is exposed in the Kalmard area of Tabas in eastern Central Iran [Aghanabati 1977]. The rock consists of plagioclase (albite-oligoclase 38 %), quartz (23 %), biotite (22 %), K-feldspar (13 %) and amphibole (4 %).

The Sefid calc-alkaline granite is a light coloured biotite granite and consists of quartz (20-25 %), alkali feldspars (mostly microcline, rarely orthoclase; 50-60 %), plagioclase (An 15-30, 10-20 %) and biotite (5 %). It seems that in the Chapedony area of Central Iran, the Sefid granite with some pegmatites represent the last phase of the late Precambrian calc-alkaline intrusive activity. An Rb/Sr absolute age of 681 Ma (whole-rock analysis) is given for the Sefid granite. A biotite of the same sample gave an age of 318 Ma, which may be due to partial recrystallization of the rock in later deformational phase(s) which altered the biotite to chlorite. Almandine-rich garnet, epidote, clinozoisite, sericite, sphene, apatite, chlorite and opaque minerals are found in the Sefid intrusives as secondary or accessory minerals [Haghipour 1974, 1978, Haghipour et al. 1977].

Muteh Calc-Alkaline Granite. This is a biotite-muscovite granite exposed in the Muteh and Hassan Robat area of the Golpaygan region of Central Iran (Fig. 2). It has invaded the Golpaygan Muteh metamorphics and seems to be older than the Upper Precambrian Soltanieh dolomite. The granite is intersected by aplitic veins [Thiele et al. 1968].

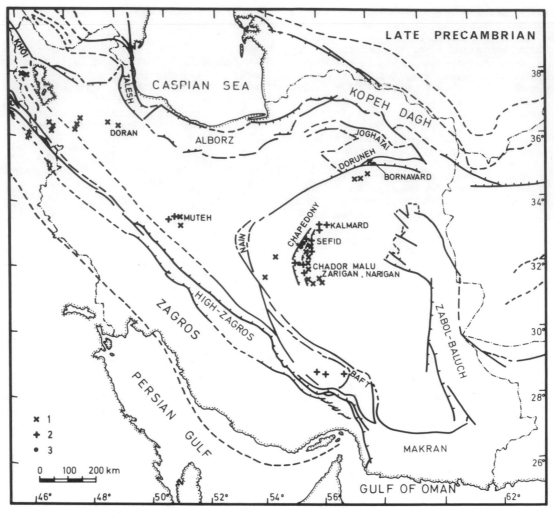

Fig. 2. Precambrian intrusive rocks exposed at the surface.
Major deep seated basement faults limiting the Iranian tectono-sedimentary units are shown by thick lines (cf. Fig. 1). Chapedony and Posht-e-Badam faults (cf. Fig. 1) presumably delineate the possible Precambrian fossil island-arc in east-Central Iran (for discussion see the text). No Precambrian intrusives are known from the Mesozoic ophiolite belts of High Zagros, Khoi, Doruneh-Joghati, Nain-Baft, Zabol-Baluch and Makran (cf. Fig. 1).
1 - Alkali granites, 2 - Calc-alkaline intrusives (gabbro to granite), 3 - Anatectic granite and diorite.
Principal source of data by which the map is constructed are given in the text. Lambert Conformal Conic Projection.

Late Precambrian Alkali Granite

Different names are given to the late Precambrian-Cambrian alkali granites in Iran, but their petrographic characters and stratigraphic positions are closely comparable to each other (Fig. 2). The rocks of this group usually lack mafic minerals and form intrusive bodies in the late Precambrian metamorphic rocks. They can be roughly divided into two types, Doran and Zarigan (Fig. 2).

Doran Type late Precambrian Alkali Granite. The late Precambrian Doran type alkali granites are apparently older than the Upper Precambrian (infracambrian) sedimentary rocks. They are intruded into the Upper Precambrian metamorphics and are covered by the Upper Precambrian-Cambrian sedimentary rocks. The granites of Doran, Bornavard and Muteh are the major representatives of the Doran type granite (Fig. 2).

Doran Alkali Granite. The Doran granite is a

coarse-grained to slightly porphyritic, white to pinkish granite characterized by the almost complete lack of macroscopically recognizable dark minerals except for sporadic biotite. The rock is composed predominantly of alkali feldspars (albite, microcline, orthoclase) with minor amounts of quartz and plagioclase and rare muscovite, biotite and sphene. At the type locality (Doran in Zanjan area) [Stocklin et al. 1964, Stocklin and Eftekhar-nezhad 1969] the granite has intruded a complex of epimetamorphic schists (Kahar Formation) [Dedual 1967] and is truncated and transgressively overlain by the late Precambrian Bayandor Formation [Stocklin et al. 1964]. The granite is poor in rubidium and attempts to date it by using total rock samples and plagioclase has failed. Biotites from this granite gave an age of 175±5 Ma [Crawford 1977]. The absolute age of Middle Jurassic for this Upper Precambrian granite indicates the effect of repeated orogenic movements which have complicated and reset the isotopic history of the rock.

Bornavard Alkali Granite. The rock is composed of K-feldspar, quartz, some albite-oligoclase and sparse biotite flakes. The granite is believed to be of roughly the same age as the Upper Precambrian rhyolites of the Taknar Formation into which it is intruded (Fig. 2). Both the granite and the Taknar Formation are affected by low-grade regional metamorphism [Razzaghmanesh 1968, Stocklin 1972]. The Bornavard granite is poor in Rb and its radiometric age of about 540 Ma is not meaningful [Crawford 1977].

Muteh Alkali Granite. The rock is rich in quartz and alkali feldspars but poor in mica. It has invaded the Golpaygan Muteh metamorphics (Fig. 2) and seems to be older than the late Precambrian Soltanieh Dolomite [Thiele et al. 1968].

Zarigan Type late Precambrian-Cambrian Alkali Granite. Like the Doran type, the Zarigan type granites are characterized by the scarcity or absence of mafic minerals and have a quartz porphyritic marginal facies. Unlike the Doran, the granites of this type have invaded the lower parts of the Upper Precambrian beds and seem to represent the plutonic equivalents of the Upper Precambrian-Cambrian Gharadash-Rizu rhyolitic lava flows and tuffs. The granites of Zarigan, Narigan and Chador Malu are the major representatives of the Zarigan type granites (Fig. 2).

Zarigan Alkali Granite. The rock is characterized by a conspicuously low content of mafic minerals. It is intruded into the Precambrian metamorphics and igneous rocks as well as Upper Precambrian sediments. The dolomites of the Upper Precambrian Rizu Formation [Huckriede et al. 1962] are the youngest rocks which show thermal contact with the Zarigan granite in the Chapedony area (Fig. 2). In the type locality, pebbles of granites similar to the Zarigan granite

have been found in the basal conglomerate of the Cambrian sequence. The granite consists mostly of alkali feldspars (40-50 %), quartz (25 %) and plagioclase (An 15, 15-25 %). Accessories and secondary minerals are green biotite (altered to chlorite), rutile, apatite, calcite, muscovite and sericite [Haghipour 1974, 1978, Haghipour et al. 1977].

Narigan Alkali Granite. The rock is a pale-pink, medium-grained biotite granite intruded into the Precambrian metamorphic rocks of Central Iran (Fig. 2). The granite has a granite-porphyry to quartz-porphyry marginal facies and has produced thermal contact effects on sedimentary rocks of the Rizu Series [Huckriede et al. 1962]. It consists of quartz (39 %), orthoclase (29 %), plagioclase (31 %) and chlorite (0.9 %).

Chador Malu Alkali Granite. This is an aplitic granite conspicuously lacking in mafic minerals except for accessory opaque minerals. The granite forms an intrusive body in metamorphic rocks of the Chador Malu area (Fig. 2), south of Chapedony [Huckriede et al. 1962, Stocklin 1972]. It consists of orthoclase (51 %), quartz (32 %), plagioclase (14.3 %) and sphene (0.6 %).

Discussion Of The Precambrian Plutonism

The Precambrian crust of Iran was partly thickened by the vast Precambrian plutonic activity (some of which reached the surface), and by compressive tectonic movements which produced folding, faulting and regional metamorphism. The original crystallization ages of the Precambrian plutonic rocks are difficult to determine since their isotopic systematics have been commonly disturbed by post-crystallization orogenic events. The ages that have been determined range from 1100 to 560 Ma. Presumably the younger ages are due to resetting by orogenic movements.

Due to the lack of paleomagnetic data it is not clear if Iran was a unified continental mass during Precambrian time or not. The Precambrian rocks of eastern Central Iran (Chapedony and Posht-e-Badam Complex) [Hushmandzadeh 1969, Stocklin 1972, Haghipour 1974, 1978, Haghipour et al. 1977] were originally composed of greywackes, calc-alkaline volcanics of intermediate composition, volcano-detritics, and rare carbonates. They were transformed to gneiss, amphibolite, schist, migmatite, and anatectite, and all have been invaded by calc-alkaline intrusions of Precambrian age (Fig. 2). The rocks in part show a tectonic melange of Precambrian metamorphics, associated with pyroxenite and pyroxene-olivine rocks partly transformed to serpentinite.

The association of metagraywacke, metadiorite, amphibolite, pyroxenite, serpentinite and calc-alkaline intrusives in the Chapedony area of eastern Central Iran may have evolved in an intraoceanic island-arc environment of comagmatic volcanism and intrusion above a subduction zone.

Petrographically the rocks are similar to rocks deposited in intraoceanic island-arcs formed near convergent boundaries of oceanic plates [Dickinson 1970, Mitchell and Reading 1971, Dickinson 1974]. The pyroxenite and serpentinite between Chapedony and Posht-e-Badam faults (Figs. 1 and 2) may be the remnants of the Precambrian oceanic crust caught up between Central Iran and east Central Iran along the Chapedony fossil island-arc zone. The metamorphics, migmatites and calc-alkaline intrusives are all elongated along an arcuate belt (presumably indicating the original shape of the island-arc) east of the Chapedony fault and follow the same trend (Fig. 2). Detailed mapping of the region, together with the systematic study of the $K_2O/(Na_2O+K_2O)$ ratios, of trace elements of the intrusives, and radiometric age dating are needed to understand the nature of the basement cratonization. Petro-chemical analyses may throw light on the direction of dip of the possible Precambrian subduction zone in the Chapedony area. Based on surface geology and intrusion of magmatic rocks into the overriding plate, it seems that the Precambrian subduction zone in the Chapedony area dipped eastwards. Owing to the lack of radiometric age dating, the time of the possible island-arc cratonization of the Iranian Precambrian basement is not clear. From stratigraphic evidence the process might have been completed before deposition of the Upper Precambrian (infracambrian) transgressive deposits (about 600 Ma) over the stabilized Precambrian craton. Presence of almandine-rich garnets in the Precambrian plutonic rocks of the Chapedony-Posht-e-Badam area of Central Iran may indicate that partial melting of the subducted oceanic crust took place at depths greater than 70 km, where the breakdown of amphibole (70-100 km) and eclogite fractionation (100-150 km) are the dominant factors [Ringwood 1974] and could lead to the formation of the Precambrian plutonic bodies. In the absence of age-controlled geochemical studies, it is hard to assume that the presence of garnet in the plutons of the Chapedony implies subduction-magma genetic processes.

The uppermost Precambrian intrusive and volcanic rocks of Iran (mostly alkaline), could be the products of the localized partial fusion of the lower crust, during the late Precambrian (Katangan) compressional movements (about 600 Ma) which thickened and shortened the crust.

The Precambrian basement rocks of Iran seem to be a continuation of the Arabian shield and they both appear to have formed a united Arabian-Iranian platform, as a part of Gondwanaland, at least from Upper Precambrian to Middle Paleozoic time [Berberian and King 1981]. The Arabian Precambrian basement appears to have been progressively cratonized by plutonism, volcanism and attendant sedimentation in an intra-oceanic island-arc environment during episodic subduction of oceanic crust through three phases of the Hijaz tectonic cycle (Table 1) over 1100 to 500 Ma [Greenwood et al. 1975, 1976, Neary et al. 1976,

Gass 1977, 1979, Hadley and Schmidt 1978, Schmidt et al. 1978]. Similar cratonization of the basement could have been responsible for the consolidation of the Iranian Precambrian basement. Further research is needed to clarify the structural evolution of this complex region during the Precambrian.

Late Paleozoic (Hercynian) Intrusives

Following the late Precambrian orogenic movements and consolidation of the Precambrian basement, the region became a relatively stable area with platform type deposits during the Paleozoic [Stocklin 1968a, 1977, Berberian 1976]. No important tectonic or plutonic activity is known to have taken place during the early and Middle Paleozoic in Iran. The region underwent some epeirogenic movements during the early (Caledonian) and the late (Hercynian) Paleozoic compressional phases. The Hercynian movements were stronger in the northeastern part of the country (southern margin of the Turan plate; Figs. 1 and 3) and were associated with two phases of granitic intrusions and emplacement of ultrabasics in the Mashhad area [Majidi 1978]. The area seems to be the western continuation of the northern Afghanistan Hercynian orogenic belt [Stazhilo-Alekseev et al. 1972, Sborshchikov et al. 1972; Fig. 3]. The Hercynian intrusive rocks of the northern Afghanistan Hercynian fold belt (gabbro to plagiogranite and ultrabasics) usually cut through Lower Carboniferous volcanics and are transgressively overlain by the Middle Carboniferous-Permian deposits. The detrital rock fragments of the Hercynian intrusives are found in the Permian beds [Stazhilo-Alekseev et al. 1973]. Similar intrusives are also reported from the Turan plate in NE Iran [Knyazev and Shnip 1970] and from the Caucasus, northwest Iran [Zaridze 1958, Azizbekov and Dzotsenidze 1970, Shevchenko et al. 1973, and Khain 1975] (see Fig. 3). Probable Hercynian syenite is reported from Julfa, Mahabad, and the Mero area (Fig. 3), northwest Iran [Eftekhar-nezhad, 1973, 1975; J. Eftekhar-nezhad, M. Qoraishi and S. Arshadi pers. comm.]. Basic plutonic bodies composed of diorite, gabbro and peridotite have been found in the Masuleh area of the Talesh mountains (SW of the Caspian Sea; Fig. 3) are thought to be of probable Paleozoic age, although the evidence is not conclusive in every case [Clark et al. 1975].

According to Majidi [1978] the first phase of the late Paleozoic (Hercynian) magmatism in the Mashhad area (Fig. 3) is characterized by intrusion of granite porphyry, biotite granite, granodiorite and hornblende-biotite tonalite. K/Ar age determinations of biotite samples from the granite porphyry of Mashhad, gave 256±10 and 215±9 Ma. The second phase is a leuco-granite with a radiometric age of 245±10 Ma (K/Ar on one biotite sample) cutting the intrusive rocks of the first phase. It is mainly composed of quartz (30-35 %), orthoclase, microcline, albite, muscovite and rare

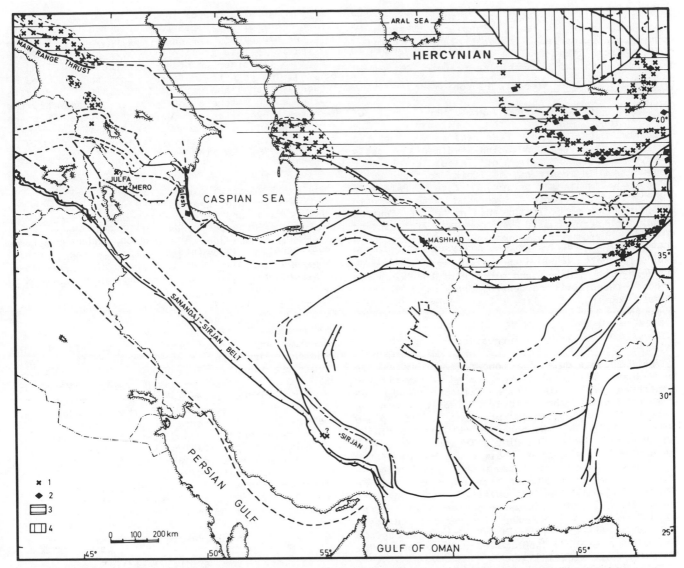

Fig. 3. Hercynian intrusive rocks of Iran and the neighbouring regions exposed at the surface.
The intrusives are mainly developed along the Hercynian belt. The Caucasian Hercynian granitization together with the tectonic lenses of serpentinites are presumably related to the closure of
the Great Caucasian ocean along the Main Range thrust [Khain 1976]. The age of the Iranian intrusives and ultrabasic rocks are not certain.
1 – Intrusive rocks, 2 – Hercynian ultrabasics, 3 – Areas with early and late Hercynian basement,
4 – Caledonian belt.
Data outside Iran are mainly compiled from Garkovets [1964], Andreyev et al. [1966], Stazhilo-
Alekseev et al. [1972, 1973], Shevchenko et al. [1973], Portnyagin et al. [1974], Khain [1975],
Burtman [1975] and Desio [1977]. Lambert Conformal Conic Projection.

biotite. Accessories are garnet, tourmaline, apatite, zircon, rutile and some opaque minerals. K/Ar ages obtained by Alberti et al. [1973] from the micas of the Mashhad granite range from 146±3 to 120±3 Ma (Upper Jurassic to Lower Cretaceous). This must either be due to rejuvenation and potassium metamorphism or from an error in K/Ar ages given by Majidi [1978].

A series of intrusive rocks ranging from lherzolite to granite is associated with the pre-Permian metamorphic rocks between Quri and Beshneh, SW of Sirjan [Sabzehei et al. 1970, Ricou 1974] (Fig. 3). The whole sequence shows some affinities with those of the island-arcs or active continental margins but unfortunately the age of the series is not known. Nabavi [1976] and

Hushmandzadeh [1977] related them to the early Paleozoic (Caledonian) orogenic movements, but the region is far from the Caledonides [Berberian and King 1981] (Fig. 3). Further research is needed to clarify the ambiguities.

Discussion Of The Paleozoic Plutonism

Alborz, Central Iran and Zagros were parts of Gondwanaland during the Paleozoic. The Precambrian basement of these regions were covered by platform-cover Gondwanian sediments (mostly detritics) with no sign of Paleozoic orogenic movements. In contrast to this southern quiet tectono-sedimentary regime, late Paleozoic (Hercynian) orogenic movements with magmatic activity took place in the area north of Central Iran (southern Russia and northern Afghanistan; Fig. 3). Exposure of numerous late Paleozoic-early Mesozoic calc-alkaline magmatic rocks and ophiolite belts between these two regions indicate a Hercynian suture and closure of the Hercynian ocean between northern Central Iran and southern Russia during the late Paleozoic-early Mesozoic time (Fig. 3).

Triassic Intrusives

The Paleozoic quiet tectono-sedimentary regime of Iran with platform cover sedimentation continued until the Lower Triassic and was disrupted in the Middle Triassic, when an important compressional movement took place approximately at the boundary between the Middle and Upper Triassic. The Middle Triassic movements remodelled the tectono-sedimentary regime of the country, and were associated with major folding , faulting, regional metamorphism and magmatism. The folded, metamorphosed and uplifted rocks, together with the related intrusive bodies, were covered unconformably by the Rhaetic-Liassic sediments indicating the end of the Middle Triassic orogenic movements [Stocklin 1968a, 1977, Berberian 1976, Berberian and King 1981].

In the western part of the Alborz mountains (Figs. 1 and 4) Triassic plutonism is manifested by the Lahijan biotite granite and granodiorite. The rock intrudes the Carboniferous metasediments and pebbles of it have been found within the Jurassic conglomerates [Annells et al. 1975]. In the western Alborz, the Masuleh diorite and gabbro of the Talesh mountains, SW of the Caspian Sea (Figs. 1 and 4) cut the Upper Paleozoic clastics and are overlain by the Rhaetic-Liassic Shemshak Formation. The rocks have suffered some late-stage metamorphism [Davies et al. 1972, Clark et al. 1975].

In the Aghdarband area of the Kopeh Dagh fold belt (Fig. 4), the sediments below the Anisian nodular limestone are cut by a great number of Triassic dykes which never enter into the Anisian nodular limestone [Seyed-Emami 1971]. The Triassic sediments and dykes are covered unconformably by the Rhaetic-Liassic Kashafrud (Shemshak) Formation

in the Aghdarband area [Madani 1977]. In the junction zone of the Kopeh Dagh belt and Central Iran in the northeast (Mashhad region, Fig. 4) the Triassic aplitic granite with K/Ar dating of 211 ± 8 Ma cuts the older granites of the area [Majidi 1978]. The granite is fine-grained and composed of quartz, orthoclase, microcline, albite, muscovite with minor amounts of biotite. Presumably the Triassic granite of Mashhad is the western continuation of the northern Afghanistan granite belt (Fig. 4) [Stazhilo-Alekseev et al. 1972, Desio 1975, Boulin and Bouyx 1977, Debon et al. 1978].

Along the Sanandaj-Sirjan belt of south and southwest Central Iran (Figs. 1 and 4) the Triassic granite, granodiorite, monzonite, diorite and gabbro are intruded through metamorphosed Paleozoic-Triassic rocks which are unconformably overlain by the Jurassic conglomerates [Sabzehei and Berberian 1972, Berberian and Nogol 1974, Sabzehei 1974, Alric and Virlogeux 1977]. The Sanandaj-Sirjan Triassic intrusives are mainly exposed in the Esfandagheh and the area west and south of Sirjan (Fig. 4). Several intrusive bodies ranging from gabbro to granite cut the metamorphosed Permian rocks in the Quri and Cheshmeh Anjir region west of Sirjan (Fig. 4). The contact metamorphic rocks related to these intrusives are found in the conglomerates of Jurassic age [Sabzehei et al. 1970, Ricou 1974]. The Sikhoran intrusive complex of the Esfandagheh area (southeastern extremity of the Sanandaj-Sirjan belt; Fig. 4) is reported to be a layered intrusion characterized by a differentiation suite from ultrabasic to granite, and which originated from a basaltic tholeiitic magma poor in alkali and rich in CaO [Sabzehei 1974; see discussion of the Mesozoic plutonism]. They produced a metamorphic aureole of pyroxene hornfels facies and are overlain by the Jurassic sediments.

Late Jurassic Intrusives

Following the Middle Triassic compressional movement, Central Iran, Alborz and Kopeh Dagh were transgressively covered by the Rhaetic-Liassic coal-bearing sandstone and shale of the Shemshak Formation [Assereto 1966]. The shallow water deposits of Lower Jurassic were followed by the Middle to Upper Jurassic marine carbonate rocks. Due to the late Jurassic movements the sea regressed from most parts of Central Iran and Alborz and the whole region underwent folding, faulting, magmatism and uplifting before deposition of the lower Cretaceous sediments [Stocklin 1968a, 1977, Berberian 1976].

The late Jurassic plutonism in the central Alborz mountains in north Iran (Fig.1) is indicated by olivine-biotite lamprophyre dykes which cut the Jurassic Shemshak Formation in Baijan, east of Damavand (Fig. 5). The olivine-biotite lamprophyres contain biotite (35 %) and olivine (15 %) phenocrysts, in a groundmass composed of plagioclase and biotite. A sample of

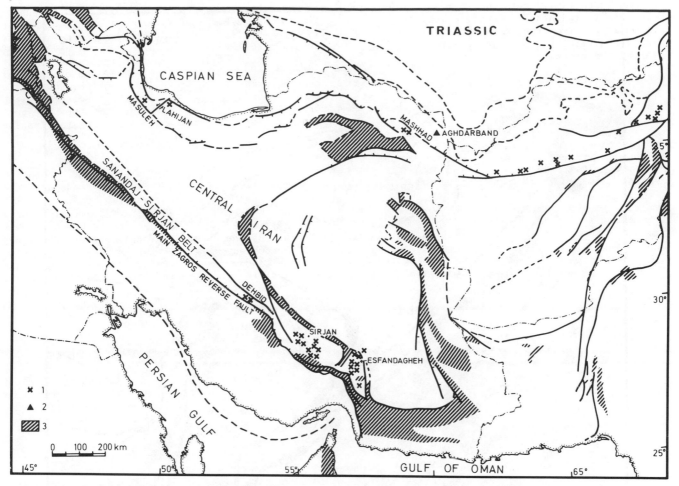

Fig. 4. Distribution of the Triassic plutonic rocks of Iran and the neighbouring regions. Presumably the plutonic belt developed along the southeastern segment of the Sanandaj-Sirjan (the Central Iranian active continental margin) represents the magmatic-arc due to subduction of oceanic crust underneath Central Iran along the present Main Zagros reverse fault line. The absence of plutonic outcrops along the middle and northwestern segments of the belt may partly be due to the sedimentary cover and/or that the subduction started from the SE.
1 – Triassic intrusives exposed at the surface, 2 – Triassic dykes, 3 – Areas with oceanic sedimentation (radiolarian chert, etc.).
Data outside Iran are mainly based on Stazhilo-Alekseev et al. [1972, 1973] and Desio [1977]. Lambert Conformal Conic Projection.

olivine-biotite-augite lamprophyre is also reported from the same area. The rock consists mainly of augite (30 %) and olivine (20 %) phenocrysts in a groundmass of biotite (25 %), plagioclase (15 %) and opaque minerals (10 %), [Allenbach 1966].

Tourmaline-bearing pegmatite granites are reported from the area east of Masuleh and west of Lissar in the Talesh mountains southwest of the Caspian Sea (Fig. 5). The rocks are definitely pre-Maastrichtian [Clark et al. 1975] and Rb/Sr dating from a muscovite sample gave an age of 175±10 Ma [Crawford 1977].

Several late Jurassic intrusive bodies are reported from Central Iran, the Sanandaj-Sirjan belt (southwestern active margin of Central Iran) and the Lut Zone (eastern part of Central Iran). The major intrusive bodies in Central Iran are the Esmailabad, Airakan, Shirkuh and Kolah Qazi granites (Fig. 5).

Esmailabad Biotite Granite. The Esmailabad granite is a pink biotite granite exposed in the eastern Central Iran (Fig. 5). It cuts the Permian and Jurassic rocks and is transgressively covered by the basal conglomerate of the Cretaceous. The rock is mostly composed of alkali feldspars (microcline, orthoclase 50 %), interstitial quartz

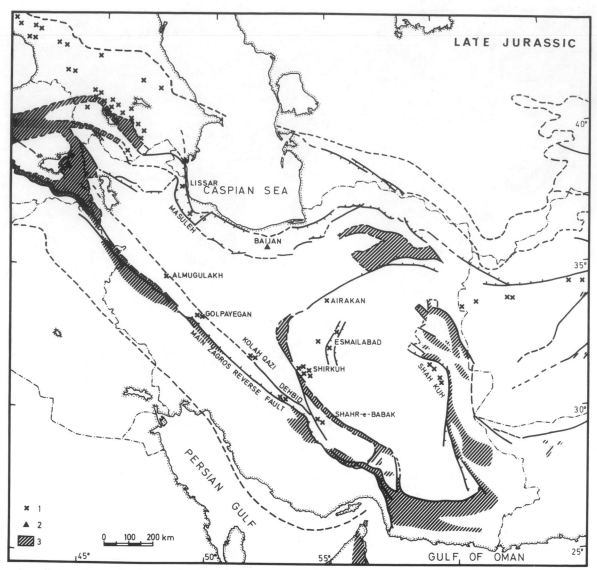

Fig. 5. Upper Jurassic-Lower Cretaceous intrusives of Iran and the neighbouring region.
Note the close relation between plutonic activity and areas of oceanic character. The subduction along the Main Zagros reverse fault line is still continuing and the plutonic activity along the Sanandaj-Sirjan belt seems to be shifting towards the NW.
1 - Intrusives, 2 - Dykes, 3 - Oceanic areas with deep water sediments (radiolarian chert, thick flysch type deposits, pelagic limestone), sheeted diabase complex and submarine pillow lavas.
Data outside Iran are mainly based on Stazhilo-Alekseev et al. [1972, 1973], Shevchenko et al. [1973], Khain [1975] and Tvalchrelidze [1975]. Lambert Conformal Conic Projection.

(20–30 %), plagioclase (An23–38, 10 %), biotite (5–10 %) and muscovite (less than 5 %). The whole rock Rb/Sr age determination of this granite in the Posht-e-Badam region gave 268, 267 and 240 Ma, which are much older than indicated by its stratigraphic position [Haghipour 1974, 1978, Haghipour et al. 1977].

Airakan Granite. The rock is exposed in the eastern Central Iran (Fig. 5). Rb/Sr dating gave

an age of 165±8 Ma (Middle Jurassic). Dating by the K/Ar method on the biotites and the feldspars of the granite gave ages of 113±9 Ma (Barremian-Albian) and 74±13 Ma (Senonian-Paleocene) respectively [Reyre and Mohafez 1972].

Shirkuh Granite. The Shirkuh granite batholith is exposed in the area south of Yazd and is covered by the basal conglomerates of the lower Cretaceous which contain pebbles of the granite [Nabavi 1970] (Fig. 5).

Kolah Qazi Granodiorite. The Kolah Qazi granodiorite of the Esfahan region in Central Iran (Fig. 5) is intruded into Jurassic shales and is covered by the Barremian-Aptian basal conglomerates [Zahedi 1976].

Intrusives along the Sanandaj-Siarjan Belt. The Almugulakh diorite of northwest Hamadan is intruded through the Jurassic sediments in the northwestern part of the Sanandaj-Sirjan belt (Fig. 5). A Rb/Sr radiometric age gave 144±17 Ma (with 87Sr/86Sr initial ratio = 0.7088), which is near the Upper Jurassic/Lower Cretaceous boundary [Valizadeh and Cantagrel 1975]. The Golpaygan granodiorites, southeast of the Almugulakh diorite (Fig. 5) have invaded the Jurassic sediments and their pebbles are found in the basal conglomerate of the Middle Cretaceous [Thiele et al. 1968]. Southeast of the belt, small stocks of diorite in the northern area of Deh Bid (Fig. 5) cut the Jurassic limestone. They are highly sheared and have undergone light dynamothermal metamorphism [Taraz 1974].

Some intrusive bodies, ranging from gabbro to granite, are exposed in Chah Dozdan and Chah Bazargani area, southwest of Shahr-e-Babak, along the Sanandaj-Sirjan belt (Fig. 5). Owing to the lack of good exposure, the stratigraphic position of these intrusives is not known. The rocks can only be grouped among the pre-Cretaceous intrusives since they are transgressively covered by the Cretaceous sediments. K/Ar radiometric age determination has resulted in 118±10 Ma (muscovite in muscovite pegmatite, Kuh-e-Chah Dozdan) to 164±4 Ma (biotite in granite gneiss, Chah Bazargani) for these rocks. The K/Ar dates probably give the approximate age of intrusion, although the actual age could be somewhat older if argon loss has occurred [Sabzehei et al. 1970].

Shah Kuh Granite of the Lut Zone. In the Lut zone of eastern Iran, some granite batholiths are exposed in Shahkuh and Chahar Farsakh (Fig. 5). The granite is intruded through the lower Jurassic shales and is transgressively covered by the Aptian Orbitolina limestone. It is a coarse-grained biotite granite with large pink crystals of K-feldspars. Quartz, K-feldspars (microcline), biotite and plagioclase are the main constituents [Stocklin et al. 1972].

Late Cretaceous Intrusives

The deformed Jurassic rocks together with the late Jurassic intrusives of Central Iran and Alborz are transgressively covered by the Cretaceous detrital sediments and carbonate rocks. Towards the end of the Cretaceous, the whole region underwent strong orogenic movement and magmatism [Stocklin 1968a, Berberian 1976].

The late Cretaceous intrusive bodies are mostly developed along the active continental margin of Central Iran (the Sanandaj-Sirjan belt; Fig. 6). In the only outcrop of the western Alborz area,

the Sardeh monzonite, hornblende porphyry and syenite (Fig. 6) crop out as small plugs and dykes, cutting the Jurassic Shemshak Formation [Annells et al. 1975].

The Upper Cretaceous intrusive bodies along the Sanandaj-Sirjan belt are mostly developed in the Hamadan and Golpaygan area with a range in composition from gabbro to granite and pegmatite (Fig. 6). The existence of inclusions of basic intrusives in the acid rocks indicates their relatively older age than the granites. Radiometric age determination of the Alvand norite in the Hamadan area gave 88.5 Ma, 78.5 Ma (Rb/Sr method) and 89.1±3 Ma (K/Ar method). The Alvand granite of the same region gave ages between 65±2 (Rb/Sr) to 81±3 (K/Ar) Ma [Valizadeh and Cantagrel 1975].

The Bazman batholith in southeast Iran (Fig. 6) is mainly composed of porphyritic granites with some minor augite-hornblende gabbros and hornblende- biotite diorites. Evidently the intrusion did not occur as one event but in successive stages from gabbro to granite. The gabbroic and dioritic bodies are cut by the main granitic intrusion. Aplitic veins are intruded into gabbros, diorites and granites of the area, and are themselves cut by the younger diabasic dykes. The batholith shows characteristics of a zoned pluton with dominance of granite intrusion. The basic and intermediate rocks are not abundant and only fringe the main plutonic body. Stratigraphic data give their age as post Triassic – pre Neogene, since the intrusive rocks cut the Permian Jamal and the Triassic Shotori Formations and are overlain by the Neogene red beds [Berberian 1974]. The Bazman batholith has a critical siting. Its position along the southeastern extremity of the Jebal-e-Barez Oligo-Miocene plutonic belt (Fig. 8) may indicate its belonging to the same episode. On the other hand, the batholith is located in the northern part of the coastal-Makran subduction zone [White and Klitgord, 1976] which has been active since Cretaceous [Farhoudi and Karig 1977, Berberian and King 1981] and may be related to its arc-magmatic activity (Figs. 1 and 6).

Biotite and hornblende-biotite granite are the common rocks of the Bazman batholith. The approximate mode of the biotite granite is orthoclase (30 %), oligoclase (30 %), quartz (25 %), biotite (10 %) and muscovite (<5 %). The rocks usually show a sort of protoclastic texture which may have been caused by slight crushing before consolidation [Holzer and Samimi 1970].

The initial 87Sr/86Sr ratio of the Bazman rocks [0.70564±0.00006; Berberian 1981, Berberian et al. 1981] is typical of the values found in the predominantly I-type magmas, characteristic of destructive continental margins [Faure and Powell 1972, Chappell and White 1974, Hedge and Peterman 1974, Faure 1977, Ishihara 1977, Beckinsale 1979]. This, together with the situation of the calc-alkaline rocks of the Bazman batholith above the Makran subduction zone (Figs. 1 and 6), the coincidence of some trace element results with the fields of the island-arc rocks, and the late

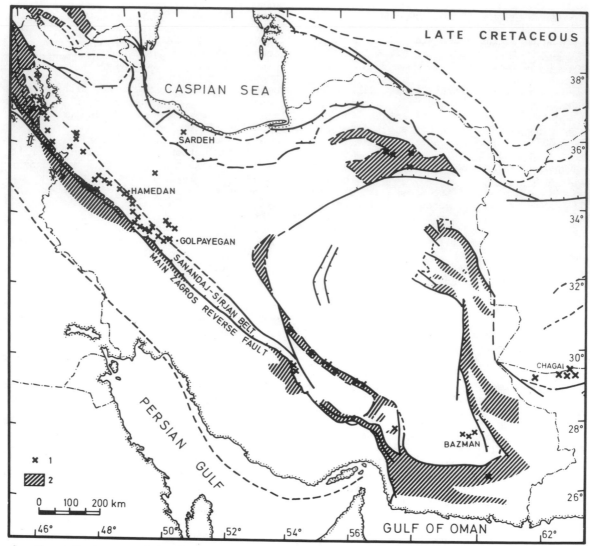

Fig. 6. Upper Cretaceous intrusive bodies exposed at the surface.
Note the complete shift of magmatic activity towards the northwestern segment of the Sanandaj-Sirjan belt. The plutonic belt seems to be the magmatic-arc activity related to the subduction of the oceanic crust underneath Central Iran. Due to the late Cretaceous collisional orogenies the ophiolites were emplaced along the continental margins. The emplaced ophiolites were covered by Paleocene-Eocene detritics.
1 - Intrusive rocks, 2 - Continental margin areas with obducted oceanic crust material (ophiolites, radiolarian cherts, pillow lavas, etc.). Lambert Conformal Conic Projection.

Cretaceous-early Paleocene radiogenic age (Rb-Sr dating, 74 to 64 Ma) [Berberian 1981, Berberian et al. 1981], may indicate that the batholith was related to the early stage of the subduction. The emplacement of the Bazman granite (the youngest phase) during the late Cretaceous, presumably reveals that the northward subduction of the Arabian (Oman) oceanic crust underneath the southeastern boundary of the Central Iranian continental margin was well established prior to the Cretaceous and gave rise to the formation of the batholith. The Bazman batholith is not the

only evidence of magmatic activity above the Makran subduction zone. Widespread E-W trending intrusion of the late Cretaceous-Paleogene granite, granodiorite and diorite batholiths, together with the late Cretaceous-Paleogene calc-alkaline volcanic rocks [Jones 1960, Abu-Bakr and Jackson 1964, Arthurton et al. 1979] are exposed northeast of the Bazman batholith in western Pakistan, all lie above the Makran subduction zone.
Obduction and emplacement of the ophiolites underneath the Makran accretionary flysch wedge,

south of the Bazman batholith (Figs. 1 and 6), could be taken as an indication of intensive subduction of the Arabian oceanic crust underneath the continental margin of Central Iran during Maastrichtian. These ophiolites are not interpreted as a continental closure (Alpine-type) since subduction of oceanic crust preceded and succeeded its late Cretaceous emplacement and the area has been and still is an active convergent plate margin since at least post Cretaceous times [Berberian and King 1981]. Therefore the Bazman calc-alkaline intrusives could be a manifestation of an Andean-type tectonic regime. The results of the analyses of the Bazman batholith [Berberian 1981] now reveals that the Maastrichtian obduction of ophiolites (remnants of the Arabian oceanic crust) presumably took place during a northward subduction phase; the Arabian oceanic crust was much older than Cretaceous and the northward subduction underneath the Central Iranian continental margin was initiated much before the Cretaceous. Assuming that the magma was produced by melting initiated by a descending slab, with an inclination of about 30° at a rate of 3 cm/yr, the Arabian oceanic crust must have reached to depth ranges of 100-150 km at least 10 My prior to the oldest intrusive rock (gabbro) of the Bazman complex. Moreover, the crystallization and ascending time of the calc-alkaine magma, generated from that depth [Green and Ringwood 1968, Oxburgh and Turcotte 1970, Marsh and Carmichael 1974], should be added to this rough estimate.

Discussion Of The Mesozoic Plutonism

Apparently Central Iran, which was connected to Arabia via the Zagros belt during the Paleozoic was detached from the latter during the late Permian and attached to Turan platform (southern Eurasia) before the late Triassic. Due to northeastward movement of Central Iran, a new ocean was formed along the present Main Zagros reverse fault line. The Middle Triassic movements, which were associated with regional metamorphism and magmatism mainly along the Sanandaj-Sirjan belt of southwestern Central Iran (Fig. 4), were interpreted as the onset of subduction of the oceanic crust underneath Central Iran [Berberian and King 1981]. The metamorphics may result from the frictional heating and pressure effects created during the descent of subducting oceanic crust underneath the active continental margin of Central Iran (the Sanandaj-Sirjan belt). Triassic ophiolites, gabbro, granite, amphibolite and metamorphosed volcano-detritic sediments exposed in the southeastern part of the belt seems to form the Triassic magmatic-arc which is partly covered by the Lower Jurassic volcano-detritic sediments. The regional calc-alkaline magmatism and metamorphism along the Sanandaj-Sirjan belt during Mesozoic time (Figs. 4 to 6) may represent other parts of this magmatic-arc, developed during subduction episodes since Triassic time.

Apparently the ocean was consumed towards the Cretaceous and final closure took place during late Cretaceous-Paleocene times, when the ophiolites were obducted onto the continental margin [Stocklin 1974, 1977, Berberian and King 1981]. However, petrochemical studies and isotopic age dating of the Oligo-Miocene plutons [Berberian 1981] disagree with the above assumption.

The distribution of Mesozoic plutonic bodies in Iran (Figs. 4 to 6) is mostly restricted to regions close to the eventual active plate margins marked by the ophiolite-melange belts (Fig. 1). They appear to have been generated extensively along and above the early Mesozoic subduction zone of the Sanandaj-Sirjan belt (Figs. 4 to 6) during 150 My of active Mesozoic subduction (from 220 to 65 Ma). During this period, active plutonic events seem to be episodic with climaxes around Middle Triassic (Fig. 4), late Jurassic (Fig. 5) and late Cretaceous (Fig. 6). This may indicate episodic plate motion and changes in consumption of oceanic crust and melting. The Triassic plutonic belt is well exposed in the southeastern part of the Sanandaj-Sirjan belt in Sirjan and Esfandagheh area (Fig. 4). The absence of Triassic intrusive rocks along the central and northwestern segments of the belt is either due to their being covered by Jurassic and Cretaceous sediments or because subduction started from the southeast and propagated northwestward. According to this view, the Triassic calc-alkaline plutonic activity along the Sanandaj-Sirjan magmatic-arc, could belong to the early tholeiitic stage of arc development [Ringwood 1974].

Sabzehei [1974] stated that fractional crystallization of the tholeiitic basaltic parental magma resulted in the Triassic granitic intrusion of the Esfandagheh area (Fig. 4). There is a large amount of experimental data tending to support the fractional crystallization model for the production of granitic magma from a basaltic parent [Bowen 1914, 1915; Andersen 1915; Bowen and Schairer 1938; Schairer and Bowen 1938, 1947; Schairer and Yoder 1960] but all of these data are for simplified systems that do not exactly duplicate natural magmas. Fractional crystallization of basaltic parental magma is now believed to be an unlikely mechanism for the generation of granitic batholiths during a late-stage crystallization residue, in the orogenic belts [Presnall 1979, Kushiro 1979]. Therefore, Sabzehei's interpretation [1974] supporting the proposition that the granitic and the layered ultrabasics of the Esfandagheh area are co-magmatic and are derived from a common basaltic magma (formed by fractional crystallization) is questionable. As in the Skaergaard [Presnall 1979] and Bushveld [Davies et al. 1970], separate origins could be adopted for the granitic and ultrabasic rocks of the Esfandagheh region, SE of the Sanandaj-Sirjan belt.

The Triassic plutonic activity (Fig. 4) was followed by the late Jurassic-early Cretaceous (Fig. 5), and the late Cretaceous intrusives (Fig.

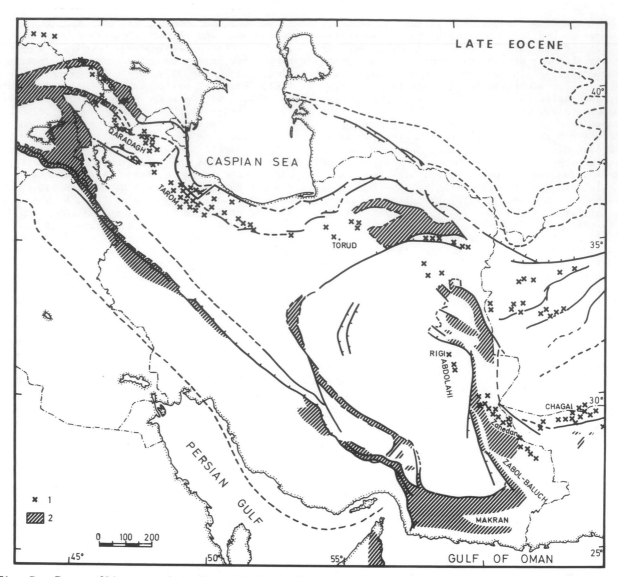

Fig. 7. Eocene-Oligocene plutonic activity in Iran.
1 - Intrusive bodies, 2 - Late Cretaceous ophiolite-radiolarite belts thrust along the continental margins. Lambert Conformal Conic Projection.

6) mainly developed along the Sanandaj-Sirjan belt (the destructive plate margin of SW Central Iran) and along the northern Makran of southeastern Central Iran (the Bazman batholith). Comparison of the distribution of the Triassic to late Cretaceous regional plutonism along the belt at the surface (Figs. 4 to 6) shows a progressive northwestward migration of plutonic emplacement in time along the active continental margin, which may support the idea of an oblique subduction episode. Alternatively, the absence of exposed plutons in some parts of the belt may be due to solidification of some plutons at the root of the magmatic-arc or they are covered by younger sediments.

Low 87Sr/86Sr initial ratios and trace-major element studies of the late Cretaceous batholith of Bazman in southeast Iran (Fig. 6) indicate that the magmas were derived by melting of a mantle or oceanic crust source above the Makran subduction zone in the Gulf of Oman [Berberian 1981]. Interpretation of the late Cretaceous batholith of Bazman in terms of rapid motion of Arabian oceanic crust beneath the Central Iranian continental margin in southeast Iran, is not strange, as subduction has been active at least since Cretaceous. Detailed study of the batholith [Berberian 1981] confirms this link and suggests that the subduction may have started much earlier than Cretaceous.

The upper Cretaceous plutonic bodies (Fig. 6) are surrounded by contact metamorphism of

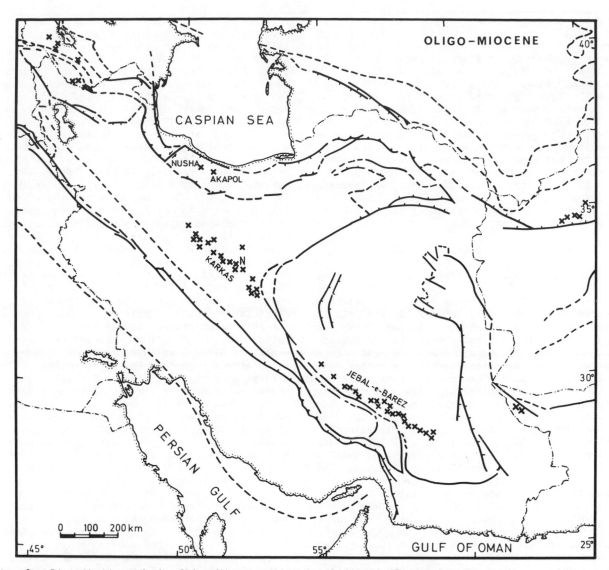

Fig. 8. Distribution of the Oligo-Miocene intrusive bodies in Iran.
Position of the Natanz intrusive complex is shown by 'N' along the Karkas intrusive belt. Data outside Iran are mainly based on Tvalchrelidze [1975]. Lambert Conformal Conic Projection.

pyroxene-hornfels facies and are associated with the regional metamorphism of the lower greenschist facies along the northwestern part of the Sanandaj-Sirjan belt [Berberian 1976, Berberian and Alavi-Tehrani 1977, Berberian and King 1981]. Therefore the regional metamorphism (presumably related to ascent of thermal flux derived from the descending slab) and the emplacement of the abundant intrusives (Fig. 6) are directly linked. In many places the regional metamorphism precedes the intrusion of granites and granodiorites [Berberian and Alavi-Tehrani 1977]. The development of plutonic rocks and associated regional metamorphism appears to decrease inland from the active plate margin, and also decreases towards the southeastern part of the belt (Fig. 6). The

garnet crystals in granite, granodiorite, pegmatite and aplite of the late Cretaceous plutonic rocks of the Hamadan area [Zarayan et al. 1972, Majidi and Alavi-Tehrani 1972] may be the result of the eclogite-controlled fractionation and /or partial melting of mantle pyrolite [Ringwood 1974] developed during the subduction of oceanic crust underneath Central Iran. Further radiometric age dating and detailed petrochemical study is needed to observe the change in K/Na and K/Rb ratios along a trend from the supposed continental margin to inland.

The intensive coeval Mesozoic volcanism is missing beyond the Sanandaj-Sirjan belt and the plutons have formed close to the coeval trench, the position of which is represented by the ophiolites emplaced along the Main Zagros reverse fault (Figs. 1 and 4 to 6). In modern arc-trench systems there are gaps between the trench and the volcanic fronts of the arcs, characteristically 125±50 km for island-arcs and 225±50 km for continental margin areas [Dickinson 1971]. The narrow arc-trench gap zone in this belt may be due to crustal shortening caused by the post-collisional convergence during several major compressional movements since the late Cretaceous orogeny, 65 Ma. Alternatively, the narrow arc-trench gap may represent either a steep subduction zone towards the northeast (Mariana type subduction) [Isacks and Barazangi, 1977] or indicate that the subduction started far away from the belt. Further detailed research is needed to resolve the ambiguities.

Late Eocene-Oligocene Intrusives

Following the late Cretaceous orogenic movements, immense volumes of dacite, andesite and basaltic lavas with tuffaceous and other clastic sediments were deposited during Eocene times, forming the Karaj Formation [Dedual 1967] in Central Iran and Alborz. During Upper Eocene-Oligocene times, the Karaj Formation and other rocks were cut by several intrusive bodies. The intrusives are mainly coarse-to medium-grained biotite granite, hornblende-biotite granodiorite, monzonite and diorite exposed mainly in the northwest of the country (Fig. 7), from Qaradagh to Tarom [Hirayama et al. 1966, Stocklin and Eftekhar-nezhad 1969, Clark et al. 1975, Didon and Gemain 1976]. The intrusives are usually elongated in the direction of folding. The Qaradagh-Tarom intrusive belt is the southeastern continuation of the Upper Eocene Little Caucasian intrusives [Khain 1975] (Fig. 7). The second important intrusive belt is composed of granite and diorite which are intruded through Eocene flysch in the southeastern Iranian Zabol-Baluch belt near Zahedan (Fig. 7).

The granite, granodiorite and diorite intrusions in the Torud area (Fig. 7) cutting the Eocene volcanics [Hushmandzadeh et al. 1978] could belong to this phase. The Rigi and Abdolahi granodiorite and diorite in the eastern Lut zone (Fig. 7) are intruded into the Aptian sediments and Eocene volcanics respectively [Stocklin et al. 1972]. They belong either to the Upper Eocene-Oligocene or to the Oligo-Miocene magmatic phase and their exact age is not clear.

Oligo-Miocene Intrusives

The last major intrusive episode in Iran is that of the Oligo-Miocene which is mainly developed in Central (Karkas belt) and southeastern (Jebal-e-Barez belt) Iran (Fig. 8).

The Karkas belt is composed of gabbroic to granitic intrusives. The acid intrusives cut the basic phase and show their relative younger age. The rocks are intruded through the folded Eocene

volcanics and have developed a thermal metamorphism [Reyre and Mohafez 1972, Amidi 1975, 1977]. The Rb/Sr age determination of the Karkas granodiorite has given 78 Ma [Reyre and Mohafez 1972] which is definitely not the age of intrusion since the rock cuts the folded Eocene volcanics. Two more ages have given 38-33 and 18 to 16 Ma by Reyre and Mohafez [1972] and 17 to 19 Ma (Rb/Sr) by Amidi [1975, 1977]. A detailed petrochemical and isotope study of the Natanz batholith along the Karkas belt [Berberian 1981] indicates that all rocks of this calc-alkaline batholith are peraluminous, possibly derived from a high-Al parental magma (the Natanz complex is indicated by 'N' in Fig. 8). The low concentration of Nb, the behaviour of K/Na and Mg/Fe ratios in intermediate rocks, the wide range of the Rb/Sr ratio, and the coincidence of the field of the Natanz calc-alkaline rocks with the known Andean-type subduction related plutons, all suggest a probable arc-magmatic origin for this suite [Berberian 1981]. The $^{87}Sr/^{86}Sr$ initial ratio obtained for this complex (0.70524 to 0.70573) [Berberian 1981] suggests an upper mantle or oceanic crust origin for this Oligo-Miocene batholith with Rb-Sr age dating of 33 Ma (for gabbro) and 25 Ma (for granite). The granitoid rocks of the Natanz batholith along the Karkas belt (Fig. 8) do not present much higher $^{87}Sr/^{86}Sr$ initial ratios and therefore, cannot be related to crustal melts.

The Jebal-e-Barez belt is exposed in southeastern Central Iran and is mainly composed of granite and granodiorite (Fig. 8). At Kuh-e-Lalezar, they were intruded through Eocene and Miocene limestones (Qom Formation) and the overlying volcanics. Their relationship with the late Miocene Upper Red Formation is not clear, but the Middle-Upper Miocene sequence of the conglomerates and pyroclastics with lava flows overlies the Jebal-e-Barez granodiorites along the northern flanks of Kuh-e-Lalezar. Therefore the intrusion of the Jebal-e-Barez rocks could not be younger than Tortonian [Dimitrijevic 1973]. K/Ar age determination of a diorite sample from the southern part of this intrusive belt gave an age of 12.9±0.1 Ma and three samples from granite gave 15±0.2, 18±0.1 and 24±0.1 Ma [Conrad et al. 1977]. The Jebal-e-Barez Oligo-Miocene plutonic activity locally reached near surface and produced the Kuh-e-Panj type subvolcanic rocks. These rocks differ from the main plutons mainly in their more pronounced porphyritic character. The general composition of these rocks is similar to that of the Jebal-e-Barez type [Dimitrijevic 1973].

The Akapol and Nusha granodiorites in the western Alborz mountains [Annells et al. 1975] possibly belong to this intrusive phase (Fig. 8). The Akapol batholith is mainly composed of quartz-monzonite and granodiorite. The essential minerals are alkali- feldspar, oligoclase, hornblende and biotite. Accessories are apatite, sphene and opaque minerals. Numerous aplite, diabase, and lamprophyric dykes cut the Akapol batholith. The diabasic dykes seem to be younger than the lamprophyres and aplites [Gansser and Huber 1962].

Pliocene Intrusives

Few granite, diorite and gabbroic intrusive bodies seem to have been emplaced during Pliocene time (Fig. 9). The Alamkuh leucogranite cutting the Paleogene volcanics in the Alborz mountains of northern Iran belong to this phase [Gansser and Huber 1962, Annells et al. 1975]. The rock is mainly composed of perthitic alkali-feldspar, quartz and biotite (partly chloritized). The accessory minerals are sphene and opaque grains with occasional apatite and zircon. Zeolite, calcite and zoisite are found as secondary minerals. The Alamkuh batholith is cut by aplitic dykes rich in tourmaline, granophyric veins, muscovite quartz-porphyry, biotite-porphyry, hornblende (biotite) diabase and hornblende-dolerite dykes. The Baneh quartz porphyry

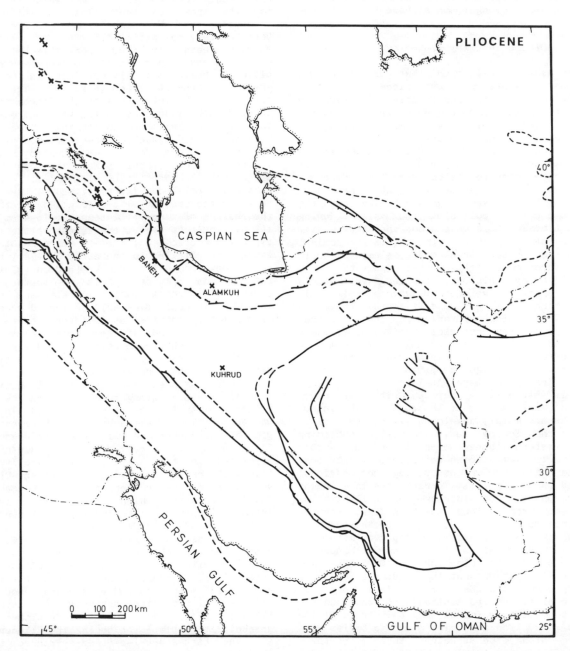

Fig. 9. Isolated granodiorite plutons of Pliocene age indicating the last phase of continental plutonic activity in Iran.
Data outside Iran are mainly based on Kornev [1961], Khain [1975], Yegorkina et al. [1976].
Lambert Conformal Conic Projection.

southwest of the Caspian Sea (Fig. 9) cutting the Neogene red beds [Clark et al. 1975] seems to belong to this tectono-plutonic episode.

In Central Iran the Kuhrud granite, granodiorite and micro-gabbro (Fig. 9) cut the Pliocene tuffs and lava flows in the area north of Soh [Zahedi 1973]. Some diorite and granodiorite-porphyry dykes, which have been injected into the Middle-Upper Miocene and Neogene conglomerates in the Jebal-e-Barez area of southeastern Central Iran (Fig. 8), could belong to this phase [Dimitrijevic 1973].

Discussion Of The Tertiary Plutonism

It is usually explained that during the late Cretaceous movements the last piece of oceanic crust between Arabia and Eurasia was presumably consumed along the present Main Zagros reverse fault and the ophiolites were emplaced along the continental margins [Stocklin 1974, 1977, Berberian and King 1981]. If this assumption is accepted, then the Tertiary plutonic activity is considered as post collision magmatism. However, some other workers argue the Late Cretaceous closure of the Tethyan ocean and believe that the continent-continent collision took place during Miocene times. It is understood that the initial 87Sr/86Sr ratio together with the petrochemical analysis of the trace-major elements would probably throw some light on this argument.

The Tertiary plutonic bodies were mainly intruded during Upper Eocene-Oligocene (Fig. 7), Oligo-Miocene (Fig. 8) and Pliocene (Fig. 9) epochs. The Tertiary plutonism in Iran was not restricted to regions close to eventual plate margins and the bulk of the magmatism seems to have occurred within continental regions as well.

The late Eocene-Oligocene plutonic activity of northern Iran (the Qaradagh-Tarom plutonic belt; Fig. 7) is spatially associated with the major fault and fold trends at the surface. The intrusive rocks are mainly developed along the axial belt of the Eocene fault-controlled subsiding basin of western Alborz, which allowed more than four kilometers of volcano-detritic sedimentation. The whole region underwent strong compressional movements and magmatism during the late Eocene orogenic phase. Considering the collision of Arabia and Iran (along the Main Zagros reverse fault line) took place in late Cretaceous, it would be difficult to assume a deep generation for magma as a result of a descending lithospheric slab to be responsible for the Qaradagh-Tarom plutonic belt. At present the plutonic belt is about 250 km northwest of the late Cretaceous collision site, and its southeastern continuation does not completely follow the NW-SE trend of the Mesozoic subduction zone. The plutonic bodies are elongated along the structural trend of northern Iran and follow the trend of the mountain range (Figs. 1 and 7) which may indicate that their alignments were influenced and controlled by the major deep seated longitudinal faults. These,

together with the emplacement of plutonic bodies after the regional uplift of the late Eocene compressional phase, may indicate that movements along the deep-seated faults, brought up hot mantle material against the lower crust (displacement of Moho) [Krestikov and Nersesov 1964, Leake 1978] leading to partial melting of the lower crust and generation of granitic magma. Further petrochemical analyses and radiometric age data are needed to establish the mechanism responsible for the anomalous heat flow conditions which produced the Qaradagh-Tarom granite magma. Unfortunately no initial 87Sr/86Sr or any detailed study has been carried out along this belt.

The second Upper Eocene-Oligocene regional plutonic belt is mainly developed in the Zahedan area of southeast Iran (Fig. 7). The Zahedan plutonic belt intruded into the Paleocene-Eocene Zabol-Baluch flysch belt which was accreted into the continental margin during Upper Eocene-Oligocene and Oligocene-Miocene movements. If all the oceanic areas in Iran were destroyed in Upper Cretaceous times (about 65 Ma), the Upper Eocene-Oligocene Zahedan plutonism could not be subduction-related. Its position relative to the Upper Cretaceous ophiolites (Fig. 7) together with the absence of a coeval intense volcanism, makes it difficult to accept the Zahedan plutonism as a typical magmatic-arc formed in response to plate subduction. Similar to the early Tertiary plutonic belt along the Gulf of Alaska [Hudson et al. 1979] the Zahedan plutonic belt of southeast Iran closely followed a major accretionary episode. Therefore it could similarly be formed as a result of anatectic processes at the deeper parts within the thickened and deformed accretionary prism of the zabol-Baluch flysch belt. The idea given here is in the absence of petrochemical or radiometric data and is not a constrained model.

The Oligo-Miocene regional plutonic activity of Iran is developed along the Karkas and the Jebal-e-Barez belt of Central Iran (Fig. 8). The region was apparently a NW-SE subsiding continental region during Eocene and Oligocene time when between 11 to 18 km of volcano-detritic sediments were deposited in the basin [Dimitrijevic 1973]. The Oligo-Miocene intrusives cut the Eocene-Oligocene volcanics during convergence of the Arabian and Iranian plates which led to important intracontinental shortening and thickening during Late Eocene and Late Oligocene times. The trend of the Karkas and Jebal-e-Barez plutons is closely associated with the regional structural trend as well as with the subduction zone of the Tethyan ocean (Figs. 1 and 8). The plutonic activity in this belt seems to have started about Middle Oligocene with maximum activity in Lower-Middle Miocene (Fig. 8) and ended with isolated granodiorite emplacement during the Pliocene (Fig. 9). The Karkas-Jebal-e-Barez plutonic belt was thought to be the product of partial fusion of the lower crust due to crustal thickening and shortening. The belt is stretched along the Mesozoic convergent plate margin (Fig. 8) and the

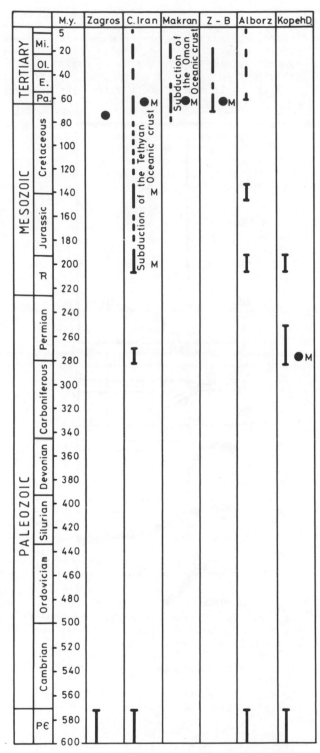

Fig. 10. Chronology of plutonic (vertical bars) and metamorphic (M) events together with ophiolite emplacement (●) in different Iranian structural units (cf. Fig. 1). Time ranges are based on data referenced and discussed in this chapter. Time scales from

Van Eysinga [1975]. C.Iran: Central Iran, Z-B: Zabol Baluch, Kopeh D: Kopeh Dagh structural units (see Fig. 1).

---

low 87Sr/86Sr initial ratio [Berberian 1981, Berberian et al. 1981] indicates that the magmas were derived by melting of a mantle or oceanic crust source. Assuming a Middle to late Tertiary continent-continent collision, the plutonism along this belt could be formed in response to plate subduction or be connected with a broken descending lithospheric slab beneath Central Iran. If we assume a late Cretaceous collision, then the time interval between the continental collision and the onset of plutonism would be about 40 My. By this time the slab must long have past the depth where calc-alkaline melts are usually generated. Detailed study of the Natanz batholith (situated along the Karkas-Jebal-e-Barez plutonic belt) cleared some of the ambiguities about the petrogenesis of this belt [Berberian 1981].

Concluding Remarks

Following the Paleozoic quiet Gondwanian tectono-sedimentary regime in Iran, tectonic activity began by continental rifting and alkali volcanism in the late Paleozoic-early Mesozoic and was followed by sea-floor spreading in the late Triassic along the Main Zagros reverse fault line. Data introduced in this study and summarized in Figures 2 to 11 show that the Mesozoic batholiths were produced above the subduction zone and are considered here as subduction by-products. Detailed isotopic and geochemical study of the late Cretaceous Bazman (Fig. 6) and the Oligo-Miocene Natanz (Fig. 8) batholiths [Berberian 1981] show that an Andean-type subduction could be satisfactorily applied as a mechanism for generating the I-type melts above subducting oceanic crust and along an active continental margin. Intrusion of granite magma took place repeatedly from Triassic to Miocene along the active continental margin of Central Iran. Apparently gaps in the Mesozoic-Tertiary plutonic activity (Fig. 10) appears to record temporary changes in spreading direction and cessation or change in rates of subduction. Features such as changes in the dip of the subducted slab, its velocity and duration in any one episode, should be determined from well-dated rocks and trace-major element studies, which unfortunately are not available for all plutonic rocks.

The late Cretaceous batholith of Bazman in southeast Iran (Fig. 6) is interpreted as subduction by-product of the Arabian oceanic crust beneath Central Iranian continental margin at the Gulf of Oman [Berberian 1981]. The study indicates that the subduction of the Oman oceanic crust may have started much earlier than Cretaceous and that the Oman oceanic crust is older than Cretaceous. Isotopic and trace-major element analyses of the Oligo-Miocene batholiths (Fig. 8) [Berberian 1981]

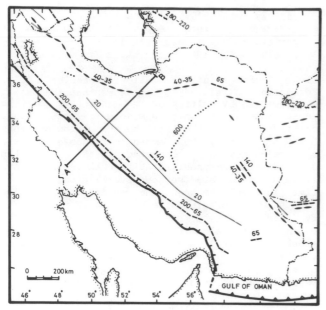

Fig. 11. Major Iranian plutonic trends since the late Precambrian with migration of plutonic fronts since Mesozoic. Numbers are in million years. A-B: Cross section perpendicular to the Mesozoic-Paleogene subduction zone (see Fig. 12).

do not correlate with the late Cretaceous closure of the Tethyan ocean between the Arabian and the Central Iranian plates. It possibly requires a much younger continent-continent collision. Despite the lack of younger pelagic sediments or of ophiolite-melange emplacement, the results of detailed study [Berberian 1981] fit with the speculative Miocene closure models of the Tethyan ocean proposed by Dewey et al. [1973], Smith [1973], Forster [1974], Haynes and McQuillan [1974], Krumsiek [1976], Kanasewich et al. [1978], Klootwijk [1979], Sengor [1979] and Powel [1979]. Therefore, the late Cretaceous ophiolites (emplaced along Zagros, Central Iranian and Makran continental margins) could have formed as arc-basin complexes above short-lived subduction zones; and their accretion to the continental margins do not indicate a continent-continent collision.

It is possible that the slope of the subducted slab underneath southwestern Central Iran decreased from the Mesozoic slope (Fig. 12) and changed the magmatic front pattern through the time (Fig. 11). A low angle subducting oceanic slab could cause the arc zone to advance northwards (extending further inland, beyond the active margin) producing the Tarom plutonic belt (Figs. 7, 11 and 12b). During the Middle Tertiary the angle of subduction zone possibly steepened back to nearly 40° situation as the arc front retreated southwestward and produced the Natanz batholith and the Karkas-Jebal-e-Barez plutonic

belt (Figs. 8, 11 and 12c). Although the scheme of changing the dip of the subduction zone and therefore, of migration of the magmatic-front was adopted by Matsuda and Uyeda [1971] for Japan, by Coney and Reynolds [1977] and Keith [1978] for the Californian Cordillera, it suffers some critisism. Apparently it is mechanically difficult to change

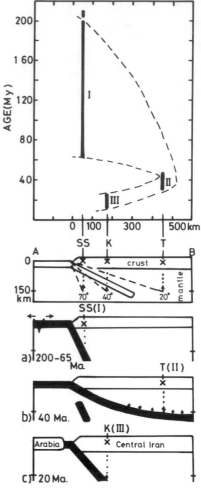

Fig. 12. Top diagram: possible trend of the Mesozoic-Tertiary plutonic rocks ploted as a function of time. Plutonic ages are projected into a line perpendicular to the Mesozoic-Paleogene subduction zone (A-B in Fig. 11).
SS: Plutonics of the Sanandaj Sirjan belt (active continental margin of Central Iran; Figs. 4, 6 and 11). T:Tarom plutonic belt (Fig. 7) with approximate age of 40-35 Ma (Fig. 11). K: Karkas-Jebal-e-Barez plutonic belt (Fig. 8) with an approximate age of 20 Ma (Fig. 11).
Cross sections: Various dip-angles of the Tethyan subduction zone underneath Central Iran with points of intersection with 150 km depth line.

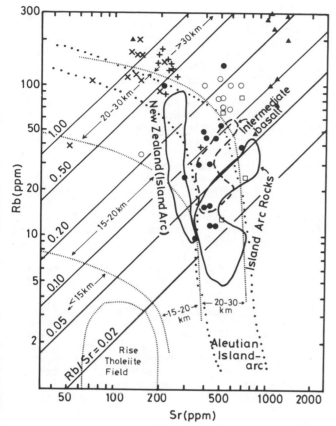

● 1, ▲ 2, + 3, × 4, ○ 5, □ 6

Fig. 13. Rb-Sr plot for the Paleogene volcanic rocks of the Natanz area along the karkas plutonic belt (cf. Fig. 8). Chemical analyses are taken from Amidi [1975, 1977].
Boundary lines of known calc-alkaline series (labelled fields) after DeAlbuquerque [1979], and the Aleutian calc-alkaline island-arc suites (Captains Bay plutons and Unalaska island-arc volcanics, shown by dotted lines) after Perfit et al. [1980]. Rb-Sr crustal thickness grid is taken from Condi [1973, 1976]. Note that almost all the Paleogene andesites fall in the field of the Aleutian island-arc rocks (dotted line).
1- Paleogene andesites, 2- Eocene shoshonites, 3- Eocene rhyolites, 4- Oligocene rhyolites, 5- Pliocene volcanics, 6- Quaternary volcanics.

the dip of the subducting oceanic crust [Elston and Bornhorst 1979].

Alternatively it is possible that the first Mesozoic Mariana-type steeply dipping slab (Fig. 12a) broke and was then overriden by a shallow dip slab during the early Tertiary due to a possible trench-continent collision (Fig. 12b; imbrication model of Lipman et al. [1972]). During this

modified early Tertiary Andean-arc stage, the hot oceanic lithosphere is being entrained beneath the Central Iranian continental crust. This could possibly be responsible for the widespread Paleogene volcanic activity of Central Iran and the plutonism along the Tarom belt (Figs. 7 and 12b).

Unfortunately no study has been carried out on the late-Eocene plutonic rocks of the Tarom belt but parts of the Paleogene andesite along its southeastern extension (from the Natanz area, 'N' in Fig. 8) have been studied. Rb/Sr and K/Rb data for these volcanics were taken from Amidi [1975, 1977] and plotted in Figure 13. Interestingly the Paleogene andesites fall in the field of the major island-arcs (especially in the field of the plutonic-volcanic rocks of the Aleutian island-arc) [Perfit et al. 1980]. This coincidence may indicate that the Paleogene volcanics, through which, the Natanz batholith was intruded, could also be subduction related. However, the volcanics were interpreted as crustal melts by Amidi [1975 and 1977].

During late Oligocene-early Miocene the dip of the subduction zone apparently steepened due to the approach of the Arabian plate to the Central Iranian continental crust (Fig. 12c), and the plutonic front shifted backwards (Fig. 11) close to the active continental margin (emplacement of the Natanz batholith and the Karkas-Jebal-e-Barez plutonic belt; Fig. 8). Development of the major Iranian porphyry copper deposits [Waterman and Hamilton 1975] in the southern segment of the Karkas-Jebal-e-Barez plutonic belt is possibly another support for their Andean-type origin. It is generally accepted that the porphyry copper deposits are characteristically associated with I-type subduction-related granites [Gustafson 1979].

Waning of calc-alkaline volcanism around 6-5 Ma (late Miocene-early Pliocene) followed by the onset of alkaline volcanism in NW Iran [Alberti et al. 1980, Innocenti et al. 1980] may indicate that by this time the last piece of broken oceanic slab(s) passed the depth where calc-alkaline melts are usually generated. During the same period the shallow water marine environment of the Zagros sedimentary basin (deposition of the early to Middle Miocene Mishan Formation) [James and Wynd 1965] was changed to synorogenic diachronous clastic deposits (Agha Jari Formation of late Miocene to Pliocene). Apparently during late Miocene and Pliocene, an Alpine continental tectonic regime was superimposed on the Mesozoic-Tertiary Andean-type and the convergence of the plates resulted in thickening and shortening of the continental crust by folding, reverse faulting and elevation of the Iranian plateau.

Acknowledgement. This work was partly supported by the Department of Earth Sciences, Cambridge University and the Geological and Mineral Survey of Iran. We would like to thank Frances Delany, Harsha Gupta, Basil King, Ian Muir, Euan Nisbet, and anonymous AGU-GSA reviewers for critically reading the manuscript and for valuable discussions.

References

Abu-Bakr, A.M., and R.O. Jackson, Geological map of Pakistan, 1:2,000,000, Geol. Surv. Pakistan, Quetta, 1964.

Aghanabati, S.A., Etude geologique de la region de Kalmard (W, Tabas): Stratigraphie et Tectonique, Geol. Surv. Iran, 35, 230p, 1977.

Alavi, M., Tectonostratigraphic evolution of the Zagrosides of Iran, Geology, 8, 144-149, 1980.

Alberti, A.A, Nicoletti, M., and C. Petrucciani, K/Ar age for micas of Mashhad granite (Khorasan, NE Iran), Period. Miner. 42(3), 483-493, Rome, 1973.

Alberti, A.A., Comin-Chiaramonti, P., Sinigoi, S., Nicoletti, M., and C. Petrucciani, Neogene and Quaternary volcanism in eastern Azerbaijan (Iran): some K-Ar age determinations and geodynamic implications, Geol. Rundsch., 69(1), 216-225, 1980.

Allenbach, P., Geologie und petrographie des Damavand und seiner umgebung (Zentral Elburz), Iran, Mitt. Geol. Inst. ETH Univ. Zurich, n.s., No. 63, 114p, 1966.

Alric, G., and D. Virlogeux, Petrographie et geochimie des roches metamorphiques et magmatiques de la region de Deh Bid-Bawanat, chaine de Sanandaj Sirjan, Iran, These, 3eme cycle, Geol. Appl., Mention Petrol. Metallogenie, Univ. Sci. Med, Grenoble (Institut Dolomieu) France, 239p, 1977.

Amidi, S.M., Contribution a l'etude stratigraphique, petrologique et petrochimique des roches magmatiques de la region Natanz-Nain-Surk (Iran Central), Thesis, Grenoble, 316p, 1975.

Amidi, S.M., Etude geologique de la region de Natanz-Surk (Iran Central), stratigraphie et petrologie, Geol. Surv. Iran, 42, 316p, 1977.

Anderson, O., The system anorthite-forsterite-silica, Am. J. Sci., 39 (4th Ser.), 407-454, 1915.

Andreyev, A.P., Brodovoy, V.V., Goldschmidt, V.I., Kuzmin, Yu.V., Morozov, M.D., and R.A. Eydlin, Deep tectonic subdivision of Kazakhstan, from geophysical data, Sovet. geol., 6, 34-47, 1966.

Annells, R.N., Arthurton, R.S., Bazley, R.A., and R.G. Davies, Explanatory text of the Qazvin and Rasht quadrangles map, Geol. Surv. Iran, E3 and E4, 94p, 1975.

Arthurton, R.S., Sarwar Alam, G., Arisuddin-Ahmad, S., and S. Iqbal, Geological history of the Alamreg-Mashki Chah area, Chagai district, Baluchestan, In Geodynamics of Pakistan, edited by A. Farah, and K.S. deJong, Geol. Surv. Pakistan, Quetta, 325-331, 1979.

Assereto, R., The paleozoic formations in Central Elburz (Iran), preliminary note, Riv. Ital. Paleont. Strat., 69, 503-543, 1963.

Assereto, R., The Jurassic Shemshak Formation in Central Elburz (Iran), Riv. Ital. Paleont. Strat., 72(4), 1133-1182, 1966.

Azizbekov, S.A., and G.S. Dzotsenidze, Magmatism in the Caucasus, Iran and Turkey, AN SSSR Izve. ser. geol. 12, 15-24, 1970, (English translation in Internat. Geology Rev., 13(10), 1464-1470, 1971).

Beckinsale, R.D., Granite magmatism in the Tin Belt of South-east Asia, In Origin of Granite Batholiths, Geochemical Evidence, edited by M.P. Atherton and J. Tarney, Shiva Publishing Ltd., U.K., 34-45, 1979.

Berberian, F., Petrogenesis of Iranian plutons: a study of the Natanz and Bazman intrusive complex, Ph.D. Thesis, University of Cambridge, 300p, 1981.

Berberian, F., Pankhurst, R.G., Muir, I.D., and M. Berberian, Late Cretaceous and early Miocene Andean-type plutonic activity in northern Makran and Central Iran (in preparation).

Berberian, M., Structural history of the Southern Lut Zone (northern highlands of Jaz Murian Depression-Baluchestan), a preliminary field note, Geol. Surv. Iran, Int. Rep., 1974., 21p.

Berberian, M., Contribution to the Seismotectonics of Iran (part II), Geol. Surv. Iran, 39, 1976.

Berberian, M., Active faults and tectonics of Iran (this volume).

Berberian, M., and N. Alavi-Tehrani, Structural analyses of Hamadan metamorphic tectonites: a paleotectonic discussion, In Contribution to the Seismotectonics of Iran (part III), edited by M. Berberian, Geol. Surv. Iran, 40, 263-279, 1977.

Berberian, M., and G.C.P. King, Towards a paleogeography and tectonic evolution of Iran. Can. J. Earth. Sci., 18(2), 1981.

Berberian, M., and M. Nogol, Preliminary explanatory text of the geology of Deh Sard and Khalor maps with some remarks on the metamorphic complexes and the tectonics of the area (two geological maps 1:100,000 from the Hajiabad quadrangle map), Geol. Surv. Iran, Int. Rep., 60p, 1974.

Boulin, J., and E. Bouyx, Introduction a la geologie de l'Hindou Kouch occidental, Mem. h. ser. geol. France, 8, 87-105, 1977.

Bowen, N.L., The ternary system: diopside-forsterite-silica, Am. J. Sci., 38 (4th Ser.), 207-264, 1914.

Bowen, N.L., The crystallization of haplobasaltic, haplodioritic, and related magmas, Am. J. Sci., 40 (4th Ser.), 161-185, 1915.

Bowen, N.L., and J.F. Schairer, Crystallization equilibrium in nepheline-albite-silica mixtures with fayalite, J. Geol., 46, 397-411, 1938.

Brown, G.F., Tectonic map of the Arabian Penensula, 1:4,000,000, Arabian Peninsula Series, Map AP-2, Directorate General of Mineral Resources, Jiddeh, 1972.

Burtman, V.S., Structural geology of Variscan Tien Shan, USSR, Am. J. Sci., 275-A, 157-186, 1975.

Chappell, B.W., and A.J.R. White, Two contrasting grnite types, Pacific Geology, 8, 173-174, 1974.

Clark, G.C., Davies, R.G., Hamzepour, B., and C.R.

Jones, Explanatory text of the Bandar-e-Pahlavi quadrangle map 1:250,000, Geol. Surv. Iran, D3, 198p, 1975.

Condie, K.C., Archean magmatism and crustal thickening, Geol. Soc. Am. Bull., 84, 2981-2992, 1973.

Condie, K.C., Magma associations, In Plate Tectonics and Crustal Evolution, edited by K.C. Condie, Pergamon Press Inc., 145-174, 1976.

Coney, P.J., and S.J. Reynolds, Cordillern Benioff zones, Nature, 270, 403-406, 1977.

Conrad, G., Conrad, J., and M. Girod, Les formation continentales tertiaires et quaternaires du bloc du Lout (Iran): importance du plutonisme et du volcanisme, Mem. h. ser. Soc. geol. France, no. 8, 53-75, 1977.

Crawford, A.R., A summary of isotopic age data for Iran, Pakistan and India, Mem. h. Ser. Soc. geol. France, 8, 251-260, 1977.

Davies, R.D., Allsopp, H.L., Erlank, A.J., and W.I. Manton, Sr- isotopic studies on various layered mafic intrusions in southern Africa, in Symposium on the Bushveld igneous complex and other layered intrusions, Geol. Soc. S. Africa, Spec. Pub. 1, 576-593, 1970.

Davies, R.G., Jones, C.R., Hamzepour, B., and Clark G.C., Geology of the Masuleh sheet 1:100,000, northwest Iran, Geol. Surv. Iran, 24, 110p, 1972.

DeAlbuquerque, Origin of the plutonic mafic rocks of southern Nova Scotia, Geol. Soc. Am. Bull., 1, 90, 719-731, 1979.

Debon, F., Le Fort, P., and Sonet, J., Des caracteres geochimiques de deux provinces plutoniques d'Afghanistan: Hindou-Kouch et Montagnes Centrales, 6eme Reun. Ann. Sci. Terr, Orsay, 135p, 1978.

Dedual, E., Zur Geologie des mittleren und unteren Karaj-Tales, Zentral-Elburz (Iran), Mitt. Geol. Inst. ETHu. Univ. Zurich, n.s., 76, 123p, 1967.

Desio, A., Geology of central Badakshan (northeast Afghanistan) and surrounding countries, In Italian Expedition to Karakorum and Hindu-Kush, Rep. III, 3, Brill, Leiden, 1975.

Desio, A., Correlation entre les structures des chaines du nurd-est de l'Afghanistan et du nord-ouest du Pakistan, Mem. h. ser. Soc. geol. France, 8, 179-188, 1977.

Dewey, J.F., Pitman, W.C., Ryan, W.B.F., and J. Bonnin, Plate tectonic and the evolution of the Alpine system, Geol. Soc. Am. Bull. 84, 3137-3180, 1973.

Dickinson, W.R., Relations of andesites, granites and derivative sandstones to arc-trench tectonics, Rev. Geophys. Space Phys., 8(4), 813-860, 1970.

Dickinson, W.R., Plate tectonic models of geosynclines, Earth Planet. Sci. Lett., 10, 165-174, 1971.

Dickinson, W.R., Subduction and oil migration, Geology, 2(9), 421-424, 1974.

Didon, J., and Y.M. Gemain, Le Sabalan, volcan Plio-Quaternaire de l'Azarbaijan oriental (Iran), Etude geologique et petrographique du l'edifice et de son environment regional, These, 3eme cycle, Grenoble, France, 1976.

Dimitrijevic, M.D., Geology of Kerman region, Geol. Surv. Iran, Yu/52, 334p, 1973.

Eftekhar-nezhad, J., Geological map of the Mahabad quadrangle, 1:250,000, Geol. Surv. Iran, B4, 1973.

Eftekhar-nezhad, J., Brief description of tectonic history and structural development of Azarbaijan, Geodynamics of South West Asia Symposium, Geol. Surv. Iran, Sp. Pub., 469-478, 1975.

Elston, W.E., and T.J. Bornhorst, The Rio Grande rift in Context of regional post-40 M.Y. volcanic and tectonic events, In Rio Grand Rift, Tectonics and Magmatism, edited by R.E. Riecker, 416-438, 1979.

Farhudi, G., and D.E. Karig, Makran of Iran and Pakistan as an active arc system, Geology 5(11), 664-668, 1977.

Faure, G., Principles of Isotope Geology, John Wiley and Sons, Inc., 419p, 1977.

Faure, G., and J.L. Powell, Strontium isotope geology, Springer-Verlag, 188p, 1972.

Fitch, F.H., Informal lithostratigraphic lexicon for the Arabian Shield, Technical Record TR., 1, Dir. Gen. Min. Res., Kingdom of Saudi Arabia, 166p., 1978.

Forster, H., Magmentypen und erzlagerstatten im Iran, Geol. Rundschau, 63(1), 276-292, 1972.

Garkovets, V.G., Structural and metallogenic relationship between Tien Shan and Ural, Sovet. geol. 11, 72-83, 1964.

Gannsser, A., and H. Huber, Geological observations in Central Elburz, Iran, Schweiz. Min. Petr. Mitt., 42(2), 583-630, 1962.

Gass, I.G., The evolution of the Pan African crystalline basement in NE Africa and Arabia, J. Geol. Soc. London 134, 129-138, 1977.

Gass. I.G., Evolutionary model for the Pan African crystallien basement, In Evolution and Mineralization of the Arabian-Nubian Shield, edited by A.M.S. Al-Shanti, IAG. Bull. 3, 11-20, 1979.

Green, T.H., and A.E. Ringwood, Genesis of calc-alkaline igneous rock suite, Contrib. Min. Pet., 18, 105-162, 1968.

Greenwood, W.R., Hadley, D.G., Anderson, R.E., Fleck, R.J., and D.L. Schmidt, Late Protozoic cratonization in southwestern Saudi Arabia, U.S. Geol. Surv. Saudi Arabian Project Rep. 196, 23p, 1975.

Greenwood, W.R., Hadley, D.G., Anderson, R.E., Fleck, R.J., and D.L. Schmidt, Late Proterozoic cratonization in southwestern Saudi Arabia, Phil.Trans. R. Soc. London, A.280, 517-527, 1976.

Gustafson, L.B., Porphyry copper deposits and calc-alkaline volcanism, In The Earth: Its Origin, Structure and Evolution, edited by M.W. McElhinny, Academic Press, 427-468, 1979.

Hadley, D.G., and D.L. Schmidt, Proterozoic sedimentary rocks and basins of the Arabian shield and their evolution, Precambrian Res., 6, A23 (Abs.), 1978.

Haghipour, A., Etude geologique de la region de Biabanak-Bafq (Iran Central); petrologie et tectonique du socle Precambrian et de sa couverture, Thesis, Grenoble, 403p, 1974.

Haghipour, A., Etude geologique de la region de Biabanak-Bafq (Iran Central); Petrologie et tectonique du socle Precambrian et de sa couverture, Geol. Surv. Iran, 34, 403p, 1978.

Haghipour, A., and G. Pelissier, Geology of the Posht-e-Badam/Saghand area (East-Central Iran), Geol. Surv. Iran, 48, Internal Rep., 144p, 1968.

Haghipour, A., Valeh, N., Pelissier, G., and M. Davoudzadeh, Explanatory text of the Ardekan quadrangle map, 1:250,000, Geol. Surv. Iran, H8, 114p, 1977.

Haynes, S.J., and H. McQuillan, Evolution of the Zagros suture zone, southern Iran, Geol. Soc. Am. Bull., 85, 739-744, 1974.

Hedge, C.E., and Z.E. Peterman, Strontium: Isotopes in Nature, In Handbook of geochimistry, Springer-verlag, N.Y., II/4, 38B1-38B14, 1974.

Hirayama, K., Samimi, M., Zahedi, M., and A. Hushmandzadeh, Geology of the Tarom district, western part (Zanjan are, Northwest Iran), Geol. Surv. Iran, 8, 31p, 1966.

Holzer, F., and M. Samimi, Geological investigations in the Puzeh Bagh-Tanak area, Bazman district, Baluchestan, Granites and mineral occurrences, Geol. Surv. Iran, Int. Rep., 1970.

Huber, H., Geological map of Iran, 1:1,000,000, with explanatory note, NIOC, Exploration and Production Affairs, Tehran, 1978.

Huckriede, R., Kursten, M., and H. Venzlaff, Zur geologie des gebietes zwischen Kerman und Saghand (Iran), Beih. Geol. Jahrb., 51, 197p, 1962.

Hudson, T., Plafker, G., and Z.E. Peterman, Paleogene anatexis along the Gulf of Alaska margin, Geology, 7, 573-577, 1979.

Hushmandzadeh, A., Metamorphism et granitisation du massif Chapedony (Iran Central), These, Univ. Grenoble, 242p, 1969.

Hushmandzadeh, A., Ophiolites of south Iran and their genetic problems, Geol. Surv. Iran, Int. Rep., 89p, 1977.

Hushmandzadeh, A., Alavi-Naini, M., and A. Haghipour, Geological evolution of Torud area (Precambrian to Recent), Geol. Surv. Iran, H5, 138p (in Farsi), 1978.

Innocenti, F., Mazzuoli, R., Pasquare, G., Serri, G., and L. Villari, Geology of the volcanic area north of Lake Van (Turkey), Geol. Rundsch., 69(1), 292-323, 1980.

Isacks, B., and M. Barazangi, Geometry of Benioff zones: lateral segmentation and downwards bending of the subducted lithosphere, In Island Arcs, Deep Sea Trenches and Back-Arc Basins, Maurice Ewing Series 1, Am. Geophys. Union, 99-114, 1977.

Ishihara, S., The magnetite-series and ilmenite-series granitic rocks, Ming. Geol., 27, 293-305, 1977.

James, G.A., and J.G. Wynd, Stratigraphic nomenclature of Iranian Oil Cosortium Agreement area, Am. Assoc. Petrol. Geol. Bull., 49(12), 2182-2245, 1965.

Jaubert, A., Voyage en Armenie et en Perse, suivi d'un memoire sur le Ghilan et le Masenderan par M.Trezel, Paris, 1821.

Jones, A.G., (ed.), Reconnaissance geology of part of West Pakistan, a Columbo Plan Cooperative Project, Ottawa, Government of Pakistan, 550p (Hunting Survey Report), 1960.

Kanasewich, E.R., Havskov, J., and M.E. Evans, Plate tectonics in the Phanerozoic, Can. J. Earth Sci., 15, 919-955, 1978.

Keith, S.B., Paleosubduction geometrics inferred from Cretaceous and Tertiary magmatic patterns in southwestern North America, Geology, 6, 516-521, 1978.

Kent, P.E., The salt plugs of the Persian Gulf region, Leicester Literary and Philos. Soc. Trans., 64, 56-88, 1970.

Kent, P.E., The emergent Hormuz Salt plugs of southern Iran, J. Petrol. Geol., 2(2), 117-144, 1979.

Khain, V.E., Structure and main stages in the tectono-magmatic development of the Caucasus: an attempt at geodynamic interpretation, Am. J. Sci., 275.A, 131-156, 1975.

Khain, V.E., The new international tectonic map of Europe and some problems of structure and tectonic history of the continent, In Europe from Crust to Core, edited by D.V. Ager and M. Brooks, Wiley, London, 1976.

Klootwijk, C.T., India's and Australia's pole path since the late Mesozoic and The India-Asia collision, Nature, 286, 605-607, 1979.

Knyazev, V.S., and O.A. Shnip, Magmatic rocks in the base of the Turanina plate, Sovets. Geol., 5, 69-82, 1970.

Kornev, G.P., Minor intrusions and sills in the eastern Nakhichevan A.S.S.R and the tectonics of their formations, Sovet. Geol. 7, 34-45, 1961.

Krestikov, V.N. and I.L. Nersesov, Relation of the deep structures of the Parmirs and Tien Shan to their tectonics, Tectonophys. 1, 183-193, 1964.

Kroner, A., M.J. Roobol, C.R. Ramsay and N.J. Jackson, Pan African ages of some gneissic rocks in the Saudi Arabian Shield, J. Geol. Soc. London, 136, 455-461, 1979.

Krumsiek, K., Zur bewegung der Iranisch-Afghanischen platte (palaeomagnetisch ergebnisse), Geol. Rundsch., 65(3), 909-929, 1976.

Kushiro, I., Fractional crystallization of basaltic magma, In The Evolution of the Igneous Rocks, Fiftieth Anniversary Perspectives, edited by H.S. Yoder, Jr. Princeton University Press, Princeton, New Jersey, 171-203, 1979.

Leake, B.E., Granite emplacement: the granites of Ireland and their origin, In Evolution in NW Britain and adjoining region, edited by D.R. Bowes and B.E. Leake, Geophys. J. Sp. Iss. 10, 221-248, 1978.

Lipman, R.W., Prostka, H., and R.L. Christiansen, Cenozoic volcanism and plate-tectonic evolution of the western United States, part 1, Early and

Middle Cenozoic, Phil. Trans. Roy. Soc. London, A.271, 217-248, 1972.

Madani, M., A study of the sedimentology, stratigraphy and regional geology of the Jurassic rocks of Eastern Kopet Dagh, NE Iran, Ph.D. Thesis, Petroleium Geology Section. I.C. London Univ., 1977.

Majidi, B., Etude petrostructurale de la region de Mashhad (Iran), les problemes des metamorphites, serpentinites, et granitoides hercyniens, Thesis, Universite de Grenoble, 277p, 1978.

Majidi, B., and N. Alavi-Tehrani, Geological report on the igneous and metamorphic rocks of the Hamedan Quadrangle, Geol. Surv. Iran, Int. Rep., 1972.

Marsh, B.D., and I.S.E. Carmichael, Benioff zone magmatism, J.Geophys. Res., 79, 1196-1206, 1974.

Matsuda, T., and S. Uyeda, On the Pacific-type orogeny and its model-extension of the paired belts concept and possible origin of marginal seas, Tectonophys., 11, 5-27, 1971.

Mitchell, A.H., and H.G. Reading, Evolution of island arcs, J. Geol. 79(3), 253-284, 1971.

Nabavi, M.H. (compiler), Geological map of the Yazd quadrangle, 1:250,000, Geol. Surv. Iran, H9, 1970.

Nabavi, M.H., An introduction to the Iranian geology, Geol. Surv. Iran, 110p (in Farsi), 1976.

Neary, C.R., Gass, I.G., and B.J. Cavanagh, Granitic association of northern Sudan, Geol. Soc. Am. Bull., 87, 1501-1512, 1976.

NIOC, Geological map of Iran, 1:2,500,000, with explanatory text, National Iranian Oil Company, 1959.

Oxburgh, E.R., and D.L. Turcotte, Thermal structure of island arcs, Geol. Soc. Am. Bull., 81, 1665-1688, 1970.

Perfit, M.R., Brueckner, and J.R. Lawrence, Trace element and isotopic variations in a zoned pluton and associated volcanic rocks, Unalaska Island, Alaska: a model for fractionation in the Aleution Calc-alkline suite, Contrib. Min. Pet., 73, 69-87, 1980.

Portnyagin, E.A., Koshlakov, G.V., and Ye. S. Kuznetsov, Problem of mutual relationships between the deep seated Paleozoic structures of Southern Tien Shan and the buried Tadzhik-Afghan Massif, Moskov. Obshch. Ispytaletey, Prirody Byull., otd. geol., 49(3), 18-23, 1974.

Powell, C. McA., A speculative tectonic history of Pakistan and surroundings: Some constraints from the Indian Ocean, In Geodynamics of Pakistan edited by A. Farah and K.A. DeJong, Geol. Surv. Pakistan, Quetta, 5-24, 1979.

Presnall, D.C., Fractional Crystallization and partial fusion, In The Evolution of the Igneous Rocks: Fiftieth Anniversary Perspectives edited by H.S. Yoder Jr., Princeton Univ. press, Princeton, New Jersey, 59-75, 1979.

Rassaghmanesh, B., Die Kupfer-blei-zink-erzlagerstatten von Taknar und ihr geologischer rahmen (Nordost-Iran), Thesis Rhein.-Westph. Techn. Hochsch., Aachen, 130p, 1968.

Reyre, D., and S. Mohafez, A first contribution of the NIOC-ERAP agreements to the knowledge of Iranian geology, Edition Technip, Paris, 58p, 1972.

Ricou, L.E., L'evolution geologique de la region de Neyriz (Zagros iranien) et l'evolution structurale des Zagrides, These, Univ. Orsay, no. AD, 1269, 1974.

Ringwood, A.E., The petrological evolution of island arc sustems, J. Geol. Soc. London, 130, 183-204, 1974.

Sabzehei, M., Les melange ophiolitiques de la region d'Esfandagheh, These, Univ. Grenoble, 306p, 1974.

Sabzehei, M., and M. Berberian, Preliminary note on the structural and metamorphic history of the area between Dowlatabad and Esfandagheh, southeast Central Iran, Geol. Surv. Iran, Int. Rep., 30p; and 1st Iranian Geol. Symp., 1973, Iranian Petrol. Inst., Tehran (Abst.), 1972.

Sabzehei, M., Majidi, B., Alavi-Tehrani, N., and H. Etminan, (compiled by W.A. Watters and M. Sabzehei), Preliminary report, geology and petrography of the metamorphic and igneous complex of the Central part of Neyriz Quadrangle, Geol. Surv. Iran. Int. Rep, 1970.

Sborshchikov, I.M., Dronov, V.I., Denikaev, Sh. Sh., Kafarskiy, A.Kh., Karapetov, S.S., Akhmedzynanov, F.U., Kalimulin, S.M., Slavin, V.I., Sonin, I.I. and K.F. Stazhilo-Alekseev, Tectonic map of Afghanistan, 1:2,500,000, Ministry of Mines and Industries, Department of Geology and Mines, Kabul, Afghanistan, 1972.

Schairer, J.F., and N.L. Bowen, The system, leucite-diopside-silica, Am. J. Sci., 35A, 289-309, 1938.

Schairer, J.F., and N.L. Bowen, The system anorthite-leucite-silica, Bull. Soc. Geol. Finlande, 20, 67-87, 1947.

Schairer, J.F., and H.S. Yoder, The system albite-forsterite-silica, in Carnegie Inst. Washington Year Book 59, 69-70, 1960.

Schmidt, D.L., Hadley, D.G., Greenwood, W.R., Gonzalez, L., Coleman, R.G., and G.F. Brown, Stratigraphy and tectonism of the southern part of the Precambrian shield of Saudi Arabia, Dir. Gen. Min. Res., Kingdom of Saudi Arabia, 8, 13p., 1973.

Schmidt, D.L., Hadley, D.G., and D.B. Stoeser, Late proterozoic crustal history in the southern Najd province, Saudi Arabia, Precambrian Res., 6, A35 (Abst.), 1978.

Sengor, A.M.C., Mid-Mesozoic closure of Permo-Triassic Tethys and its implications, Nature 279, 590-593, 1979.

Seyed-Emami, K., A summary of the Triassic in Iran, Geol. Surv. Iran, 20, 41-53, 1971.

Shevchenko, V.I., Stankevich, Ye.K., and I.A. Rezanov, Pre-Cenozoic magmatism in Caucasus and western Turkmenia in relation to deep-seated sutures, Vyssh. Ucheb. Zavedniy Izv., Geologiya i Razvedka, 2, 3-12, 1973.

Smith, A.G., The so-called Tethyan ophiolites, In Implication of Continental Drift to the Earth

Sciences, edited by D.H. Tarling and S.K. Runcorn, Academic Press, London, V.2, 977-986, 1973.

Stazhilo-Alekseev, K.F., Chmyriov, V.M., Mirzad, S.H., Dronov, V.L., and A.Kh. Kafarskiy, The main features of magmatism of Afghanistan, Geol. Surv. Afghanistan, 31-43, 1973.

Stazhilo-Alekseev, K.F., Dronov, V.I., Kalimulin, S.M., Kafarskiy, A. Kh., Katchetkov, A.Ya., Mirzad, S.H., Salah, A.S., Sborshchikov, I.M., Feoktistov, V.P., Chmyriov, V.M., and M.A. Chatvan, Scheme of magmatic complexes of Afghanistan arrangement, 1:2,500,000, Ministry of Mines and Industries, Geol. Surv., Afghanistan, 1972.

Stocklin, J., Structural history and tectonics of Iran; a review, Am. Assoc. Petrol. Geol. Bull., 52(7), 1229-1258, 1968a.

Stocklin, J., Salt deposits of the Middle East, Geol. Soc. Am. Sp. Paper 88, 157-181, 1968b.

Stocklin, J., Iran Central, septentrional et oriental, In Lexique Stratigraphique International, III, Fasc. 9b, Iran, CNRS, Paris, 1-283, 1972.

Stocklin, J., Possible ancient continental margins in Iran, In The geology of continental margins, edited by C.A. Burk and C.L. Drake, New york, Springer, 873-887, 1974.

Stocklin, J., Structural correlation of the Alpine ranges between Iran and Central Asia, Mem. h. ser. Soc. Geol. France, no.8, 333-353, 1977.

Stocklin, J., Eftekhar-Nezhad, J., and A. Hushmandzadeh, Geology of the Shotori Range (Tabas area, East Iran), Geol. Surv. Iran, 3, 69p, 1965.

Stocklin, J., and J. Eftekhar-Nezhad, Explanatory text of the Zanjan Quadrangle Map, 1:250,000, Geol. Surv. Iran, D4, 59p, 1969.

Stocklin, J., Eftekhar-Nezhad, J., and A. Hushmand-zadeh, Central Lut reconnaissance, east Iran, Geol. Surv. Iran, 22, 62p, 1972.

Stocklin, J., and M.H. Nabavi, Tectonic map of Iran, Geol. Surv. Iran, 1973.

Stocklin, J., Ruttner, A., and M.H. Nabavi, New data on the Lower Paleozoic and Pre-Cambrian of north Iran, Geol. Surv. Iran, 1, 29p, 1964.

Taraz, H., Geology of the Surmaq-Deh Bid area, Abadeh region, Central Iran, Geol. Surv. Iran, 37, 148, 1974.

Thiele, O., Alavi, M., Assefi, R., Hushmandzadeh, A., Seyed-Emami, K., and M. Zahedi, Explanatory text of the Golpaygan quadrangle map, 1:250,000, Geol. Surv. Iran, E7, 24p, 1968.

Tvalchrelidze, G.A., Comparative metellogenic description of the pyritic and copper porphyry associations, Geologiya Rudnykh Mestorozhdeniy, 3, 3-18, 1975.

Valizadeh, M.V., and J.M. Cantagrel, Premieres donnees radiometriques (k-Ar et Rb-Sr) sur les micas du complexe magmatique du Mont Alvand, Pres Hamadan (Iran occidental), C.R. Acad. Sc. Paris, D, 281, 1083-1086, 1975.

VanEysinga, F.W.B., Geological Time Table, 3rd edition, Elsevier, 1975.

Waterman, G.C., and R.L. Hamilton, The Sar Cheshmeh porphyry copper deposit, Economic Geology, 70, 568-576, 1975.

White, R.S., and K. Klitgord, Sediment deformation and plate tectonics in the Gulf of Oman, Earth Planet. Sci. Lett., 32, 199-209, 1976.

Yegorkina, G.V., Sokolova, I.A., and L.M. Yegorova, Deep-seated structure of Armenian ultrabasic belts, Sovet. Geol., 3, 127-134, 1976.

Zahedi, M., Etude geologique de la region de Soh (W de l'Iran Central), Geol. Surv. Iran, 27, 197p, 1973.

Zahedi, M., Explanatory text of the Esfahan Quadrangle Map, 1:250,000, Geol. Surv. Iran, F8, 49p, 1976.

Zarayan, S., Forghani, A.H., and H. Fayaz, Le massif granitique d'Alvande et son aureole metamorphique (3eme Partie), Bull. Faculty of Science, Tehran Univ. vol. 4(3), 82-90, 1972.

Zaridze, G.M., Association of or beds in granitoids of the Caucasus and genesis of these rocks, Sovet. Geol., 4, 81-97, 1958.

ACTIVE FAULTING AND TECTONICS OF IRAN

Manuel Berberian

Department of Earth Sciences, Bullard Laboratories, University of Cambridge,
Madingley Rise, Madingley Rd., Cambridge CB3 OEZ, U.K.

Abstract. Study of earthquake faulting in Iran has shown that the early Quaternary tectonic history as well as the pre-Quaternary geological record is very important in understanding the present-day continental deformation during earthquakes. Iranian seismic activity, which is intimately related to reactivation of the existing faults, can be separated into three categories: 1) in the Zagros active fold-thrust belt, where shortening along 'longitudinal high-angle reverse basement faults' spread over the entire belt, is absorbed by ductile layers of the top sedimentary cover (usually no earthquake rupture is observed at the surface), 2) in the Central Iranian plateau where the earthquakes are accompanied by surface faulting along 'mountain-bordering reverse faults', and 3) in the Makran accretionary flysch wedge where the oceanic crust of the Gulf of Oman is subducting underneath southeastern Central Iran. In general seismicity is discontinuous and earthquakes show seismic gaps of several years to several centuries. The tectonic movements during the late Neogene-early Quaternary, and the mechanism of the recent active fault motions show that the Iranian plateau is a 'broad zone of compressional deformation'. The plateau is a relatively weak belt affected by several collisional orogenic movements and is being compressed between two blocks of greater rigidity (Arabia and Eurasia), since 65 Ma. The compressional motion between these blocks resulted in a continuous 'thickening and shortening' of the continental crust by reverse faulting and folding in a NE-SW direction. Entrapment of the Iranian plateau between two impinging zones of the Arabian plate in the west and the Indian plate in the east, has provided a unique 'constrained convergent zone' along the Alpine-Himalayan belt, where the crust has to undergo only shortening and thickening along several reverse faults. Although the present deformation is inhomogeneous, it is not confined to simple rigid plate boundaries or concentrated along major strike-slip faults predicted by the slip-line theory. The deformation mainly takes place along many 'mountain-bordering reverse faults' and very few strike-slip faults are involved. The active faults are deep-seated multi-role structures inherited from previous deformational phases; some of them possibly influenced sedimentation during different geological periods.

Introduction

Accurate knowledge of historical and modern seismicity as well as recent faulting is an important tool in understanding active continental tectonics. Epicentre maps of Iran, especially those which include both instrumental and pre-instrumental data [Berberian 1976a, 1977a] show that the country may be considered as a broad seismic zone over 1,000 km wide and extending from the Turan platform (southern Eurasia) in the northeast to the Arabian plate in the southwest (Fig. 1). The Kopeh Dagh and the Zagros, the two active fold-thrust mountain belts, respectively in NE and SW Iran, form the active border belts at the contacts with the aseismic cratons of Turan and Arabia (Figs. 1 and 2).

The Iranian plateau is characterized by active faulting, recent volcanics and high surface elevation along the Alpine-Himalyan mountain belt [Berberian 1976a]. Tectonic studies indicate that the Iranian plateau has a very high density of active and recent faults and that reverse faulting dominates the tectonics of the region [Berberian 1976a, 1979a]. Unlike northern Asia-Minor [Mckenzie 1972], central Asia [Molnar and Tapponnier 1975, 1978, Tapponnier and Molnar 1977, 1979] and Pacific margin of North America [Atwater 1970, Anderson 1971, Jones et al. 1977], where large scale horizontal motion occurs on narrow zones of major strike-slip faults, the active deformation in Iran is spread out over a large area along many reverse faults. This is presumably due to the fact that the Iranian plateau is, on the one hand, confined between the convergent movements of the Arabian and the Eurasian plates, and on the other hand, is laterally trapped between the Arabian plate with eastern Asia-Minor in the west and the Indian plate with Eurasia in the east (Fig.3). Because of this entrapment, none of the continental blocks forming the Iranian plateau can easily move sideways from the colli-

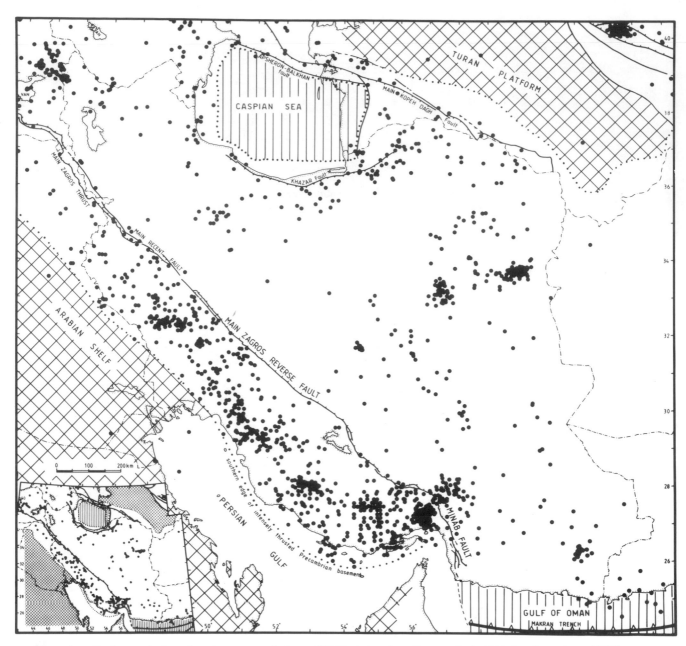

Fig. 1. Instrumental epicentre map of Iran (USGS data for the period 1961 to November 1980). It shows the high shallow inhomogeneous seismic activity of the Iranian plateau (cf. Fig. 12), and the two inactive platforms of Turan in the NE and Arabia in the SW (marked by cross hatching; cf. Fig. 2). Most activity is concentrated along the Zagros active fold-thrust belt (with a sharp break in seismicity to the northeast of the Main Zagros reverse fault line; cf. Fig. 2) at the northern margin of the Arabian plate (characterized by buried faults); and less activity in Central and east Iran (characterized by surface ruptures with earthquakes larger than Mb=5.5). Location error varies from 30 km in 1963 to 10 km in 1980 [Berberian 1979c]. No earthquake larger than Ms=7.0 has been experienced in the Zagros during the 20th century, but shocks of magnitude over Ms=7.0 have occurred in Central Iran. The seismicity follows the pattern of the fault map (Fig. 4), the documented seismically active fault map (Fig. 5) and the young fold belts of the mountain system in different structural units (Figs. 2 and 3). Apparently the recent deformation stops in eastern Central Iran and does not continue into western Afghanistan. Areas with oceanic crust (Makran in southeast and South Caspian in north) are indicated by vertical hatching. Lambert Conformal Conic Projection.

sion zone, along a major strike-slip fault. Since the Asia-Minor block is free to move westwards, the convergent movements of the Arabian plate, west of the impinging zone, are simply taken up by sideways motion along two major strike-slip faults (North and East Anatolian). Because of entrapment of the Iranian plateau, the motion in the eastern part of the Arabian impinging zone is first taken up by several reverse faults with large strike-slip component (dextral-reverse oblique slip close to the impinging zone) and then by pure thrust faulting towards Central Iran (Fig. 3).

The convergent movements between the Arabian and the Eurasian plates, with an estimated rate of about 4.7-5.1 cm/yr [LePichon 1968, McKenzie 1972, Jacob and Quittmeyer 1979], are principally taken up by folding and reverse faulting along inherited structures within the Iranian continental crust (Fig.3). Only part of this rate is taken up by subduction of the Oman oceanic crust and deformation of the accretionary wedge in the Makran ranges of southeast Iran (Figs. 2 and 3). Wrench faulting is not the dominant tectonic regime in present-day Iran as has been previously suggested by Wellman [1966], Nowroozi [1971], Cummings [1975], Nogole-Sadate [1978], Holcombe [1978] and Mohajer-Ashjai and Nowroozi [1979], and the slip-line theory [Cummings 1975, Molnar and Tapponnier 1975] is not applicable to the Iranian plateau. As with the Pacific margin of North America [Jones et al. 1977, Irving 1979, Coney et al. 1980], large scale stike-slip motion may have occurred along the major Iranian faults before the crust became trapped between Arabian, Eurasian and Indian plates but this is not detectable from presently available paleogeographic data. Lack of oceanic magnetic anomaly pattern in present-day Iran (due to consumption of the Tethyan oceanic crust) and limited paleomagnetic data makes it difficult to illustrate the nature of the old motions along the major Iranian faults and blocks.

Some of the Iranian Quaternary faults are directly associated with known large magnitude earthquakes, and are capable of generating future earthquakes [Berberian 1976b, Berberian and Mohajer-Ashjai 1977]. Some other Quaternary faults, not directly related to historical (pre-1900) and 20th century earthquakes, should also be considered as potential sources of future earthquakes. Due to the considerable location error in the instrumental and relocated epicentres [Ambraseys 1978, Berberian 1979b, c] and lack of adequate micro-earthquake surveys, it is impossible to deduce the activity of most Iranian faults from the data supplied by national and international seismic agencies. The mean error in the instrumental and relocated epicenters of the

medium to large magnitude Iranian earthquakes range from 500 km (1903) to 300 km (1918), 30 km (1963) and 15 km (1977) which is not acceptable for detailed seismotectonic and seismic risk studies [see Berberian 1979c for discussion]. Moreover, poor azimuthal coverage combined with systematic and random reading errors and bias due to the difference between the real earth and the assumed earth model has caused considerable error in the focal depth estimation of the Iranian earthquakes. The correlation of Iranian instrumental seismic data with faults, requires location accuracies which unfortunately are not obtainable with the present limited number of regional seismological stations in the country. Inaccurate fault maps together with mislocated instrumental epicentres were presumably the main factors responsible for failure to detect many seismic faults in Iran. On the other hand, the absence of Quaternary faulting on many geological maps of Iran can be attributed to the fact that most geologists have been mapping the bedrocks and were not interested in recent faulting in alluvium. To avoid any confusion, an up-to-date fault map of the country has been prepared (Fig. 4).

There is a need to identify and delineate accurately the active faults and to develop realistic values for the amount, distribution and likelihood of surface fault displacement. The basic premise applied in the assessment of future surface faulting during earthquakes is that its type, amount and location will be similar to that which has occurred in the recent geological past, as evidenced in the early Quaternary geological record [Albee and Smith 1966, Allen 1975]. The active fault map of Iran (Fig. 5) which has been prepared on the basis of extensive field work and bibliographic research, shows that recent Quaternary faults bordering the major physiographic features, i.e. mountain and plains, are the most important active faults. Although the whole country is highly faulted (Fig. 4), the major destructive earthquakes with surface ruptures take place along these "mountain bordering reverse faults" (Fig. 5 and Table I). Usually the Neogene border folds of the mountains are thrust over the Quaternary alluvial deposits of the nearby compressional grabens (depressions limited by reverse faults) by these 'frontal' active faults (Fig. 5a). Despite much evidence showing their activity in late Quaternary and recent periods, most of these faults had not been recognized or mapped prior to the destructive earthquakes. In a case like the Ferdows earthquake of 1968.09.01, the fault was not discovered even after a destructive earthquake had taken place

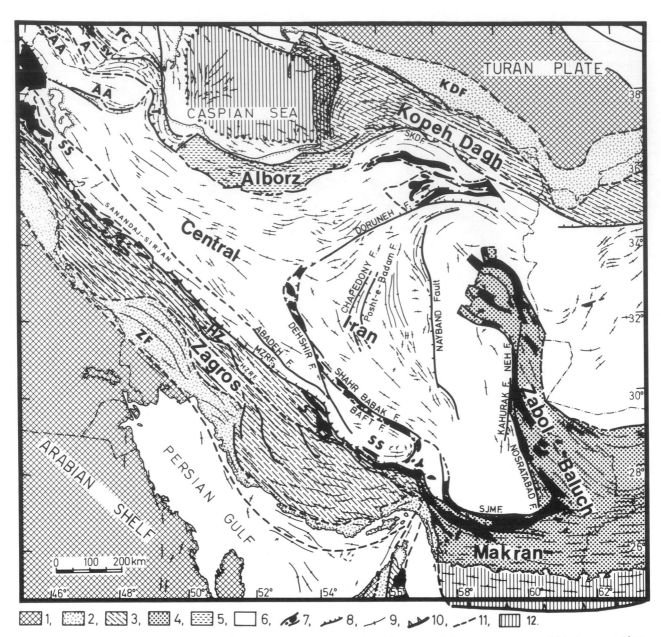

Fig. 2. Iranian major tectono-sedimentary units illustrating an inhomogeneous collision region between Arabia and Eurasia (cf. Fig. 3).

1. Stable areas: Arabian Precambrian shelf in the SW and Turanian Hercynian platform in the NE. The low dipping, relatively flat lying beds south and southwest of the Persian Gulf comprise the Arabian shelf over the buried Precambrian stable shield. 2. Neogene-Quaternary foredeeps, transitional from unfolded forelands to marginal fold zones, with strong late-Alpine subsidence. ZF: Zagros foredeep in the SW; KDF: Kopeh Dagh foredeep in the NE. 3. Main sector of the marginal active fold-thrust belt peripheral to the stable areas (Zagros and High Zagros -HZ- in the SW, and Kopeh Dagh in the NE). 4. Zabol-Baluch (east Iran) and Makran (southeast Iran) post-ophiolite flysch belts. Late Tertiary seaward accretion and landward underthrusting seem to be responsible for the formation of the present Makran ranges. 5. Alborz Mountains, bordering the southern part of the Caspian Sea. 6. Central Iranian Plateau (Central Iran) lying between the two marginal active fold-thrust belts. In northwestern part of the country, Central Iran wedges out and joins the Transcaucasian early Hercynian median mass (TC), the Sevan-Akera ophiolite belt (SV), and the Little Caucasus [A: Armenian (Miskhan-Zangezurian) late Hercynian belt, with a possible continuation to the Iranian Talesh Mountains (T) along the western part of the Caspian Sea; AA: the

Araxian-Azarbaijan zone of Caledonian consolidation, with the Vedi (V) ophiolite belt]. SS: Sanandaj-Sirjan belt, a narrow intracratonic mobile belt (during Paleozoic) and active continental margin (Mesozoic), forming the southwestern margin of Central Iran in contact with the Main Zagros reverse fault (MZRF). The belt bears the imprints of several major crustal upheavals (severe tectonism, magmatism and metamorphism). The Central Iranian tectonic unit widens eastwards and joins Central Afghanistan in the east. 7. Late Cretaceous High Zagros-Oman ophiolite-radiolarite (75 Ma) and the Central Iranian ophiolite-melange belts (65 Ma), with outcrops indicated in black. The southeastern parts of the Middle Cretaceous (110 Ma) Sevan-Akera and Vedi ophiolites of the Little Caucasus are shown in the NW of the country. The extensive belts of ophiolites mark the original zone of convergence between different blocks. The position of ophiolites are modified by post emplacement convergent movements. 8. Major facies dividing basement faults, bordering different tectono-sedimentary units. Contrasting tectono-sedimentary regimes, belts of ophiolites and associated oceanic sediments, together with paleogeographic contrasts along the Main Zagros (MZRF) and the High Zagros (HZRF) reverse faults in southwest, and the South Kopeh Dagh fault (SKDF) in northeast, indicate existence of old geosutures along these lines. The Chapedony and Posht-e-Badam faults delineate the possible Precambrian island-arc in eastern Central Iran. SJMF: South Jaz Murian Fault in SE. 9. Late Alpine fold axes. 10. Active subduction zone of Makran in the Gulf of Oman. 11. Province boundary. 12. Areas with oceanic crust (South Caspian Depression and Gulf of Oman). Lambert Conformal Conic Projection. Based on data from Gzovsky et al. [1960], Abu-Bakr and Jackson [1964], Stocklin [1968a, 1977], Adamia [1968], Sborshchikov et al. [1972], Brown [1972], Tectonic Map of Caucasus [1974], Berberian [1976a and c], White and Klitgord [1976], Huber [1978], and Shikalibeily and Grigoriants [1980] after modification and correction.

[Berberian 1979a]. A glance at the past geological history of the active mountain-bordering reverse faults shows a history of repeated movements of different character along some of these deep-seated sedimentation controlling faults. The Tabas earthquake of 1978.09.16 is an interesting case in point, indicating the activity of a multi-role basement fault during different orogenic movements [see Berberian 1979a and 1981a for discussion].

Zagros-Type Earthquakes (i.e., earthquakes along buried faults)

Allen [1975] emphasized the importance of employing geological evidence of late Quaternary movement to determine the sites of future seismic slip. This is valid where the seismic faults are not covered by ductile sediments. However, in the Zagros (the active fold-thrust mountain belt of southern-southwestern Iran, northern Iraq, northern Syria and southern Asia-Minor; Fig. 2) the earthquakes can not generally be correlated with the known surface faults, because of the special salt-tectonics character of the region [Berberian and Tchalenko 1975, Berberian 1976c, Berberian and Papastamatiou 1978]. The apparent lack of agreement between earthquakes and surface faults in the Zagros may arise from two factors: deficiency of seismic data and lack of knowledge of the relation between the Precambrian metamorphosed basement at depth (where most earthquakes occur) and the top of the sedimentary cover, where the geological structures are observed.

In the Zagros belt the metamorphosed Precambrian basement is covered by 8 to 14 kilometers of sedimentary rocks which were folded during late Miocene-Pleistocene time. A highly plastic salt layer (the Hormoz salt) [Stocklin 1968b] of late Precambrian age (Fig. 6), acts as a 'zone of slippage', disconnecting the sedimentary cover structures from those in the basement. Movements along basement faults can cause earthquakes in the Zagros active fold-thrust belt without producing surface faulting in the overlying sedimentary cover, presumably due to the presence of the plastic Hormoz deposits [Berberian 1976c]. Usually the aftershock distribution of large magnitude earthquakes in the Zagros active fold-thrust belt spreads out over 40 km [NEIS, ISC, Dewey and Grantz 1973, North 1973, 1974, Jackson and Fitch 1979]. It is unlikely that faulting of this dimension associated with large magnitude earthquake (Fig. 7) reactivates in the top sedimentary cover without surface faulting.

Within the sedimentary cover are plastic layers other than the Hormoz salt (Fig. 6): Permian evaporites (Nar member) of the Dalan Formation, Lower Triassic Dashtak evaporites, Lower Jurassic Adaiyah and Alan anhydrite, Upper Jurassic Gotnia and Hith anhydrite, Lower Eocene Kashkan-Sachun gypsum, Lower Miocene Kalhur gypsum and finally the Miocene salt-gypsum layers of the Gachsaran Formation (Lower Fars in Iraq and Kuwait) [James and Wynd 1965, Setudehnia 1972, Szabo and Kheradpir 1978]. Each of these incompetent evaporite horizons, together with other semi-ductile beds (such as the Jurassic Neyriz and the Eocene Pabdeh Formations), mechanically decouples the overlying structures from the underlying ones, behaving as zones of decollement. These tectonically incompetent units, characterized by extreme mobility, are responsible for remarkable structural changes and complications: synclines at the surface correspond to anticlines at one

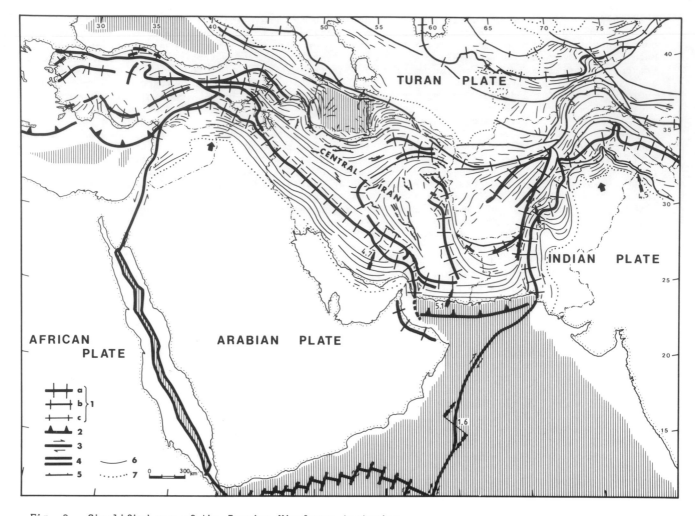

Fig. 3. Simplified map of the Iranian-Himalayan tectonics.
Collusion and convergence of African, Arabian and Indian plates with Eurasia (Turan plate in this figure) has caused major crustal shortening and thickening along the Iranian-Himalayan active fold-thrust mountain belt. Note that the Iranian plateau with an extensive area of deformation is trapped between two rigid old cratons in the north and the south and the two impinging zones (marked by thick small arrows) in the east and the west. There is an apparent widening of the deformed zones towards east Iran and wedging out towards NW. Large wedges of continental material in eastern Afghanistan and Asia-Minor, are being pushed away from the impinging zones. Whereas active frontal reverse faulting is the dominant mode of deformation in the trapped continental crust of Iran.
1. Collision zone of continental plates (a: Alpine, b: Triassic and Hercynian, c: Caledonian). 2. Subduction site. 3. Strike-slip and transform faults. 4. Rift area. 5. Major fault (known earthquake faults in Iran; triangle: thrust, bars: high-angle reverse). 6. Fold axes. 7. Edge of the late-Alpine fold-thrust belts and the Neogene-Quaternary foredeeps. Arrows indicate vectors of relative plate motion (in cm/yr) between Arabia, India and Eurasia across the assumed plate boundaries. Areas with oceanic crust are indicated by vertical hatching.
Based on data from: Gzovsky et al. [1960], Abu-Bakr and Jackson [1964], Gansser [1964, 1966], Matthews [1966], Laughton et al. [1970], CIGMEMR [1971], McKenzie and Sclater [1971], Brown [1972], McKenzie [1972], Sborshchikov et al. [1972], Minster et al. [1974], Molnar and Tapponnier [1975], Schlich [1975], Owen [1976], Berberian [1976a], Stocklin [1977], Biju-Duval et al. [1977], Bergougnan et al. [1978], Ketin [1978], Huber [1978], Desio [1979], Kravchenko [1979], Jacob and Quittmeyer [1979], Peive and Yanshin [1979b], and Shikalibeily and Grigoriants [1980] after modification and correction.

Fig. 4. Fault Map of Iran and the neighbouring regions.
The map represents the major lines of weakness, most of which have been inherited from the
previous orogenic phases. Except for the recent playas and deserts which are covered by annual
salt-mud deposits or contemporary sand dunes, active and weak features occur throughout the
country including the previously supposed stable blocks. The faults follow the pattern of the
young fold axes (cf. Figs. 2 and 3). Documented seismically active faults are shown as heavy
lines. Areas with oceanic crust are indicated by vertical hatching, and inactive platforms by
cross hatching. Lambert Conformal Conic Projection.
Mainly based on data from Berberian [1976a, f, 1979a], Huber [1978], all the published and
unpublished geological maps of the Geological and Mineral Survey of Iran up to 1981, Abu-Bakr and
Jackson [1964], Amurskiy [1971], Sborshchikov et al. [1972], Tectonic map of Caucasus [1974],
Toksoz et al. [1977], Peive and Yanshin [1979b], Kazmi [1979], and Shikalibeily and Grigoriants
[1980] after modification and correction.
Inset: Horizontal component of the compressional axes (shown by short lines) determined from the
focal mechanism solutions (cf. Fig. 5).

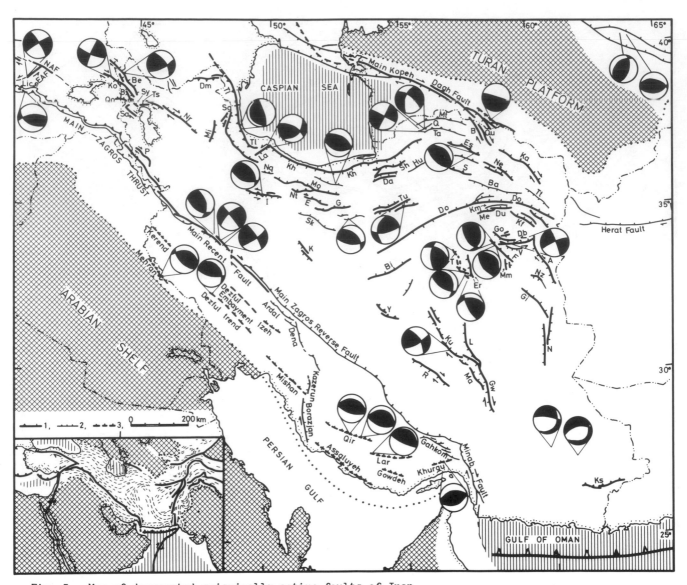

Fig. 5. Map of documented seismically active faults of Iran.
The map shows faults which are either observed on the surface and are known to have been
associated with earthquakes (Table I), or are lineaments in the seismicity that are not associated
with surface features (this applies only to the Zagros and they are thought to be faults in the
basement which fail to reach surface because of the salt tectonics; see Table II). Geometry of the
active faults and the mechanism of the large magnitude earthquakes along them clearly show that
the active tectonics of the region is dominated by frontal reverse faulting at the foot of the
fold-thrust mountain belts (Tables I, II and III). Predominant right-lateral reverse faulting in
NW Iran indicates sideways motion of blocks being pushed away from the impinging zone of the
Arabian plate with eastern Asia-Minor (NAF: North Anatolian fault, EAF: East Anatolian fault).
Deformation and seismicity is spread out over a broad zone of the continent along several mountain
bordering reverse faults and the continental crust is thickening and shortening (vertical expansion) in a NE-SW direction [Berberian 1976a, 1979a]. Letters indicate entries in Table I. Only
selected fault plane solution are shown. The compressional axes deduced from the fault plane solutions (Fig. 4 inset) often conform with the maximum horizontal shortening deduced from the Neogene
axes of the fold systems [Berberian 1976a]. Many more faults than those documented and shown in
this figure must be active, therefore, this figure alone should not be used as a guide to seismic
risk.
1. Documented active faults associated with the known damaging or destructive earthquakes
(triangle for thrust; short lines at right angle for high-angle reverse fault; see also Table I).

2. Segments of the High Zagros deep-seated reverse fault system in the Zagros (Table III) and the possible active faults in the Central Iran. 3. Buried seismic trends (basement reverse faults) in the Zagros active fold-thrust belt (see Table II). Active subduction zone of the Arabian oceanic crust (the Makran trench) is shown by a thick line with solid toothmarks in the Gulf of Oman. The southern edge of the Zagros folds indicating the boundary line between the northern active (top sedimentary fold-thrust belt with intensely thrusted Precambrian basement) and the southern rigid continental crust of Arabia (slightly folded and tilted top sedimentary cover with a non-thrusted basement; marked by cross hatching) is shown by a dotted line. Areas with oceanic crust are indicated by vertical hatching. Lambert Conformal Conic Projection.
Inset left: Entrapment of the Iranian plateau between Africa, Arabia, India and Turan (all indicated by cross-hatching; cf. Fig. 3).
Fault plane solutions after Shirokova [1962], McKenzie [1972, 1976], Berberian et al. [1979a, b], Kim and Nuttli [1977], Jackson and Fitch [1979], Sengor and Kidd [1979], Jacob and Quittmeyer [1979], Jackson [1980a], Kristy et al. [1980], Jackson and Fitch [1981], and Berberian [1981a, b]. Fault map is mainly based on Berberian [1976a, b and f, 1977d, 1979a, 1981b] , Ambraseys [1975], Toksoz et al. [1977], Tchalenko [1977] and Berberian et al. [1979a] after modifications and corrections (also see Table I).

kilometre depth; surface faults change from high-angle reverse to near horizontal thrust (listric), or die out altogether.

The recent microearthquake surveys in Kermanshah, Qir and Bandar Abbas [Asudeh 1977, Savage et al. 1977, Von Dollen et al. 1977, Niazi et al. 1978] and the results of SNAEOI [1976, 1977, 1978] show that the focal depths of the earthquakes in the Zagros active fold-thrust belt fall in the overlying sedimentary cover (0-12 km; usually micro-earthquake activity) as well as in

Fig. 5a. Active fault features across a section of the Sabzevar mountain-bordering active reverse fault in NE Central Iran (marked by 'S' in Fig. 5).
The northern folded Neogene clay deposits (left) are thrust over the southern horizontal Quaternary alluvial deposits of the Sabzevar plain (right) along this WNW-ESE recent frontal reverse fault. No large magnitude earthquake has occurred along this fault since the Beihaq destructive earthquake of 1052.06.02 which destoyed the town of Beihaq (the present Sabzevar; Table I) and the area remained relatively quiescent since then (cf. Fig. 10).

the Precambrian basement along longitudinal basement reverse faults (below 8 or 12 km; usually large magnitude earthquakes). These results indicate that recent plate tectonic models suggesting active subduction of the lithosphere towards the northeast [Nowroozi 1971, 1972, Berry 1975, Bird et al. 1975, Bird 1978] cannot be based on earthquake data. Intermediate or subcrustal depth location of some damaging earthquakes in the Zagros [USGS, ISC, Nowroozi 1972] which caused concentrated destruction within a small area are also doubtful [Berberian 1979b, c]. A detailed field study of the Khorgu 1977.03.21 earthquake of Ms=7.0 in the southeastern part of the Zagros belt (Table II and Fig. 5), and the preliminary processing of the macroseismic data indicated that the focal depth of the earthquake was between 10 and 20 km, and that the seismicity was the result of the reactivation of the basement faults [see Berberian and Papastamatiou 1978 for discussion]. This idea was later confirmed by P-wave modelling of the Zagros earthquakes [Jackson 1980a, Jackson and Fitch 1981]. The aftershock sequences of two large magnitude earthquakes in the Zagros belt also failed to demonstrate subcrustal seismic activity [Jackson and Fitch 1979].

Fault plane solution of earthquakes in the Zagros [Shirokova 1962, 1967, Balakina et al. 1967, Canitez and Ucer 1967, Canitez 1969, McKenzie 1972, Nowroozi 1972, North 1972, Dewey and Grantz 1973, Kim and Nuttli 1977, Jackson 1980a, Jackson and Fitch 1981, Jackson et al. 1981] show thrusting with high-angle planes at depth (dipping NE or SW) with a NW-SE strike roughly parallel to the fold axes of the sedimentary cover. Intense seismicity spread over the entire belt (Fig. 1) indicates that a large number of reverse basement faults are being reactivated in the Zagros, and that the basement in the folded part of the belt is 'intensely thrusted' (Figs. 1 and 5). Transverse basement faults revealed by aero-magnetics such as Hendurabi, Khark, Abadan, and basement controlled features and paleostructural lineaments like the

TABLE I – Summary of the data on the Iranian active faults (Fig. 5) and their related major earthquakes.

	FAULT NAME	FT	FL	EQ. DATE	Ms	Mb	MMI	EFL	MO.	SD	REF*
A	Abiz	R	0080	1936.06.30	6.0	6.1	VII				1,2
		SS		1979.11.14	6.6	6.6	VII	020			3
B	Badalan	SS	0040	1970.03.14		5.3	VII				4,5
Ba	Balher ?	R	0200	1940.05.04	6.6		VIII				1,2
Be	Bedavli	SS	0100	1319.00.00							4,5
		R		1696.05.00							
				1968.04.29		5.3	VII				
Bg	Baghan-Germab	SS	0080	1929.05.01	7.4		X	>50			2,4,6,
				1929.07.13		5.7	VII				7,8,9
Bi	Biabanak	R	0150	1969.11.11		5.0	VI				
Da	Damghan	R	0080	0856.12.22							4,9,10,
											11
Db	Dasht-e-Bayaz	SS	0110	1968.08.31	7.5	6.0	X	080	1800F	037	2,4,8,
									0860S		9,12,
									0480P		13,14,
											15
		SS		1979.11.27	7.1	6.1	VIII	060			3
Dm	Dasht-e-Moghan?	R	0090	1924.02.19		5.9	VII				16
Do	Doruneh	R	0650	1923.05.25		5.7	VIII				2,4,9,
											10,17
Du	Doughabad	R	0060	1619.05.00			VIII				1,2,18
				1962.10.05		5.0	VI				
E	Eivanekei	R	0080	0743.00.00							10,16,18,
				0855.00.00							19
				0864.01.00							
				0958.02.23							
				1384.00.00							
Er	Esfandiar	R	0060	1973.05.05		4.6	VI				2,9
				1973.05.11		5.1	VI				
Es	Esfarayen	R	0120	1695.05.11			VIII				2,15,18
				1963.03.31		4.9	VI				
				1969.01.03	6.0	5.4	VII		0200s		
F	Ferdows	R	0100	1947.09.23	6.9		VIII				1,2,15
				1968.09.01	6.3	5.9	VII	030	0085s		
				1968.09.04		5.4	VI				
G	Garmsar	R	0040	1945.05.11		4.7	VI				4,16,20
Gi	Giv	R	0060	1946.02.10	5.2	5.0	VII				1,2
Go	Gonabad	R	0080	1238.00.00			VII				1,2,4
				1678.00.00			VIII				
Gw	Gowk	R	0150	1877.00.00			VII				9,21
		SS		1909.10.27	5.5		VII				
				1911.04.29	5.6		VII				
				1948.07.05	6.0	5.8	VII				
				1969.09.02	5.2	5.3	VI				
	Herat	R		0849.00.00			VIII				22,23
Hu	Hunestan ?	R	0250	1971.02.14	5.3	5.3	VI				
				1980.07.23		4.9	V				
I	Ipak	R	0110	1962.09.01	7.2		IX	100	0630s	3.4	4,8,9,
											15,16,
											24,25
K	Kashan	R	0070	1574.08.00							9,10,11
				1755.06.07							
				1778.00.00							
				1794.03.14							
				1844.05.11							
Ka	Kashafrud ?	R	0180	1673.07.30			VIII				2,18,19
		SS		1687.04.00			VII				26
Kf	Khaf	R	0130	1336.10.21				>20			1,2,4,
											8,9,27
Kh	Khazar	R	0210	0874.11.00			VII				11,16,18
				1436.00.00			VI				19,20

TABLE I. Cont.

FAULT NAME	FT	FL	EQ. DATE	Ms	Mb	MMI	EFL MO.	SD	REF*
			1805.00.00						
			1809.12.07						
			1944.04.05		5.0	VII			
			1953.04.18		4.8				
Km Kashmar	R	0090	1903.09.25		6.2	VIII			2,4,17
Ko Khoy	SS	0120	1843.00.00						5,28,29
			1881.00.00						30
			1883.00.00						
			1896.00.00						
			1976.11.24	7.3		IX	050		
			1977.05.26		5.2	VI			
Ks Kishi ?	R	0140	1979.01.10		5.4	VII			
Ku Kuh Banan	SS	0260	1854.11.00			VIII			4,9,10,
			1875.05.00			VIII			21,31
			1897.05.22			VIII			
			1933.11.28		6.2	VIII			
			1977.12.19	5.8	5.5	VII	020		
L Lakar Kuh	R	0130	1911.04.18	6.5		VIII			9,21
La Lahijan ?	R	0070	1678.02.03			VII			11,16,
			1956.04.12		5.5				18,19
Ma Mahan	R	0090	1934.01.02	5.6		VII			21
Me Mehdiabad	R	0060	1925.05.02	4.5		V			1,2
Mi Mianeh	R	0060	1879.03.22			VIII			4,8,9,
									10,11,16
									18
Mm Mohamadabad			1941.02.16	6.4	6.2	VIII	>15		1,2,27
Mo Mosha	R	0190	1665.06.00						4,9,11,
			1802.00.00						16,19,32
			1811.06.20						
			1830.05.09						
			1930.10.02		5.5	VII			
			1955.11.24		4.0	V			
Mt Maraveh Tappeh?	R		1978.06.14		4.8	VI			16
Main KopehDadg	R	0600	1895.07.09	8.2					7
			1946.11.04	7.0					
			1963.09.18						
			1963.09.20						
			1964.02.12						
			1966.06.29						
			1968.07.14						
			1969.03.26						
Main Recent	R	0700	0913.00.00						4,6,8,
	SS		1008.04.27						9,10,11
			1107.00.00						33
			<1889						
			1909.01.23	7.3		X	>40		
			1909.04.11		5.6	VII			
			1909.11.01		5.1	VI			
			1957.12.13	7.0		IX	>05		
			1958.08.16	6.7		VIII	>20		
			1961.10.14			VI			
			1961.10.28		5.0	VI			
			1963.03.24		5.9	VII			
			1975.09.01		4.9	VI			
N Neh	R	0100	1928.03.08		5.2	VII			1,9,19
Ne Neyshabur	R	0100	1270.10.07			VIII			2,18,19,34
			1389.02.00			VIII			
			1405.11.23			VIII			
			1673.07.30			VIII			
			1928.08.21		5.2	VII			
Nq North Qazvin ?	R	0060	1119.12.10			VIII			11,16,18
			1639.00.00			VII			19
			1808.12.16						

TABLE I. Cont.

	FAULT NAME	FT	FL	EQ. DATE	Ms	Mb	MMI	EFL	MO.	SD	REF*
Nr	North Tabriz	R	0200	0858.00.00			VII				9,16,18
				1042.11.04			VIII				19,35
				1304.11.07			VII				
				1345.00.00							
				1459.00.00			VII				
				1550.00.00			VII				
				1641.04.05			VIII				
				1650.00.00							
				1657.00.00							
				1664.00.00							
				1721.04.26			VIII				
				1727.11.18			VIII				
				1779.12.27			VIII				
				1856.10.04							
				1965.02.10		5.1	VII				
Nt	North Tehran	R	0090	1665.06.00							9,10,11
				1802.00.00							18,19
				1811.06.20							
				1830.05.09							
				1830.10.02							
				1955.11.24		4.0	VI				
				1970.10.03		4.1	V				
Nz	Nozad	R	0060	1493.01.10			VIII	>10			1,2,8,
											9,27
P	Pasveh ?		0060	1970.10.25		5.5	VII				
Q	Qarnaveh ?	R	0050	1970.07.30	6.7	5.7	VIII		0120s		6,16,36
Qo	Qotur ?	R	0040	1715.03.08			VII				28
				1969.05.14		4.5	VI				
Qu	Quchan	SS	0050	1871-1872							2,7
				1893.11.17							
R	Rafsanjan	R	0110	1923.09.22	6.7		VIII				4,9
S	Sabzevar	R	0070	1052.06.02			VIII				2,18,19
Sa	Salmas	SS		1930.05.06	7.4		X	>30			2,37,38
Sg	Sangavar	R	0070	1863.12.30							11,16
				1864.01.02			VIII				
				1896.01.02			VIII				
				1969.03.10			V				
Sh	Shavar	R	0080	1890.07.11			VIII				11,16
Sk	Siahkuh ?	R	0140	1977.05.25		5.4	VI				
Sy	Shekaryazi	R	0015	1930.05.08	6.2		VII				2
T	Tabas	R		1939.06.10		5.5	VII				2
				1974.06.17		4.8	VI				
				1974.06.24							
				1978.09.16	7.7	6.5	X	085	1300F	025	
									0440p	017	
Ta	Takal Kuh ?	R	0100	1974.03.07		5.2	VI				16
Tl	Talesh	R	0120	1913.04.16		5.2					9,16
				1978.11.04		6.0	VII				
Ts	Tasuj	R	0030	1857.10.27							11
Tt	Torbat Jam	R	0080	1918.03.24	5.8		VII				1,2
Tu	Torud	R	0120	1953.02.12	6.4		VIII				8,9,16
V	Vondik ?	R	0055	1976.11.07	6.2	5.6	VII				
Y	Yazd	R	0070	1975.11.15		4.2	V				
				1976.02.15		4.4	V				
				1978.02.16		4.5	V				

* References include:

1 Ambraseys and Melville 1977
2 Berberian 1979a
3 Haghipour and Amidi 1980
4 Berberian 1976c
5 Berberian 1977b

6 Tchalenko et al. 1974
7 Tchalenko 1975a
8 Berberian 1976b
9 Berberian 1976a
10 Berberian 1976g

11 Ambraseys 1974
12 Ambraseys and Tchalenko 1969
13 Tchalenko and Berberian 1975
14 Hanks and Wyss 1972
15 North 1973
16 Berberian 1981b
17 Tchalenko et al. 1973
18 Melville 1978 (used only for
     correcting the European dates
     of few historical earthquakes)
19 Nabavi 1972
20 Tchalenko 1974
21 Ambraseys et al. 1979
22 Heuckroth and Karim 1970
23 Poirier and Taher 1980

24 Ambraseys 1963
25 Berberian 1976d
26 Ambraseys et al. 1969
27 Ambraseys 1975
28 Tchalenko 1977
29 Toksoz and Arpat 1977
30 Toksoz et al. 1977, 1978
31 Berberian et al 1979a
32 Berberian 1976e
33 Tchalenko and Braud 1974
34 Melville 1980
35 Berberian and Arshadi 1976
36 Ambraseys et al. 1971
37 Tchalenko and Berberian 1974
38 Berberian and Tchalenko 1976c

Abbreviations used in the Table:

EFL=Earthquake fault length in km, EQ.=Earthquake, FL=Fault length in km,
FT=Fault type (R:reverse, SS:strike-slip), Mb=Body-wave magnitude,
MMI=Modified Mercalli Intensity, MO.=Seismic moment :x10^{24}dyne-cm (values
calculated from: f=field data, P=P-wave spectrum, p=long period P-waveforms,
S=S-wave spectrum, s=Surface wave determination), Ms=Surface-wave magnitude,
R-Reverse fault, SD=Stress drop in bars, SS=Strike-slip fault, ?=Possible
active fault related to the mentioned earthquake. First column letters indicate
official fault names (given in the second column) shown in Fig. 5.

Haftgel and Hendijan highs [see Morris 1977] seem to be inactive today (Fig. 12).

The Zagros earthquakes have shown that, with the information presently available, it is not possible to use seismicity to establish smaller seismotectonic units. Neither the mountain/depression division, nor the large scale fault-flexure zones or strongly folded arcs, nor the surface faults (except major boundary faults cutting basement and the sedimentary cover) can be correlated easily with the earthquake locations. It therefore seems that the Zagros belt should neither be divided into several seismotectonic zones as suggested by Nowroozi [1976], nor that expected maximum intensities for future earthquakes should be assigned to different zones. The large magnitude basement earthquakes in the Zagros showed that earthquakes may occur anywhere. Since no correlation can be established between the seismicity (reactivation of basement faults) and the tectonic features at the surface, it should be assumed at present that an event has the same probability of occurrence anywhere in the Zagros. That is to say, the seismic risk level will be the same everywhere along the belt until we recognize the exact position of the longitudinal reverse active faults in the basement.

Compilation of accurate macroseismic data of the belt would show the trend and the location of the buried basement reverse faults in the Zagros. Unfortunately the seismic history of the Zagros (especially the historical earthquakes) is not documented well enough to be used for detecting all seismic trends in the belt. The available seismic data together with preliminary conclusions are presented in Figures 5 and 7 (see also Table II). The data seems to suggest a few distinct seismic trends which are given below. Some of the longitudinal seismic trends apparently coincide with sedimentary facies changes in the Zagros sedimentary basin. It is difficult to understand this because the detachment and southward displacement of the sedimentary cover must have severed the facies boundaries from underlying basement faults.

a. Assaluyeh Trend: This trend contained 10 damaging and destructive earthquakes (Figs. 5 and 7; Table II). There are several indications that the Assaluyeh seismic trend (buried seismic reverse fault) coincides with sedimentary facies changes. The Assaluyeh seismic trend is near the southwestern boundary between the Jurassic shallow marine carbonates (Surmeh Formation in the NE) and the lagoonal-reefal carbonates (Sargelu Formation in the SW) [Mina et al. 1967]; it lies along the change between the Eocene neritic to basinal marls of the Pabdeh Formation in the south and the transitional shallow marine carbontes and neritic to basinal marls (Jahrom-Pabdeh Formation) in the north [James and Wynd 1965, Mina et al. 1967]; it is approximately the southern limit of the early to Middle Miocene Mishan sea with shale and marl depositions [Bizon et al. 1972, Ricou 1974]; it lies parallel to the main Tertiary foredeep [Kamen Kaye 1970]; and finally it corresponds to the present dividing line between the Zagros fold-thrust belt in the NE and the Arabian platform in the SW, and also corresponds to the boundary between the mountain front and the foothills of the Zagros belt.

b. Lar Trend: This trend produced 7 destructive shocks (Table II) one of which had a NW-SE thrust mechanism (Figs. 5 and 7). The Lar trend also coincides with some sedimentary facies changes. The trend is situated in the zone of the approximate southern and southeastern limit of the lower Cretaceous Gadvan Formation and corresponds with the southwestern boundary of the upper Cretaceous-lower Eocene Sachun evaporites [James and Wynd 1965]; is in the zone of the southwestern boundary of the Eocene shallow marine carbonates in the north (Jahrom Formation) and the transitional shallow marine carbonates and neritic to basinal marls of the Eocene Pabdeh Formation in the south [James and Wynd 1965]; and finally lies in the boundary region of the lower to middle Miocene Razak red beds in the north and the northeastern limit of the Gachsaran evaporite facies in the south [James and Wynd 1965, Stocklin 1968a].

c. Qir Trend: This trend is based on 5 destructive earthquakes (Table II) two of which with NW-SE thrust mechanisms (Figs. 5 and 7). The trend is located along the transition between the lower Miocene Razak red beds in the northeast and the Miocene Gachsaran evaporites in the south and southwest [James and Wynd 1965, Stocklin 1968a].

d. Mishan Trend: Two shocks (Table II), one with a NW-SE thrust mechanism fall on this trend. It follows the surface break of the Mishan fault [Berberian and Tchalenko 1976b] (Fig. 5). The trend could possibly be the southeastern continuation of the Dezful trend (see Fig. 5).

e. Izeh Trend: This trend is based on four earthquakes (Table II and Fig. 5) and follows the northern boundary of the Dezful Embayment [Morris 1977].

f. Kerend Trend (near Iraqi-Iranian border zone): Four destructive and damaging earthquakes (Table II) occurred along this trend (Fig. 5). The trend coincides with the southwestern boundary zone of the upper Cretaceous-Paleocene Amiran flysch; southern boundary of the lower Eocene Kashkan evaporites; and the northern limit of the neritic to basinal Eocene Pabdeh marl in south, and the southwestern limit of the shallow marine carbonates of the Jahrom Formation in the north [James and Wynd 1965, Ricou 1976].

g. Mehran Trend (along the Iraq border): This trend contained five shocks (Table II) three of which had a NW-SE thrust mechanism (Fig. 5). The trend is situated along the southern edge of the Zagros folds.

The damage zone of a few earthquakes in the Zagros corresponds, at the surface, to major boundary faults (Table III). Usually the Precambrian Hormoz salt is intruded along these faults, reaching the surface. This indicates that

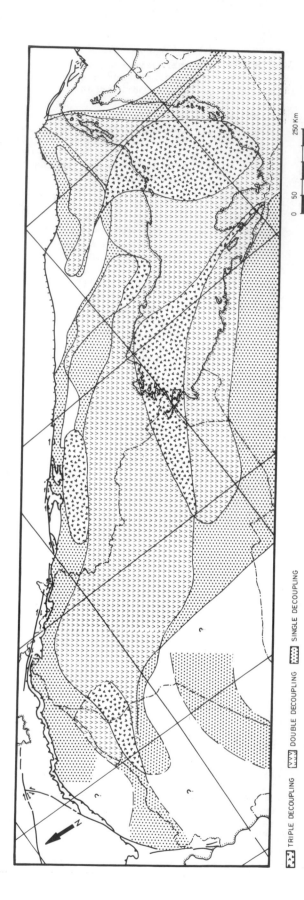

TRIPLE DECOUPLING   DOUBLE DECOUPLING   SINGLE DECOUPLING

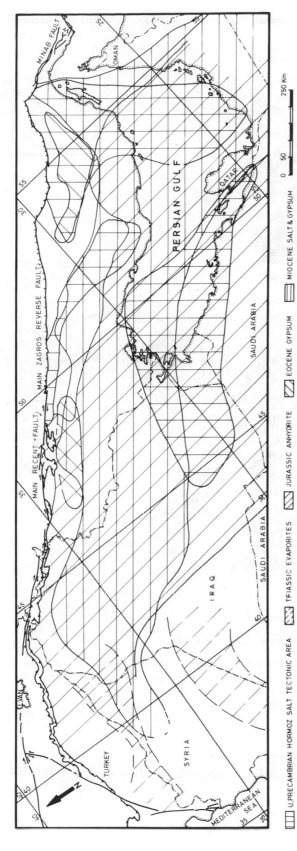

⊞ U.PRECAMBRIAN HORMOZ SALT TECTONIC AREA. ⊠ TRIASSIC EVAPORITES ⊟ JURASSIC ANHYDRITE ⊘ EOCENE GYPSUM ⊞ MIOCENE SALT & GYPSUM.

Fig. 6. Distribution of evaporite (lower map) and zones of single, double and triple decoupling (the upper map) in the Zagros active fold-thrust belt. Note that almost all the belt has at least one evaporite layer and in many places two or three such horizons. The highly mobile salt and other evaporitic horizons at the base and within the sedimentary cover mechanically decouple the overlying structures from those at depth. This mechanical decoupling gives rise to the characteristic fold pattern of Zagros, and prevents any surface expression of the longitudinal reverse basement faults. Evaporites stop in NE at the line of the Main Zagros reverse fault, and the active Main Recent fault falls at the edge, along the junction of the Central Iran and the Zagros. The difference in seismic faulting and folding behaviour between the Zagros active fold-thrust belt and the rest of the country (Figs. 1, 2, 4 and 5) can be attributed to the salt tectonics. Lambert Conformal Conic Projection.
Based on data from James and Wynd [1965], Wolfart [1965], Stocklin [1968b], Brown [1972], Bizon et al. [1972], Ricou [1974, 1976], and Berberian [1976c].

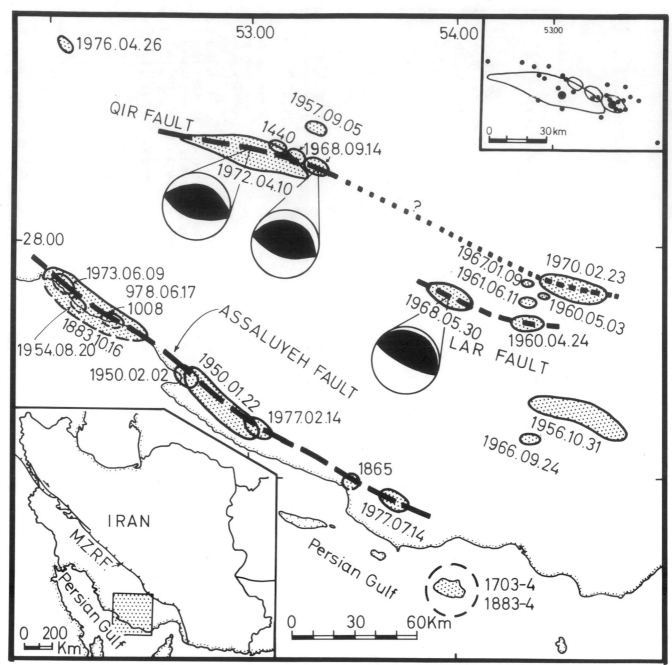

Fig. 7. Qir, Assaluyeh and Lar buried seismic trends in the Zagros active fold–thrust belt representing the approximate position of the longitudinal seismic reverse faults in the Precambrian basement (see Table II). Epicentral regions (stippled areas) after Berberian [1976a, 1977a], Haghipour et al. [1972], Ambraseys et al. [1972], Berberian and Tchalenko [1976a, b]. Fault plane solutions after McKenzie [1972], Nowroozi [1972], and North [1972]. Lambert Conformal Conic Projection.

Inset top right: Relocated epicentres of 1968 and 1972 earthquakes relative to 1972.04.10 earthquake which was used as master event [Jackson and Fitch 1979]. The relocation pattern is positioned here by setting the 1968.09.14 event over the localized damage zone.

TABLE II – Seismic trends (buried high-angle reverse faults) in the Zagros active fold-thrust belt (Fig. 5) and their related major earthquakes.

FAULT NAME	FT	FL	EQ. DATE	Ms	Mb	MMI	MO.	SD	EFL	Ref*
Assaluyeh	R	0200	0978.06.17			VIII				1
			1008.00.00			VII				
			1865.00.00							
			1883.10.16							
			1950.01.22		5.7	VII				
			1950.02.02		4.7	VI				
			1954.08.20							
			1973.06.09		4.5	VI				
			1977.02.14		4.6	V				
			1977.07.14		4.3	V				
Dezful	R		1977.06.05		5.5					
Gowdeh	R		1956.10.31		6.0	VII				2
Izeh	R		1885.00.00							3
			1929.07.15		6.5	VIII				
			1972.08.06		5.0					
			1978.12.14		5.5	VII				
Kerend	R	0100	0872.06.22			VIII				4,5,6
			0958.04.00			VIII				
			1150.04.01			VII				
			1972.06.08		4.9	VI				
Khorgu	R		1977.03.21	7.0	6.2	VIII				7,8
Lar	R	0050	1400.00.00			VII				2,5,6,
			1566.00.00							9
			1677.00.00			VII				
			1766.00.00							
			1960.04.04		6.0	VII				
			1966.09.18	6.2	5.9		04.2			
			1968.05.30		5.2	VII				
Mehran	R	0100	1864.12.20							10
			1917.07.15		6.0	VIII				
			1972.06.12		5.3	VI				
			1972.06.13		5.1					
			1972.06.14		5.3					
Mishan	R	0040	1971.08.27		5.0	VI				1
			1972.07.02		5.4	VII		010		
Qir	R	0140	1440.00.00			VII				2,6,9
			1903.11.15							
			1968.09.14	6.0	5.8	VII	0029			
			1968.09.14		5.1	VI				
			1972.04.10	6.9		IX	0150s	3.6	085	

* References include:

1	Berberian and Tchalenko 1976b	6	Melville 1978
2	Berberian 1976c	7	Berberian 1977d
3	Ambraseys 1979	8	Berberian and Papastamatiou 1978
4	Ambraseys 1968	9	Ambraseys et al. 1972
5	Nabavi 1972	10	Ambraseys 1975

the faults are deep-seated and their activity can extend through the sedimentary sequences. Three examples are cited here:

a. Gahkom Fault: The Gahkom earthquakes of 1962.11.06, 1963.07.29, 1964.08.27 and the Sarchahan earthquakes of 1965.06.21 and 1974.12.02 [Berberian and Tchalenko 1976a] which took place along the Gahkom deep-seated high-angle reverse fault system, may indicate the activity of the fault at depth (Fig. 5 and Table III). The fault plane solutions of the 1965.06.21 [McKenzie 1972] and 1963.07.29 [Canitez 1969] earthquakes show thrusting at depth parallel to the surface trend of the fault. The fault forms the southeastern segment of the High Zagros fault system, the

TABLE III - Deep-seated boundary faults cutting top-sedimentary cover as well as the basement of the Zagros active fold-thrust belt (Fig. 5) with the closest seismic activity.

FAULT NAME	FT	FL	EQ. DATE	Mb	Ms	MMI	MO.	EFL	Ref*
Ardal ?	R	0100	1666.00.00			VII			1,2
			1880.00.00			VII			
			1922.03.21		5.5	VII			
			1958.07.26		4.2	V			
			1960.09.21		5.0	VI			
			1977.04.06		5.5	VII			
Borazjan ?	R	0170	1925.12.18		5.5	VII			3
Dena ?	R	0200	1934.03.13		5.3	VII			1,2
			1975.05.09		4.9	VI			
			1975.09.21		5.2	VII			
Gahkom ?	R	0080	1962.11.06		5.8	VII			4
			1963.07.29		5.2	VII			
			1964.08.27		5.3	VII			
			1965.06.21	5.5	5.7	VII	07.7		
			1974.12.02		5.4	VII			
Kazerun ?	R	0050	1967.01.15		4.7	V			3
			1968.06.23		5.2	VI			
			1971.10.23		4.5	V			
Main Zagros ?	R	2250	1965.06.21		5.0	VI			5
			1973.08.28		4.8	VI			
			1973.11.11		5.5	VII			
			1975.09.06	6.8	6.1	VIII		030	

* References include:

1 Berberian and Navai 1977
2 Ambraseys 1979
3 Berberian and Tchalenko 1976b
4 Berberian and Tchalenko 1976a
5 Toksoz and Arpat 1977

southern margin of the High Zagros belt and the northeastern limit of the Paleocene Sachun evaporites. The fault has been an active facies divider at least since the Paleozoic [Szabo and Kheradpir 1978]. Thrusting of the High Zagros Silurian shales over the Quaternary alluvial deposits and intrusion of the Precambrian Hormoz salt along the Gahkom fault indicate that the fault is deep-seated and penetrates at least the base of the sedimentary cover.

b. Dena Fault: The damage zone of three 20th century earthquakes (Table III) lie along the Dena deep-seated reverse fault (Fig. 5). No fault plane solution is available for these shocks. The fault forms the central segment of the High Zagros fault system, and has been an active facies divider at least since the late Precambrian time, when it formed the northern limit of the late Precambrian Hormoz salt basin. During the late Cretaceous orogenic movements, it formed the southern boundary of the middle part of the High Zagros belt. Finally during Miocene, the fault formed the northeastern limit of the Gachsaran evaporite facies [Stocklin 1968b, Bizon et al. 1972, Ricou 1976].

c. Ardal Fault: This trend corresponds to the epicentral areas of five destructive earthquakes (Table III) and follows the surface break of the Ardal high-angle reverse fault [Berberian and Navai 1977]. Along this deep-seated fault a linear Precambrian Hormoz salt plug reaches the surface (Fig. 5).

Partly Covered Active Faults Known From Surface Geology In Central Iran

Faults without established Quaternary movements or faults partially covered by a thin layer of Quaternary alluvial-fluvial deposits and lava flows, are usually ranked in the "pre-Quaternary" fault category [Berberian 1976e]. They are considered aseismic or have been given a low seismic risk value. Lack of historical seismic damage record and the relatively low level of instrumentally located earthquakes along some faults in Central Iran should not be taken as indicating that the region is aseismic. Study of earthquake faulting and seismotectonics of Iran has shown that some of these faults are as hazardous as the faults which clearly cut the Quaternary alluvial deposits. Some major Iranian earthquake faults partly covered by fluvial deposits in plains, lowlands, and salt-mud playas, could have been recognized prior to large magnitude earthquakes, if the whole length of these faults, together with their associated physiographic features, had been analysed carefully. The Dasht-e-Bayaz earthquake fault of 1968.08.31 (80 km long) [Ambraseys and Tchalenko 1969, Tchalenko and Berberian 1975], the Salmas earthquake faults of 1930.05.06 (30 km long) [Tchalenko and Berberian 1974, Berberian and Tchalenko 1976c] and the 1978.09.16 Tabas earthquake fault (85 km long) [Berberian 1979a] are typical examples of active faults which were not recognized as active prior to the large magnitude earthquakes (Fig. 5 and Table I). As mentioned earlier ,in one case, the Ferdows fault [Berberian 1979a] was not even recognized and mapped after the destructive earthquake of 1968.09.01.

The Chalderan destructive earthquake of 1976.11.24 (Ms=7.3) on the northwestern Iranian border is an interesting case. The earthquake was associated with 50 km of fresh faulting in an area where no large earthquakes had been felt for at least a few generations [Toksoz and Arpat 1977, Toksoz et al. 1977, 1978]. In the southeastern continuation of the Chalderan earthquake fault, east of the Iranian border, there is a major recent and seismic fault (the Khoy fault; Fig. 5 and Table I), along which some sagponds, recent landslides, sharp topographic lineaments and a river can be easily recognized on the aerial photographs. At least four destructive earthquakes are known to have taken place along this fault (1843, 1881, 1883, 1896) [Tchalenko 1977, Berberian 1977b]; two damaging shocks took place on 1970.03.14 (Mb=5.3) and 1977.05.26 (Mb=5.2). Since the northwestern extension of this fault is covered by Quaternary lava flows in Asia-minor and fluvial deposits in Iran, this part of the fault was not recognized as such, and the area was not considered a major seismic zone on the seismic risk maps of the two countries.

Change In Fault Behaviour

A comparison between the character of earlier geological deformations and those of the present day earthquakes shows that the tectonic pattern did generally not change between the late Tertiary-early Quaternary and the present [Tchalenko and Berberian 1975, Berberian 1976d, 1979a]. However in a few interesting cases the fault behaviour changed between the early Quaternary and the Present.

The Bob-Tangol earthquake of 1977.12.19 in southeastern Iran resulted from the reactivation of an early Quaternary high-angle reverse fault,

the Kuh Banan fault, but surprisingly the surface rupture over 20 km length during the earthquake, and the fault plane solution (Fig. 5), showed a predominant right-lateral movement [Berberian et al. 1979a]. A similar case has been noticed for the Bedavli earthquake of 1968.04.29. The earthquake took place along the Bedavli early Quaternary high-angle reverse fault in NW Iran [Berberian 1977b], while the fault plane solution [Mckenzie 1972] shows a prevalent right-lateral strike-slip movement (Fig. 5).

Another example of changing slip vector from early Quaternary to present is the case of the Main Recent Fault [Tchalenko and Braud 1974, Tchalenko et al. 1974] in the junction zone of the Zagros active fold-thrust belt and Central Iran (Fig. 5). The Main Recent Fault behaved as a predominantly high-angle reverse fault during late Tertiary-early Quaternary orogenic movements (thrusting of the lower Cambrian sandstones over the Mesozoic and Tertiary rocks), while recent earthquakes showed a prevailing right lateral strike-slip movement. Similar time-dependent changes of fault character have been documented along the Alpine fault in New Zealand [Scholz et al. 1973].

These events show that slip-vectors on individual faults can change and that the evidence of early Quaternary movement cannot always be extrapolated for the present day continental deformation. Change in fault behaviour and slip-vector presumably indicate that the contemporary stress field and regions of deformation could be different from those which prevailed in the Neogene, and that they are being changed with time. It is therefore concluded that in a structurally complex and inhomogeneous collision region such as the Iranian plateau, short term behaviour is not obviously related to interplate slip-vectors derived from time-averaged data [Berberian et al. 1979a].

Sympathetic Faulting

The main shock or a series of aftershocks can trigger a chain reaction of seismic events on nearby faults that readjusts the disturbed mechanical state of the region. Two examples, illustrated in Figs, 8 and 9, show sympathetic faulting at distance from the primary event with change in mechanism. This indicates that in a compresional deformation zone, large crustal earthquakes involving considerable lateral-slip may trigger activity on neighbouring reverse faults of different trend. In the case of Dasht-e-Bayaz and Ferdows earthquakes (Fig. 8) the focal mechanism of the main shock and the surface break show strike-slip motion, while two of the large magnitude aftershocks were produced by thrusting at the western end of the strike-slip fault. Previous activity along both fault sets, especially along the Ferdows fault (Fig. 8), shows that its reactivation (and the difference in the mechanism with the main shock) is not just an

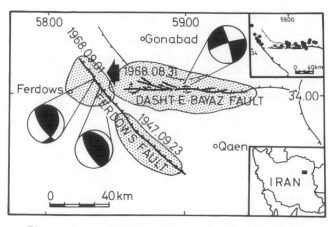

Fig. 8. Reactivation of the Ferdows frontal reverse fault during the largest aftershock of the 1968 Dasht-e-Bayaz earthquake with stike-slip faulting. Dasht-e-Bayaz earthquake fault associated with the main shock of magnitude Ms=7.5 had a vertical left-lateral mechanism striking E-W [Ambraseys and Tchalenko 1969, Tchalenko and Berberian 1975]. After 20 hours the readjustment of the local stress system triggered a shock of magnitude Ms=6.3, 60 km west of the main shock damage zone. Left-lateral strike-slip motion on the northern block along the Dasht-e-Bayaz fault (arrow) produced compression at its western boundary near the Ferdows frontal reverse fault and reactivated the latter [Berberian 1979a]. Stippled areas are the epicentral regions of the large magnitude earthquakes. Transverse Mercator Projection.
Inset top right: Relocated epicentres of 1968 teleseismically recorded earthquakes relative to the 1968.09.01 earthquake which was used as master event [Jackson and Fitch 1979]. The relocation pattern is positioned here by siting the earthquakes relative to the dips of the Dasht-e-Bayaz (high-angle reverse dipping north with major left-lateral strike-slip component) and the Ferdows (frontal thrust dipping NE) faults.

aftershock occurrence. The Dasht-e-Bayaz E-W fault is terminated at both ends by major NW-SE structures transverse to the strike of the fault (Figs. 4 and 5). The fault-blocks, produced by these faults, could be partially decoupled from each other and therefore might have moved without major internal deformation. It seems that after a considerable lateral motion and relative westward movement of the northern block, the Ferdows frontal reverse fault in the west, served as a location for high stress concentration [Berberian 1979a]. Similarly, the 1979.11.14 earthquake (Ms=7.1) and its aftershocks along the Abiz fault (Fig. 5 and Table I), triggered the eastern seg-

ment of the Dasht-e-Bayaz fault and caused surface ruptures associated with the 1979.11.27 (Ms=6.6) earthquake [Haghipour and Amidi 1980].

Seismic Creep And Interseismic Stage

Unlike oceanic plate boundaries, the sum of the displacements determined from seismic moment in Iran is very much less than the rate of plate interaction estimated on a long-term basis from sea floor magnetic lineaments [North 1974, 1977]. This may imply that seismic faulting is not the primary deformation and a substantial portion of deformation is taken up aseismically by fault creep, bedding-plane slip or other forms of ductile deformation. Since the continental deformation along the convergent zones is distributed and controlled by velocity boundary conditions of plate interactions described by the plate tectonics theory, the continental faults would possibly reactivate and move at slower rates than the plate rates. No evidence of fault creep has been documented so far along the faults in

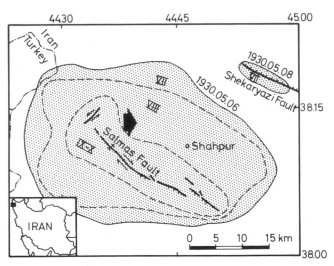

Fig. 9. Reactivation of the Shekaryazi reverse fault during the largest aftershock of the 1930 Salmas earthquake. The Salmas earthquake fault associated with the main-shock of magnitude Ms=7.4 had a right-lateral displacement and normal mechanism at the surface. Ground displacements associated with this earthquake were interpreted in terms of a regional exten-sion and east-northeastward motion of crustal material (arrow) [Tchalenko and Berberian 1974, Berberian and Tchalenko 1976c]. After 41 hours, the readjustment of the local stress system triggered the largest shock of magnitude Ms=6.2 along the Shekaryazi reverse fault about 30 km northeast of the main-shock epicentral zone [Berberian 1979a]. Intensities are on the Modified Mercalli scale. Transverse Mercator Projection.

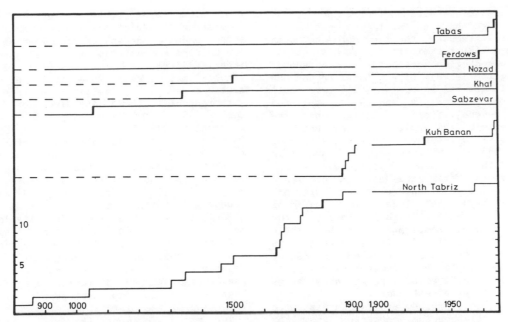

Fig. 10. Cumulative diagram of the major earthquakes along some active Iranian faults indicating the long recurrence period of large magnitude earthquakes (events with Ms>6.0 are shown as heavy lines) in Central Iran (cf. Table I and Fig. 5). Despite low magnitude earthquakes which occur frequently in Central Iran, large magnitude shocks are discontinuous and have long periods of relative quiescence. Note the difference in the recurrence period of earthquakes along the dominant strike-slip faults (North Tabriz and Kuh Banan) and the pure thrust faults (Tabas, Ferdows, Nozad, Khaf and Sabzevar). The difference may be due to the fact that thrusts are high-stress seismic faults and the minimum differential stress required to initiate sliding along the thrust faults is higher than stresses required for strike-slip or normal faults [Sibson 1974, 1975, 1977]. Lack of historical seismic damage record over the past 10 centuries along the Tabas and the Sabzevar frontal active reverse faults indicate that there was no adequate assessment of the seismicity and seismic hazard on this basis. This illustrates the importance of recognizing the active fault features in the field [Berberian 1979a]. Time in years is plotted as abscissa and number of earthquakes as ordinate. Historical data after Berberian [1976a, 1979a], Ambraseys [1975], Berberian and Arshadi [1976], and Ambraseys et al. [1979].

Iran [King et al. 1975, Tchalenko 1975b]. Does the regional geology (e.g. presence of ophiolite belts and ductile beds) contribute to this? Of course, the seismic data are only available for a very short time span, and seismicity may be abnormally low at present. Limited historical (non-instrumental) earthquake records show long periods of quiescence followed by periods of large magnitude seismic activity (Fig. 10) [Ambraseys 1971, 1974, 1975, Ambraseys and Melville 1977, Berberian 1979a]. The absence of seismicity along major plate boundaries enables the prediction of large magnitude earthquakes on the basis of the seismic gap method [Sykes 1971, Kelleher et al. 1973, Kelleher and Savino 1975]. Migration of large magnitude shallow earthquakes along active plate margins [Mogi 1968, McCann et al. 1979, 1980, Sykes et al. 1980] suggests that these earthquakes are dependent events. This sort of simple migration has not been observed in the complex continental convergent zones. Only in one

case a migration of seismic activity towards nucleation zone of a large magnitude earthquake along the Tabas active fault (Fig. 5, Table 1) is documented in Central Iran [Berberian 1979a]. Available data shows that large magnitude earthquakes in the Iranian plateau occur in regions with histories of both long- (11 centuries or more in Tabas) and short-term (11 years in Dasht-e-Bayaz) seismic quiescence (Table I, Figs. 5 and 10). The limited historical data and the variable long and short recurrence periods of the large magnitude earthquakes makes it difficult to predict earthquakes in the trapped Iranian convergent zone as well as other active continental areas.

Revised Seismic Risk Map Of Iran

The first preliminary seismic risk map of Iran [Berberian and Mohajer-Ashjai 1977] was found to be incomplete after the Tabas-e-Golshan destructive earthquake of 1978.09.16 [Berberian 1979a]. Since the Tabas earthquake fault was not known prior to this earthquake, the region had not been recognized as a high seismic zone. It is essential to use geological data and seismic records for evaluating the seismic risk. Based on the new detailed fault map and an up-to-date active fault map of the country (Figs. 4 and 5), and the results of several field investigations, the preliminary seismic risk map was revised and all the available data were used to prepare the new version (Fig. 11). In this version, the country is divided into three zones of minor, moderate and major damage zones. Until more data is available on the exact age and nature of the faults and their relationships to local and regional seismicity, and on the relation between the surface structures and those at depth and their relation to the seismic activity in Zagros, this revised version of the seismic risk map may serve as a general and simplified seismic zoning map. It still should be considered as a preliminary stage in the preparation of a better map.

Comparison of the Iranian seismic zones with those of the neighbouring countries, show that they do not fit each other along the frontiers. This inconsistency should be resolved by a joint seismic risk study in the countries of the region.

Recent Continental Movements In Iran

Iran is the site of convergent plate collisions along the Alpine-Himalayan active mountain belt (Fig. 3). As the African-Arabian plate approached Central Iran, subduction of the Mesozoic ocean crust beneath Central Iran apparently occurred along the 'Zagros suture zone', the location of which is indicated by a belt of deformed ophiolites, pelagic sediments and turbidities with clastic blocks (trench sediments; see Figs. 2 and 3). Distribution of the convergent petrotectonic assemblages mark the contact between two different continental blocks (Zagros in the south and Central Iran in the north) along the present line of the Main Zagros reverse fault. The ophiolites and pelagic sediments in Iran were emplaced in the form of a stack of thrust sheets during the late Cretaceous (65 Ma). This emplacement together with the associated change of sedimentary conditions from oceanic to shallow marine, and the absence of a younger ophiolite sequence in Central Iran presumably indicate that since collision (probably 65 Ma.) [Berberian and King 1981] the geological and tectonic history of the Iranian plateau has been typically 'intracontinental'. Subsequent sedimentation was controlled by major multi-role basement faults which have clearly been inherited from previous orogenic phases. Post-collisional convergence resulted in progressive crustal thickening and shortening by folding, reverse faulting and the gradual rise of the mountain belts above sea level.

As in other active sialic regions, both deformation and seismicity is spread out over a broad

zone along several active faults (Figs. 1 to 5) suggesting a complex mode of deformation. Unlike young oceanic crust, the continental crust has a non-uniform composition, and usually contains many inherited planes of weakness. These planes (faults) may be reactivated if the stress field is suitable and the faults are oriented favourably with respect to the stress direction [Jaeger and Cook 1969; Mckenzie 1969, 1972; Molnar and Tapponnier 1975; Bird et al. 1975; Sykes 1978]. Because of its heterogeneous nature, one would tend to view the Iranian plateau from a mechanical point of view as a jumble of large crustal fragments of various rigidity and history. Each fragment is limited by major, deep-seated, old boundary faults; the geological structure, deformational style and seismicity in each fragment has a special and consistent pattern, different from those of the adjacent units (Figs. 1 and 2). These major boundary faults, which are usually old, deep-seated multi-role fractures reactivated during different orogenic periods, have played an important part in the structural evolution of the plateau.

The major fragments (tectonic units) indicated by the available seismo-tectonic data are the two fold-thrust border belts of Zagros (in the SW) and Kopeh Dagh (in the NE). They rest on the Precambrian Arabian and the Hercynian Turan platforms respectively. The third fragment, Central Iran, lies between them and the fourth unit is the accretionary flysch belt of east (Zabol-Baluch) and southeast Iran (Makran; see Fig. 2). The axes of maximum crustal shortening in the different units, based on the Neogene fold axes of the mountain belts (Fig. 3), together with the horizontal component of the compressional axes determined from the focal mechanism solutions (Fig. 4 inset) indicate a NE-SW direction of shortening [Berberian 1976a]. The drift of the Arabian plate towards the north-northeast with respect to Eurasia, which is roughly perpendicular to the Neogene fold axes, is presumed to be responsible for this shortening (Fig. 3). The character of each tectonic unit is summarized below.

A) Zagros Active Fold-Thrust Belt

This active border belt lies on the northeastern margin of the Arabian continental crust, on a Precambrian metamorphic basement. It is a young fold-thrust mountain belt currently shortening and thickening as a result of the collision of the Arabian and Central Iranian plates and their continuing convergent movements (Figs. 2 and 3). Although the suture zone between the Zagros active fold-thrust belt and Central Iran lies along the present Main Zagros reverse fault line (Fig. 2), Alavi [1980] suggested that it lies to the northeast of the Sanandaj-Sirjan belt of southwest Central Iran (see Fig. 2).

The geological history of the belt is comparatively simple one of continuous sedimentaion (with

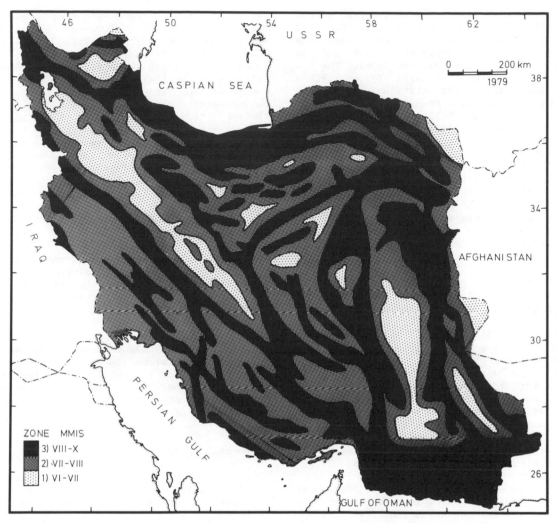

Fig. 11. First revised seismic risk map of Iran.
Three seismic zones are considered of which zone 3 (the major damage zone) covers the zones of the
Quaternary faults and the areas associated with the past destructive earthquakes. The intensities
shown are not the maximum possible intensity, but only probable. Based on the first preliminary
seismic risk map of Iran [Berberian and Mohajer-Ashjai 1977]. Lambert Conformal Conic Projection.

minor disconformities) on a subsiding basement
from the late Precambrian to the Miocene time. The
sedimentation was of platform-cover type during
the Paleozoic; miogeosynclinal from Middle
Triassic to Miocene, and synorogenic with
diachronous conglomerates in late Miocene to
Pleistocene times [Stocklin 1968a, James and Wynd
1965, Berberian 1976c]. The belt was a 'passive
continental margin' undergoing thinning,
subsidence and extension with deposition of shalow
water shelf carbonates during the Mesozoic
[Berberian and King 1981]. The subsidence was the
effect of the late Paleozoic intracontinental
rifting along the present Main Zagros reverse
fault which preceded the Triassic oceanic
spreading. The crust was subsided at least for
170 My with apparently no major tectonic activity

other than subsiding along normal faults which was
probably an isostatic response to crustal
thinning.
   The belt was uniformly folded into a NW–SE fold
system due to the NE–SW compression applied from
the Arabian plate (Figs. 2 and 3) mainly during
the late Miocene to Pleistocene orogenic movements
(the High Zagros marginal belt, however, had
already been deformed during the late Cretaceous
collisional orogeny). Telescoping of the whole
belt and rethickening of its crust during
collisional orogenies could have been carried out
by reversal of fault motion along the Mesozoic
normal faults [Helwig 1976, Jackson 1980, Jackson
and Fitch 1981, Berberian and King 1981, Jackson
et al. 1981]. The intensity of deformation and
the amplitude of the folds decreases

southwestwards with distance from the High Zagros in the northeast. The southern and southwestern limbs of the Zagros folds (steeper than the norheastern limbs) are locally overturned, and in places overthrusts have been developed. The recent age of the folding is implied by anticlines forming topographic highs with juvenile erosion patterns superimposed upon them. Upper Precambrian salt has risen diapirically through the whole sedimentary sequence and several salt horizons decouple the upper structures from those at depth (Fig. 6). At present the Zagros active fold-thrust belt becomes more compressed and narrower in the NW (near the impinging zone of the Arabian plate with the northwestern part of the Iranian plateau) and has a width of about 200 km. The belt widens towards southeast to about 350-400 km. Apparently this pattern is influenced by the shape of the Arabian-plate margin and the geometry of the collision zone along the northern margin of Arabia.

The isodepth contour map of the Moho surface in the Zagros belt [Peive and Yanshin 1979a] shows a 50-55 km thick crust along the High Zagros (the northern margin of the belt). This gradually thins out towards southwest to a crust of about 35 km thick along the Zagros foothills at the northern shorelines of the Persian Gulf. A thicker crust along the High Zagros marginal belt could be due to the fact that this marginal belt has undergone two major collisional orogenies during the late Cretaceous and the late Miocene-Pleistocene times; whereas the rest of the belt was only affected by the latter phase. Moreover, the High Zagros was the northern edge of the Mesozoic passive margin of the Arabian-plate and it underwent larger amount of subsidence and sedimentation.

The belt is a broad zone of continuing compressional deformation which is marked by a high level of shallow crustal seismicity (Fig. 1). Most of the earthquakes in the Zagros active fold-thrust belt are shallower than 30 km (Fig. 1) and their hypocentres do not appear to increase in depth towards the northeastern margin of the belt. Some of the deeper focal depths reported for this belt (Fig. 12) are either real deep shocks or are poorly recorded events [Berberian 1979b, c; Jackson 1980b]. The entire belt is intensely seismic and there is no evidence for seismic shortening on a single northeast dipping plane. The intense seismicity is limited to the Main Zagros reverse fault in the northeast and the southwestern edge of the Zagros folds in southwest (Figs. 1 and 2). The Main Zagros reverse fault is believed not to be seismically active, but the local damage zones of the earthquakes of 1965.06.21 (at Hajiabad; 28.30N, 55.91E), 1973.08.28 (at Gondoman; 31.87N, 51.15E), 1973.11.11 (at Qeshlag; 30.56N, 53.03E); and finally the Lice earthquake in south Asia-Minor on 1975.09.06 with a 30 km surface rupture and thrust mechanism [Toksoz and Arpat 1977], apparently belie this judgment (Fig. 5 and Table III).

The belt is truncated to the SE by the Minab fault, north of the Straight of Hormoz, but the seismicity stops west of this fault (Fig. 1), along the Hormoz-Jiroft seismic trend [Berberian 1976c]. The Minab fault (Fig. 5) has been an important tectonic structure at least since the late Cretaceous time, separating the continent-continent collision in the west (Zagros) from the oceanic-continent plate boundary in the east (Makran; Figs. 3, 4 and 5). The belt is characterized by a high proportion of low to medium magnitude shallow earthquakes and very few earthquakes of magnitude 7. Series of earthquakes resembling swarms are also characteristic of this belt.

Fault plane solutions of the 20th century earthquakes in the Zagros are consistent with predominantly high-angle faulting; this indicates that the present deformation appears due to NE-SW thrusting along several NW-SE trending buried reverse basement faults. Recent folding and uplifting of the belt is already evident in the deep river valleys cutting the anticlines, fossil (raised) beaches, the height of Quaternary alluvial terraces and uplift of historical canals [Lees and Falcon 1952, Vita-Finzi 1979a]. The minimum average rate of uplift of the Zagros is estimated to be about 1 mm/yr since the early Pliocene time [Falcon 1974, Vita-Finzi, 1979a]. Based on surface geology, the thrusting and folding of the Zagros mountains represent a shortening of about 20 per cent across the present 250 km width of the belt since the early Pliocene time.

B) Kopeh Dagh Active Fold Belt

This northeastern border fold belt of the Iranian plateau (Figs. 3 and 4) is formed on the southwestern margin of the Turan continental crust (southern Eurasia) and is presumably underlain by Hercynian metamorphosed basement. The sedimentary cover consists of about 10 km of Mesozoic-lower Tertiary sediments deposited in a subsiding sedimentary basin during the Mesozoic extensional phase. Although the Kopeh Dagh basin did not evolve into an oceanic crust, the thickness of the Mesozoic sediments implies that its crust was strongly stretched and thinned along normal faults at that time. Like the Zagros belt, the sediments were folded into long linear NW-SE folds perpendicular to the trend of the relative motion between Arabia and Eurasia, during the last phase of the Alpine orogeny. The minimum horizontal crustal shortening during Neogene, was about 15 per cent on the Iranian side of the fold belt (with the present width of 70 km). Folds in western Kopeh Dagh have a gently dipping northern flank whereas the southern flank is steep and thrust faulted. The folds in eastern Kopeh Dagh are symmetric and cut by numerous transverse conjugate shear faults. These shear faults displace some longitudinal reverse faults and folds, and a few (like the Baghan-Germab fault; Fig. 5) have been recently active. The belt is

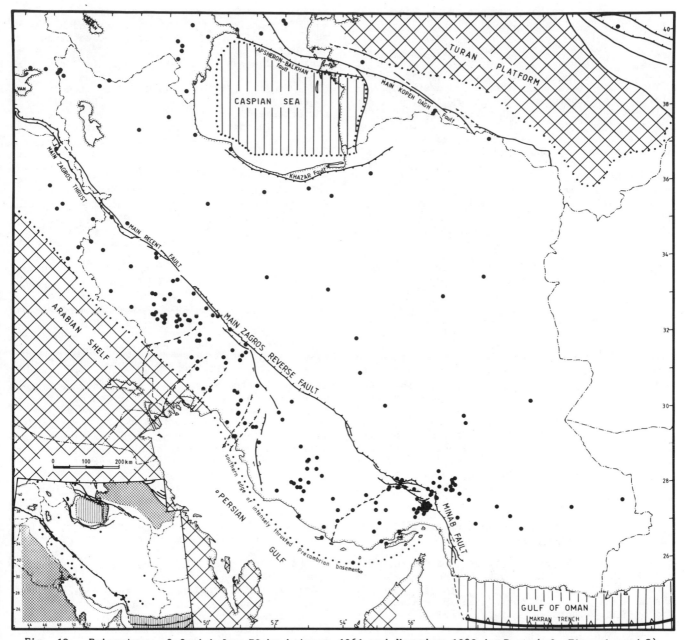

Fig. 12. Epicentres of foci below 50 km between 1961 and November 1980 in Iran (cf. Figs. 1 and 2). The depth of foci is given by USGS. There is great error in focal depth estimation and large portion of them may not represent the real depth of the earthquakes. Lambert Conformal Conic Projection.

Inset left: Distribution of epicentres of events with foci below 50 km depth and with magnitude Mb>5.0. Comparison with the main figure shows that a large number of reported deeper events are small magnitude earthquakes (Mb<5.0) which were recorded by few stations. Therefore, their calculated focal depths are not reliable.

bordered on the north by the Main Kopeh Dagh reverse fault and is separated in the south from Central Iran by several reverse faults (Fig. 2). Salt horizons which could produce major decollement, decoupling different horizons, are absent in the Kopeh Dagh. Minor plastic horizons are the gypsiferous mudstone of the Paleocene Pesteh-leigh Formation and the upper Jurassic-lower Cretaceous thin gypsum layers of the Shurijeh Formation [Afshar-Harb 1969, 1970, 1979]. The isodepth contour map of the Moho surface in the Kopeh Dagh belt [Peive and Yanshin 1979a] shows a 45 km thick crust in northern Kopeh Dagh. Apparently the crust thickens towards the south and becomes 50 km thick along its southern border zone, where joins Central Iran along several reverse faults.

The Kopeh Dagh fold-thrust mountain belt is elevated by northward thrusting in the north (along the Main Kopeh Dagh reverse fault; Fig. 5) and southward thrusting in the south of the range. This geological observation is also supported by the fault plane solutions of the earthquakes along the northern (1948.10.05 Eshghabad earthquake) and the southern (1963.03.31 Esfarayen; see Table I and Fig. 5) margins of the belt. The recent continental deformation in the Kopeh Dagh consists of movements along the longitudinal reverse and transverse conjugate shear faults (Fig. 5). Aftershock sequence of the Eshghabad 1948.10.05 earthquake along the northeastern margin of the Kopeh Dagh belt illustrated listric thrust fault(s) dipping southwards underneath the Kopeh Dagh fold-thrust belt [Rustanovich 1957] (Fig. 5).

The belt is characterized by a high proportion of large magnitude shallow earthquakes [Tchalenko 1975a, Berberian 1976a, c]. Although the damage zones of two earthquakes (1970.07.30 and 1974.03.07) coincide at the surface with two longitudinal thrust faults (Qarnaveh and Takal Kuh; see Table I and Fig. 5) along the Kopeh Dagh belt, fault plane solutions of these earthquakes [Jackson and Fitch 1979] indicate a left-lateral motion along a NE-SW transverse shear fault. Right-lateral displacement of a fold axis by the Baghan-Germab active fault in eastern Kopeh Dagh (Fig. 5) amounts to about 9 km; the average displacement rate is therefore about 4-5 cm/yr since the early Pliocene time [Tchalenko 1975a, Afshar-Harb 1979].

C) Central Iranian Fold-Thrust Belts And Compressional Depressions

Central Iran (Fig. 2), between the Zagros and the Kopeh Dagh fold-thrust border belts, has undergone several major orogenic phases, and is characterized by various syntectonic metamorphic and magmatic events, especially during the late Paleozoic, Middle Triassic, late Jurassic and late Cretaceous phases along its southwestern margin [Berberian 1976c].

Central Iran was a part of the Arabian plate during the Precambrian-Paleozoic time, and was separated from Eurasia by the Hercynian Ocean in the north [Berberian and King 1981]. Late Paleozoic rifting in the Arabian-Iranian platform along the present line of the Main Zagros reverse fault, seems to be responsible for separation and northward motion of Central Iran from Arabia, and opening of the High Zagros Alpine Ocean in the south. Subduction of the High Zagros Alpine oceanic crust underneath southern active margin of Central Iran (Sanandaj-Sirjan belt in Fig. 2) produced 'Andean-type magmatic-arc' during

Mesozoic and possibly Tertiary time. Continuous northward drift of the Arabian plate and northward subduction of the oceanic crust, led to closure of the High Zagros Alpine Ocean and the Arabian-Central Iranian collision [Berberian and King 1981].

The post-collisional history of Central Iran (possibly since Eocene or Miocene) is characterized by extensive magmatism and regional uplift. Paleogeographically, the Central Iranian Cainozoic magmatism [see Berberian 1976a, Berberian and King 1981] has no clear association with single plate boundary (Fig. 3) and is possibly derived from a mixing process between primary magma with basaltic affinities and crustal material. Clearly a close look at the tectonic-petrologic setting of the magmatic rocks is required. If the Mesozoic oceanic crust between Zagros and Central Iran was consumed during the late Mesozoic time, then post collisional convergent movements could be responsible for regional uplift of Central Iran (by folding and reverse faults) and partial melting of the lower crust and upper mantle. A detailed trace-major element and isotopic study of the Tertiary magmatic rocks of Central Iran is needed to understand the petrogeneses of these rocks and the structural evolution of the region. Evolution and structure of the active continental plateaus are accepted to be a complicated subject and different mechanisms could be involved during evolution of different regions [Powel and Conaghan 1973, Bird et al. 1975, Toksoz and Bird 1977, McGetechin and Merrill 1979]. Due to deficiency of detailed geological, structural and geophysical data about the crust and upper mantle properties, no constrained model can be proposed for the evolution of the region.

The isodepth contour map of the Moho surface in Central Iran [Peive and Yanshin 1979a] shows that the thickest part of the crust (50 km thick) is located along the Sanandaj-Sirjan belt (southwestern active continental margin of Central Iran during Mesozoic) and also along the northeastern part of the zone (Paleozoic continental margin in contact with the Kopeh Dagh belt). Apparently the rest of the region has a crustal thickness of about 45 km.

Decollement tectonics is documented in two places in Central Iran. In the Kuh Banan fold-thrust belt (between Kuh Banan and Lakar Kuh faults in Fig. 5) the upper Precambrian Hormoz salt acts as a zone of slippage between the metamorphosed Precambrian basement and the top sedimentary cover. Kuh Banan and Lakar Kuh faults, bordering this decollement zone, are documented seismically active faults of this small decollement zone in Central Iran. Detailed study of the best located aftershocks of the 1978 Tabas-e-Golshan earthquake demonstrated an active imbricate listric thrust system with fault planes flattening into a basement decollement zone beneath the Shotori fold-thrust mountain belt (east of the Tabas active frontal fault in Fig. 5) [Berberian 1981a].

Dominance of 170 My or 210 My of Andean-type tectonic regime (with intrusion of voluminous volcanics and plutonic rocks) in Central Iran has led to development of a very hot uppermost mantle underneath Central Iranian crust. Presumably existence of large number of recent and active volcanoes in Central Iran and their absence in Zagros and Kopeh Dagh, indicate that uppermost mantle underneath Central Iranian crust has a higher temperature than those of Zagros and Kopeh Dagh. Therefore the brittle-ductile boundary in Central Iran could be shallower than the one at Zagros, and that most earthquakes in Central Iran are possibly shallower than events in Zagros or Kopeh Dagh belts.

Seismically, Central Iran is characterized by 'sporadic and discontinuous' earthquake activity with large magnitude shallow earthquakes along several 'mountain-bordering reverse faults' (Figs. 1 and 5) which have been formed and reactivated during previous orogenic phases. Unlike the ideas presented by Ambraseys and Melville [1977] and Shirokova [1977], the earthquakes in Central Iran are not thought to be related to the "minor fault zones cutting across the major faults". Almost all the major earthquakes in Central Iran have occurred along the frontal reverse faults separating mountain belts from compressional depressions [Berberian 1979a]. This is not surprising, since the late Alpine movements responsible for the present day physiographic features took place mainly along these mountain-bordering reverse faults. Some of these frontal reverse faults occur along old normal faults which were responsible for the formation of sedimentary basins during older tensional phases [Berberian 1979a, 1981a and b, Berberian and King 1981]. This reversal of fault motion during different tensional and compressional phases [Mercier 1976] indicates that the present active faults are deep-seated and have a long geological history. It also demonstrates that the intracontinental sedimentary basins were formed and deformed entirely within continental crust. Despite the dominant active compressional regime, normal faulting is observed in one locality in the northwestern Central Iran [Berberian and Arshadi 1977]. The question of how a single stress field can simultaneously account for parallel normal and reverse faulting within a narrow zone remains unanswered until a detailed study is carried out in this region.

Large magnitude earthquakes in Central Iran are usually characterized by long recurrence periods along frontal reverse faults and the recent aseismic period along some major recent Iranian faults presumably indicates an interseismic period of quiescence after a period of seismic activity (Fig. 10). Due to poor seismic station coverage in Iran, it is not clear whether after the aftershock activity of large magnitude earthquakes in Central Iran has decreased, the fault zone becomes completely quiescent until the next earthquake, or if at a certain stage, the fault becomes quiescent as a part of a process

premonitory to a large magnitude earthquake [Kelleher and Savino 1975]. It is accepted that the interseismic stage, which occurs during the period between large earthquakes, is characterized by the steady accumulation of elastic strain. It has been observed that in Central Iran this interseismic stage of quiescence varies from 11 years (in the case of the Dasht-e-Bayaz active fault; Fig. 5 and Table I) to 11 centuries or more (in the case of the Tabas active fault; Figs. 5, 8 and 10, Table I). Apparently this stage is shorter along the strike-slip faults than along the thrust faults (Fig. 10). This might be due to the fact that the reverse faults are high-stress seismic faults and the minimum differential stress required to initiate sliding along the reverse faults is higher than stresses required for strike-slip or normal faults [Sibson 1974, 1975, 1977]. Apparently the deformation is characteristically more intense for a given displacement around the thrust faults [Hills 1965]. Moreover, where there is evidence for shear heating around faults, it is almost invariably associated with thrust faults [Scholz, 1980]. Since the recurrence intervals at a given location in Central Iran appear to be variable and usually long, the late Neogene-Quaternary geological history along the recent faults provides important information about the potential activity of the faults [Berberian 1979a].

Some large magnitude earthquakes in Central Iran are characterized by relatively little or no foreshock-aftershock activity. The Siahbil earthquake of 1978.11.04, which took place along the Talesh active fault (Fig. 5 and Table I) [Berberian 1981b], is a case which may indicate a sharp dislocation in which most of the seismic energy is released during the main faulting stage.

Recent movements along active faults (Figs. 4 and 5; Table I), fault plane solutions of earthquakes and post-Neogene fold axes (Fig. 3) indicate that regional NNE-SSW compression and shortening dominates the tectonic pattern of the Iranian Plateau [Berberian 1976a, 1979a]. There is an apparent widening of Central Iran towards east and wedging out towards the impinging zone of the Arabian plate in the northwest. Eastern Central Iran has a general low elevation whereas northwestern Central Iran has high elevations with widespread high altitude active volcanoes. The total amount of crustal shortening by folding and reverse faulting during the Plio-Pleistocene orogenic phase is unknown. The horizontal shortening is estimated to be about 25 per cent in the Shotori mountains of eastern Central Iran (east of the Tabas active fault in Fig. 5). The measured vertical movement across two mountain bordering active reverse faults in Central Iran (Tabas and Ferdows; Fig. 5) is at least some 1,000 and 500 m respectively since the late Neogene-early Quaternary compressional phase [Berberian 1979a]. Quartzo-feldspathic mylonites and pseudotachylites are found to be associated with still active fault of Salmas in the northwestern

Central Iran [R.H. Sibson personal communication] (Fig. 5). This could be considered as evidence for a long continued activity and a large degree of differential uplift and erosion (exposure of the Precambrian metamorphic rocks with mylonites) along the fault.

D) Accretionary Flysch Belt

The Makran ranges of southeast Iran, and the Zabol-Baluch ranges of east Iran along the Afghanistan-Pakistan border, are post-Cretaceous flysch belts which join together in SE Iran and continue into the Pakistani Baluchestan ranges (Fig. 2). The flysch deposits are believed to have been deposited on upper Cretaceous ophiolitic-melange in a continental slope environment.

The Makran ranges are a zone of east-west Tertiary folds overthrust south with north-dipping longitudinal imbricate reverse faults extending from the Minab fault in the west to the Ornach Nal and Chaman faults of Pakistan in the east (Figs. 2 and 3). The flysch belt is shortened by a north-south compression system into steep E-W folds and north-dipping thrusts. The isodepth contour map of the Moho surface in the Makran ranges [Peive and Yonshin 1979a] shows a thick crust of about 50 km thick in the northern Makran (southeastern active continental margin of Central Iran). Apparently it gradually thins southwards to a crust of 40 km thick along the coastal Makran. However recent investigations along offshore Makran indicated a much thinner crust [Niazi et al. 1980, White and Louden 1981].

The Makran region is an unusually wide deformed accretionary sediment prism formed from material scraped off the northerly subducting oceanic lithosphere of the Gulf of Oman (Arabian plate) underneath the Central Iranian continental margin, at a rate of about 5 cm/yr [Stoneley 1974, White and Klitgord 1976, Farhoudi and Karig 1977, Jacob and Quittmeyer 1979, Quittmeyer et al. 1979]. The ophiolitic melange of the Makran ranges cannot be interpreted as an indication of continental closure, since subduction of oceanic crust preceded and succeeded its late Cretaceous emplacement [Berberian and King 1981]. The area has been an active convergent plate margin at least since Cretaceous. Folding and stacking of the surface sediments of the Oman abyssal plain is thought to be due to the ongoing subduction of the oceanic crust along the Makran continental margin [White and Klitgord 1976, White 1977, White and Ross 1979]. Apparently the continental slope of the Gulf of Oman off western Makran is underlain by an oceanic lithosphere with thick pile of stacked sediments. The subducting oceanic crust has a thickness of about 6.7 km and is carrying about 7 to 12 km of sediments into the accretionary wedge [Niazi et al. 1980, White and Louden 1981]. Presumably the subducting oceanic lithosphere underneath the Makran ranges has a gentle dip (less than 2 degrees) from the coastal

trench to the hingeline (approximately 170 km inland) and deepens northward [Jacob and Quittmeyer 1979, White and Louden 1981]. Apparently the Bazman-Taftan-Soltan active volcanic-arc, north of Makran, is associated with this subduction zone. The presence of Pleistocene and Holocene raised beaches and marine terraces along the Makran coast indicates a tectonically emerging coast [Falcon 1947, Snead 1969, 1970, Little 1972, Vita-Finzi 1975, 1979b, Ghorashi 1979, Page et al. 1979]. Episodic uplift of the convergent continental margin of the Makran coast could be the result of the large-magnitude earthquakes [Page et al. 1979]. The ^{14}C and U-Th dates on shells from the raised beaches of Iranian Makran, range from 114,000 to 156,000 YBP, and the Holocene rate of uplift on the coast is about 0.01-0.02 cm/yr increasing towards east [Page et al. 1979].

The zone is characterized by a low and scattered seismicity compared with the Zagros active fold-thrust belt [Berberian 1976a; Fig. 1]. Lack of seismicity along the subduction zone is not clear; it could be due to the very shallow angle of subduction [White and Klitgord 1976]. The depth of earthquakes in Makran apparently increases from very shallow at the coast to 80 km inland [Jacob and Quittmeyer 1979, Quittmeyer and Jacob 1979]. Fault plane solutions of the Makran ranges of southeast Iran (Fig. 5) show that the region is characterized by two seismic regimes: i) shallow earthquakes with thrust mechanism (reactivation of the reverse faults at the surface), and ii) moderate to deep earthquakes with normal faulting mechanism (deformation of the subducting oceanic crust) [Jacob and Quittmeyer 1979]. No normal fault is observed at the surface in the Makran ranges. The Jaz Murian Depression, which is situated in the northern part of the Makran ranges, is considered to be downthrown along normal faults [Farhoudi and Karig 1977]. The South Jaz Murian fault, bordering the southern part of the depression and the northern part of the Makran ranges (Fig. 1), has a high-angle reverse character at the surface.

The area has experienced two large magnitude earthquakes (1945.11.27, Ms=8.0 and 1947.08.05, Ms=7.3) along the Makran active subduction zone, off the coast of Makran [Sondhi 1947, Quittmeyer et al. 1979]. Recent low level seismicity in the region west of that affected by 1945 event, suggests that the western part of the coastal Makran region may be the site of the next large magnitude earthquake [Quittmeyer 1979].

Recent Tectonics Models

Plate tectonics has provided a good description of the active deformation of oceanic lithosphere which is characterized by narrow belts of seismicity separating larger aseismic rigid plates [Isacks et al. 1968]. In contrast, continental lithosphere in convergent zones with diffuse seismicity behaves differently [McKenzie 1972,

1977, Sykes 1978] and the deformation is mainly controlled by reactivation of the old deep-seated faults. Based on the plate tectonics concept, three regional recent-tectonic models were used by Nowroozi [1972] , McKenzie [1972, 1977] and Dewey et al. [1973] to study the complex behaviour of the continental crust in Iran and the Middle East.

The tectonics models of Nowroozi [1972] and McKenzie [1972, 1977] were both based on a very short term seismicity, fault plane solutions of the large magnitude earthquakes and limited structural data. The differences between the two models are minor. There is no Lut plate and an area with extensional features (marked by Quaternary and recent volcanoes in the Nowroozi model) in McKenzie's model. The sense of motion of the Caspian plate in the two models is opposite to each other (northeast in McKenzie's model and northwest in Nowroozi's model) and the Persian or Iranian plate moves northward and northeastward according to Nowroozi and McKenzie respectively.

In order to assess this sort of approach, the model of Nowroozi [1972] is projected on the documented active fault map together with the trend map of the post Neogene-early Quaternary fold axes (Fig. 13). Nowroozi [1972] introduced four rigid plates in Iran with vague and hypothetical boundaries for which there is very little or no geological and seismotectonic field evidence. His proposed Lut and Caspian plates, have imaginary boundaries, neither tectonic nor seismic. The Lut plate moving faster northward than the Persian plate, should produce left-lateral motion along the Kuh Banan fault. However, the 1977.12.19 earthquake along this fault produced a right-lateral motion [Berberian 1977c, Berberian et al. 1979a] (Fig. 5). The eastern boundary of the Lut plate cuts all structural trends, near the Pakistan border (Fig. 13) especially the major recent faults and the post Neogene - early Quaternary fold axes. The disadvantage of this approach is that with each large magnitude earthquake (like the Dasht-e-Bayaz earthquake of 1968.08.31 for which the seismogenic fault became the northern limit of the Lut plate in Nowroozi's model) a new plate boundary must be drawn. A broad zone of recent faulting (Fig. 4) makes it impractible to recognize several rigid microplates with well defined boundaries in Central Iran. The model of Dewey et al. [1973] faces similar problems; it is not supported by the existing field evidence. Some of the features introduced in the recent tectonic models of Iran, such as the boundaries of the Caspian plate in the three mentioned models [see Berberian 1981b], are clearly unrelated to geological and structural data. Major fault zones exist within all proposed plates and the earthquakes occur on large number of faults spread over a wide area. This indicates that rigid plate models do not adequately represent the active regions of the continents involved in crustal compression (especially in trapped convergent zones like the Iranian Plateau).

Concluding Remarks

Evaluation of seismological and tectonic data coupled with seismotectonics and field investigations of medium to large magnitude earthquakes and study of aerial photographs and Landsat imagery has led to the discovery of several formerly undetected active faults. These faults are deep-seated structures inherited from previous orogenies and play an important role in the present-day tectonics of the region. Faults, some of which had controlled sedimentary facies distribution and the location of basins, became responsible for the formation of the present physiographic features, and are now the site of the seismic activity. As with the folding and faulting, active deformation indicated by seismicity is widespread throughout the entire Iranian plateau, suggesting a complex mode of deformation. The data presented show that most of the active faults in Iran are 'frontal reverse faults' bordering the mountains along which the continental crust is being thickened and shortened.

At present the width of the deformed zone in northwest Iran (close to the impinging zone of the Arabian plate with the western Iranian plateau) is about 850 km. This region is characterized by high elevation and widespread high altitude volcanoes. The same zone widens in the east and becomes 1450 km wide and has a lower elevation (Fig. 3). This pattern, which is influenced by the shape of the Arabian plate margin, may indicate that the continental crust of the Iranian plateau becomes slightly thicker in the northwest (due to larger amount of crustal shortening) than in the east. However, this is not observed on the isodepth contour map of the Moho surface [Peive and Yanshin 1979a]. Widespread active volcanoes in Central Iran (especially in northwest and east Central Iran) may indicate a higher temperature for the lower crust and uppermost mantle of the region. Whereas no volcanic activity is recorded in the Zagros and the Kopeh Dagh active border fold-thrust mountain belts.

The Iranian plateau with a relatively weak continental crust lies in a wide inhomogeneous collision zone between two continental blocks of greater rigidity. The collision and the convergent movements of Eurasian and Afro-Arabian blocks are responsible for the uplift and deformation of the mountain belts, the formation of the present physiographic features and the active tectonics of Iran. The present-day deformation is mainly taken up by reverse faulting within the Iranian continental crust and to a lesser extent by strike-slip motion (sideway movements of the blocks) in the northwest, and consumption of the Oman oceanic crust along the Makran trench in the southeast.

The Iranian plateau is folded and densely faulted, and unlike California, Asia-Minor and Central Asia, linear narrow zones of more intense deformation do not occur. It is therefore

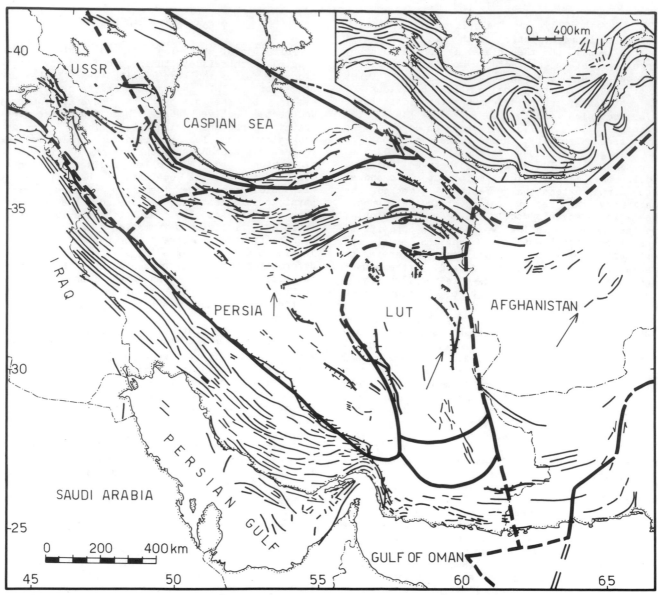

— Documented Seismic Fault ; —— Post—Neogene Fold Axis ; ▬ Nowroozi's Plate Boundaries(1972).

Fig. 13. Appraisal of Nowroozi's active plate boundaries [1972], by comparison with the geometry and character of the documented recent active faults and the post Neogene—early Quaternary fold axes (cf. Figs. 3 to 5). The comparison demonstrates the inapplicability of the rigid plate concept on the active continental areas like Iran. The recent fold and fault pattern demonstrates that the active tectonics of Iran are dominated by reverse faulting and the deformation throughout the country is dominated by NE—SW compression and shortening. The active tectonics (seismicity, recent faulting and folding) are not resticted to, or even concentrated on, the boundaries of the proposed subplates. Inset: Late Alpine structural trends in Iran and the neibouring countries.

difficult to subdivide the region into simple rigid plates and microplates with well defined boundaries. However, observations indicate that the deformation is certainly not homogeneous. Intense seismicity occurs in the Zagros fold-thrust belt whereas in Central Iran and Alborz the seismicity is scattered and less intense. Seismic boundaries between the major tectonic units are fairly wide and are not well defined.

The 'scattered seismic pattern' in most parts of the Iranian continental crust (Figs.1 and 5) [also see Berberian 1976c, 1977d, 1979a] may be due to the fact that the crust consists of an agglomeration of different continental fragments separated from the Arabian shield and accreted to Eurasia during several collisional orogenies [Berberian and King 1981]. Apparently in this crust some older and stronger Precambrian shield-like cores have not remobilized during younger deformational phases and resisted deformation; although they were subject to the same forces as the adjacent mobile belts. Comparison of the pre-Quaternary and the recent continental deformation indicates that the present activity is the continuation of the long-established tectonic regime that has resulted in uplifting of the mobile fold-thrust mountain belts along 'frontal reverse faults', and downthrusting of possibly more rigid blocks which are not deformed (apparently represented by some compressional depressions). The difference in elevation between the compressional depressions and the bordering active fold-thrust mountain belts is possibly caused by differences in crustal structure and reversal of fault motions during a dominant compressional tectonic regime.

The seismic activity is mainly concentrated along the 'frontal reverse faults' of the mobile belts (mountain-front and foothill thrusts), and earthquake faulting follows exactly the obvious fault scarps created by the late Quaternary faulting. These frontal active reverse faults are the 'youngest sets of imbricate thrusts' in the young fold-thrust mountain belts, and have been reactivated extensively during the late Alpine orogenic movements.

Acknowledgement. I would like to acknowledge all the help and facilities which I received during the last ten years of field work and research in Iran from the Geological (and Mineral) Survey of Iran. Were it not for understanding the urgency of the seismotectonics investigations in Iran shown by N. Khadem and R. Assefi the former Managing Directors, J. Eftekhar-nezhad Deputy Director, and M.H. Nabavi Head of the Geological Department of the Geological and Mineral Survey of Iran, and for their constant help and encouragement, the present knowledge of the seismotectonics of the region would probably have not been achieved. I am grateful to Frances Delany, Harsh Gupta, Dimitri Papastamatiou, Chris Scholz, John Tchalenko and anonymous AGU and GSA reviewers for their comments, corrections and help during the preparation of the present paper. Gratitude is also expressed to the Department of the Armenian Affairs of the Galuste Gulbenkian Foundation (Lisbon), the British Petroleum and the British I.B.M. for donating separate small grants during the course of this study. This paper is dedicated to the memory of all those who have lost their lives during earthquakes.

References

Abu-Bakr,A.M. and R.O. Jackson, Geological map of Afghanistan, 1:2,000,000, Geological Survey of Pakistan, Quetta, 1964.

Adamia, S.A., The pre-Jurassic formations of the Caucasus, Academy of Sciences of the Georgian SSR, Geological Institute, Tbilisi (in Russian), 1968.

Afshar-Harb, A., History of oil exploration and brief description of the geology of the Sarakhs area and the anticline of the Khangiran, Iran Petrol. Inst. Bull., 37, 86-149, 1969.

Afshar-Harb, A., Geology of Sarakhs area and Khangiran gas field, N.I.O.C., and 8th ECAFE Bandung, 1970.

Afshar-Harb, A., The stratigraphy, tectonics and petroleum geology of Kopet Dagh region, northern Iran, Ph.D. Thesis, Petroleum Geology Section, Royal School of Mines, Imperial College, London, 1979.

Alavi, M, Tectonostratigraphic evolution of the Zagrosides of Iran, Geology, 8, 144-149, 1980.

Albee,A.L. and J.L. Smith, Earthquake characteristics and fault activity in southern California, In: Lung, R. and Proctor, R. (eds.), Engineering Geology in Southern California, Glendale, Calif. Assoc. Eng. Geologists, 9-33, 1966.

Allen, C.R., Geological criteria for evaluating seismicity, Geol. Soc. Am. Bull., 86, 1041-1057, 1975.

Ambraseys, N.N., The Buyin-Zara (Iran) earthquake of September 1962: a field report, Bull. Seism. Soc. Am., 53(4), 705-740, 1963.

Ambraseys, N.N., Early earthquakes in north-central Iran, Bull. Seism. Soc. Am., 58(2), 485-496, 1968.

Ambraseys, N.N., Value of historical records of earthquakes, Nature, 232(5310), 375-377, 1971.

Ambraseys, N.N., Historical seismicity of north-central Iran, Geol. Surv. Iran, 29, 47-96, 1974.

Ambraseys, N.N., Studies in historical seismicity and tectonics, Geodynamics Today, Royal Society, London, 7-16, 1975.

Ambraseys, N.N., The relocation of epicentres in Iran, Geophys. J.R.astr.Soc., 53, 117-121, 1978.

Ambraseys, N.N., A test case of historical seismicity: Isfahan and Chahar Mahal, Iran, Geogr. J., 145(1), 56-71, 1979.

Ambraseys, N.N. and C.P. Melville, The seismicity of Kuhistan, Iran, Geogr. J., 143(2), 179-199, 1977.

Ambraseys, N.N. and J.S. Tchalenko, The Dasht-e-Bayaz (Iran) earthquake of 31 August 1968: a field report, Bull. Seism. Soc. Am., 59(5), 1751-1792, 1969.

Ambraseys, N.N., A.A. Moinfar and J.S. Tchalenko, The Karnaveh (northeast Iran) earthquake of 30th July 1970, Ann. di Geofis., 24(4), 475-495, 1971.

Ambraseys, N.N., A.A. Moinfar, and J.S. Tchalenko, Ghir earthquake of 10 April 1972, UNESCO, SN 2789/RMO,RD/SDE, Paris, 1972.

Ambraseys, N.N., A. Arsovski and A.A. Moinfar, The Gisk earthquake of 19 December 1977 and the seismicity of the Kuhbanan fault-zone, UNESCO, FRM/SE/GEO/79/192, Paris, 1979.

Ambraseys, N.N., G. Anderson, S. Bubnov, S.

Grampin, M. Shahidi, T.P. Tassios, and J.S. Tchalenko, Dasht-e-Bayaz earthquake of 31 August 1968, UNESCO 1214/BMS. RD/SCE, Paris, 59p, 1969.

Amurskiy, G.I., The deep structure of the Kopetdag, Geotectonics, 1, 34-40, 1971.

Anderson, D.L., The San Andreas fault, Scientific America, 225(5), 52-68, 1971.

Asudeh, I., Depth of seismicity and a crustal model for the velocity of seismic P and S waves in Zagros, M.Sc. Thesis, School of Science, Mashhad Univ., Mashhad, Iran, 1977.

Atwater, T., Implications of plate tectonics for the Cenozoic tectonic evolution of western North America, Geol. Soc. Am. Bull., 81, 3513-3536, 1970.

Balakina, L.M., A.V. Vvedenskaya, L.A. Misharina and E.I. Shirokova, The stress state in earthquake foci and the elastic field of the Earth, Izv. Earth Physics, 6, 3-15, 1967.

Berberian, M., Seismotectonic map of Iran, Geol. Surv. Iran, 39, 1976a.

Berberian, M., Documented earthquake faults in Iran, Geol. Surv. Iran, 39, 143-186, 1976b.

Berberian, M., An explanatory note on the first seismotectonic map of Iran; a seismotectonic review of the country, Geol. Surv. Iran, 39, 7-142, 1976c.

Berberian, M., The 1962 earthquake and earlier deformation along the Ipak earthquake fault (Iran), Geol. Surv. Iran, 39, 419-428, 1976d.

Berberian, M., Pre-Quaternary faults in Iran, Geol. Surv. Iran, 39, 259-270, 1976e.

Berberian, M., Generalized fault map of Iran, Geol. Surv. Iran, 39, 1976f.

Berberian, M., Quaternary faults in Iran, Geol. Surv. Iran, 39, 187-258, 1976g.

Berberian, M., Macroseismic epicentres of Iranian earthquakes, Geol. Min. Surv. Iran, 40, 79-100, 1977a.

Berberian, M., An introduction to the seismotectonics of the Maku region (NW Iran), Geol. Min. Surv. Iran, 40, 151-202, 1977b.

Berberian, M., Against the rigidity of the Lut Block; a seismotectonic discussion, Geol. Min. Surv. Iran, 40, 203-228, 1977c.

Berberian, M, Contribution to the seismotectonics of Iran (part III), Geological and Mineral Survey of Iran, 40, 279p, 1977d.

Berberian, M., Earthquake faulting and bedding thrusts associated with the Tabas-e-Golshan (Iran) earthquake of 16 September 1978, Bull. Seism. Soc. Am., 69(6), 1861-1887, 1979a.

Berberian, M., Discussion on the paper by Nowroozi, A.A., 1976, seismotectonic provinces of Iran, Bull. Seism. Soc. Am., 69(1), 293-297, 1979b.

Berberian, M., Evaluation of the instrumental and relocated epicentres of Iranian earthquakes, Geophys. J. R. astr. Soc., 58, 625-630, 1979c.

Berberian, M., Aftershock tectonics of the 1978 Tabas-e-Golshan (Iran) earthquake sequence; a documented active 'thin- and thick-skinned tectonic' case, Geophy. J. R. Astr. Soc. (submitted 1981a).

Berberian, M., The Southern Caspian: a compressional depression floored by a modified oceanic crust, Can. J. Earth Sci. (submitted 1981 b).

Berberian, M. and J.S. Tchalenko, On the tectonic and seismicity of the Zagros active folded belt, Geodynamics of SW Asia, Tehran Symp., Geol. Surv. Iran, 1975.

Berberian, M. and S. Arshadi, On the evidence of the youngest activity of the North-Tabriz fault and the seismicity of Tabriz city, Geol. Surv. Iran, 39, 397-418, 1976.

Berberian, M. and J.S. Tchalenko, Earthquakes of Bandar Abbas-Hadjiabad region (Zagros-Iran), Geol. Surv. Iran, 39, 371-396, 1976a.

Berberian, M. and J.S. Tchalenko, Earthquakes of the southern Zagros (Iran): Bushehr region, Geol. Surv. Iran, 39, 343-370, 1976b.

Berberian, M. and J.S. Tchalenko, Field study and documentation of the 1930 Salmas (Shahpur-Azarbaidjan) earthquake, Geol. Surv. Iran, 39, 271-342, 1976c.

Berberian, M. and S. Arshadi, The Shibly rift system (Sahand region, NW Iran), Geol. Min. Surv. Iran, 40, 229-235, 1977.

Berberian, M. and A. Mohajer-Ashjai, Seismic risk map of Iran, a proposal, Geol. Surv. Iran, 40, 121-150, 1977.

Berberian, M., and I. Navai, Naghan (Chahar Mahal Bakhtiari-High Zagros, Iran) earthquakes of 6 April 1977: a preliminary field report and a seismotectonic discussion, Geol. Min. Surv. Iran, 40, 51-77, 1977.

Berberian, M., and D. Papastamatiou, Khurgu (north Bandar Abbas, Iran) earthquake of 21 March 1977: a preliminary field report and a seismotectonic discussion, Bull. Seism. Soc. Am., 68(2), 411-428, 1978.

Berberian, M., I. Asudeh and S. Arshadi, Surface rupture and mechanism of the Bob-Tangol (SE Iran) earthquake of 19 December 1977, Earth Planet. Sci. Lett., 42(3), 456-462, 1979a.

Berberian, M., I. Asudeh, R.G. Bilham, C.H. Sholz, and C. Soufleris, Mechanism of the main shock and the aftershock study of the Tabas-e-Golshan (Iran) earthquake of September 16, 1978: a preliminary report, Bull. Seism. Soc. Am., 69(6), 1851-1859, 1979b.

Berberian, M., and G.C.P.King, Towards a paleogeography and tectonic evolution of Iran, Can.J.Earth Sci., 18(2), 1981.

Bergougnan, H., J.H. Brunn, C. Fourquin, P.C.De Graciansky, M. Gutnic, J. Marcoux, O. Monod, and A. Poisson, The Alpine folded chains of Asia Minor, In: Lemoine (ed.), Geological Atlas of Alpine Europe and Adjoining Alpine Areas, Elsevier, 509-544, 1978.

Berry, R.H., Evidence for two different basement types beneath the Zagros range, Iran, Geodynamics of Southwest Asia, Tehran Symposium, Geol.Surv. Iran, sp.pub., 70-88, 1975.

Biju-Duval, B., Dercourt, J., and X. Le Pichon, From the Tethys ocean to the Mediterranean seas: a plate tectonic model of the evolution of the western Alpine system, In: Biju-Duval, B. and L. Montadert (eds.), Structural History of the Mediterranean Basins, Editions Technip, Paris, 143-164, 1977.

Bird, P., Finite element modelling of lithosphere deformation: the Zagros collision orogeny, Tectonophysics, 50, 307-336, 1978.

Bird, P., M.N. Toksoz and N.H. Sleep, Thermal and mechanical models of continent-continent convergence zone, J. Geophys. Res., 80(32), 4405-4416, 1975.

Bizon, G., J.J. Bizon and L.E. Ricou, Etude stratigraphique et paleogeographique des formations tertiaries de la region de Neyriz (Fars interne, Zagros iranien), Rev. Inst. Fr. Petr., 27(3), 369-405, 1972.

Brown, G.F., Tectonic map of the Arabian Peninsula, 1:4,000,000, Ministry of Petroleum and Mineral Resources, Directorate General of Mineral Resources, Kingdom of Saudi Arabia, Jiddah, 1972.

Canitez, N., The focal mechanisms in Iran and their relations to tectonics, Pure Apple. Geophys., 75(IV), 76-87, 1969.

Canitez, N. and B. Ucer, Computer determinations for the fault-plane solutions in and near Anatolia, Tectonophysics, 4(3), 235-244, 1967.

CIGMEMR, International geological map of Europe and the Mediterranean region, 1:5,000,000, International Geological Congress, Commission for the Geological Map of the World, Bundesanstalt fur Bodenforschung/UNESCO, 1971.

Coney, P.J., Jones, D.L. and J.W.H. Monger, Cordilleran suspect terranes, Nature, 228, 329-333, 1980.

Cummings, D, Theory of plasticity applied to faulting, Lut area, East-Central Iran, Geodynamics of Southwest Asia, Tehran Symposium, Geol.Surv. Iran, sp.pub., 115-136, 1975.

Desio, A., Geologic evolution of the Karakorum, In: Farah, A. and K. De Jong (eds.), Geodynamics of Pakistan, Geol. Surv. Pakistan, 111-124, 1979.

Dewey, J.W. and A. Grantz, The Ghir earhtquake of 10 April 1972 in the Zagros mountains of southern Iran: Seismotectonic aspects and some results of a field reconnaissance, Bull. Seism. Soc. Am., 63(6), 2071-2090, 1973.

Dewey, J.F., W.C.Pitman, W.B.F. Ryan and J. Bonnin, Plate tectonics and the evolution of the Alpine system, Geol. Soc. Am. Bull., 84, 3137-3180, 1973.

Falcon, N.L., Raised beaches and terraces of the Iranian Makran coast, Geogr. J. 109, 149-151, 1947.

Falcon, N.L., Southern Iran: Zagros Mountains, In : Spencer, A.M. (ed.), Mesozoic-Cenozoic Orogenic Belts, Geol.Soc. London, 199-211, 1974.

Farhoudi, G. and D.E. Karig, Makran of Iran and Pakistan as an active arc system, Geol., 5(11), 664-668, 1977.

Gansser, A., Geology of the Himalayas, Interscience, London, 289p, 1964.

Gansser, A., The Indian Ocean and the Himalayas, Eclog.Geol.Helv., 59, 831-848, 1966.

Geological and Mineral Survey of Iran, Published and unpublished geological quadrangle maps of Iran.

Ghorashi, M., Late Cainozoic faulting in SE Iran, Ph.D. Thesis, Univ. College, London, 289p. 1978.

Gzovsky, M.V., Krestnikov, N.N., Leonov, I.A., Rezanov and G.I. Reisner, Map of the youngest tectonic movements in Central Asia, Izv. Geophys. Ser., 1168-1172, 1960.

Haghipour, A., and M. Amidi, The November 14 to December 25, 1979 Ghaenat earthquakes of northeast Iran and their tectonic implications, Bull. Seism. Soc. Am., 70 (5), 1751-1757, 1980.

Haghipour, A., M.H. Iranmanesh and M. Takin, The Ghir earthquake in southern Persia (a field report and geological discussion), Geol. Surv. Iran, Int. Rep., 52, 1972.

Hanks, T.C. and M. Wyss, The use of body-wave spectra in the determination of seismic-source parameters, Bull. Seism. Soc. Am., 62(2), 561-589, 1972.

Helwig, J., Shortening of continental crust in orogenic belts and plate tectonics, Nature, 260, 768-770, 1976.

Heuckroth, L.E., and R.A. Karim, Earthquake history, seismicity and tectonics of the regions of Afghanistan, Kabul University, 102p., 1970.

Hills, E.S., Elements of structural geology, Methuen and Co. Ltd., 502p, 1965.

Holcombe, C.J.H., Interplate wrench deformation in Iran, Afghanistan and western Pakistan, Geol. Runds., 67(1), 37-48, 1978.

Huber, H., Geological map of Iran, 1:1,000,000, with explanatory note, National Iranian Oil Company, Exploration and Production Affairs, Tehran.

Irving, E., Paleopoles and paleolatitudes of North America and speculations about displaced terrains, Can. J. Earth Sci., 16, 669-694, 1979.

Isacks, B., Oliver, J., and L.R. Sykes, Seismology and the new global tectonics. J.Geophys.Res., 73, 5855-5899, 1968.

ISC, International Seismological Centre, Edinburgh, Scotland, Earthquake Bulletins.

Jackson, J.A., Reactivation of basement faults and crustal shortening in orogenic belts, Nature 283, 343-346, 1980a.

Jackson, J.A., Errors in focal depth determination and the depth of seismicity in Iran and Turkey, Geophys. J. Roy. astr. soc., 61, 285-301, 1980b.

Jackson, J.A. and T.J. Fitch, Seismotectonic implications of relocated aftershock sequences in Iran and Turkey, Geophys. J.R. astr. Soc., 57, 209-229, 1979.

Jackson, J.A. and T.J. Fitch, Basement faulting and focal depths of the larger earthquakes in the Zagros mountains (Iran), Geophys. J. Roy. astr. Soc., 64, 561-586, 1981.

Jakson, J.A., T.J. Fitch, and D.P. McKenzie, Active tectonics and the evolution of the Zagros fold belt, Geol. Soc. Lond., sp. Pub. 9, 1981.

Jacob, K.H. and R.C. Quittmeyer, The Makran of Pakistan and Iran: trench-arc system with active plate subduction, In: Farah, A. and K. De Jong (eds.), Geodynamics of Pakistan, Geol. Surv. Pakistan, 305-317, 1979.

Jaeger, J.C. and N.G.W. Cook, Fundamentals of rock mechanics, Methuen, London, 1969.

James, G.A. and J.G. Wynd, Stratigraphic nomenclature of Iranian Oil Consortium Agreement Area, Am. Assoc. Petrol. Geol. Bull., 49(12), 2182-2245, 1965.

Jones, D.L., Silberling, N.J., and J. Hillhouse, Wrangellia-a displaced terrane in northwestern North America, Can.J.Earth Sci., 14, 2565-2577, 1977.

Kamen-Kaye, M., Geology and productivity of Persian Gulf synclinorium, Am. Assoc. Petrol. Geol., 54(12), 2371-2394, 1970.

Kazmi, A.H., Active fault systems in Pakistan, In: Farah, A. and K.De Jong (eds.), Geodynamics of Pakistan, Geol. Surv. Pakistan, 285-294, 1979.

Kelleher, J., L. Sykes and J. Oliver, Possible criteria for predicting earthquake locations and their application to major plate boundaries of the Pacific and the Caribbean, J. Geophys. Res., 78, 2547-2585, 1973.

Kelleher, J. and J. Savino, Distribution of seismicity before large strike slip and thrust type earthquakes, J. Geophys. Res., 80, 260-271, 1975.

Ketin, I., Main orogenic events and paleogeographic evolution of Turkey, Bull. Min. Res. Expl. Inst., Turkey, 88, 4p, 1978.

Kim, S.G. and O.W. Nuttli, Spectral characteristics of anomalous Eurasian earthquakes, Bull. Seism. Soc. Am., 67(2), 463-478, 1977.

King, G.C.P., R.G. Bilham, J.W. Campbell, D.P. McKenzie, and M. Niazi, Detection of elastic strainfields caused by fault creep events in Iran, Nature, 253(5491), 420-423, 1975.

Kravchenko, K.N., Tectonic evolution of the Tien Shan, Pamir and Karakorum, In: Farah, A. and K. DeJong (eds.), Geodynamics of Pakistan, Geol. Surv. Pakistan, 25-40, 1979.

Kristy, M.J., Burdick, L.J. and D.W. Simpson, the focal mechanisms of the Gazeli, USSR, earthquakes, Bull. Seism. Soc. Am., 70(5), 1737-1750, 1980.

Laughton, A.S., Whitmarsh, K.B., and M.T. Jones, The evolution of the Gulf of Aden, Phil. Trans. Roy. Soc. London, A.267 (1181), 227-266, 1970.

Lees, G.M. and N.L. Falcon, The geographical history of the Mesopotamian plains, Geogr. J., 118, 24-39, 1952.

LePichon, X.L., Sea floor spreading and continental drift, J. Geophys. Res., 73, 3661-3697, 1968.

Little, R.D., Terraces of the Makran coast of Iran and parts of West Pakistan, M.A. Thesis, Univ. Southern California, 151p, 1972.

Matthews, D.H., The Owen fracture zone and the northern end of the Carlsberg ridge, Phil. Trans. Royal. Soc. London, A, 259, 172-186, 1966.

McCann, W.R., Nishenko, S.P., Sykes, L.R., and J. Krause, Seismic gaps and plate tectonics: seismic potential for major boundaries, Pageoph., 117, 1082-1147, 1979.

McCann, W.R., Perez, O.J., and L.R. Sykes, Yakataga gap, Alaska: seismic history and earthquake potential, Science, 207, 1309-1314, 1980.

McGetchin, T.R., and R. Merrill (eds.), Plateau uplift: mode and mechanism, Tectonophysics (Special Issue), 61, 1-3, 1979.

McKenzie, D.P., Speculations on the consequences and causes of plate motions, Geophys. J.R. astr. Soc., 18, 1-32, 1969.

McKenzie, D.P., Active tectonics of the Mediterranean region, Geophys. J.R. astr. Soc., 30, 109-158, 1972.

McKenzie, D.P., The East Anatolian fault: a major structure in eastern Turkey, Earth Planet. Sci. Lett., 29, 189-193, 1976.

McKenzie, D., Can plate tectonics describe continental deformation?, Int. Symp. Structural History of the Mediterranean Basins, Split (Yugoslavia); In: B. Biju-Duval and L. Montadert (Eds.), Editions Technip, Paris, 189-196, 1977.

McKenzie, D.P., and J.G. Sclater, The evolution of the Indian Ocean since the late Cretaceous, Geophys. J. Roy. astr. Soc., 24, 437-528, 1971.

Melville, C.P., Arabic and Persian source material on the historical seismicity of Iran from 7th to the 17th centuries A.D., Ph.D. Thesis, Cambridge Univ., 1978.

Melville, C.P., Earthquakes in the history of Nishapur, Iran, 18, 103-120, 1980.

Mercier, J.L., La neotectonique, ses methodes et buts. Un example: l'arc Egian (Mediterranee orientale), R.Geogr.Phys.Geol.Dyn., (2), XVIII (4), 323-346, 1976.

Mina, P., T. Razaghnia and Y. Paran, Geological and geophysical studies and exploratory drilling of the Iranian continental shelf, Persian Gulf, Proc. 7th World Pet. Cong., Mexico, 870-903, 1967.

Minster, J.B., Jordan, T.H., Molnar, P., and E. Haines, Numerical modelling of instantaneous plate tectonics, Geophys. J. Roy. astr. Soc., 36, 541-576, 1974.

Mogi, K., Migration of seismic activity, Bull. Earth Res. Inst., 46, 53-74, 1968.

Mohajer-Ashjai, A., and A.A. Nowroozi, The Tabas earthquake of September 16, 1978 in east-Central Iran, a preliminary field report, Geophys. Res. Lett., 6(9), 689-692, 1979.

Molnar, P. and P. Tapponnier, Cenozoic tectonics of Asia: effects of a continental collision, Science, 189(4201), 419-426, 1975.

Molnar, P. and P. Tapponnier, Active tectonics of Tibet, J. Geophys. Res., 83(B11), 5361-5375, 1978.

Morris, P., Basement structure as suggested by aeromagnetic surveys in southwest Iran, 2nd Iranian Geol. Symp., The Iranian Petroleum Institute, 1977.

Nabavi, M.S., Seismicity of Iran, M.Ph. Thesis, Univ. London, 273p, 1972.

NEIS, National Earthquake Information Service, Preliminary Deterimination of Epicentres, Montly listing. U.S. Geological Survey.

Niazi, M., I. Asudeh, G. Ballard, J. Jackson, G. King, and D. McKenzie, The depth of seismicity in the Kermanshah region of the Zagros mountains (Iran), Earth Planet. Sci. Lett., 40, 270-274, 1978.

Niazi, M., Shimamura, H., and M. Matsu'ura, Microearthquakes and crustal structure of the Makran coast of Iran, Geophys. Res. Lett., 7(5), 297-300, 1980.

NOAA, National Oceanic and Atmospheric Administration, Environmental Data Service, U.S.A, Earthquake Bulletins.

Nogole-Sadate, M.A.A., Les zones des decrochement et les virgations structurales en Iran, Consequences des resullats de l'analyse structurale de la region de Qom, These, 3 eme cycle, Univ. Sci. Med. Grenoble, France, 1978.

North, R., A thrust event in southern Iran, Lincoln Lab., Semi-Annual Technical Summary, 60-61, M.I.T., 1972.

North, R., Seismic slip rates in the Mediterranean and Middle East, Nature, 252, 560-563, 1974.

North, R., Seismic source parameters, Ph.D. Thesis, Churchill College, Cambridge, U.K. 132p, 1973.

North, R., Seismic moment, source dimensions and stresses associated with earthquakes in the Mediterranean and Middle East, Geophys. J.R. astr. Soc., 48, 137-161, 1977.

Nowroozi, A.A., Seismo-tectonics of the Persian Plateau, eastern Turkey, Caucasus and Hindu-Kush regions, Bull. Seism. Soc. Am., 61(2), 317-341, 1971.

Nowroozi, A.A., Focal mechanism of earthquakes in Persia, Turkey, West Pakistan and Afghanistan and plate tectonics of the Middle East, Bull. Seism. Soc. Am., 62(3), 823-850, 1972.

Nowroozi, A.A., Seismotectonic provinces of Iran, Bull. Seism. Soc. Am., 66(4), 1249-1276, 1976.

O'Brien, C.A.E., Tectonic problems of the oilfield belt of southwest Iran, Rept. 18th Int. Geol. Cong., London, 6, 45-58, 1950.

Owen, H.G., Continental displacement and expansion of the Earth during the Mesozoic and Cenozoic, Phil. Trans. Roy. Soc., A.281, 223-291, 1976.

Page, W.D., J.N. Alt, L.S. Cluff and G. Plafker, Evidence for the recurrence of large-magnitude earthquakes along the Makran coast of Iran and Pakistan, Tectonophysics, 52, 533-547, 1979.

Peive, A.V., and A.L. Yanshin (eds.), Scheme of distribution of continental crusts of various ages in northern Eurasia and relief of the Moho surface, scale 1:5,000,000, Geological Institute, USSR, 1979a.

Peive, A.V., and A.L. Yanshin (eds.), Tectonic map of northern Eurasia, scale 1:5,000,000, Geological Institute, USSR, 1979b.

Poirier, J.P., and M.A. Taher, Historical seismicity in the Near and Middle East, north Africa and Spain from Arabic documents (VIIth-XVIIth century), Bull.Seism.Soc.Am., 70(6), 2185-2201, 1980.

Powell, C.McA., and P.J. Conaghan, Plate tectonics and the Himalayas, Earth Planet. Sci. Lett., 20, 1-12, 1973.

Quittmeyer, R.C., Seismicity variations in the Makran region of Pakistan and Iran: relation to great earthquakes, Pageoph., 117, 1212-1228, 1979.

Quittmeyer, R.C. and K.H. Jacob, Historical and modern seismicity of Pakistan, Afghanistan, northwestern India, and southeastern Iran, Bull. Seism. Soc. Am., 69(3), 773-823, 1979.

Quittmeyer, R.C., Farah, A., and K.H. Jacob, The seismicity of Pakistan and its relation to surface faults, In: Farah, A., and K. DeJong (eds.), Geodynamics of Pakistan, Geol. Surv. Pakistan, 271-284, 1979.

Ricou, L.E., L'evolution geologique de la region Neyriz (Zagros iranien) et l'evolution structurale des Zagrides, These, Univ. Orsay, no AD, 1269, 1974.

Ricou, L.E., Evolution structural des Zagrides, la region clef de Neyriz (Zagros Iranien), Soc. Geol. Fr. Bull., 55(125), 1-140, 1976.

Rustanovich, O.N., Some problems of the investigation of the seismic activity of the Ashkhabad region, Izv. Akad. Nauk SSSR (geofiz. ser), 1, 10, 9-21, 1957.

Savage, W.U., J.N. Alt and A. Mohajer-Ashjai, Microearthquake investigations of the 1972 Qir, Iran, earthquake zone and adjacent areas, Geol. Soc. Am. Abst. J., 496, 1977.

Sborshchikov, I.M., V.I. Dronov, Sh.Sh. Denikaev, A.Kh. Kafarskiy, S.S. Karapetov, F.U. Akhmedzyanov, S.M. Kalimulin, V.I. Slavin, I.I. Sonin, and K.F. Stazhilo-Alekseev, Tectonic map of Afghanistan, 1:2,500,000, Ministry of Mines and Industries, Department of Geology and Mines, Kabul, Afghanistan, 1972.

Schlich, R., Structure et age de l'ocean Indian occidental, Soc. Geol. France, Mem-hors serie, 6, 103p., 1975.

Scholz, C.H., J.M.W. Rynn, R.W. Weed, and C. Frohlich, Detailed seismicity of the Alpine fault zone and Fiordland region, New Zealand, Geol. Soc. Am. Bull., 84, 3297-3316, 1973.

Scholz, C.H., Shear heating and the state of stress on faults, J. Geophys. Res., 85 (B11), 6174-6184, 1980.

Sengor, A.M.C. and W.S.F. Kidd, Post-collisional tectonics of the Turkish- Iranian plateau and a comparision with Tibet, Tectonophysics, 55, 361-376, 1979.

Setudehnia, A., Stratigraphic lexicon of southwest Iran, Lexique Stratigraphique International, III, Asie, F. 9b Iran, 289-376, 1972.

Shikalibeily, E.Sh., and B.V. Grigoriants, Principal features of the crustal structure of the South-Caspian basin and the conditions of its formtion, Tectonophysics, 69, 113-121, 1980.

Shirokova, E.I., Stresses effective in earthquake foci in the Caucasus and adjacent districs,

Bull. Acad. USSR, Geophys. Ser., 10,809-815, (Engl. Transl.), 1962.

Shirokova, E.I., General features in the orientation of principal stresses in earthquake foci in the Mediterranean-Asian seismic belt, Bull. Acad. USSR Earth Physics, 1, 22-36, 1967.

Shirokova, E.I., Changes in the focal mechanisms and their relation to revived faults in the Middle and Far East, Izvestia Earth Physics, 13(9), UDC 550, 341, 621:626, 1977.

Sibson, H.R., Frictional constraints on thrust, wrench and normal faults, Nature, 249(5457), 542-544, 1974.

Sibson, H.R., Generation of pseudotachylyte by ancient seismic faulting, Geophys. J. R. astr. Soc., 43, 775-794, 1975.

Sibson, H.R., Fault rocks and fault mechanisms, J. Geol. London, 133, 191-213, 1977.

SNAEOI, Bulletin of the Seismographic Network, Bushehr region, Atomic Energy Organization of Iran, Site and Environmental Management, Tehran, 1976, 1977, 1978.

Snead, R.E., Physical geography reconnaissance: West Pakistan coastal zone, Univ. New Mexico Publ. in Geography, I, Albuqureque, 55p, 1969.

Snead, R.E., Physical geography of the Makran coastal plain of Iran, Rep. Off. Nav. Res., Contract N00014 66 C D104, Task order NR 388 082, Univ. New Mexico, Albuqureque, 715, 1970.

Sondhi, V.P., The Makran earthquake, 28th November 1945, Indian Minerals, 1, 147-154.

Stocklin, J., Structural history and tectonics of Iran: a review, Bull. Am., Assoc. Petrol. Geol., 52(7), 1229-1258, 1968a.

Stocklin, J., Salt deposits of the Middle East, Geol. Soc. Am. Sp. Paper, 88, 159-181, 1968b.

Stocklin, J., Structural correlation of the Alpine ranges between Iran and Central Asia, Mem. h. ser. Soc. Geol. France, 8, 333-353, 1977.

Stonely, R., Evolution of the continental margins bounding a former Tethys, In: C.A. Burk and C.L. Drake (eds.), The Geology of Continental Margins, Springer, N.Y., 889-903, 1974.

Sykes, L.R., Aftershock zones of great earthquakes, seismicity gaps, and earthquake prediction for Alaska and the Aleutions, J. Geophys. Res., 76, 8021-8041, 1971.

Sykes, L.R., Intraplate seismicity, reactivations of preexisting zones of weakness, alkaline magmatism, and other tectonism postdating continental fragmentation, Rev. Geophys. Space Physics, 16(4), 621-688, 1978.

Sykes, L.R., Kisslinger, J.B., House, L., Davies, J.N., and K.H. Jacob, Rupture zones of great earthquakes in the Alaska-Aleutian arc, 1784 to 1980, Science, 210, 1343-1345, 1980.

Szabo, F. and P. Kheradpir, Permian and Triassic stratigraphy, Zagros basin, south-west Iran, J. Petrol. Geol., 1(2), 57-82, 1978.

Tapponnier, P. and P. Molnar, Active faulting and tectonics in China, J. Geophys. Res., 82, 2905-2930, 1977.

Tapponnier, P., and P. Molnar, Active faulting and Cenozoic tectonics of the Tien Shan, Mongolia,

and Baykal regions, J. Geophys. Res., 84(B7), 3425-3425, 1979.

Tchalenko, J.S., Recent destructive earthquakes in the Central Alborz, Geol. Surv. Iran, 29, 97-116, 1974.

Tchaleko, J.S., Seismisity and structure of Kopet Dagh (Iran, USSR), Phil. Trans. Roy. Soc. London, 278(1275), 1-25, 1975a.

Tchalenko, J.S., Strain and deformation rates at the Arabian/Iran plate boundary, J. Geol. Soc. London, 131, 585-586, 1975b.

Tchalenko, J.S., A reconnaissance of the seismicity and tectonics at the northern border of the Arabian plate (Lake Van region), Rev. Geogr. Phys. Geol. Dyn., XIX(2), 189-208, 1977.

Tchalenko, J.S. and M. Berberian, The Salmas (Iran) earthquake of 6 May 1930, Ann. di Geofis., 27(1-2), 151-212, 1974.

Tchalenko, J.S. and M. Berberian, Dasht-e-Bayaz fault, Iran: earthquake and earlier related structures in bedrock, Geol. Soc. Am. Bull., 86, 703-709, 1975.

Tchalenko, J.S., M. Berberian and H. Behzadi, Geomorphic and seismic evidence for recent activity of the Doruneh fault (Iran), Tectonophysics, 19, 333-341, 1973.

Tchalenko, J.S., J. Bruad and M. Berberian, Discovery of three earthquake faults in Iran, Nature, 248, 261-263, 1974.

Tchalenko, J.S. and J. Braud, Seismicity and structure of the Zagros (Iran): the Main Recent fault between 33 and 35 N., Phil. Trans. Roy. Soc. London, 227(1262), 1-25, 1974.

Tectonic Map of Caucasus, 1:1,000,000, Academy of Sciences, USSR, 1974.

Toksoz, M.N. and E. Arpat, Studies of premonitory phenomena preceeding two large earthquakes in eastern Turkey (abst.), EOS, Am. Geophys. Union Trans., 58(12), 1195, 1977.

Toksoz, M.N., E. Arpat and F. Sargolu, East Anatolian earthquake of 24 November 1976, Nature, 270, 423-425, 1977.

Toksoz, M.N., and P. Bird, Formation and evolution of marginal basins and continental plateaus, In: Talwani, M., and W. Pitman (eds.), Island Arcs, Deep Sea Trenches and Back-Arc Basins, Maurice Ewing Series 1, Am. Geophys. Union, 379-393, 1977.

Toksoz, M.N., J. Nabelek and E. Arpat, Source properties of the 1976 earthquake in east Turkey, a comparision of field data and teleseismic results, Tectonophysics, 49, 199-205, 1978.

USGS, United States Geological Survey, Preliminary and monthly determinations of epicentres.

Vita-Finzi, C., Quaternary deposits in the Iranian makran, Geogr. J., 141, 415-420, 1975.

Vita-Finzi, C., rates of Holocene folding in the coastal Zagros near Bandar Abbas, Iran, Nature, 278(5705), 632-634, 1979a.

Vita-Finzi, C., Contributions to the Quaternary geology of southern Iran, Geol. Min. Surv. Iran, 47, 52p, 1979b.

Von Dollen, F.J., J.N. Alt, D. Tocher and A.

Nowroozi, Seismological and geological investigations near Bandar Abbas, Iran, <u>Geol. Soc. Am., Abst.</u>, <u>9</u>, 521, 1977.

UNESCO, International geological map of Europe and the Mediterranean region, 1:5,000,000, International Geological Congress, Commission for the Geological Map of the World, Bundesanstalt fur Bodenforschung/<u>UNESCO,</u> 1971.

Wellman, H.W., Active wrench faults of Iran, Afghanistan and Pakistan, <u>Geol. Rundsch.</u>, <u>55(3)</u>, 716-735, 1966.

White, R.S., Seismic bright spots in the Gulf of Oman, <u>Earth Planet. Sci. Lett.</u>, <u>37</u>,29-37, 1977.

White, R.S. and K. Klitgord, Sediment deformation and plate tectonics in the Gulf of Oman, <u>Earth Planet. Sci. Lett.</u>, <u>32</u>, 199-209, 1976.

White, R.S. and D.A. Ross, Tectonics of the western Gulf of Oman, <u>J. Geophys. Res.</u>, <u>84(B7)</u>, 3479-3489, 1979.

White, R.S. and K.E. Louden, The Makran continental margin: structure of a thickly sedimented convergent plate boundary, <u>Am. Assoc. Petrol. Geol. mem.</u> (Hedberg Conf., 1981; submitted).

Wolfart, R., Tektonik und palaogeographische Entwicklung des mobilen Schelfes im Bereich von Syrien und dem Libanon, <u>Zeitsch. Deutsch. Geol. Gesell.</u>, <u>117</u>, 544-589, 1965.

A BRIEF REPORT ON GEODYNAMICS IN IRAN

Jovan Stöcklin

Huebstrasse 9a, 9011 St. Gallen, Switzerland.

Introduction

The political events in Iran in 1978/79 were a
severe drawback to scientific research in the
country. One of the consequences was a virtual
breakdown of all activities under the Iranian programme of the International Geodynamics Project.
Since late 1978 the contacts between the National
Iranian Geodynamics Committee and Working Group 6
of the Inter-Union Commission have been interrupted,
and it was not possible to obtain a contribution
to this Report from the Iranian Committee. This is
all the more regrettable as Iran had been very
actively engaged in an extensive and promising
study programme within the framework outlined by
Working Group 6 at its session in Hyderabad in
March 1973.

The following notes are not to substitute for an
official Iranian report. The writer was not a member
of the Iranian Geodynamics Committee and not directly involved in its activities. He, however, was
engaged in exploration work in Iran and maintained
personal contacts with many Iranian colleagues till
early 1979. These notes are fragmentary personal
recollections intended mainly to explain the situation in Iran with regard to the completion of
geodynamic study programmes. Included is a brief
summary of a few recent publications, which, in the
writer's opinions are particularly relevant to the
work and purposes of the International Geodynamics
Project.

Geodynamics Activities

One outcome of the 1973 Hyderabad meeting of
Working Group 6 (then WG 3b) was the creation of a
National Iranian Committee on Geodynamics in 1974.

The Committee comprised representatives of the
Geological Survey of Iran, the National Iranian Oil
Company, several Iranian Universities, the mining
industry, and other institutions and individuals
involved in geodynamic work. For the execution of
programmes the Committee was strongly dependent on
the Geological Survey of Iran. At that time, the
Survey was engaged in an extensive, systematic mapping programme with which were related numerous investigations of high interest to the Geodynamics
Project, in particular structural and tectonic in-

vestigations, studies of magmatism and metamorphism,
of the classical Iranian ophiolite and ophiolitic
mélange complexes, stratigraphic basin studies,
deep crustal studies by geophysical methods, and
study of the seismotectonic and neotectonic phenomena.

None of these studies became ever an official
part of the International Geodynamics Project. The
role of the National Geodynamics Committee in these
Geological Survey programmes was purely consultat-
ive and intended to establish and maintain a link
with Working Group 6, and to make the results
available to the Working Group as a contribution
to the International Geodynamics Project. The
Committee was, however, instrumental in the organization of the Tehran Symposium on the Geodynamics
of Southwest Asia (8 to 15 September 1975). The
results of the Symposium have been collected in
the "Proceedings" which were issued in 1977 as a
Special Publication of the Geological Survey of
Iran.

Aeromagnetic Survey

One of the programmes of the Geological Survey
closely related to geodynamic problems was an
airborne magnetic survey of Iran. It initially
focussed on one of the two geotraverses across
Iran recommended by the 1973 Hyderabad meeting but
was eventually extended to cover the whole country.
It resulted in the preparation and printing of a
geomagnetic map series of Iran at the scale of
1:250,000 and an overall map at the scale of
1:1,000,000. Sample sheets were demonstrated in
March 1978 at the Kathmandu Conference on Geodynamics of the Himalayan region. To the writer's
knowledge, the maps have not been officially
released for publication but are in principle
available for inspection at the Geological Survey
Institute in Tehran.

Geotraverses

Of the nine geotraverses across the Iranian-
Himalayan foldbelt selected by the 1973 Hyderabad
meeting, two were placed in Iran: one from the
southern Caspian shore across the Alborz Range,
Central Iran and the Zagros to the Persian Gulf;

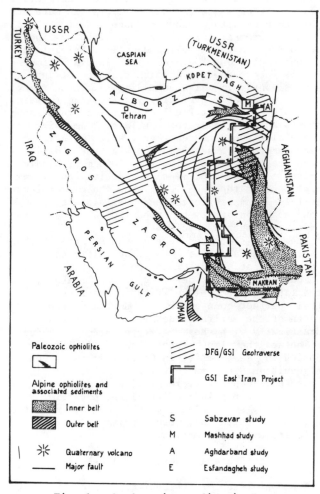

Fig. 1.   Geodynamic studies in Iran.

Legend:

Paleozoic ophiolites

Alpine ophiolites and associated sediments
- Inner belt
- Outer belt

☀ Quaternary volcano

— Major fault

DFG/GSI Geotraverse

GSI East Iran Project

S   Sabzevar study

M   Mashhad study

A   Aghdarband study

E   Esfandagheh study

---

the other from Sarakhs in Northeast Iran, at the Soviet border, across the eastern Kopet Dagh and the Doruneh Fault to the northern Lut block.

A specific project combining various geological and geophysical investigations along these two traverses was planned as a joint venture of the Geological Survey of Iran and the Scientific Research Council of West Germany ('Deutsche Forschungsgemeinschaft'). Scientists from the Geological Survey and other Iranian Institutions were to cooperate with research workers from several German Universities. For reasons of economy, is was decided to combine the two traverses into one, running from Sarakhs in the northeast to the central Lut, from there west to Central Iran, and from there southwest to the Persian Gulf.

Negotiations for this project began in 1975 but were delayed for more than two years, so that the actual investigations had barely started when the political disturbances interrupted them in 1978. To the writer's knowledge, no results of this short-lived cooperative effort have become avail-

able, and the prospects for an eventual resumption appear to be very uncertain.

East-Iran Project

It is most unfortunate that the political events in Iran have totally interrupted, shortly before completion, one of the most comprehensive and most promising study programmes ever undertaken in the Alpine ophiolite zones: the East-Iran project of the Geological Survey. Although this, too, was not officially a part of the Geodynamics Project, the results to be expected from it were of greatest importance to the solution of regional and global geodynamic problems.

After several years of discussions and negotiations, the East Iran project was finally launched in 1976 under the general direction and supervision of the Geological Survey of Iran. The Survey contracted for this purpose four Iranian exploration firms working in partnership with a large number of Australian, French, US-American and Canadian geologists, and several noted scientists were repeatedly consulted and asked to check and coordinate the field work. The project covered, in addition to a major portion of the Lut block, most of the vast expanses of ophiolite and flysch rocks constituting the East Iranian Ranges and the Iranian Makran, the classical 'coloured mélange' region. The immediate target was the preparation and publication of a geological map series on a scale 1:100,000, accompanied by comprehensive explanatory texts and review maps, on a scale 1:250,000. Based on the results of this first phase, systematic mineral exploration was to follow in a second phase.

In addition to mapping, in which photogeology was extensively used, the work included detailed stratigraphic and facies analyses of the thick oceanic and flysch sediments, petrographic and geochemical studies of the ophiolites and associated magmatic rocks (which had revealed complete, well-developed sequences of peridotites-gabbros-sheeted dykes pillow lavas, etc.), paleontologic and radiometric dating, investigation of the widespread blue schist metamorphism, structural analysis, and study of the geodynamic history of ophiolite emplacement and mélange formation.

By late 1978 the greater part of the field work had been completed, compilation and cartographic preparation of the maps as well as composition of the texts were far advanced, and a few maps had already been printed. At this stage, however, the rapidly deteriorating situation in Iran and related financial difficulties led to a complete deadlock in the project activities, and soon all foreign partners were obliged to leave the country before the main part of the material could be handed to the printers.

There is no doubt that the East Iran project has brought forth a great deal of those factual data which are still so badly lacking in many of the wild speculations about the geodynamic role of the Alpine ophiolites. It remains to be hoped that the

Geological Survey of Iran will be in a position to publish the results at least in their presently available fragmentary form and to fill eventually the remaining gaps.

New Data on Ophiolites and Related Phenomena

The ophiolites of Iran can be broadly grouped in into three zones (Stöcklin 1977).

1) Ophiolites associated with pre-Jurassic, largely metamorphic rocks; they are known mainly from northern Afghanistan but reappear in a few outcrops in the Binalud uplift of the eastern Alborz in Northeast Iran, and in scattered small wedges near the north foot of the western Alborz.

2) An "inner sub-belt" of Alpine ophiolites, characterized mainly by Upper Cretaceous-Lower Tertiary ophiolitic mélanges and thick flysch sequences; they occur northeast of the Main Zagros Thrust, in several more or less linear but discontinuous outcrops in Central Iran and, most extensively, in the East Iranian Ranges and the Iranian Makran.

3) An "outer sub-belt" of Alpine ophiolites, characterized mainly by large coherent peridotite sheets associated with Mesozoic (Jurassic-Cretaceous) deep-water sediments and turbidites, occurring in two major outcrops (Kermanshah and Neyriz) southwest of the Main Zagros Thrust and reappearing in the Oman Range of southeastern Arabia.

Binalud-Aghdarband (Northeast Iran)

Majidi (1978) submitted to the University of Grenoble a thesis dealing with the ophiolites, granites and metamorphic rocks of Binalud (near Mashhad) in Northeast Iran. Simultaneously, but independantly, Ruttner (1980) obtained significant stratigraphic and structural data on a thick pre-Jurassic volcano-sedimentary sequence exposed in the Aghdarband uplift east of Mashhad, as a result of detailed mapping and coal exploration. The results of both studies are closely interrelated.

Majidi found paleontological evidence for a Devonian-early Carboniferous age of the metamorphic complex of Binalud. The complex consists mainly of a thick, slightly metamorphic pelite-greywacke sequence. Minor calc-schists and limestones containing scarce Devonian conodonts, crinoids and brachiopods occur in the lower part, whereas the upper part, which has yielded traces of Carboniferous plants, contains considerable amounts of ophiolitic material as well as cherty radiolarian slates. The ophiolites occur as numerous flow sheets interbedded in the metasediments. They are mostly serpentinized but have revealed relict structures of dunite, wehrlite, dolerite, and spilitic pillow lava. Locally, mélange-like mixtures of the sedimentary and igneous material have been observed. Porphyric granites (Mashhad Granites) have intruded and altered the metamorphic sequence in at least two intrusive phases, for which K/Ar dating indicated ages ranging from 256 to 211 m.y. Ophiolites and granites were

found reworked in conglomerates of a Rhaeto-Liassic plant-bearing sequence that overlies the metamorphic complex with pronounced unconformity.

These observations are in line with the findings of Ruttner in the Aghdarband uplift. Here, late Devonian conodonts were detected in black limestones interbedded in a thick sequence of dark slates containing also some diabase. These rocks are unconformably overlain by a sequence (missing, probably eroded, at Binalud) of red shales, sandstones and conglomerates, at least 2 km thick and containing abundant reworked granite material of Mashhad type. These Permo-Scythian red clastic beds are followed by more than 1000 m of Middle Triassic volcano-sedimentary deposits containing coal beds in the upper part. The whole is intensely folded and imbricated and overlain with pronounced unconformity by subhorizontal Lower Jurassic plant-bearing sediments.

The thick slate sequence of the Devonian-Carboniferous, the associated basic and ultrabasic rocks, the red clastic beds of the Permo-Scythian, the volcano-sedimentary deposits of the Triassic, and the late Paleozoic-late Triassic granites have their perfect counterparts in northern Afghanistan, north of the Harirud lineament, and contrast drastically with the shelf-type carbonate deposition and the almost total absence of magmatism during the same period in Central Iran and Central Afghanistan. They suggest formation of oceanic crust and deep-sea sedimentation in a Paleo-Tethys north of the Binalud-Harirud line, possibly followed by subduction and collision as indicated by the granitization, metamorphism and intense late Triassic deformation.

Sabzevar (Northeast Iran)

The ophiolites of Sabzevar are typical representatives of the "inner sub-belt" of Alpine ophiolites in Iran. They were the subject of a study programme of Saarland University (Germany). Lensch et al. (1975) described the ophiolites as comprising intrusive rocks grading from dunites to tonalites and including diabasic sheeted dyke complexes, spilitic pillow lavas, and tuffs interbedded with Upper Cretaceous pelagic limestones. The rocks are tectonically associated with large slabs of harzburgites representing upper mantle material.

A avi-Tehrani (1975) studied the metamorphism of the Sabzevar ophiolites and distinguished two principal phases: 1) a pre-late Cretaceous phase of oceanic load metamorphism, expressed mainly by serpentinization and spilitization and thought to have taken place during a period of oceanic expansion; and 2) a late Cretaceous to early Eocene phase of blue schist and green schist high-pressure metamorphism related to compression.

The ophiolitic magmatites were found to belong to a special K- and Ti- poor, calc-alkaline type of ocean-ridge magmatism with only slight tholeiitic affinities, reminiscent of the Troodos massif in Cyprus. The post-ophiolitic andesitic rocks

of the region, supposed to be the products of a
subduction zone, are associated with sediments
lacking in terrigenous detritus in their lower
part and thus suggesting an island arc rather
than a continental margin position. This, too,
supports the comparison with the Troodos massif,
for which Vine (1975) concluded an origin in a
marginal basin or back arc spreading situation,
possibly in close proximity to the island arc and
subduction zone itself.

Esfandagheh (South Iran)

A geologic and petrographic study of the extremely complex Esfandagheh area in South Iran was the
subject of a thesis submitted by Sabzehei (1974) to
the Univer ty of Grenoble.

The Esfandagheh area belongs to the Sanandaj-
Sirjan zone situated immediately northeast of the
Main Zagros Thrust. It has a kind of transitional
character between Central Iran (continental crustal
nature, pronounced early and late Kimmerian deformation, magmatism and metamorphism) and the eugeosynclinal ophiolitic belt accompanying the Main
Zagros Thrust. The ophiolitic assemblages, too,
display a transitional character between the "inner
sub-belt" (Central Iran type) and the "outer sub-
belt" (Zagros type), and they seem to have formed
in a sector where the two sub-belts merge. They
occur in a number of irregularly shaped, fault-
bounded, graben-like depressions between "horsts"
of highly metamorphic rocks covered by "normal"
shallow-water Jurassic-Cretaceous sediments which,
however, reflect conditions of stronger subsidence
with more significant synsedimentary volcanism than
usually found in the more "epi-continental" environ-
ments of Central Iran. A sharp unconformity, attrib-
utable to the early Kimmerian phase of diastrophism,
separates the Jurassic-Cretaceous deposits from the
pre-Jurassic metamorphic complexes. The latter
consist mainly of marble and abundant amphibolite
in the lower part (Abshur Complex) and of thick
micaschists, green schists, gneisses, and some
quartzites and marbles in the upper part (Sargaz
complex).

The age of the metamorphic complexes was hither-
to unknown; the presence of both Precambrian and
Paleozoic formations had been suspected. Sabzehei
established a careful relative chronology of
successive phases of deformation and metamorphism.
Moreover, he discovered organic traces indicating
a Devonian to possibly Carboniferous age for the
thick micaschist/green schist sequence of the
Sargaz Complex. Petrographic and geochemical
analyses indicated for these Devonian-Carboniferous
schists a derivation from a thick shale and greywacke sequence associated with submarine tholeiitic
basalt flows and tuffs, a facies strongly contrasting with the essentially non-volcanic, shelf-type
carbonate deposits characterizing the Devonian-
Carboniferous of Central Iran. Sabzehei concluded
from this the formation of a rift-like intra-crat-
onic trough in Devonian-Carboniferous time, a kind
of fore-runner of the Mesozoic Zagros rift now
marked by ophiolites.

Most significant, however, was Sabzehei's discovery of an important phase of magmatic intrusion
and ophiolite emplacement pre-dating the Jurassic
transgression but post-dating the early Kimmerian
processes of folding and metamorphism. These late
Triassic or early Jurassic intrusions comprise
large igneous bodies displaying a magmatic differentiation covering the whole spectrum from ultrabasic to gabbroic, tonalitic and granitic fractions.
The ultrabasic portions form several huge bodies
(up to 200 sq. km) of layered, partly chromite -
bearing dunites, harzburgites and pyroxenites,
showing intrusive heat contacts with the metamorphosed Paleozoic host rocks. They are now in faulted
contact with the Mesozoic coloured mélange complexes of the Esfandagheh area and seem to have
formed their original, early Mesozoic substratum.

The mélange rocks, in which only late Cretaceous
and early Tertiary microfaunas have been found but
which are thought to include older Cretaceous and
Jurassic deposits in the more metamorphosed
elements, display a repetition of the pre-Jurassic
metamorphic events: an initial phase of static
metamorphism supposedly associated with a period of
expansional movements, followed by a phase of high
pressure (blue schist) metamorphism related to compression.

Although Sabzehei's investigations did not allow
precise dating of the late Paleozoic - early Mesozoic processes of magmatism, deformation and metamorphism, the relative sequence of these processes
and their grouping into a first, pre-Jurassic, and
a second, Mesozoic-Tertiary, cycle, have been well
demonstrated. As regards the first, pre-Jurassic
cycle, one cannot fail to notice a conspicuous
similarity in the relative succession, if not in
absolute time, with the processes described by
Majidi from the "Paleo-Tethys suture" in Northeast
Iran and briefly discussed above. They seem to
support Sabzehei's view that repeated rifting and
compression along old (? Assyntic) lineaments, with
repeated opening and closing of narrow, Red Sea-
type oceanic troughs between rigid continental
blocks may have been the fundamental mechanism in
the tectonic evolution of Iran - rather than long-
distance continental migration with subduction and
consumption of gigantic volumes of oceanic crust
as assumed in most current plate tectonic models
for the Middle East.

References

Alavi-Tehrani, N., On the metamorphism in the oph-
iolitic rocks in the Sabzevar region (NE-Iran).
In: Proceedings of Tehran Symposium on the Geo-
dynamics of Southwest Asia, 8-15th Sept. 1975.
Geol. Surv. Iran, Spec. Publ., pp.25-52, 1975.
Lensch, G., A. Mihm, E. Sadredini and F. Vaziri-
Tabar, Geology, geochemistry and petrogenesis of
the ophiolitic range north of Sabzevar (Khorassan,
Iran). In: Proceedings of Tehran Symposium on the
Geodynamics of Southwest Asia, 8-15th Sept. 1975.
Geol. Surv. Iran, Spec. Publ., pp.215-248, 1975.
Majidi, B., Etude pétrostructurale de la région de
Mashhad, Iran. Les problèmes des métamorphites,

serpentinites et granitoides 'hercyniennes'.
Thesis, Grenoble University, 277 p., 1978.

Ruttner, A.W., Sedimentation und Gebirgsbildung in
Ost-Iran: In: J. Pohlmann (ed.), Festschrift Max
Richter, Berliner geowiss. Abh. (A), 20, pp.3-20,
1980.

Sabzehei, M., Les mélanges ophiolitiques de la
région d'Esfandagheh(Iran méridional). Etude pétro-
logique et structurale - Interprétation dans le
cadre iranien. Thesis,Grenoble University,306 p,1974.

Stöcklin, J., Structural correlation of the Alpine
ranges between Iran and Central Asia, Mém. h. sér.
Soc. géol. Fr., 8, pp. 333-353, 1977.

Vine, F.J., The structural significance of the
Troodos igneous massif, Cyprus. In: Proceedings
of Tehran Sympmposium on the geodyncamics of Southwest Asia, 8-15th Sept. 1975. Geol. Surv. Iran,
Spec. Publ., pp. 400-409, 1975.

GEOLOGICAL OBSERVATIONS AND GEOPHYSICAL INVESTIGATIONS CARRIED OUT IN AFGHANISTAN OVER THE PERIOD OF 1972 - 1979

Abdullah Shareq

Afghanistan Geological Survey, Department of Geological and Mines Survey, Kabul

Abstract. The geological-geophysical investigations (1972-1979), a synthesis of all data
available on the geology, and the compilation of
the geological map of Afghanistan resulted in the
obtaining of much new data on stratigraphy, tectonics and magmatism. The geophysical investigations carried out in comparatively small areas
of the country have provided better descriptions
of some geological structures and some ideas of
the deep structure of the region.

The stratigraphy of Paleozoic and more recent
formations was worked out on the basis of fossils.
The Precambrian is subdivided on the basis of the
degree of regional metamorphism.

For the first time, the intrusive formations
have been considered as derivatives of certain
tectono-magmatic cycles; the rocks were divided
into separate phases of intrusion according to
their composition and dated according to their
relationships with the country rock.

Tectonically Afghanistan is regarded to be of
heterogeneous structure, divided by the Hari Rod
and Central Badakhshan Faults into northern and
southern blocks. The tectonic map of Afghanistan
was compiled, and the tectonic position of Afghanistan in the Mediterranean zone structures was
determined.

Brief results of the airborn geophysical investigations carried out in western Afghanistan are reported here for the first time.

Seismic observations indicate that the depth of
earthquake foci increases from west to east and
from south-west to north-east. Some deep faults
are seismically active.

Introduction

From 1968 to 1972, geological and geophysical
surveys at a scale of 1:500 000 were carried out
in Afghanistan. Field investigations covered
over 50% of the country. In 1972, all the information received in the process of these operations, as well as earlier available information
was summarized and a set of geological, tectonic
and magmatic maps of Afghanistan, at a scale of
1:1 000 000 was compiled for the XXIV International Geological Congress (Canada).

At the end of 1972, the generalized geological
map of central and south-western Afghanistan, at
a scale of 1:500 000 was compiled (Dronov et al.
1973). The map was used as a base for compiling
the geological map of Afghanistan at a scale of
1:1 00 000 and also for the preparation of the
generalized Afghanistan geological map at a scale
of 1:500 000 (Katarsky et al., 1977). The latter
was based on the materials of the Afghan-Soviet
systematic geological survey of 1968-1975, including all the work on regional geology carried
out by German, French and Italian geologists
prior to 1973. The map has been compiled in
accordance with the existing international legend;
relatively isochronous stratified and intrusive
bodies are coloured the same shade and accompanied
by international standard age indices. Regional
geological surveys and synthesis of the accumulat-
ed data gave many new details on stratigraphy because the work was conducted by a uniform method
throughout the country. Simultaneously, with the
map preparation, all the available geological
data on the stratigraphic units of Afghanistan
were revised and correlated, both with the international standard scale, and with the coeval
formations of neighbouring countries. In the
process of the geological survey, it became possible to sub-divide stratigraphically in detail
the formations of the country and to collect rich
fossil assemblages.

Based on its geological structure, the stru-
tural formational units and the geological
evolution, the country can be divided into the
northern, middle, south-western, central, western, southern and south-eastern areas. Within
the eastern portion of the country, south-eastern Hindu Kush was singled out as an independent
district, as a considerable part of it lies to
the east of the present Afghan border.

The geological survey, synthesis of all available materials and compilation of the above-
mentioned geological map of the country, resulted in accumulation of a good deal of new data
that essentially changed our views on the stratigraphy, magmatism and tectonics of Afghanistan.

General results of the geological and geophysical investigations are given below.

Stratigraphy. Prior to 1970, the stratigraphy was only known in broad outline. Many of now established units were either unknown or poorly known. It is for the first time that the crystalline complexes developed in the north-eastern part of the country, namely in south Badakhshan, are correlated (though rather conventionally) to the Archean, based on their degree of regional metamorphism and a correlation with the adjacent territories is possible. The Archean strata are granitized and now represented by various gneisses, amphibolites, marbles, calciphyres, quartzites and are metamorphosed in the hornblende-granulite and amphibolite facies (Figure 1).

The Archean formations extend into the crystalline series of the south-western part of the Soviet Pamir. Absolute age of this series has been determined (Dronov et al., 1976).

Apart from the Archean formations, some individual isolated outcrops of ancient formations have been recognized in a number of districts where they were considered to be pre-Devonian or Proterozoic. Recent investigations of the Proterozoic metamorphic formations in Afghanistan resulted in definite identification of three geographically separate complexes by the degree of metamorphism (Figure 2).

The lower complex is represented by various gneisses, crystalline schists, marbles, amphibolites and quartzites, with a total thickness of 3000-4000 m. The rocks are in the amphibolite facies, with some granitized bands.

The middle complex, consisting of metamorphic rocks of green schist and lower stages of amphibolitic facies, comprises various crystalline green coloured schists, including those developed from volcanites; amphibolized volcanitic gneisses, amphibolites, marbles and marble limestones, dolomites and quartzites. The total thickness of this complex exceeds 5000 m.

The lower portion of the upper complex comprises various slates, sandstones and aleurolites, that have been altered to phyllites and are interbedded with bands or lenses of conglomerates, dolomites, cherts and volcanites of acid to basic composition. Their total thickness is 6000 m.

The upper parts of the upper complex consist of rubefied altered volcanites of acid to intermediate composition alternating with metamorphosed series of phyllites and carbonate rocks. Their thickness is 1500-2000 m (Dronov et al., 1976).

A good deal of new data, related to the problem of stratigraphical division of the Paleozoic, has been collected. For the first time within the country, the Vendian-Cambrian and Cambrian sediments were recognized and within the Paleozoic formations, on the whole, 20 sub-divisions were reliably distinguished. In some districts the stratigraphic division is more detailed, including series, stages and zones.

Vendian-Cambrian sediments are found in central and middle Afghanistan unconformably overlying the Upper Proterozoic. In central Afghanistan, Vendian-Cambrian formations are represented by limestones and dolomites containing Vendian algae and reach 500-900 m. in thickness. In middle Afghanistan, the lower part of Vendian-Lower Cambrian formations consists of reddish sandstones and aleurolites, 1000-1500 m thick.

The upper series, probably with erosion, is represented by limestones and dolomites. Late Cambrian trilobites and brachiopods are found in the upper part.

Other sub-facies of the Paleozoic also yielded a lot of new data concerning their distribution, more detailed division, more precise definition of their limits, age and so on.

The composition and structure of the 20 abovementioned sub-facies are not absolutely identical throughout the country, or even within the limits of one region, but vary from one location to another. The Paleozoic sub-facies appears to consist of an assemblage of low grade metamorphic

STRATIGRAPHICAL SCHEME OF ARCHEAN ROCKS OF AFGHANISTAN

INTERNATIONAL STRATIGRAPHIC SCALE	THE PAMIRIAN - NURESTA MEDIAN MASS			
	REGIONAL STRATIGRAPHIC SCALE	CHARACTERISTIC SIGN OF METAMORPHISM STAGE	SOUTHERN BADAKSHAN	
ARCHEAN GROUP	SANGLECH SERIES — UPPER PART	ALMANDINE SILIMANITE SUB-FACIES, THE AMPHIBOLITE FACIES	TARASHAN SUITE. BIOTITE, GARNET-BIOTITE, SILLIMANITE BIOTITE, GNEISS, BANDS OF INJECTIBLE GNEISSES, AMPHIBOLITES AND MARBLES. 1500 M	
			SHEKRAN SUITE. AMPHIBOLE, GARNET-AMPHIBOLE GNEISS, BANDS OF BIOTITE GNEISS, AMPHIBOLITES. 600 M	
			DARMARAKH SUITE. BIOTITE GARNET BIOTITE GNEISS, BANDS OF BIOTITE – AMPHIBOLE GNEISSES, AMPHIBOLITES, MARBLES 2000 - 2500 M	
	SANGLECH SERIES — MIDDLE PART	HORNBLENCE – GRANULITE SUB-FACIES OF GRANULITE FACIES	SAKHI SUITE	ALTERNATION OF MARBLE, BIOTITE AND AMPHIBOLE GNEISSES, AMPHIBOLITES, QUARTZITES 850 - 1400 M
				BIOTITE GARNET-BIOTITE AMPHIBOLE-BIOTITE GNEISS, INTERBEDS OF AMPHIBOLITE QUARTZITES, MARBLES. 850 - 1600 M
				BIOTITE GARNET-BIOTITE AMPHIBOLE-BIOTITE AMPHIBOLE GNEISSES, BANDS OF QUARTZITES, MARBLES 600 - 1000 M
	SANGLECH SERIES — LOWER PART		VALEJ SUITE BIOTITE GARNET-BIOTITE GNEISS, BANDS, LENSES OF AMPHIBOLITES, MARBLES, CALCIPHYRES, AMPHIBOLE BIOTITE, GARNET-SILLIMAITE-BIOTITE, CORDIERITE-GARNET-BIOTITE GNEISSES. 600 M	

GEOLOGICAL STRUCTURE AND ORE DEPOSITS OF AFGHANISTAN (IN PRESS)

Fig. 1. Stratigraphic sketch of the Archean rocks of Afghanistan.

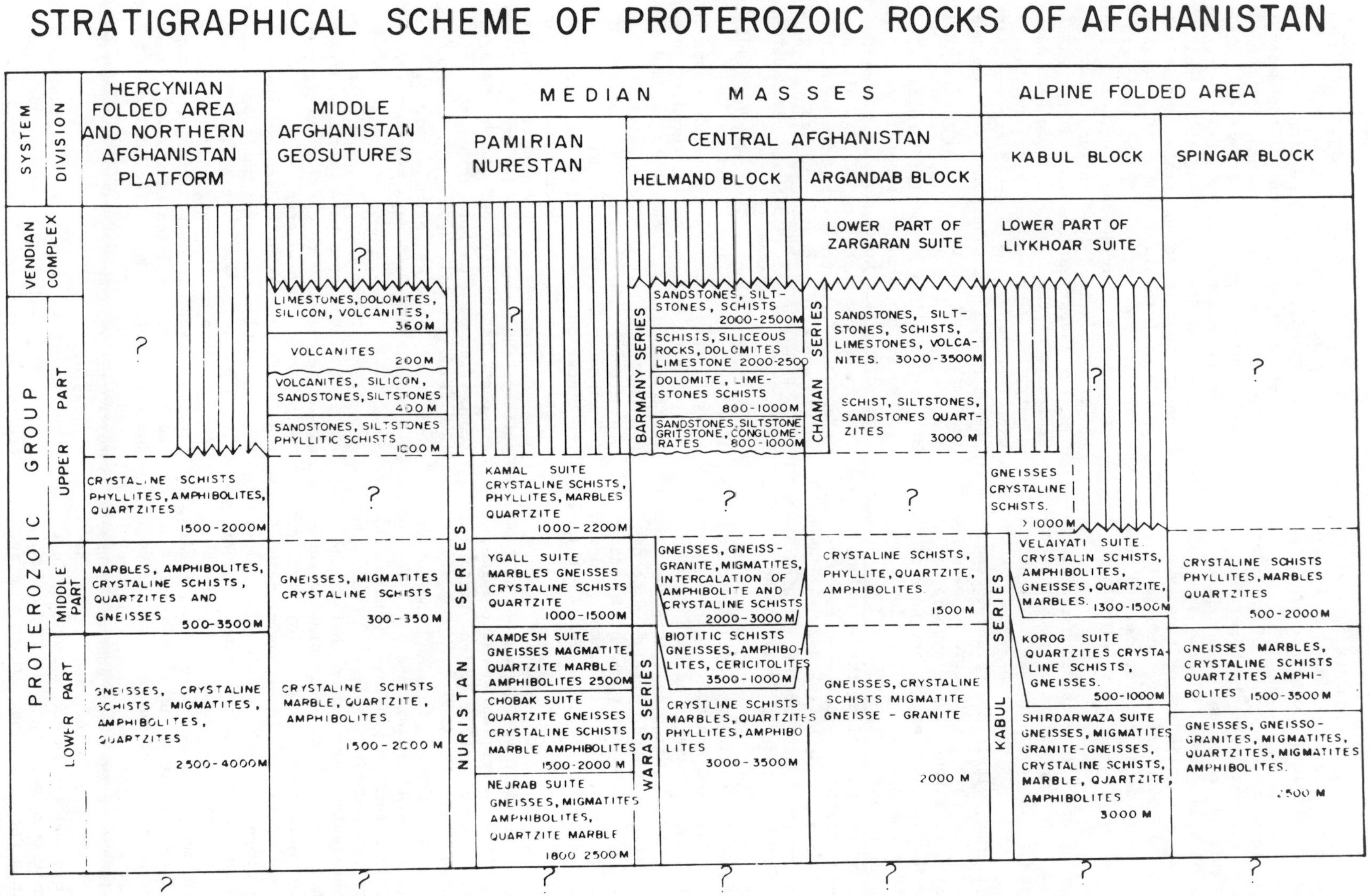

STRATIGRAPHICAL SCHEME OF PROTEROZOIC ROCKS OF AFGHANISTAN

SYSTEM	DIVISION	HERCYNIAN FOLDED AREA AND NORTHERN AFGHANISTAN PLATFORM	MEDIAN MASSES — MIDDLE AFGHANISTAN GEOSUTURES	MEDIAN MASSES — PAMIRIAN NURESTAN	CENTRAL AFGHANISTAN — HELMAND BLOCK	CENTRAL AFGHANISTAN — ARGANDAB BLOCK	ALPINE FOLDED AREA — KABUL BLOCK	ALPINE FOLDED AREA — SPINGAR BLOCK
PROTEROZOIC / VENDIAN	GROUP / COMPLEX — UPPER PART	LIMESTONES, DOLOMITES, SILICON, VOLCANITES 360 M; VOLCANITES 200 M; VOLCANITES, SILICON, SANDSTONES, SILTSTONES 420 M; SANDSTONES, SILTSTONES PHYLLITIC SCHISTS 1000 M		KAMAL SUITE CRYSTALINE SCHISTS, PHYLLITES, MARBLES QUARTZITE 1000-2200 M (NURISTAN SERIES)	SANDSTONES, SILTSTONES, SCHISTS 2000-2500 M; SCHISTS, SILICEOUS ROCKS, DOLOMITES LIMESTONE 2000-2500 M; DOLOMITE, LIMESTONES SCHISTS 800-1000 M; SANDSTONES, SILTSTONE GRITSTONE, CONGLOMERATES 800-1000 M (BARMANY SERIES)	LOWER PART OF ZARGARAN SUITE; SANDSTONES, SILTSTONES, SCHISTS, LIMESTONES, VOLCANITES 3000-3500 M; SCHIST, SILTSTONES, SANDSTONES QUARTZITES 3000 M (CHAMAN SERIES)	LOWER PART OF LIYKHOAR SUITE; GNEISSES CRYSTALINE SCHISTS	
	MIDDLE PART	CRYSTALINE SCHISTS PHYLLITES, AMPHIBOLITES, QUARTZITES 1500-2000 M; MARBLES, AMPHIBOLITES, CRYSTALINE SCHISTS, QUARTZITES AND GNEISSES 500-3500 M	GNEISSES, MIGMATITES CRYSTALINE SCHISTS 300-350 M	YGALL SUITE MARBLES GNEISSES CRYSTALINE SCHISTS QUARTZITE 1000-1500 M	GNEISSES, GNEISS-GRANITE, MIGMATITES, INTERCALATION OF AMPHIBOLITE AND CRYSTALINE SCHISTS 2000-3000 M (WARS SERIES)	CRYSTALINE SCHISTS, PHYLLITE, QUARTZITE, AMPHIBOLITES 1500 M	VELAIYATI SUITE CRYSTALIN SCHISTS, AMPHIBOLITES, GNEISSES, QUARTZITE, MARBLES 1300-1500M / >1000M (KABUL SERIES)	CRYSTALINE SCHISTS PHYLLITES, MARBLES QUARTZITES 500-2000 M; GNEISSES MARBLES, CRYSTALINE SCHISTS QUARTZITES AMPHIBOLITES 1500-3500 M
	LOWER PART	CRYSTALINE SCHISTS MARBLE, QUARTZITE, AMPHIBOLITES 1500-2000 M; GNEISSES, CRYSTALINE SCHISTS MIGMATITES, AMPHIBOLITES, QUARTZITES 2500-4000 M	CRYSTALINE SCHISTS MARBLE, QUARTZITE, AMPHIBOLITES 1500-2000 M	KAMDESH SUITE GNEISSES MAGMATITE QUARTZITE MARBLE AMPHIBOLITES 2500 M; CHOBAK SUITE QUARTZITE GNEISSES CRYSTALINE SCHISTS MARBLE AMPHIBOLITES 1500-2000 M; NEJRAB SUITE GNEISSES, MIGMATITES, AMPHIBOLITES, QUARTZITE MARBLE 1800-2500 M	BIOTITIC SCHISTS GNEISSES, AMPHIBOLITES, CERICITOLITES 3500-1000 M; CRYSTALINE SCHISTS MARBLES, QUARTZITE PHYLLITES, AMPHIBOLITES 3000-3500 M	GNEISSES, CRYSTALINE SCHISTS MIGMATITE GNEISSE-GRANITE 2000 M	KOROG SUITE QUARTZITES CRYSTALINE SCHISTS GNEISSES 500-1000M; SHIRODARWAZA SUITE GNEISSES, MIGMATITES GRANITE-GNEISSES, CRYSTALINE SCHISTS, MARBLE, QUARTZITE AMPHIBOLITES 3000 M	GNEISSES, GNEISSO-GRANITES, MIGMATITES, QUARTZITES, MIGMATITES AMPHIBOLITES 3500 M

Geological Structure and Ore Deposits of Afghanistan (In Press)

Fig. 2. Stratigraphic sketch of the Proterozoic rocks of Afghanistan.

(only occasionally to green schists facies) phyllitoid and argillaceous schists, fine- and coarse-grained terrigene and carbonate rocks of marine origin.

In some districts at the contact of Cambrian and Ordovician, presumably in the Lower Tournaisian and Namurian stages and in the Lower Permian stage as well, green, altered submarine volcanites of acid to basic composition have been found. In the Vendian-Lower Cambrian and Permian strata, in some districts, red and multi-coloured sediments, probably of continental origin, occur.

It is a fact of great interest that tectonically undisturbed Paleozoic strata overlie the Proterozoic sediments in distinct angular or azimuth unconformity.

In some districts, within the Paleozoic sequence at various stratigraphic levels, traces of a gap and unconformity were found. For example, at the bottom of the Middle Cambrian; at the base of the Devonian; before the Frasnian; in the lower portions of the Famennian; at the base and roof of the Tournaisian; at the roof of the Namurian; within the Upper Moscovian stage or before the Upper Carboniferous stage; at the base and in the upper portions of the Lower Permian and within the Upper Permian stage, etc., (Chmyriov et al., 1976b).

The Mesozoic sediments are well developed in Afghanistan. Within the Mesozoic sequence, there are 27 sub-divisions of narrow and broad age ranges singled out, including sediments that are grouped, or undifferentiated, many of which have been identified for the first time. All Mesozoic sub-divisions are, on the whole, formed predominantly by an assemblage of fine- to coarse-grained terrigene carbonate rocks.

Certain areas of Mesozoic deposition include green-stone and rubefield volcanites of the Lower and Upper Triassic, Berriasian-Hauterivian, Barremian-Aptian, Upper Aptian-Albian, Campanian and Maestrichian stages.

In middle and northern Afghanistan, the Norian-Rhaetian, Lower-Middle Jurassic and partially Lower Cretaceous strata show either continental, or mixed continental-marine facies.

Recent investigations have shown that in the northern part of the country, the Triassic sediments have been metamorphosed into clayey shale, and Jurassic and Cretaceous rocks having been lithified and compacted with practically no metamorphism. In the south of Afghanistan, however, the Triassic, Jurassic and Cretaceous rocks have been regionally metamorphosed into clayey facies, and locally, into phyllitic shale.

In different districts of the country, at various levels of the Mesozoic sequence, evidence of a stratigraphic gap or even an unconformity has been found; for example, before and within the Rhaetian, before and within the Bathonian, in the roof of the Oxfordian or within the Cimmerian, before or within the Barrimian, within the Upper Aptian, before or within the Campanian and within

the Maestrichian stages (Chmyriov et al., 1976a).

As far as the lower limit of the Mesozoic stage is concerned, it should be emphasized that in the northern part of the country, the Mesozoic sediments shown everywhere lacunae and even lie unconformably on the older formations, while in the center and south, they have only traces of lacunae or concealed hiatus. Abundant Mesozoic fossils made it possible to undertake detailed stratigraphic division in a number of districts.

Cenozoic formations of Afghanistan are represented both by marine and continental sediments and are subdivided into 16 substages.

Recent investigations showed that Paleogene sediments of the middle and southern parts of the country are represented by continental facies with rubefied, acid intermediate volcanites at the base, and red sandstones and conglomerates in the upper part of the sequence. In the northern and south-eastern districts, Paleogene strata are formed mainly of marine carbonates and terrigene rocks including a rather large percentage of acid to basic volcanites of submarine origin in a number of districts of Afghanistan.

Neogene formations everywhere of continental origin are mainly coarse-grained terrigene; their formation is related to the orogenic development of the country.

At the beginning of the Quaternary volcanic activity occurred locally.

Locally pre-Eocene stratigraphic gaps and unconformities have been discovered in some areas at various levels of the Cenozoic, and also in the uppermost part of the Oligocene and Miocene stages and at the base of the Quaternary.

Furthermore, within the Quaternary sequence, the traces of gaps are observed before each substage. The relations between Cenozoic formations and older sediments in the northern and southern parts of the country are different. In the northern part, the Neogene and Cretaceous strata are represented by a uniform carbonate series which is difficult to divide. In the southern half of the country, the Paleogene strata overlie the older sediments with gaps and unconformities.

Magmatism. Until recently, the intrusive formations of Afghanistan were poorly studied. Regional geological surveys carried out in middle and south-eastern Afghanistan, as well as the synthesis of the available data, in the process of the preparation of the Afghanistan geological map (1:500 000), (Kafarsky et al., 1977), yielded abundant data on the intrusive formations. These formations vary in composition, age, morphology and structural position as was well indicated in the sketch of the magmatic complexes of Afghanistan (1:1 000 000), (Stazilo-Aleksev et al., 1973), and at a scale of 1:2 000 000 (Abdullah et al., 1976).

The intrusive complexes in Afghanistan correspond to the following three tectono-magmatic cycles: Proterozoic, Paleozoic-early Mesozoic and Meso-Cenozoic. It has also been shown that the intrusive magmatism of each geological area was characterized by its own features (Map 1). There is a clear relationship between the intrusive magmatism, tectonics and mineral resources of the region. The intrusive formations of different ages are subdivided into rock associations, complexes and separate phases. The volcanogene and intrusive formations and complexes of different ages from Proterozoic to Quaternary (inclusive) have been described (Stazjilo-Aleksev et al., 1976).

The oldest intrusions are related to Proterozoic. Spatially closely associated with the outcrops of the Precambrian metamorphic formations, they are represented by small subconformable isometric and linear bodies consisting of two complexes. The early Proterozoic complexes include gabbro, metadiabases, orthoamphibolites, diorites and plagiogranites. The late Proterozoic complexes consist of gneiss granites, granites and plagiogranites. Their radiometric age is 575 m.y.

The Paleozoic and early Mesozoic intrusions were formed during the Hercynian tectono-magmatic cycle and are developed only in the north-eastern part of the country within the mountain systems of north-western Badakhshan and western Hindu Kush. The Paleozoic intrusions are represented by small isometric and lenticular discordant bodies of early Carboniferous age, composed of three complexes. The oldest complex is formed of dunites, peridotites and serpentinites; the youngest one, of diorites, granodiorites and plagiogranites (Stazhilo-Aleksev et al., 1976). Their age is determined by the fact that they are intrusive into the Lower Carboniferous volcanics. The early Carboniferous complexes are overlain by the Middle-Upper Carboniferous terrigene-carbonaceous formations. The radiometric age of the plagiogranites is 360 m.y. The early Triassic intrusions spatially converge with the early Carboniferous and are represented by large batholith-like granitoid bodies which were formed during the four phases of intrusion.

The first and the second phases are represented by granodiorites and granites; the third phase by granites, and the last one by granites and granosyenites. Their radiometric age is 203-216 m.y. In the north of the country, the localized Triassic intrusions form small isometric bodies constituting two complexes. The older subvolcanic complex consists of andesite-porphyrites and granite-porphyries, while the younger one is composed by granodiorites, granosyenites and granites. The radiometric age of the complex is 144-155 m.y. (Stazjilo-Aleksev et al., 1976).

The Alpine tectono-magmatic cycle started with the emplacement of late Jurassic - early Cretaceous intrusions, represented by small sub-volcanic bodies of diabases and gabbro-diorites that are spatially connected with volcanites of the same age. The early Cretaceous intrusions are formed by four complexes, of which the two older occur in central and south-western Afghanistan. These are small bodies and are spatially connected with the late Jurassic - early Cretaceous volcano - sedimentary formations (ophiolite group). The first complex consists of dunites, periodotites and serpentinites; the second complex of gabbro, diorites and plagiogranites. The third one is represented by linear and isometric bodies of gabbro, monzonites, diorites and granodiorites which were developed in Nurestan and Pagman. The fourth complex comprises batholith-like masses of granitoid composition and varying sizes and is known in middle and the north-eastern Afghanistan. The radiometric age of diorites determined by the age of the Pagman mass is 103 m.y.

The late Cretaceous - Paleocene intrusions are confined to the central and south-western regions of the country. These are isometric or linear bodies consisting of gabbro, monzonites, diorites, granites, granosyenites, syenite-porphyries, syenites and gabbro-diorites. They cut the rocks from the Proterozoic to the Lower Cretaceous period inclusive and are overlain by Eocene-Oligocene volcanites. They have been dated at 97-106 m.y.

The Eocene intrusions are known in south-eastern Afghanistan and in the southern areas of northern Afghanistan. They are small stock-like subconformable bodies belonging to two complexes. The older complex consists of diabases and diorites; the younger one is formed by dunites, peridotites and serpentinites.

The latest investigations proved that the largest Eocene ultrabasic massif of Afghanistan (Logar) is associated with a zone of abyssal fracture. It is a digitation overthrust sheet of complicated structure, split into a series of compressed layers. The Eocene-Oligocene intrusions are represented by subvolcanic dikes, small stocks and sill-like bodies of liparites, dacites and granite-porphyries. They are connected, both spatially and genetically, with rubefied volcanites of intermediate to acid composition dated Eocene-Oligocene (Stazjilo-Alekev et al., 1976).

The Oligocene intrusions are wide-spread throughout the Central Afghan and Pamirs-Nurestan median masses. These are batholith-like, isometric or elongate bodies of different sizes. Three phases of emplacement have been identified; the first consists of granodiorites, diorites and plagiogranites; the second phase of granodiorites, alaskites, granosyenites, granites; the third is represented by granites. The granitoids cut the Eocene-Oligocene volcanites and are overlain transgressively by the Neogene. The granitoids have been dated radiometrically from 17-23 to 60-80 m.y.

Two Miocene complexes include minor intrusions in the form of dikes and stock-like bodies. The older complex consists of diorite-porphyrites, granodiorite-porphyries, monzonite-porphyries and syenite-porphyries. The younger complex is composed of nepheline syenites.

The early Quaternary intrusions occur only

locally, mainly along the zones of meridional faults and are spatially connected with the development of the coeval volcanites. They have much in common with andesites, dacites and liparites. The most noteworthy is the Khanneshin volcanic structure, which is characterized by peculiar rocks such as soevites, alvikites and ankerite-barytic carbonatites (Eremenko et al., 1975).

Tectonics. New data on the country's stratigraphy and analysis of the 1977 geological map of Afghanistan (1:500 000), allowed corrections to be made, and in a number of cases, the existing views on the tectonics of the region to be revised. The latest investigations showed that Afghanistan, being one of the eastern segments of the Mediterranian fold belt, represents a typical heterogeneous structure. The Hercynian folded area and epi-Hercynian (epi-Cimmerian) plateau can be delineated in its northern part. The structures of middle Afghanistan, which had been formed at the site of the Paleozoic-Mesozoic troughs, lie further to the south. South and south east of the latter, lie the Central Afghan and Pamir-Nurestan median masses. In the east, south and west the mass is surrounded by the Alpine folded area. The north Afghan plateau is the southern margin of the Turanian platform, while all the remaining territory of Afghanistan is related to the Mediterranean fold belt (Map 1).

The Hercynian fold belt includes structures developed in north-western Badakhshan and western Hindu Kush. The stratigraphic sequence of the area includes structural-formational complexes represented by metamorphic and volcano-sedimentary strata. They are separated by gaps or angular unconformities, and cut by the following intrusive formations: Proterozoic granitoids occurring as small sub-conformable bodies in the Proterozoic metamorphic rocks; geosynclinal ultrabasites of early Carboniferous that form small lenticular bodies and stocks; early Carboniferous gabbro-diorites and geosynclinal plagiogranites occurring as stock-like discordant bodies; early and late Triassic (orogenic) granitoids characterized by large batholith-like bodies and small stocks confined to the axial part of the west Hindu Kush Range and north-east of it.

The Hercynian folded structures form the basement of the north Afghan plateau. They are mostly overlain by the platform cover, so they can be seen only in exposures within the cores of anticlinal structures and in deep valleys. In the north-eastern part of the country (south-western Hindu Kush and north-eastern Badakhshan) the basement is free of platform cover. The linear character of the main Hercynian structures is clearly expressed here where they trend northeast as a gently curved east-south-east arc. In the south-east they are limited by the Central Badakhshan fault and in the north-west, by the Khokhan Fault (Kafarsky et al., 1976). These structures plunge to the south-west, under the cover of the north Afghan Plateau, while to the north-east they pass into the Hercynian fold structures of the Soviet northern Pamir. In the south-west the structures are characterized by a sublatitudinal or east-north-eastern strike; in the center and north the strike is submeridional.

A significant part of northern Afghanistan is occupied by the north Afghan epi-Hercynian plateau which is separated from the structures of middle Afghanistan by the Hari Rod fault and in the east and south-east by the Khokhan fault - from the structures of the western Hindu-Kush and north-western Badakhshan.

The plateau shows clearly (Map 1) its cover and folded basement. The basement has already been described; the cover is represented by gently dipping Jurassic, Cretaceous and Paleogene terrigene-carbonate beds (200-3000m). The cover is cut by minor intrusions of Miocene diorite-porphyrites and granodiorites-porphyrites. The plateau has been investigated by detailed geological and geophysical surveys and drilling. It is split into the Murgab-Hari Rod, Maimana, Shibergan and Mazarif blocks, separated from each other by the Bandi-Turkestan, Andarab-Mirza-Valang and Alburz-Marmuli faults. In the north lie the Amu Darya Quaternary superimposed basin and the Afghan-Tajik basin, reworked by the Alpine movements. Within each of the blocks structures of the second and third order have been described (Chmyriov et al., 1976a).

Middle Afghanistan is bounded to the north by the Hari Rod and Central Badakhshan faults, and by the Karganaw fault in the south. The latest investigations gave much information on the geological structure of middle Afghanistan which is the most complicated area of the country. It consists of metamorphosed Precambrian-Paleozoic, Meso-Cenozoic and volcano-sedimentary complexes that are cut by the Cretaceous-Paleogene intrusions of granitoids. Lately, within the area of middle Afghanistan, more than ten structural-formational zones were distinguished. More than nine tectonic blocks, more than twelve schuppen zones and overthrusts have been described. In the western part of the area, the structures have been shown to strike sub-latitudinally, while in the eastern part their strike is north-eastern or sub-meridional. The greater majority of the recent earthquake epicenters and the chain of mineral springs are confined to the eastern portion of the area. Middle Afghanistan has an overlapped-folded block structure, formed at the site of the Paleozoic-Mesozoic marine troughs.

The Central Afghan and Pamir-Nurestan median masses are located in the inner parts of the country. The Pamir-Nurestan median mass is related to the pre-Baikalian structures, while the Central Afghan median mass is considered to represent Baikalian consolidation. The Pamir-Nurestan mass is now known to be bounded by the Tanjser-Kokche faults and by the Bagharak interformational granite massif in the northwest, and by the Kunar and Tashkapruk faults in the southeast. The Pamir-Nurestan mass is not a homogeneous structure and two blocks can be clearly defined: south Badakhshan and Nurestan. In the south Badakhshan block, large intrusive massifs have not yet been found. The block is characterized by dome-shaped, brachy-form and simple folds, with the major folds and disjunctive structures oriented about N-S.

The Nurestan block is separated from the south Badakhshan one by the Zebak fault and is characterized by the presence of intrusive formations represented by sub-conformable intrusive sheets presumably of Precambrian granite-gneisses, discordant stock-like bodies of early Cretaceous gabbro-monzonites-granodiorites and batholith-like discordant and sub-discordant bodies presumably of Oligocene granites. The block is formed mainly by steeply-dipping metamorphosed sediments with a north-eastern strike. At the south-western margin of the block, the structures veer north-west following the contour of the Kabul Zone (Chmyriov et al., 1976a).

The Central Afghan mass is not homogeneous in structure and includes the Helmend-Argandab uplift, the Farekh trough and the Seistan superimposed basin.

The Helmend-Argandab uplift is limited in the north-west by the Helmend fault, and in the south-east, by the Tarnak-Moqur fault. According to Dronov and co-authors (1973), its folded basement is formed of two metamorphic series; the lower, altered to amphibolite facies, is presumably related to the early-Middle Proterozoic; the upper series, presumably of late Proterozoic age, is altered to phyllitic facies. In the upper series, north-eastern oriented, linear harmonic and iso-clinal folds are observed.

The cover of the Helmend-Argandab uplift comprises 1000-5000 m. of metamorphosed carbonate-terrigene marine rocks of Vendian to Jurassic age. They overlie the folded basement in sharp discordance and are overlain also unconformably by 500-1000 m of coarse-grained terrigene and volcanogene continental Cenozoic formations.

The intrusions of the Helmend-Argandab uplift comprise several complexes of different ages: gabbroids, metadiabases, orthoamphibolites, diorites, and granodiorites of early-Middle Proterozoic, the emplacement of which is connected with the pre-Baikalian tectono-magmatic cycle; Upper Proterozoic diorites and plagiogranites that have been intruded during the Baikalian tectono-magmatic cycle, cut the Upper Proterozoic basement.

Early Cretaceous intrusions are small or medium sized bodies of ultrabasites, gabbrodiorites and plagiogranites.

Late Cretaceous-Paleogene stock-like intrusions are of gabbro, monzonites, diorites, granites and granodiorites. Oligocene intrusions of granitoid composition are represented by larger elongate and stock-shaped bodies. The Helmend-Argandab uplift, on the whole, is characterized by the presence of disjunctive dislocations striking north-east (i.e. sub-conformable with the general strike of the folded structures) or about N-S.

The cover is characterized by the simple, shallow, linear folds oriented north-east.

The Farakh trough lies north-west of the Helmend-Argandab uplift and is separated from it by the Helmend fault. The northern limit of the trough is the Karganaw fault; in the west it is bounded by the Dasht-i-Demdam fault; further west the Alpine structures of western Afghanistan and eastern Iran are developed. The latest investigations made it possible to delimit the trough more precisely and revealed new factors on its inner structure. The Farakh trough is wedge shaped; it occurs on the Baikalian folded basement and is filled with over 10,000 m of terrigene-volcanogene rocks of Carboniferous, Permian, Triassic, Jurassic and Lower Cretaceous ages. Within the Farakh trough, the Tithonian-Lower Cretaceous formations are predominantly developed, and due to their heterogeneity, they can be subdivided into four structural-facial zones: Zuri, Harit Rod, Anardara and Khash Rod.

The intrusive formations are of different ages. The early Cretaceous intrusions are those formed by small massifs of ultrabasic, basic and acid rocks that spatially associate with ophiolites.

Late Cretaceous-Paleocene intrusions are widely developed and represented by small stock-like bodies of basic to acid rocks. Eocene-Oligocene subvolcanic bodies of acid to intermediate rocks together with the volcanites of prophyry formation compose the Paleogene volcano-plutonic association. Oligocene intrusions are large stock-like granitoid massifs cutting Eocene-Oligocene volcanites. Miocene intrusions are represented by numerous sub-meridional dikes of diorite-porphyrites and granodiorite-porphyries that cut all the pre-Neogene formations.

On the whole, the folds show predominantly a N-E strike. The folds are of different types: those confined to terrigene rocks are linear, harmonic and even isoclinal. Carbonate rocks are crumpled into relatively simple shallow brachyform folds. The age of folding is identified as Cenomanian-Turonian. The folds of the Paleogene strata are brachyform, very rarely linear. The Neogene rocks are mainly faulted.

The most recent investigations have changed the geological contours of the Alpine fold area; and its regional extent is now better defined, and the tectonic characteristics of the area have become more completely understood (Chmyriov et al., 1980).

The Alpine fold belt includes western Afghanistan (Asparan and Kishmaran), parts of southern and south-eastern Afghanistan. Its limits are everywhere tectonic. In western Afghanistan, the Alpine structures formed on the site of a late Cretaceous trough. The main geosynclinal complex, dated Campanian-Maestrichian, lies in sharp discordance on the folded basement, formed as carbonaceous and terrigene formations regionally metamorphosed into phyllites and clayey schists.

Their total thickness is 4000-5000 m. The main geosynclinal complex consists of two formations. The lower one, 300-800 m thick, consists of volcanogene-carbonate rocks; the upper one is flyshoid, predominantly sandy-aleurolitic, with minor amounts of carbonaceous rocks and volcanites of intermediate to basic composition.

Intrusive formations are conventionally referred to as early Cretaceous and Miocene. The early Cretaceous intrusions are granites and confined to the folded basement. The Miocene intrusions are represented by stock-like and dike-like bodies of gabbro-diorites, diorite-porphyrites and granodiorite-porphyrites that cut the formations of geosynclinal complex and the Campanian-Maestrichian beds.

The folded basement is characterized by harmonic and isoclinal structures of late Jurassic - early Cretaceous age. The structures are oriented submeridionally.

In south-eastern Afghanistan, the main Paleogene geosynclinal complex is represented by marine-terrigene-flysch formation with minor limestones and volcanites, overlying the eroded folded basement in discordance.

Geophysical Investigations

From 1972 to 1979 only limited geophysical investigations were undertaken in Afghanistan. Airborne geophysical surveys in western Afghanistan and ground control within the Kabul Block are important and mentioned here.

Airborne geophysical surveys (magnetic and gammaray surveys) were conducted in the western part of Afghanistan (Knyazhez and Bukhmastov, unpublished data, 1976).

The analysis of the airborne magnetic surveys showed that the most contrasted magnetic anomalies are confined to the areas of intrusive and effusive formations. The zones of major faults can be recognized fairly distinctly. The surveys enabled a number of faults to be more precisely delineated, as well as contours of the intrusive bodies.

The analysis of such parameters as the geomagnetic field, types and size of the anomaly-causing elements, anomalous zones and the depths of magnetic masses, show the area covered by airborne magnetic investigations to be divided into four blocks, namely: northern, Ghurian, Schindan and Farakh (Figure 3).

The northern block in its south-eastern part is characterized by the presence of extensive positive anomalies of significant amplitude. In its northern part a magnetic field showing a mosaic structure is observed. The southern limit of the area is some 25 km north of the valley of the Hari Rod river, where a zone of positive magnetic field is recorded. Tectonically, the northern area includes the western part of the epi-Hercynian plateau of northern Afghanistan, while the southern is associated with the Alpine fold belt, characterized by a negative magnetic field.

The Ghurian block lies south of the northern block and is characterized by a positive magnetic field. Its southern limit passes through the village of Hasankala. Tectonically, the Ghurian block occupies the Alpine fold belt in the north and partially the Yakovlang-Davindar zone, the Hazorsangan block (middle Afghanistan) and a part of the Zori Zone (Knyazhev et al., unpublished data, 1976). Within this block, the western continuation of the Hari Rod fault also became clear through the airborne geophysical investigations. While interpreting the airborne magnetic surveys, this fault structure was considered to be of lesser significance than it was thought to at the time of the geological survey at a scale of 1:500 000.

The Shindan block lies south of the Shurian block and shows a magnetic field of complicated

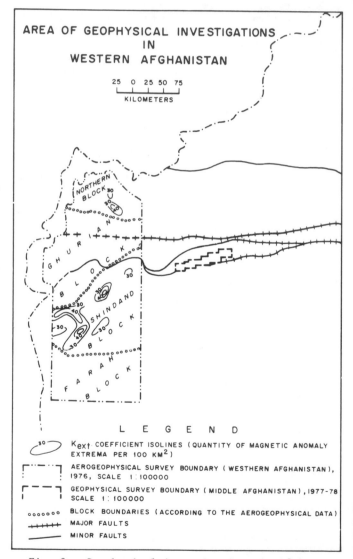

Fig. 3. Geophysical investigations in Afghanistan.

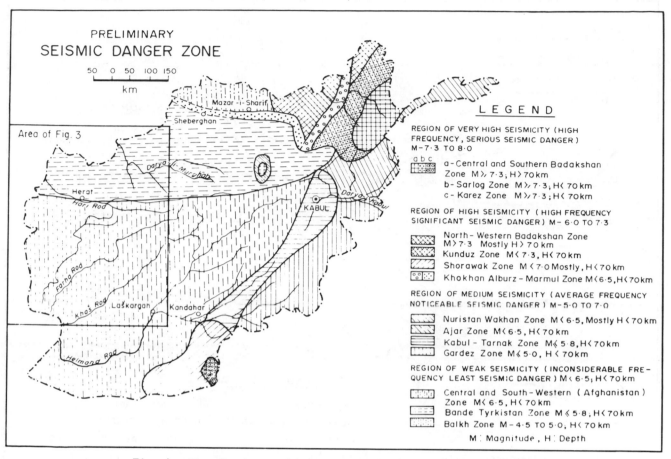

Fig. 4. Map showing preliminary seismic danger zones in Afghanistan.

mosaic structure. Its southern limit passes south of the zone of high-gradient negative magnetic anomalies that extends from north-east to south-west. Tectonically, the Shindan block occupies a considerable part of the northern Anardara zone and a certain part of the Zori zone.

The Farah block lies south of the Shindan block with its southern limit at 32°07' N. Extensive positive magnetic anomalies, frequently of isometric shape, are typical of this block. Tectonically, the Farah block occupies the Anardara zone and a marginal part of the Farah trough.

The blocks described above are defined on the basis of the differences in the magnetic field which reflects different depths to crystalline basement in western Afghanistan. The Churian block is the most uplifted. The strike of two geological structures determined during geological observations completely coincides with that of the anomalous zones. The presence of the majority of magnetic anomalies in the western part of the country can be explained by the presence of basic intrusions. At the western margin of the northern Afghanistan epi-Hercynian

plateau, where the Eocene-Oligocene effusive formations are developed, airborne magnetic surveys have revealed a north-western strike of the structures. In the northern part of the area, at a depth of 400-600 m, an intrusive massif formed by diorites and granodiorites is inferred.

In the northern part of the Ghurian block, at the latitude of the town of Herat, within the Shindan block and in some other localities, the airborne magnetic surveys have delineated a number of intrusive, effusive and subvolcanic formations of varying sizes occurring at the depths of 1 to 3,5 km.

Ground Geophysical Surveys. Since 1973, the ground geophysical investigations have been mainly restricted to the Kabul block. These investigations were aimed at helping the prospecting surveys and hydrogeological work.

The geophysical investigation included magnetic prospecting, natural electric field and vertical electric sounding. It aimed at studying the regional geological structure and discovering structures favourable for mineralization. Here, results are discussed briefly.

From 1974 to 1977, magnetic prospecting at a

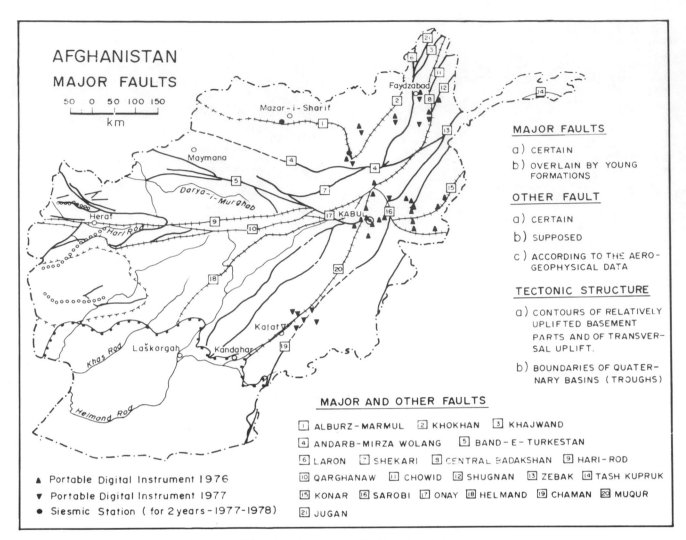

Fig. 5. Map showing major faults in Afghanistan.

scale of 1:25 000 and 1:50 000 was conducted in an area of 800 km^2 (Kubatkin, Voinov et al., unpublished data, 1978). The interpretation of the data resulted in the compilation of a structural-tectonic sketch of the area, more than 50% of which is overlain by Neogene and Quaternary sediments. The geological structure of the region was somewhat clarified and faults accompanied by zones of crumpling and exfoliation of rocks were defined; the contours of intrusive massifs, both which either crop out or are masked by unconsolidated formations were delineated.

This work covered the north-eastern part of the ultrabasite Logar massif, and yielded the first data on the magnetic properties of the rocks (Abdullah et al., 1978).

During 1977-78, systematic study of the Logar massif was carried out at a scale of 1:100 000 and at present approximately 70% (1800 km^2) of the massif is covered by the magnetic survey. There the magnetic prospecting contributed

effectively to the mapping of the Logar massif contacts with the country rock and to the study of fault zones.

According to Voinov (Abdullah et al., 1978) the ultrabasics are usually serpentinized in the endogenic contacts of the massif and in the fault zones, and these zones are characterized by a highly differentiated anomalous magnetic field (from − 30 millioersted to + 50 millioersted). As a rule, the unaltered ultrabasics and the country rocks are poorly magnetized, with a weak magnetic field.

Within the Nalbandon deposit (middle Afghanistan), the magnetic survey was found to be less effective for mapping the sedimentary rocks to a scale of 1:1 000 000 (Figure 3).

The method of natural electric field (NEF) was used in the areas of thin Neogene and Quaternary sediments within the Kabul block and in an area of more than 900 km^2 in middle Afghanistan. A great number of anomalies and anomalous fields

with intensities of up to -500 m were discovered.
A large number of these anomalies are associated
with carbonaceous material in the shallow terrigene cover; and the others correspond perhaps to
the localities with sulphide mineralization.

The method of vertical electric sounding (VES)
was used within the Kabul block to determine the
depth of unconsolidated deposits. This method
was also used during hydrogeological investigations within the populated areas and agricul-
tural land.

To date, about 1000 VES have been conducted
in the Kabul block, along with drilling. These
have been helpful in locating water reservoirs
underground.

Seismicity. Afghanistan is a seismically active
country. Epicentres are recorded in all regions,
but their energy, frequency and density as well
as focal depth vary.

The analysis of data on earthquakes registered
in Afghanistan, the data on the seismicity of the
contiguous countries in addition to the available
information on tectonics received during the geological survey at a scale of 1:500 000, improved
our understanding of the seismicity in Afghanistan
(Abdullah, 1976).

On the whole, the depth of earthquake foci and
magnitude increase from west to east and from
south-west to north-east (Slavin et al., 1970;
Heuckroth et al., 1970). The north-eastern part
of the country, particularly north-western Badakhshan, is a zone of high seismicity where during
the period 1938-1967, 1063 earthquake shocks were
registered.

On the basis of the available data on seismicity and taking into account the tectonic regions
of the country, regions of the intense, high,
medium and weak seismicity could be delineated
(Figure 4).

During 1977-78, the experts of the French Geological Mission and the MIT geophysicists, in
collaboration with Afghan geologists, conducted
short-term geophysical observations by means of
portable seismological stations in the northwestern Badakhshan, the western Hindu Kush,
Nurestan and elsewhere (Figure 5). The main purpose of these investigations was to study the regional distribution of earthquakes in connection
with regional tectonics and to determine the
plunge of the fault planes.

The microseismic investigations resulted in the
discovery of seismic activity associated with
some large faults in Afghanistan: Alburz-Marmul,
Central Badakhshan, Panjser, Pagman, Kunar and
Chaman. It was proved once again that in northwestern Badakhshan and partially western Hindu
Kush intermediate depth earthquakes occurred, and
in other areas they were shallow. From June 13
to July 14, 1977, 1300 seismic shocks were registered in the region. On the 23-24th of May,
1978, 80 shocks occurred in the Panjser valley,
and within two months their number rose to about
500. The depth of the earthquake in the Panjser
valley is estimated to be up to 40 km.

The fact that numerous earthquake hypocentres
are confined to the area north-west of the Panjser fault, indicates a north-western dip of the
plane of this structure.

At present, the seismicity of the Alburz-Marmul
fault zone in western Afghanistan is well understood. Within a year, in the Puli-Khumri region,
and south-west of Mazari Sharif, the seismic
stations registered 3000 and 1000 shocks, respectively, which confirms the high activity of
the fault. The seismic characteristics of the
fault are given by Mirzaev and co-authors (1979),
who also compiled a map of the seismic regions

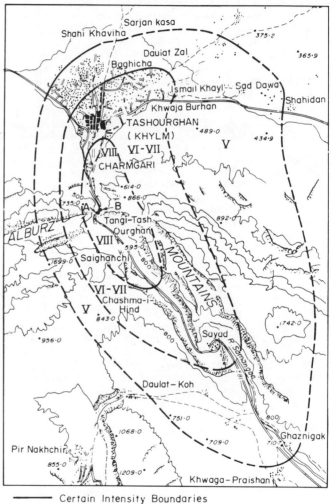

AFGHANISTAN, SAMANGAN, PROVINCE (KHULM)
Intensity Distribution of 19 March, 1976 Earthquake.

——— Certain Intensity Boundaries

– – – Uncertain Intensity Boundaries

VIII Earthquake Intensity in Ball (Gubin, 1960)
A }
B } Landslides

Fig. 6. Isoseismal map of the 19 March 1976
earthquake in Afghanistan.

of the Alburz-Marmul fault zone. Destructive
earthquakes take place here. Isoseismals for
one of them (Figure 6) were given by Abdullah
(Abdullah, 1977).

There is no doubt that numerous epicentres in
the regions of Kunduz, the western Hindu Kush,
north-western and southern Badakhshan, Nurestan,
Bakhan and Kabul are confined to deep fault zones.
The earthquakes testify to the ongoing neotectonic
movements. In this connection, we consider it
necessary to continue seismic investigations within the deep fault zones in Afghanistan in order
to describe the earthquakes and to determine the
morphology of major faulted structures.

References

Abdullah, S., Seismotectonic regioning of Afghanistan (abstract), presented in IV Naychno-
Methdicheskay, konferensia Kabylski Poletech-
nicheski Institute (in Russian), Kabul, 1976.

Abdullah, S., The Khulm (North Afghanistan)
earthquake of 19 March, 1976, (abstract) presented in International Symposium on Recent
Crustal Movement, Stanford University, Palo
Alto, California, 1977.

Abdullah, S., W. M. Chmyriov, K. F. Stazhilo-
Aleksev, V. I. Dronov, P. J. Gannon, B. K.
Lubemov, A. Kh. Kafarsky and E. P. Malyarov,
Mineral Resources of Afghanistan, 2nd Edition,
Kabul, 1977.

Abdullah, S., V. N. Voinov, E. B. Nevretdinov
and L. V. Kubatkin, The Logar ultrabasite
massif and the character of its magnetic field
(East Afghanistan), paper presented to International Geodynamics Conference (Alpine-Himalayan Region) Kathmandu (in press), 1978.

Chmyriov, V. M., A. Kh. Kafarsky, S. Abdullah,
V. I. Dronov, and K. F. Stazhilo-Aleksey,
Tectonic Regioning of Afghanistan, paper presented to Himalayan Geology Seminar, New Delhi
(in press), 1976a.

Chmyriov, V. M., S. Abdullah, V. I. Dronov,
A. Kh. Kafarsky, A. S. Salah and K.F. Stazhilo-
Aleksev, Geological Map of Afghanistan, paper
presented to Himalayan Geology Seminar, New
Delhi (in press), 1976b.

Chmyriov, W. M., V. I. Dronov, K. F. Stazhilo-
Aleksev, A. Kh. Kafarsky, E. P. Malyarov and

I. M. Sborshchikov, Geological structure and
ore deposits of Afghanistan, Moscow (in press),
1980.

Dronov, V. I., A. Kh. Kafarsky, Sh. Sh. Denikaev,
A. S. Salah, I. I. Sonin, V. M. Chmyriov and
S. Abdullah, Scheme of Stratigraphy of Afghan-
istan, pp. 86-89 edition 1 Kabul, 1973.

Dronov, V. I., S. Abdullah, A. S. Salah and V. M.
Chmyriov, The Main Stratigraphical Sub-Division
of Afghanistan, paper presented to Himalayan
Geology Seminar, New Delhi (in press), 1976.

Erikmenko, G. K., B. Ya. Vikhter, V. M. Chmyriov,
S. Abdullah, On the carbonatite volcanic formations of the Middle East, Tehran symposium on
the Geodynamics of Southwest Asia, Special
Publication of the Geological Survey of Iran,
pp. 185-199, 1975.

Heuckroth, L. E., and R. A. Karim, Earthquake
history seismicity and tectonic of the regions
of Afghanistan, Kabul, 1970.

Kafarsky, A. Kh., S. Abdullah, The tectonics of
North-East Afghanistan (Badakshan, Wakhan,
Nuristan) and relationship with the adjacent
territories, International Colloquium on the
Geotectonics of the Kashmir Himalaya Karakorum-
Hindu Kush-Pamir orogenic belts, Accademia
Nazionale dei Lincei, Rome, 1976.

Kafarsky, A. Kh., V. I. Dronov, K. F. Stazhilo-
Aleksev, I. M. Sborschikov, V. I. Slavin, S.
Abdullah, V. B. Averyanov, V. M. Chmyriov and
A. Ashmat, Geological Map of Afghanistan,
1:500 000, 1977.

Mirzaev, K. M., A. M. Babaev and S. Abdullah,
Seismichiskoe rayoni-rovanie zone sochlenenei
Severo-Afghanskovo vistypa e Afghano-Tajeksay
Depressie (in press), USSR, 1979.

Slavin, V. I., A. P. Soloveva and Y. Ya. Solovev,
Seismotectonichesko Raynirovahniya Afghanistana,
Isvestie Veshechk Ychebnichk Zavedenyi Gheolo-
ghia e Razvedko, Moscow, 1970.

Stazhilo-Aleksev, K. F., V. M. Chmyriov, S. H.
Mirzad, V. I. Dronov and A. Kh. Kafarsky, The
main features of magmatism of Afghanistan, pp.
31-43, edition 1, Kabul, 1973.

Stazhilo-Aleksev, K. F., V. M. Chmyriov and M.T.
Gerowall, The intrusive magmatism of Afghanistan, (abstract) presented in IV Naychno-
Metodicheskay Konferensia, Kabylski Poletech-
knicheski Institute (in Russian), Kabul, 1976.

TECTONICS OF THE CENTRAL SECTOR OF THE HIMALAYA

K. S. Valdiya

Kumaun University, Nainital-263001, India

Abstract. The Kumaun Himalaya is not only an epitome of the tectonic design of the whole Himalaya mountain, but also it is seismotectonically the most reponsive sector. The Lesser Himalaya is built up of a succession of three nappes overriding the autochthonous Precambrian sedimentary sequences; the sedimentary Krol (=Berinag) Nappe, the epimetamorphic Ramgarh (=Chail) Nappe, and the mesometamorphic Almora (=Jutogh) Nappe at the top. The southern frontal part of the Krol sheet in the proximity of the Main Boundary Thrust, which is registering measurable neotectonic movements, exhibits schuppen structure of overturned to recumbent folds dissected by reverse faults, giving rise to depressions and lakes. Long deep faults developed along crestal planes in the central part of the synclinorial Krol Nappe bring up the basement in tectonic windows, and have differentiated the Krol belt into two parts. The southern belt comprises the full succession of the Palaeozoic formations while the northern part has only the lower (older) half of the sequence.

The intensity of deformation in the autochthonous zone of the inner Lesser Himalaya increases progressively northward. Overturned and tight isoclinal folds beneath the Berinag (=Krol) overthrust sheet are locally split into minor tectonic schuppen with a locally inverted sequence. In the north the combined autochthonous and allochthonous formations are involved in multiple repetitions by imbricate thrusting and inversion in the zone of schuppen in the immediate proximity of the root of the crystalline nappe. The structures of this northern belt thus indicate very severe deformation in the root zone.

The asymmetry of folding of the synclinally folded thrust sheet of the mesograde crystalline rocks (Almora Nappe) progressively increases northwards so that near the root, the autochthonous formations are squeezed up as tectonic wedges between the southern isoclinal synclinal klippe and the northern homoclinal root that dips 15-20° NE at the base of the Great Himalaya. The Main Central (Vaikrita) Thrust (MC(V)T) constitutes the tectonic boundary between the strongly cataclastically deformed and mylonitized rocks of the Munsiari unit (root zone) and the higher-grade Vaikrita Group of the Great Himalaya. Delimited by the 30°-45° northeastward dipping MC(V)T at its base and the steeply hading Malari Thrust-Fault in the north, the Great Himalaya is a huge tectonic slab which is still geodynamically active, as evidenced by the tilting of recent river terraces, many geomorphic features and recurrent seismicity. The MC(V)T, associated with a line of hot springs, is an intracrustal boundary thrust registering a net vertical throw of the order of 20,000 m. The young Tertiary granites injected into the Vaikrita Group are related to the post-orogenic phase of deformation and represent presumably the fusion of the deep, sunken part of the frontal edge of the Indian shield.

Higher incidence of earthquakes in the areas affected by transverse and diagonal faults and the transverse trend of high seismicity in the quantitative seismicity map of the Himalaya, together with the pattern of vertical distribution of foci, suggest that the higher seismicity of northeastern Kumaun and adjoining Nepal is related to a large extent to strike-slip movement along some of the transverse faults. It seems that the crustal accomodation consequent on convergence of the Indian and Asian plates is being accomplished not only by left-lateral strike-slip movement along E-W faults in the Tibetan plateau and adjacent areas but also by dextral (predominant) and sinistral movement on the transverse faults of the Himalaya and the adjoining Penisular domain.

Introduction

Structural and stratigraphic studies in the last decade have demonstrated that the crucial central sector of the Himalayan arc (Figure 1)-the Kumaun Himalaya - lying between the borders of the states of Himachal Pradesh and Nepal is not only geodynamically a very responsive region but also an epitome of the entire Himalaya mountain. Exhibiting full development of all the four lithotectonic and physiographic zones, (the Siwalik, Lesser Himalaya, Great Himalaya and Tethys) the Kumaun sector registers the most frequent seismicity anywhere in the Himalaya, and provides eloquent evidence of neotectonic movements along the many thrusts and faults.

The ruggedly youthful topography of the frontal Siwalik, composed of the late Tertiary to Pleistocene molasse, is rising in relation to the flat plains of the Ganga Basin along the Himalayan Frontal Fault, and the sedimentary thrust sheet of the outer Lesser Himalaya continues to advance southwards along the Main Boundary Thrust over the Siwalik. The Himalayan Frontal Fault delimits the tectonic boundary of the hilly Siwalik against the plains of Uttar Pradesh and lies to the south of the Main Boundary Thrust. The Lesser Himalaya, comprising three thrust sheets overriding the autochthonous Riphean sediments (Figures 2 and 3), has a mild and mature topography, deeply dissected by rivers in consequence of recent rejuvenation. The thrust plane that separates the Lesser Himalaya from the extremely rugged realm of the Great Himalaya to the north has thrown up the Precambrian basement to form the highest mountain rampart characterized by sharp peaks, precipitous scarps and deep gorges. To the north yet another regional thrust fault separates the Great Himalayan block from the huge sedimentary sequence of the Tethys Himalaya. Higher seismicity and a number of geomorphic features indicate that the fault-delimited tectonic slab of the Great Himalaya is rising at a faster rate than the Lesser Himalaya. The Tethys realm, which contains an almost unbroken record of sedimentation from the late Precambrian to the Cretaceous, is severely folded and cut by imbricate thrusts in the northern belt. It is delimited and cut in the north by a deep fault that has brought up ophiolite and melange, part of which has travelled 80 kilometres south as vast thrust sheets.

There are thus four major thrust faults

(Table 1), three of which are of intracrustal or even deeper and they separate the various lithotectonic-physiographic subprovinces. The central subject of this paper is the tectonics of these intracrustal boundary thrusts and intraformational faults.

Geodynamics of the Southern Belts of Lesser Himalaya

The Lesser Himalaya is separated from the Outer Himalaya in the south by the Main Boundary Thrust (MBT). The MBT has been defined (Valdiya, 1980) as a series of four steeply inclined thrusts that separate the autochthonous Cenozoic sedimentary zone, including the Siwalik, from the old Lesser Himalayan sub-province. In Kumaun it is the Krol Thrust which serves as the MBT. It has brought one or the other of the six lithostratigraphic units of the Krol Nappe over and against the Siwalik (Figures 2 & 3; Table 1), and east of the Yamuna valley, has completely overlapped and concealed the Nahan Thrust, the original MBT separating the Eocene Subathu from the Siwalik.

The Krol Thrust is superficially a steeply inclined (50°-45°) plane. However, tunnelling in the Tons Valley has revealed that the inclination diminishes with depth to 20°-25° or even less, the surficial steepness being attributed to later movements. The occurrence of Lower Riphean turbidites (the oldest sediments) in the tectonic windows of Bidhalna and Phart in the southern belt (in Tehri District) is eloquent evidence of the flattening of the Krol Thrust. Characterized by shattering and granulation of rocks, the MBT is outlined by perennial springs and straight wide valleys with slopes scarred by landslides.

Schuppen Structure Related to MBT: One of the conspicuous features associated with the MBT is the schuppen structure in the overthrust pile,

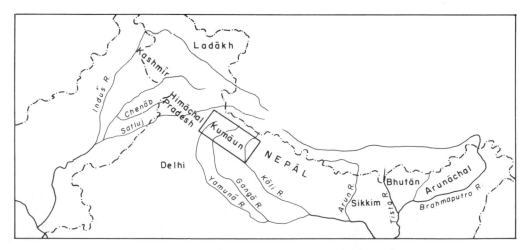

Fig. 1. Location of the Kumaun Himalaya.

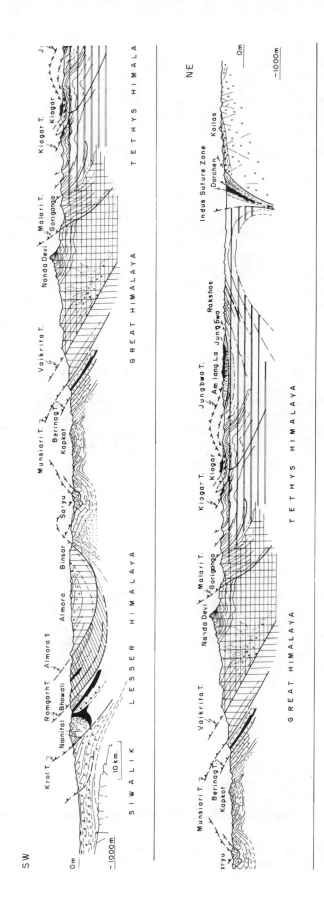

Fig. 3. The generalized cross-section of the Kumaun Himalaya, modified after Heim and Gansser (1939), (Courtesy Jour. Geol. Soc. Ind.). (Slanting lines: - Ramgarh (Chail) unit; Vertical lines: - Almora - Munsiari (Jutogh) unit.) Small circles = molasse; Large dots = quartzarenite; Small dots = sublitharenite and slates; Solid black = basic ultrabasic rocks; Crosses = granites; Fine stipples = Flysch.)

as discernible in the Nainital (Figures 4a & 10) and Lansdowne Hills (Figure 4b). Recent study in the Nainital Hills has demonstrated that the apparently conformable southern flank of the Krol Nappe is thrown into overturned isoclinal folds, with anticlinal folds being split by minor thrusts along crestal planes, resulting in the repetition and inversion of formations. Thus along one of these thrust faults (Manora T.) passing through Hanumangarhi, the Lower Krol has moved over recent talus deposits (Figure 5b). A large number of reverse faults on the northern slopes have given rise to lakes and the Nainital Fault has uplifted the northern block compared to the southern (Figure 10).

In the Lansdowne Hills, the Permian (Tal) has been thrust over the tightly folded and truncated Krol-Tal-Subathu succession (Figure 4b). In the Jaunsar Hills, the Chandpur Thrust has split the otherwise unbroken succession of the Krol belt.

Recent Tectonic Activity along the MBT: A large number of buildings in the villages lying in the proximity of the MBT in the Nainital Hills (Bhalaun-Patkot-Baldiakhan-Naikana) have developed discernible cracks parallelling the trend of the thrust, and the frames of windows and doors have been distorted. The trees are tilted (5°-10°) on the hillslope and landslides are very frequent. In the Nihal River (SW of Nainital) the recent point-bar deposit adjacent to the MBT is tilted. (Figure 5a).

Recent geodetic measurements in the Dakpatthar area in the Tons valley have shown significant ground deformation in the thrust zone. Precise measurements by Ansari and co-workers (1976) over a period of three years demonstrated that the area to the south of the MB (Krol) Thrust is moving southwest compared to the country to the north - the average southerly component being 6.38 mm/year and the westerly component 2.7 mm/year. The Nahan Thrust immediately below the MB

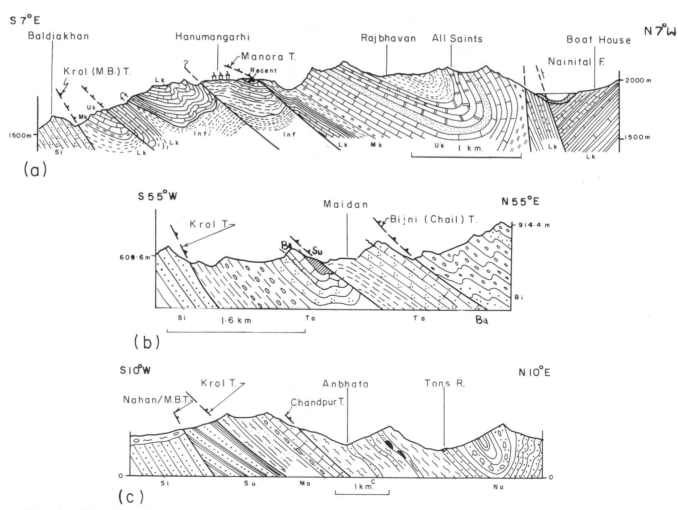

Fig. 4. The schuppen structure related to the Main Boundary (Krol) Thrust: (a) Nainital Hills and (b) Lansdowne Hills. (Inf = Infrakrol; Lk = Lower Krol; Mk = Middle Krol; Uk = Upper Krol; Ba = Bansi; Ta = Tal; Su = Subathu (Lr Eocene); Si = Siwalik; Bi = Bijni (= Ramgarh).

(Krol) T at Kalawar, likewise registered measurable movement; the horizontal component being of the order of 9.02 mm/year in the direction 132° E and the strike-slip component 0.38 mm/year (Sinvhal et al., 1973). In the neighbouring locality, the Subathu shales have been uplifted at the rate of 5.2 mm/year with respect to the Siwalik sandstone across the Nahan Thrust (Jalote, 1978). A sample of the carbonaceous gouge from the thrust zone near Dhamaun gave the age of 38,270 ± 2480 years. Krishnaswamy et al., (1970, see Figure 8) state that the thrust movement of the Eocene (Subathu) over Recent scree and outwash deposits in the Dopaharia stream in the Tons is in the range 75 to 400 metres.

In the Mussoorie Hills, according to Nosin (1971), a number of terraces have been uplifted 150 to 200 metres with respect to the Dun fans across the MBT. Earlier Jalote (1966) had demonstrated southward advance (135 m) of the Lower Palaeozoic Chandpur phyllites over the Upper Pleistocene to subrecent Dun gravels throughout the tract between Dehradun and Rishikesh (Jalote and Mithal, 1971).

Further data on the dynamic activity of the MBT is given in the section on seismicity.

Sedimentary Nappes and Differentiation of the Krol Belt

Existence of the Krol Nappe: The MB (Krol) Thrust has brought a thick succession of Palaeozoic sedimentary formations - Mandhali-Chandpur-Nagthat-Blaini-Krol-Tal (Table 2) - against and over the Siwalik and Tertiary sediments of the Outher Himalaya. The thrust is synclinally folded and its northern flank is known as the Tons Thrust (Auden, 1934, 1937) in the Jaunsar area; in the west, Dharkot Thrust in northern Tehri (Saklani, 1971), and Srinagar Fault in the Alaknanda valley (Mehdi et al., 1972; Kumar and Agarwal, 1975). East of the Dudhatoli massif in northeastern Pauri-Garhwal, the 60-65° SW-dipping North Krol Thrust (NKT) vanishes under the North Almora Thrust delimiting the base of the crystalline nappe. Like its southern flank, the NKT has brought various Palaeozoic formations such as the Mandhali, Chandpur, Nagthat and Blaini against and over the different Precambrian formations - the Chakrata, Rautgara, Deoban and Berinag of the autochthonous zone to the north. The contact is marked by: pronounced structural discordance across the Thrust discernible throughout the western sector (Figure 6), tremendous crushing and granulation of the quartzites, marmorization of the carbonate bands (in Jaunsar and southwestern Tehri), conspicuous crumpling and chevron-folding of the phyllites in the Bhagirathi and Alaknanda valleys, truncation of folds and related structural trends of the autochthon in the Nagun valleys in southwestern Uttarkashi (Jain, 1971) and the occurrence of greatly deformed wedges of the Eocene rocks between the

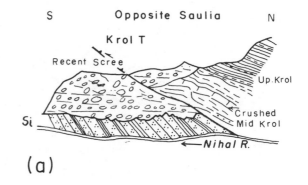

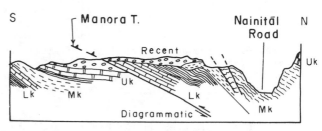

Fig. 5. The Krol Thrust near Saulia, SW of Nainital is active as evidenced by southward advance of Middle Krol over recent point-bar deposit which is tilted. Similarly the Monora Thrust-Fault of the Nainital Hills is active as evidenced by advance of the Lower Krol upon recent scree. (Lk = Lower Krol; Mk = Middle Krol; Uk = Upper Krol; Inf = Infrakrol; Si = Siwalik.)

autochthonous Chakrata Formation and the overthrust Mandhali in the Jaunsar area.

In recent years Ranga Rao (1968), Agarwal and Kumar (1973), Rupke (1974) and Kumar et al., (1974) have raised strong doubts about the very existence of the Krol Nappe, for they see lithological gradation between the Chakrata rocks of the autochthonous and the Chandpur of the Krol unit. However, not only is there a distinct difference in lithology, but the styles and orientations of folds in the two units are at variance, indicating that the two units belong to different structural regimes. The discordance and disharmony of structural trends (Figure 7b, c) are apparent in every section including the valleys of Nayar, Ganga, Nagun, Bhadri, Yamuna and Tons and in the Bidhalna-Phart windows. Cappings of the Subathu upon the Chakrata but below the Chandpur (Figures 6a, 7a) and the evidence of southward-movement of the overlying Chandpur in the Bidhalna and Phart windows in southern Tehri (Jain, 1972) bear further testimony to the validity of Auden's concept (1937) of the allochthonous nature of the Krol Belt and indicate post-early Eocene thrust-

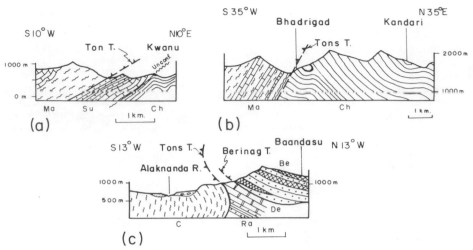

Fig. 6. Pronounced discordance of structures across the North Krol (=Tons) Thrust in (a) Kwanu area, (b) Bhadrigad valley, east of the Yamuna and (c) Alaknanda valley. Different formations of the Krol Nappe have been thrust up against and over various units of the autochthon. (Ma = Mandhali; Su = Subathu; Ch = Chakrata; C = Chandpur; Ra = Rautgara; De = Deoban; Be = Berinag).

ing. In the Jajal area on the Rishikesh-Chamba-Tehri road (Figure 7a), it is the Blaini which rests upon the Chakrata, the intervening Changpur and Nagthat being completely eliminated by thrusting.

Northerly Extension of the Krol - the Berinag Sheet: In the inner Lesser Himalaya the Precambrian argillo-calcareous rocks of the Deoban and/or Mandhali formations are overlain by a thick sequence of quartzarenites with penecontemporaneous basaltic volcanics constituting the Berinag unit. The Berinag Formation of the Purola-Sandra belt extends westward across the Tons valley and skirting around the eastern slopes of the Chaur massif, joins up with the Nagthat Formation of the Krol Nappe (Rupke, 1974; Sharma, 1972). In the Krol Nappe, the Nagthat is underlain successively by the Chandpur and the Mandhali (Table 1) but in the inner Lesser Himalaya its extension, the Berinag, is underlain by the Mandhali without the intervening Chandpur (Figure 8a). Locally even the Mandhali is missing so that the Berinag rests upon the still older Deoban (Figure 8b). In the Someshwar-Swarahat, Rudraprayag and Uttarkashi areas the Berinag succeeds the oldest rock formation - the Rautgara - (Figure 8c) with the Deoban-Mandhali and Chandpur eliminated from the sequence. In eastern Kumaun, the Deoban and Mandhali rocks are inverted and partly truncated (Figure 8d, e) under the Berinag quartzites (Misra and Valdiya, 1961; Valdiya, 1962; 1968) - obviously a result of folding and thrusting of the latter. Shattering and mylonitization of quartz-arenite and basalt corroborate this deduction (Saklani, 1971; Jain, 1971; Pachauri, 1972). Likewise the Blaini, the Krol and the Tal formations that normally succeed the Nagthat

(=Berinag) (Table 2) in the southern tectonic unit are completely absent in the north, as pointed out by Valdiya (1977, 1979a).

It is thus evident that the Berinag sheet represents the extension of the Krol Nappe of which the base and top have been removed. The Krol-Berinag sheets cover a very vast part of the Kumaun Lesser Himalaya.

Faulting of the Krol Nappe and Differentiation of the Krol Belt: The southern flank of the Krol Nappe is considerably affected by long, deep faults, usually developed along the axial or crestal planes of the tight anticlines. The E-W trending, deep Nayar Fault (Auden, 1951) coincides with the crest of an anticline developed in the Krol synclinorial nappe (Figure 9a) which has brought up even the lower Riphean Chakrata turbidites of the autochthon against the various units of the Krol Nappe and has also faulted down the plane of the Krol Thrust along with the Lansdowne syncline in the south. It has thus given rise to the tectonic window of the Nayar-Phart belt. The throw is of the order of 1500 to 3000 metres. A very similar situation exists in the Mussoorie Syncline to the northwest. Here the ESE-WNW trending Aglar Fault (Rupke, 1974) has uplifted the northern block compared to the synclinal Mussoorie Hills and brought constrasting lithologies with different structural disposition in juxtaposition (Figure 9b). Another fault in the Jajal-Ampata area to the northeast has brought up the Chakrata rocks of the basement against the Blaini of the allochthonous unit.

The Nayar-Aglar faults split the Krol Nappe into two distinct halves - the southern Lansdowne-Mussoorie-Jaunsar synclines comprising the full succession (Mandhali-Chandpur-Nagthat-

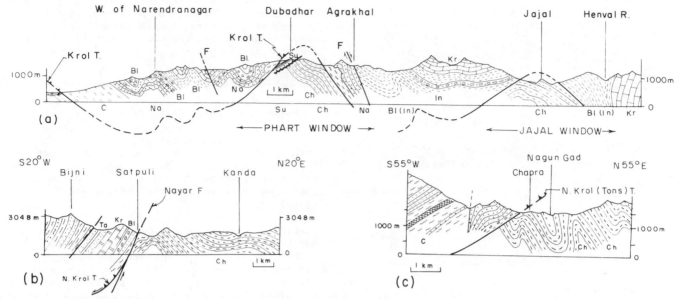

Fig. 7. The evidence for the existence of the Krol Nappe. (a) The Phart and Jajal windows on the Rishikesh-Tehri section, (b) Nayar valley, near Satpui, showing discordance of structural orientation, (c) Nagun valley, S. of Dharasu, District Uttarkashi. (C = Chandpur; B = Blaini; Na = Nagthat; Ch = Chakrota; Kr = Krol; Ta = Tal.)

Blaini Krol-Tal) and the northern Pauri-Nagtibba synclines made up only of the Chandpur and Nagthat formations. The Krol basin was thus divided into two parts as early as the beginning of the Blaini sedimentation, and these faults today delimit the two parts. One of the inferences could be that faulting started in the early Blaini times; that is the faulting was syn-sedimentary. However, I regard the Nayar-Aglar faults as having formed as a consequence of

Table 1. Tectonic Succession in Kumaun Lesser Himalaya

Tectonic Unit	Outer Belt		Inner Belt	
Almora (=Jutogh) Nappe	Almora Group		Vaikrita Group — Main Central (Vaikrita) Thrust Munsiari Formation	
	Almora Thrust		Munsiari Thrust	
Ramgarh (=Chail) Nappe	Ramgarh Group		Bhatwari - Barkot Formations	
	Ramgarh Thrust		Bhatwari - Barkot Thrusts	
Krol Nappe	Sirmur Group	(Subathu Fm (Bansi Fm		
	Mussoorie Group	(Tal Fm (Krol Fm (Blaini Fm		
	Jaunsar Group	(Nagthat Fm (Chandpur Fm	Berinag Formation	
	Mandhali Formation			
	Krol Thrust		Berinag Thrust	
Allochthon	Subathu Fm		Tejam Group	(Mandhali Fm (Deoban Fm
	Damtha Group	(Rautgara Fm (Chakrata Fm	Damtha Group	(Rautgara Fm (Chakrata Fm

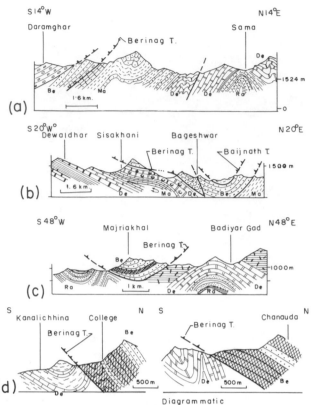

Fig. 8. The Berinag Formation rests on different rock-formations of the autochthon and in some places discordantly. (Ra = Rautgara; De = Deoban; Ma = Mandhali; Be = Berinag.)

strong later folding along the axes and crests of the tightly compressed anticlines and as involving the basement. The differentiation of the basin at the beginning of the Blaini times may be due to folding (not faulting) and to the uplift of the Pauri-Nagtibba upwarp. The considerable thinning of the various units in the northern flank of the Lansdowne-Mussoorie synclines is attributed to the evolution of this upwarp (Rupke, 1974). The windows exposing the autochthon, in the author's opinion, represent anticlinal culminations.

The Nainital Hills along the line of the Mussoorie-Lansdowne Hills are likewise bewilderingly faulted by long, deep reverse faults (Figures 4a and 10). The Nainital Fault has caused the uplift, of the order of 30 m, of the northern Naina-Skerkadanda range compared to the southern Ayarpata block which is an overturned south-vergent syncline. This phenomenon has given rise to the lake. Like the Nayar Fault, the Nainital Fault also joins up with the Krol Thrust. A number of faults seemingly diverge from it, rendering the hillsides very unstable. Similar conditions obtain in the adjoining Bhimtal area. The uplift of the blocks, the fold pattern and the trend of the numerous tear

faults in the region suggest that the southward-directed pressure consequential to the movements along the Krol Thrust, has been responsible for the upward thrusting (Figure 10) and for the great height of the Krol belt mountain rampart overlooking the plains.

Deformation in the Inner Sedimentary Belt

The southern part of the inner sedimentary zone is constituted of autochthonous Precambrian sediments that can be divided into four litho-stratigraphic formations: The Chakrata-Rautgara-Deoban-Mandhali (Table 2). The larger northern part is covered by the allochthonous Berinag sheet, which represents the northerly extension of the Krol Nappe.

Folding in Autochthonous Zone: The autochthon-our belt shows open, upright to overturned folds, locally tightly closed or even isoclinal. In the proximity of the North Almora Thrust the much deformed rocks are compressed into fan-shaped folds. All along the belt from the Kali River to the Tons the folded autochthonous sedimentary succession has been abruptly elevated to a ruggedly high range – much higher than the thrust sheets that rest upon the autochthon. This fact, together with the backfolding of the North Almora Thrust, such as seen in the Saryu valley (Figure 11c) in eastern Kumaun (Valdiya, 1963), suggests that even the autochthon has moved southwards and upward, after the emplacement of the crystalline nappe.

Away from the thrust zone, the folds gradually open up and become asymmetric to overturned or even upright. However in the vicinity of the Berinag Thrust they show even greater deformation. Under the Berinag quartzites and volcanics, the Deoban-Mandhali rocks are tightly or even iso-clinally folded, and locally split up into minor tectonic scales along axial planes. This phenomenon, discernible in the Chandaak-Gangolihat-Ganai belt in the Pithoragarh District and Kanda-Sisakhani belt in the Almora District, has given rise to inversion and repetition of beds (Figures 8b, 11a, b) over tens of kilometres of strikewise distance (Misra and Valdiya, 1961; Valdiya, 1962, 1968).

Like the Krol Nappe, the autochthon is also affected by long, deep, reverse faults. One of the most significant faults recognized and mapped by Kumar (1970) and Kumar and Agarwal (1975) follows the Alaknanda valley between the tract south of Nandprayag and north of Rudraprayag and trends almost E-W (Figure 2). It serves as the boundary fault between the schuppen zones to the north and the overturned isoclinal fold in the Berinag formations to the south. Offsetting indicates that the northern block has moved eastward along the Alaknanda Fault, which is thus a dextral wrench fault, presumably related to the Chamoli schuppen structure.

Table 2. Lithostratigraphy of the Autochthonous Unit and the Krol Nappe in Kumaun

	Sirmur Group	(Subathu Fm (Lower Eocene) (~~~~~~~~~~~~~~~~~~~~~~
		(Bansi (Singtali) Fm (? Up. Cretaceous to Palaeocene) ~~~~~~~~~~~~~~~~~~~~~~~~~~~
Krol Nappe	Mussoorie Group	(Tal Formation (? Permian) (Krol Formation (? Carboniferous) (Blaini Formation
	Jaunsar Group	(Nagthat (= Berinag) Formation (Chandpur Formation
		Mandhali Formation (sliced off from the autochthon)
		Krol Thrust
	Tejam Group	(Mandhali Formation (Up. Riphean to Vendian) (Deoban Formation (Middle Riphean)
Autochthon	Damtha Group	(Rautgara Formation (Lower Riphean) (Chakrata Formation
		(Base Not Seen)

Folding Pattern in the Berinag Nappe: In the inner Lesser Himalaya the Berinag sheet is concordantly folded with the autochthonous formations. Generally the concentric folds are tight

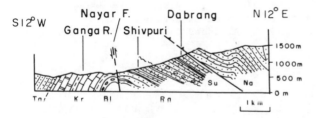

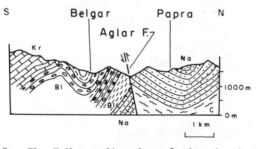

Fig. 9. The E-W trending deep faults in the Nayar and Aglar valleys, developed along the axial crestal plane, have brought up the basement or older rocks in juxtaposition to the different rock formations of the Krol Nappe, and thus pushed down in the southern side of the bounding Krol Thrust. (Symbols as in Figure 7.)

and overturned. The monotonous lithology makes these folds difficult to recognize leading to wrong estimates (Valdiya, 1965) of the great thickness of the Berinag. Another notable feature is that the southern flanks of the folds are vastly thicker than the northern limbs. This is possibly due to repeatedly tight folding of the southern flank in consequence of movement along the Berinag (=Krol) Thrust. In northern Chamoli and western Uttarkashi the deformation has been so intense that the folds were split up into imbricate slabs (Figure 12) along with the concordantly underlying units giving rise to the schuppen structure.

Shuppen Zones Under the Overthrust Sheets: Below the crystalline thrust-sheets (Munsiari or Jutogh unit) the sedimentary succession is involved in multiple, imbricate thrusting giving rise to schuppen structure in the Purola (in western Uttarkashi) and Bhatwari (in northwestern Chamoli) areas as already pointed out by Valdiya (1978). The Purola-Barkot schuppen zone (Figure 12a) comprises five tectonic schuppen and the Chamoli schuppen zone, seven schuppen (Valdiya, 1978; see Figure 6). One of the most striking features of these tectonic schuppen is the occurrence within a succession of epimetamorphic slates, phyllites, quartzites and marble of highly sheared and mylonitized granitic quartz-porphyry, clearly pointing to the great tectonic deformation these granitic rocks have suffered, presumably as a result of multiple thrusting in the root zone of the crystalline

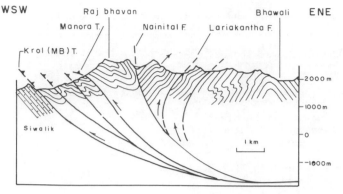

Fig. 10. Diagrammatic section showing that the movement along the Krol Thrust may have been responsible for the formation of the faults of the Nainital Hills, and consequent evolution of the lakes.

nappes. One of the tectonic schuppen of the Purola area extends across the Tons River into the Chail Nappe recognized by Pilgrim and West (1928) in Himachal Pradesh just as the Berinag sheet is relayed across the valley by the Jaunsar tectonic unit of the Chaur-Simla belt (Figure 2).

The lithology, the presence of mylonites and the tectonic position of the Barkot unit of the Purola schuppen (Figure 2 and map) are exactly like those of the Bhatwari and Pokhri sheets of the Chamoli schuppen (also see Valdiya, 1978, Figure 8). The Pokhri sheet, with its metaflysch associated with mylonitized granite-porphyry, extends across the Alaknanda valley into the Pindar valley where it comprises the Rautgara-Deoban-Mandhali succession - the Precambrian formations of the autochton. The lithology of the Pokhri-Bhatwari-Barkot units indeed bears very strong resemblance with that of the Rautgara, the oldest exposed formation of the autochthon in central and eastern Kumaun. Significantly, at the top of the Pokhri succession there are lenses of crystalline magnesite recalling those of the Deoban which normally succeeds the Rautgara in the autochthon. It thus seems that the lithologically identical Pokhri-Bhatwari sheets are constituted of the early Riphean Rautgara rocks that were isoclinally overfolded, split up by imbricate thrusting along axial planes, and pushed up along with the granite-porphyroid emplaced in the base of the formation (Valdiya, 1978). Alternatively, the sheared, mylonitized porphyroid represents the basement that has been caught up in the imbricate thrusting of the schuppen.

Similar schuppen structure, but involving only the younger formations of the autochthon (Deoban and Mandhali) and the Berinag is discernible to the north of Pipalkoti (Pakhi-Tangani-Gulabkoti) in the Alaknanda valley, and also in the Kali Ganga valley (Ghes-Balan), a south-flowing tributory of the Pindar. In the Dhauli-Kali valleys in the extreme east, the Mandhali and Berinag rocks have been thrust up as a wedge (Sirdang Zone) between the isoclinally folded Chhiplakot klippe and its root (Figure 12b) (Valdiya and Gupta, 1972).

It thus seems that the belt immediately under the Main Central Thrust, forming the junction of the Great and Lesser Himalaya, experienced the highest degree of deformation so that the rocks were intensely and tightly overfolded, split up along axial planes into packs of tectonic schuppen as noticeable in the belt west of the Alaknanda, or were cut by conjugate-

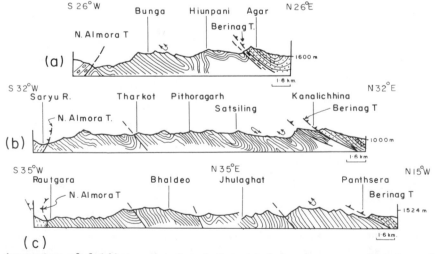

Fig. 11. The intensity of folding and related thrusting increases northward towards the Berinag Thrust in the autochthon. Note the imbrication and inversion of rocks in the proximity of the Berinag Thrust and also the back-folding of the North Almora Thrust and related uplift of the autochthon in comparison to the crystalline of the allochthon. Pithoragarh area, northeastern Kumaun.

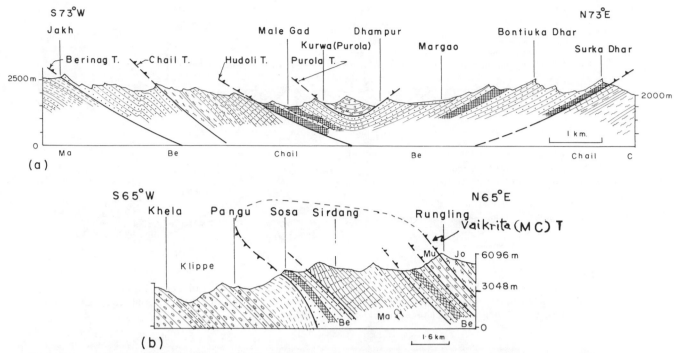

Fig. 12. (a) The schuppen zones of Purola (western Uttarkashi) (b) In northeastern Kumaun the Sirdang zone represents the wedged up autochthon between the overturned isoclinal klippe of the thrust sheet and its much-attenuated to wholly eliminated root, the Munsiari Fm. (Ma = Mandhali; Be = Berinag; Mu = Munsiari; Jo = Joshimath (Vaikrita Group).

ly paired transverse tear faults, as seen east of the Saryu River. The Sirdang zone is really the wedged-up autochthon between the folded crystallines in a partially developed schuppe.

Ramgarh Nappe and Bijni Klippe: In the outer Lesser Himalaya (Dudhatoli-Ranikhet-Almora-Champawat) the vast Krol Nappe is overriden by a thick pile of epimetamorphic flysch, characterized by a voluminous body of much mylonitized granitic quartz-porphyry (Figure 13a, b; see Valdiya, 1978, Figure 3, 4). This epimetamorphic sheet - the Ramgarh - lies under and frames the southern flank of the Almora Nappe of the crystallines. Likewise, in the Lansdowne Hills the crystallines are underlain by a sheet of epimetamorphics, but without the porphyroid (see Valdiya, 1978, Figure 5). It is known as the Bijni nappe (Auden, 1937). In lithology and structural position the Ramgarh and Bijni resemble the Bhatwari-Barkot units. It is therefore deduced (Valdiya, 1978) that the Ramgarh unit represents the far-travelled part of the Chail Nappe, and the Bijni sheet is the distal klippe of this nappe.

In recent years many workers such as Mehdi et al. (1972), Kumar et al. (1974), Saxena (1974), Raina and Dungrakoti (1975), and Shah and Merh (1978) have questioned the existence of the Ramgarh Thrust (Pande, 1950) and its easterly extension (Figure 2 and map) the Ladhiya Thrust

(Valdiya, 1963). The existence of this delimiting thrust is borne out (Valdiya, 1978) by the pronounced discordance of structures of the underlying Nagthat and the Ramgarh unit, discernible throughout the belt (Valdiya, 1978; Figures 5, 9), the persistently wide zone of shattering and crushing of the Nagthat quartzites and pronounced mylonitization and related cataclastic deformation of the granitic porphyroid of the Ramgarh unit throughout its extent, the truncation of the synclinally overfolded Palaeozoic Nagthat-Blaini-Krol succession (Table 2) by the early Riphean porphyroid (1170 ± 20 m.y.; Rhanot et al., 1976) of the Ramgarh as discernible in the Nandhaur valley and the completely different lithologies on the two limbs of the anticlinally folded Nagthat in the Nainital District - the southern limb consists of Nagthat, Blaini and Krol while in the northern limb the same Nagthat is succeeded by a succession radically different from the Blaini-Krol lithology. Likewise in the Lansdowne Hills the older Bijni rests upon the Permian and/or Palaeocene formations (Figure 4b).

It may be emphasized that the attitude of the Ramgarh Thrust varies from place to place. Over a vast tract in southeastern part, it is a thrust plane of low inclination but in the central sector between Kala-agar and Bhatronjkhan, it is a reverse fault between the 20-35° NNE/NE-dipping Ramgarh rocks and 60-80° SSW dipping Nagthat quartzites.

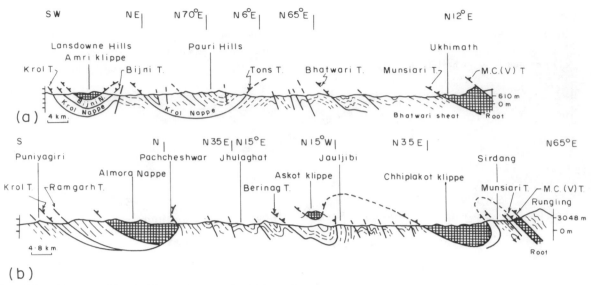

Fig. 13. The synclinal Almora Nappe (gridlines) its klippen and the root in eastern Kumaun Himalaya. The Ramgarh sheet (inclined lines) lies below the Almora Nappe, and is presumably derived from the Rautgara of the autochthon (blank).

Crystalline Nappe and Its Root

East of the Alaknanda valley, the larger part of the Lesser Himalaya in Kumaun is covered with a thick synclinal sheet of mesograde metamorphics intruded by trondhjemitic suite of granites (Figure 2). This is the Almora Nappe (Figure 13) and its three rows of klippen concordantly folded with, and thus preserved in, the synclinal cores of the Berinag sheet; the Amri Klippe (Auden, 1937; Valdiya, 1975) of the Lansdowne Hills in the frontal parts, the Askot-Baijnath-Nandprayag klippen (Heim and Gansser, 1939; Valdiya, 1977; 1979) on the back of the Almora Nappe and the overturned Chhiplakot klippe in the northeastern corner (Figures 2, 13b) almost connected with the root (Gupta and Valdiya, 1972; Valdiya, 1977, 1979a). The folded Almora Thrust delimits the boundary of the crystalline sheet, occupying the highest topographic level (Figure 3) and therefore represents the last large-scale thrust movement. Since the Amri klippe has ridden over the Lower Eocene Subathu (Figure 4b), the age of thrusting is post-Lower Eocene. The Munsiari Formation underlying the Vaikrita Group of the Great Himalaya, and which joins up with the Jutogh of Himachal (Pilgrim and West, 1928), represents the root of the Almora Nappe and its detached pieces, that is, the klippen.

The Almora and Munsiari Thrusts: The Almora Thrust defining the base of the southeasterly plunging nappe is an asymmetrically folded plane. The southern flank, called the South Almora Thrust by Heim and Gansser (1939), dips 20°-30° NNE/NE, while the northern limb is inclined 45-70° SSE/SE or even vertical to overturned as discernible in the Saryu valley (Figures 3, 14)

in the extreme east (Valdiya, 1963). The klippen are likewise delimited by asymmetrically folded thrust planes. The asymmetry of folding increases progressively northward towards the root (Figures 14) so that near the root, the Chhiplakot unit is an overturned isoclinal klippe (Figure 13b). The steepening or overturning of the thrust plane is a consequence of later movements which pushed up the underlying autochthonous sedimentaries against or above the crystallines as seen in the Sirdang Zone.

The steeply inclined North Almora Thrust - which Mehdi et al., (1972), Misra and Sharma (1972), Saxena (1974) and Merh (1977) regard as a reverse fault, is marked by a zone of chaotically crushed and complexly folded Rautgara rocks of the autochthon. A persistent band of phyllonite and mylonite with pronounced retrograde changes, forms the base of the overthrust sheet. In the klippen too, the rocks of the crystalline sheet as well as of the Berinag unit exhibit cataclastic deformation, including mylonitization accompanied by retrograde metamorphism. The existence of the South Almora Thrust has been doubted by Saxena (1974), Saxena and Rao (1975) and Kumar et al., (1974). However, the folowing facts provide proof of the existence of the South Almora Thrust: (i) the change from the green schist facies of the Ramgarh unit to the epidote-amphibolite facies and augen gneiss of the Almora, (ii) the development of three generations of folding (namely, isoclinal, reclined-recumbent NNE-SSW trending earliest fold overprinted by NW-SE open, asymmetic to slightly overturned folds, and NE-SW trending upright plunging folds of the last generation) in the Almora sheet compared to

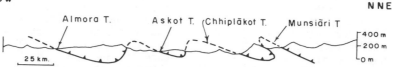

SSW   NNE

Fig. 14. The intensity of folding, reflected in the asymmetry of the crystalline thrust sheet, progressively increases northward until near the root the folds are isoclinal and wedged up along imbricate thrust planes.

only later two generations in the Ramgarh-Berinag unit, (Valdiya, 1963; Das, 1966, Vashi and Merh, 1965; Powar, 1970; Vashi and Laghate, 1972), (iii) the morphological contrast in the garnets of the Ramgarh and Almora rocks (compared to the syntectonically formed garnet with a helicoidal pattern of inclusions of the Almora schists, the Ramgarh garnet is euhedral and undeformed), (iv) the presence of a band of phyllonite along the thrust zone, (v) structural discordance seen in the Askot-Baijanath klippen and, (vi) in the case of Amri klippe of the Lansdowne Hills, the overstepping of the crystallines on the Permian and/or Tertiary sediments.

The Root Zone: The low-angle (10-30°) tectonic boundary plane between the sedimentary rocks (Berinag or Mandhali) of the parautochthon and the crystalline rocks of the Munsiari Formation at the base of the Great Himalaya (Figures 3, 13) was described as the Main Central Thrust by Heim and Gansser (1939). However, Munsiari Thrust is a better name for it, for not only it does not mark the boundary of the Lesser and Great Himalaya (Valdiya, 1979b) but it extends westwards into the Jutogh Thrust of Himachal Pradesh, and is certainly not the continuation of the MCT identified by Bordet (1973), Hashimoto et al., (1973) and Pecher (1975, 1977) in the adjoining western Nepal. The real MCT, which I locally and tentatively designated as the Vaikrita Thrust (Valdiya, 1977, 1979), constitutes the boundary between the mesometamorphic Munsiari (=Jutogh) and the higher grade Vaikrita Group forming the bulk the Great Himalaya. Sandwiched between the Vaikrita (MC) Thrust above and the Munsiari Thrust below, the crystalline rocks of amphibolite facies of the Munsiari Formation are intruded by porphyritic granite. The wholesale and intense cataclastic deformation has converted a sizeable part of the Munsiari into mylonitized rocks, including augen gneiss with sheared and fragmented felspars particularly in the proximity of the Munsiari Thrust.

As already pointed out, below the Munsiari Thrust the sedimentary sequence is intensely deformed, giving rise to schuppen structure as prominently seen in western Uttarkashi and northwestern Chamoli and in the Pipalkoti area. These schuppen also involve the Munsiari crystalline rocks in the imbricate repetitions of the formations. The Munsiari Thrust is also responsible

for the truncation, attenuation and finally elimination of the Berinag Formation in the northern belt.

Geodynamics of the Root Zone: Judging from the scale and intensity of deformation suffered by the rocks in the proximity of the Munsiari Thrust, and appreciating the fact that its squeezed out and far-travelled part - the Almora Nappe and the chains of klippen - register a width of thrusting of the order of 120 km or so, it is quite obvious that the root zone had experienced severe compression and tectonic dynamism. The age of the granite gneiss of the Munsiari Formations near Kalamuni has been dated at 1895 ± 100 m.y. and of the Askot crystallines near Didihat at 1960 ± 100 m.y. (Bhanot, V. P., et al., 1977) and the leucocratic granite of the Almora unit near Almora, approximately at 700 m.y. (per. com. V. B. Bhanot). The granitic rocks thus point not only to the Almora-Munsiari rocks being of early Precambrian age (considerably older than the oldest of the sedimentary formations the Lower Riphean Chakrata and Rautgara of the autochthon) but also to the lithotectonic unity of the Munsiari and Askot-Baijnath crystallines. Possibly these crystallines constitute the basement on which the sedimentary rocks of the Lesser Himalaya were deposited. Strong deformation in the northern extremity of the basin and consequent thrusting along the Munsiari and Vaikrita (MC) thrusts may have uplifted and pushed southward the superficial part of the postulated basement. It must be admitted, however, that there is no evidence in support of this speculation, and hence it remains in the realm of possibility.

Junction of Great and Lesser Himalaya: It has already been stated that the Vaikrita Thrust recognized by the author (Valdiya, 1977, 1979a) constitutes the real boundary between the Great and Lesser Himalaya. There is no perceptible discordance in the dip of the strata, although the style and orientations of the first-generation folding are quite different across the dividing plane. However the change in the grade of metamorphism is abrupt. The epidote-almandine-amphibolite facies rocks of the Munsiari give way to higher-grade metamorphics characterized by such minerals as kyanite and sillimanite in the psammitic gneisses and diopside, vesuvianite and idocrase, in the calc-granulites. The 30-45°

northward heading thrust is associated with a line of hot springs in the region, and this testifies to its being a deep thrust. In my opinion (Valdiya, 1979b) the Vaikrita Group represents the basement of the Tethyan sedimentary pile, just as the Munsiari rocks possibly formed the infrastructure of the Lesser Himalayan sediments. Part of the seismicity of the belt may be attributed to movement along the MC(V)T, and partly (to a much smaller extent) to that along the Munsiari Thrust.

Tectonics of the Great Himalaya

The Great Himalaya (Figure 15) consists of katazonal high-grade metamorphics (Vaikrita Group) intruded by Tertiary granite, and is characterized by flowage folds of usually intrastratal dimension. The Great Himalaya represents a huge (~10,000 m thick) homoclinal tectonic slab (Figure 3) demarcated by the moderately dipping (30-45°) Main Central (Vaikrita) Thrust in the south and the steeply dipping Malari Thrust in the north (Valdiya, 1979a). Between the two tectonic planes the Vaikrita Group represents the upthrusted basement of the Tethyan sediments, the net throw being of the order of 20 km.

<u>Southern Tectonic Boundary</u>: As already demonstrated, the MCT demarcates the boundary between the two radically different lithotectonic realms,

the Lesser Himalaya and the Great Himalaya (Figure 3). The junction is marked by a thick zone of extreme deformation and mylonitization, and development of schuppen structure as witnessed in the Alaknanda-Bhilangana (in northern Chamoli) and Yamuna-Tons (western Uttarkashi) triangles (Figure 2). Significantly, the hot springs of Kumaun lie in the proximity of the MCT, such as seen at Dar (Dhauli River), Madhyamaheshwar, Gaurikund (Mandakini R.), Yamunotri (Yamuna R.), Karcham (Sa Huj R.), etc. (Figure 15). A similar situation exists in north-central Nepal. The thermal springs are further pointers to the deep-seated nature of the MCT.

<u>Northern Boundary</u>: The Malari Thrust (Figure 3, 15) sharply cuts the Great Himalayan Vaikrita rocks from the Tethyan sedimentary succession, Kumar et al., (1972) and Shah and Sinha (1974) describe this as a fault. Normally the Vaikrita metamorphics transitionally grade upwards into Proterozoic sediments of the Tethyan succession in the Himalaya (e.g. in Spiti (Himachal), Dhaulagiri (Nepal), etc.). However, in the Kali valley, the steeply dipping thrust has considerably attenuated and sheared off the Martoli flysch of the sedimentary unit. In the western Dhauli valley the Martoli flysch rests discordantly over the sheared migmatized Vaikrita rocks.

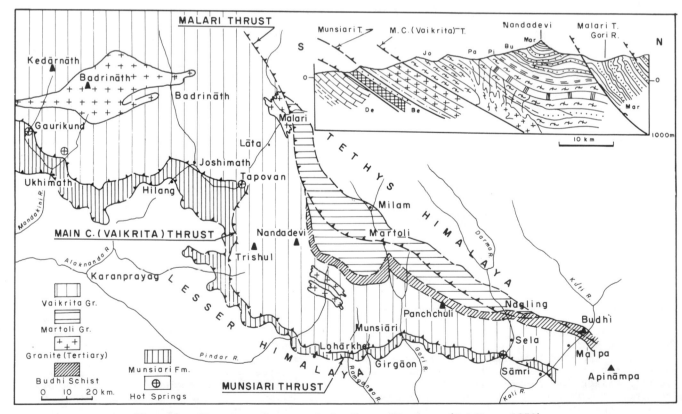

Fig. 15. The tectonic map of the Great Himalaya (Valdiya, 1979).

A couple of sympathetic faults and thrusts have developed in the overthrust succession in the Girthi valley. Gravitational sliding along these faults have given rise to spectacular north-vergent back-folds. In the Gori valley the Martoli shows sigmoidal folding of the most complicated type in the proximity of the Malari Thrust.

Neotectonic Activity: Between the two delimiting thrust planes the Great Himalayan block is rising in elevation as evidenced by the tilting of recent river deposits such as at Malpa in the Kosi valley and near Sela and Nagling in the Dhauli valley (Figure 15), convex to vertical slopes of the valley walls, and rapids and cascades in the river beds, etc. Interestingly, the maximum strain-energy release related to the MCT is observed only in Kumaun (Verma et al., 1977). The seismic plane, with an inferred dip of 40° NE, which Kaila and Narain (1976) have recognized in Kumaun sector, coincides with MC (Vaikrita) T and not with the Munsiari Thrust (15-20°) as surmised by them. It thus seems that the MC(V)T is serving as the plane of resurrection movement of the Great Himalaya. The movement is however slow compared to that on the MBT.

The deep-seated nature of the Vaikrita rocks, and the great depth of the MCT is also indicated by the occurrence of post-tectonic leucocratic young Tertiary (20-25 m.y.) granite which has not only extensively injected and migmatized the Vaikrita, but occur as great batholithic bodies also, such as the Badrinath Granite. The granite is post-orogenic and may represent the fusion of the deep, sunken part of the frontal edge of the Indian shield which became involved in the Himalayan revolution.

Transverse Structures

One of the remarkable features of the tectonic architecture of the Kumaun Himalaya is the existence of transverse folds and faults, affecting all the four lithotectonic subprovinces. They have been described comprehensively by the author (Valdiya, 1976). It would suffice here to highlight the salient features.

Transverse Folds: Transverse (upright folds plunging and oriented NE/NNE-SW/SSW) are developed in all the four lithotectonic realms, and thus constitute the youngest set of folds (Figure 16). They have affected even the MBT, which has brought the Lesser Himalayan rocks over sediments as young as the Upper Siwalik. These upright folds are unrelated to the older set of recumbent to reclined folds trending almost in the same direction, namely NNE/NE/ENE-SSW/SW/WSW, but which are discernible only in the overthrust crystalline sheet, the Almora Nappe, its klippen and the root. The younger folds are also unrelated to the "Aravalli Structures" in the pre-

Krol rocks pointed out by Auden (1935).

The younger transverse folds of larger dimension have given rise to low-amplitude domal and basinal structures in the autochthonous belt in the inner Lesser Himalaya, immediately south of the root of the crystalline nappe. The examples of low-amplitude domes are seen in the valleys of Mandakini, Alaknanda (Pipelkoti area), Kali Ganga-Gyanganga (N of Pindar), Saryu (Kapkot area) and Eastern Ramganga (Tejam area). The changes in the strike of the Chakrata rocks (the base of the autochton) in the Nayar - Ganga valleys in southern Pauri are attributed to the superimposition of the younger folds on the earlier set.

The transversal structure of the Badarinath area in the Great Himalaya and in Malla Johar in the Tethys zone involving the obducted ultrabasic sheet and melange (Heim and Gansser, 1939) are manifestations of this youngest folding in the far northern belts.

Faults and Fractures: Much more conspicious than the folds are the transverse (NNE/N-SSW/S) and conjugately related oblique (NNW/NW-SSE/SE) faults (Figure 16) and fractures that dissect the Lesser Himalaya rather extensively (see Figures in Valdiya, 1976). Cutting across and dextrally offsetting the thrust planes, including the young M.B.T., the faults have caused a considerable crushing, shearing and displacement of rock formations, particulary in the southern belt. The faults that have right-laterally displaced the M.B.T. by as much as 10 to 12 km have dissected the Siwalik belt as well. In the outer Lesser Himalaya some of the faults coincide with the thrust planes delimiting the crystalline Almora Nappe and its klippen, such as the Jamarcheura Fault in the Ladhiya valley, the Raintoli Fault in the southern reaches of the Saryu valley, the Chaukhutia Fault in the Dwarahat area, the Bhikiasen Fault in the Binau-Naurar valley, the Raitpur Fault in the Medi Gad valley in the Lansdowne Hills, etc. This fact has let many a worker to deduce that the thrust planes defining the base of the crystalline sheets are deep faults.

In the autochthonous belt in the north, not only the overthrust crystallines have been truncated by the NNW/NW-SSE/SE trending faults, but there are bodies of very coarse, porphyritic granite within the autochthon in the proximity of the faults. The granite bodies are highly sheared and locally crushed as a result of movement along the fault, as witnessed in the Gori valley between Toli and Chipaldera and between Baram and Baikot, and in the Senduna valley, south of Gopeshwar in Chamoli (Figure 2). Possibly these faults brought up the granites of the basement against the younger rocks of the autochthon.

In the northwestern corner, the Arakot Fault in the Pabar valley registering right-lateral

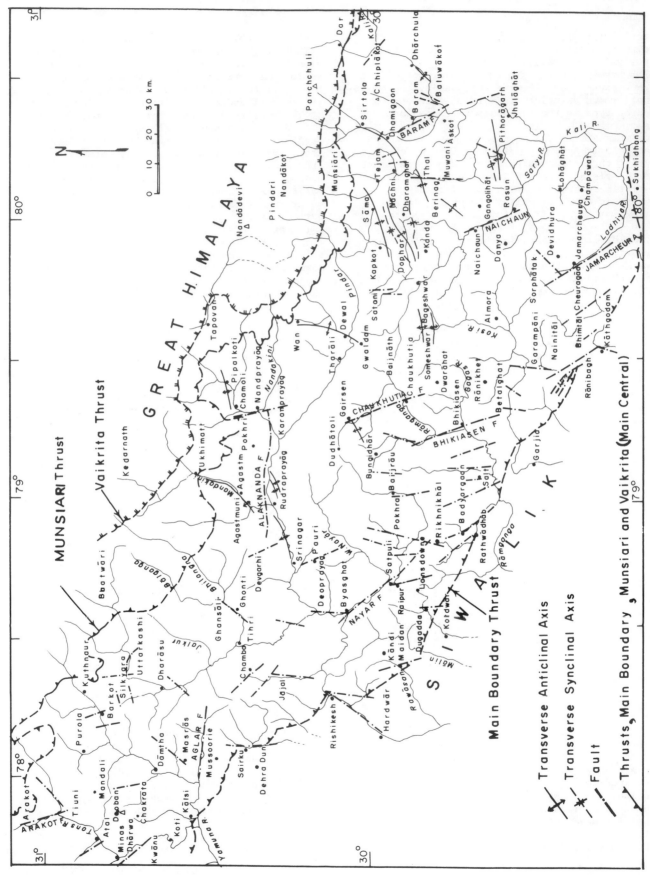

Fig. 16. The distribution of transverse and conjugately related tear faults registering dominant dextral displacement (Valdiya, 1976).

shear movement is responsible for the evolution of a mini syntaxial bend.

Parallelism with the Structures of the Peninsular India Block: Significantly, the transverse faults and fractures of the Lesser and Outer Himalaya demonstrate notable parallelism with the basement of the Ganga Basin and in the adjoining Peninsular Indian block. The Muradabad Fault, the Tilhar-Datarganj Fault, and the Sohna-Sonipat faults that frame the Aravalli horst may be mentioned (Figures 18).

The Himalayan transverse folds show broad parallelism with the hidden ridges recognized in the basement of the Ganga Basin, such as the subterranean Bundelkhand-Faizabad Ridge and the Delhi-Haridwar Ridge which is the underground prolongation of the Aravalli Range.

The remarkable parallelism of the folds and faults in the Lesser Himalayan and the basement of the Ganga Basin taken in conjunction with stratigraphical similarity of the Lesser Himalaya and the Vindhyan Basin (Valdiya, 1964, 1969, 1975) cannot be dismissed as fortuitous. They point to the stratigraphic and structural unity of the two geological subprovinces.

Seismicity of the Kumaun Himalaya

Northeastern Kumaun: The northeastern part of Kumaun (Dharchula-Kapkot belt) and adjoining northwestern Nepal (Bajang) (Figure 17) are frequently rocked by earthquakes of magnitudes between 5 and 6. The earthquakes on May 21, 1979 and July 29, 1980, were of the scale 5 to 6,

while the strongest so far recorded was 7.5 on October 28, 1916, all located in the Bajang area in NW Nepal. The quantitative seismicity map of the Himalaya (Kaila and Narain, 1976) shows that this part of the Himalaya has the highest seismicity anywhere in the Himalayan arc with a value above 6 (Figure 18). The strain-energy release map based on the data concerning the period between 1900 and 1970 (Verma et al., 1977) confirms this fact. The general scatter of the epicentres of the shallow earthquakes over a wide belt (Figure 17a) trending NW-SE precludes the possibility of the relation of the earthquakes with any of the thrust planes, such as the Munsiari Thrust or the Main Central (Vaikrita) Thrust. Since there is no suggestion of progressive northward increase in the depth of foci (Figure 17b) as one would expect if the movement were to take place on the N-dipping plane of the MCT or for that matter on any other thrust, it can be surmised that the high seismicity of northeastern Kumaun is not related much to the movement along the MCT or Munsiari Thrust. This is in contradiction of the deduction of Srivastava (1973), Rastogi (1974) and Molnar et al., (1977) whose fault-plane mechanism solutions indicate predominant thrust faulting with pressure directions acting perpendicular to the orographic trend, and tensions oriented parallel to it. While the earthquake of 3rd March, 1969, resulted from dip-slip faulting, the 27th June, 1966 earthquake originated from a fault that registered strike-slip movement as well. It is noticed that there is a predominance of earthquakes resulting from strike-slip faulting

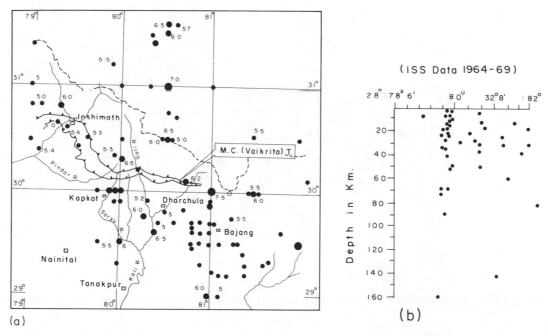

Fig. 17. (a) Spatial distribution of spicentres of (shallow) earthquakes; (b) Vertical distribution of foci (after Chaudhary et al., 1974).

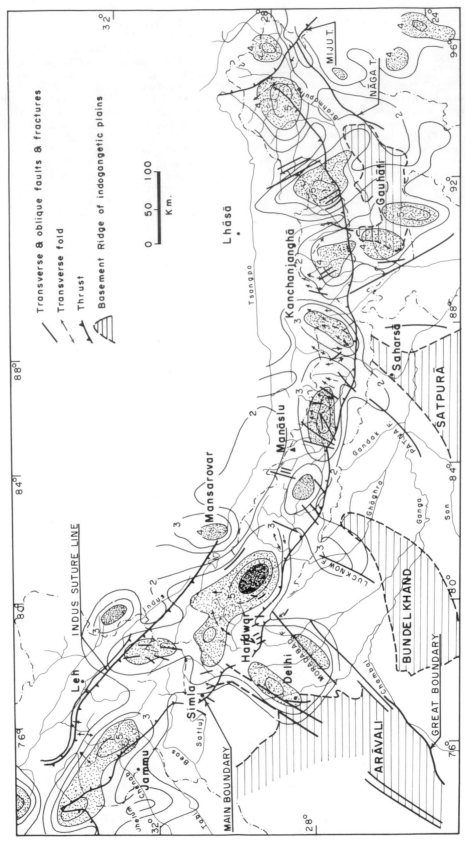

Fig. 18. Quantitative seismicity map of the Himalaya showing transversal trends. Note that Kumaun is seismically the most active part of the Himalayan arc (after Valdiya, 1976, and Kaila and Narain, 1976).

throughout the Himalaya (Srivastava, 1973), which is inconsistent with the notion of the Lesser Himalayan plate slipping under the Great Himalayan mass. This is the situation in the inner (northern) belt of the Himalaya. Possibly both strike-slip and dip-slip movements are taking place (Chauhan, 1975) along the Himalayan faults and thrusts.

The quantitative seismicity map of Kaila and Narain (1976) shows a conspicuous transversal northeasterly trend spanning the high seismicity belt of Delhi with that of northeastern Kumaun (Figure 18). The unmistakably linear distribution in a northerly direction of the epicentres in the Dharchula area is suggestive of tear movement along the transverse faults, presumably concomitant with the strike-slip movement along the thrust planes. Not only the number of earthquakes is higher but the depth of foci is also comparatively greater (33 to 60 km) in the Bajang-Dharchula area, characterized by the tightly compressed synclinal Chhiplakot crystalline klippe - and wedging up to the autochthonous base in the Sirdang belt. And this zone is cut by a number of transverse and oblique faults (Figure 16) such as the one along the Gori River from Baram to Baikot (Valdiya, 1976) and near Baluwakot in the Kali valley. It is significant that this area lies along the line of the active Muradabad-Datarganj faults in the basement of the Ganga Basin and that the December, 1966 earthquake, originating in the Dharchula area, badly shook the townships of Muradabad and Pilibhit in the Plain.

Humla valley in Tethys Himalaya: Another high seismicity area is discernible in the Gurla Mandhata area in the India-Nepal-Tibet trijunction over the terrain drained by the Humla Karnali (see Figure, Valdiya, 1976). The river follows a wide gravelly alluvial-filled valley dissected by the NW-SE trending tear fault that has caused a considerable left-lateral offsetting of the MC (Vaikrita) Thrust (Valdiya, 1976). The faulting is possibly related to the schuppen structure in the Tethyan sedimentary succession of the Kuti-Kalapani belt. Interestingly, the alluvial gravels have been perceptibly displaced (A. Gansser, per. com., 1978), presumably by the fault movement. I attribute the high seismicity of this area to the movement along this tear fault.

Seismicity of the Southern Border: The occurrence of earthquakes is less frequent in the outer belt of the Himalaya, but their magnitude is comparatively higher. Fault-plane mechanism solutions of six earthquakes in the whole of Outer Himalaya indicate that the movement is mainly dip-slip, the dip angle ranging from 30° to 70° (Chaudhury and Srivastava, 1976). The movement may be associated with the Main Boundary Thrust, which as already demonstrated, shows signs of recent tectonic activity. This

corroborates the deductions of Le Fort (1975) and Molnar et al., (1977) that the movement of the plates has now largely shifted from the MCT to the MBT.

Thrust Movement Versus Tear Movement: To those who ascribe the evolution of the Himalaya to the convergence of the Indian and Asian continuents, it has been tempting to associate the seismicity of the Himalayan region with movement along either the Main Central Thrust (Kaila and Narain, 1976) or the Main Boundary Thrust (Le Fort, 1975; Molnar et al., 1977). Pointing out the remarkable parallelism of the transverse structures of the Himalaya (Valdiya, 1973) with the subsurface ridges and faults of the basement of the adjoining Ganga Basin, the author suggested that the seismicity of the Himalayan region, particularly Kumaun, is related to the strike-slip movement along some of the transverse faults of these two lithotectonic provinces (Valdiya, 1976). Striking confirmation has come from the quantitative seismicity map prepared by Kaila and Narain (1976). Impressed by the conspicuous transverse trends of high seismicity, they too are of the opinion that the occurrence of earthquakes of the eastern Himalaya is most probably controlled by the transverse faults and fractures that extend far into Tibet. Therefore in my opinion the origin of the recurring and shallow earthquakes should be attributed to the neotectonic activity of the many transverse and oblique faults registering predominantly dextral displacement in Kumaun.

Impact of the Drifting Indian Block: The central sector of the Himalayan arc - the Kumaun-Nepal border - appears to be experiencing maximum pressure of the drifting Peninsular India, as is manifest in the higher seismicity of this part compared to the other. That this sector has borne the maximum brunt of the drifting Peninsular block is also evident from the fact that the crystalline basement has been warped up giving rise to the Gurla Mandhata massif and Rakshas Tal high amidst Tethyan sediments (Heim and Gansser, 1939). The obducted ultrabasic rocks and melange of the Kailas-Mansarovar belt in Tibet have travelled over 80 km to the south from their source (Gansser, 1974).

Between the long strike-slip faults that limit the Himalayan arc to the east and to the west and which show respectively dextral and sinistral movement, the fault-dissected Indian block together with its Himalayan front is moving northwards - the various segments registering different rates of movement. Simultaneously, there is underthrusting of the southern belts along the Main Boundary Thrust. Thus the crustal accommodation consequent on the convergence of the Indian and Asian plates is being accomplished not only by left-lateral strike-slip movement along the many E-W trending faults of the Tibetan plateau as demonstrated by Molnar and Tapponier

(1975) but also by dominantly strike-slip movement on many transverse fault of the Himalayan
region.

Acknowledgement. The work, started during my
tenure at the Wadia Institute of Himalayan
Geology, was continued with the extremely gener-
ous financial support of the Department of
Science and Technology, Government of India (on
Great Himalaya) and the University Grants
Commission (on Nainital Hills). I am deeply
grateful to the DST and the UGC for their kind
assistance. The draft was critically reviewed
by Dr. F. Delany, to whom I am very grateful.
The manuscript was typed by P. B. Ghansyal.

References

Agarwal, N. C. and Gopendra Kumar, Geology of
the upper Bhagirathi and Yamuna valleys,
Uttarkashi District, Kumaun Himalaya,
Him. Geol., 3, 1-23, 1973.

Ansari, A. R., R. S. Chugh, H. Sinvhal, K. N.
Khattri and V. K. Gaur, Geodetic deter-
mination of earth strain and creep on the
Krol Thrust in the Dakpathar area, Dehradun
District, U. P., Him. Geol., 6, 323-337,
1976.

Auden, J. B., The Geology of the Krol Belt,
Rec. Geol. Surv. Ind., 67, 357-454, 1934.

Auden, J. B., Transverses in the Himalaya,
Rec. Geol. Surv. Ind., 69, 123-167, 1935.

Auden, J. B., The structure of the Himalaya
in Garhwal, Rec. Geol. Surv. Ind., 71,
407-433, 1937.

Auden, J. B., The bearing of geology on multi-
purpose projects. Proc. Ind., Sci. Congr.,
Section 5, 109-153, 1951.

Bhanot, V. P., A. K. Bhandari, V. P. Singh
and A. K. Geol, The petrographic studies and
the age determination of Koidal gneiss,
Kumaun Himalaya, Current Science, 45, 18,
1976.

Bhanot, V. P., B. K. Pandey, V. P. Singh and
V. C. Thakur, Rb-Sr whole-rock age of the
granitic gneiss from Askot area, eastern
Kumaun and its implication on tectonic
interpretation, Him. Geol., 7, 118-122,
1977.

Bordet, P., On the position of the Himalayan
Main Central Thrust within Nepal Himalaya,
Proc. Seminar Geodynamics Himalayan Region,
N.G.R.I. Hyderabad (India) 148-155, 1973.

Chaturvedi, A., L. S. Srivastava and R. S.
Mithal, Tectogenesis and seismicity of the
Kumaun Himalaya, Him. Geol., 3, 336-344,
1973.

Chaudhury, H. M., and B. N. Srivastava,
Seismicity and focal mechanism of some recent
earthquakes in northeast India and neighbour-
hood, Annali di Geofisica, 29, 41-57, 1976.

Chauhan, R. K. S., Seismotectonics of Delhi
region, Proc. Ind., Nat. Sci. Acad., 41, 429-
447, 1975.

Das, B. K., The study of metamorphics and struc-
ture of the Chaukhutia area, Almora District,
U.P., Ph. D. Thesis Banaras Hindu University
Press, Varanasi, 1966.

Gansser, A., The ophiolitic melange, a worldwide
problem on Tethyan examples, Eclog. Geol. Helv.
67, 479-507, 1974.

Hashimoto, S., Y. Ohta and C. Akiba, (Eds) Geology of the Nepal Himalayas, Himalayan Comm.,
Hokkaido Univ., Sapporo (Japan), 286p, 1973.

Heim, A., and A. Gansser, Central Himalaya,
Geological Observation of the Swiss Expedition
in 1936, Mem. Soc. Helv. Sci. Nat., 73, 1-245,
1939.

Jain, A. K., Stratigraphy and tectonics of
Lesser Himalayan region of Uttarkashi, Garhwal
Himalaya, Him. Geol., 1, 25-58, 1971.

Jain, A. K., tructure of Bidhalna-Pharat windows Thrust Unit Garhwal, U. P., Him. Geol.,
2, 188-205, 1972.

Jalote, P. M., Some observations on recent move-
ments along the Krol Thurst, Rajpur, Dehradun
District. Proc. Third Symp. Earthquake
Engineering, Roorkee Univ., 455-458, 1966.

Jalote, P. M., In: Geological Survey of Indian
News, 9(8), 5, 1978.

Jalote, P. M., and R. S. Mithal, Geological and
tectonic evolution of the recent activity
along the Krol Thrust in the Dun valley,
Jour. Engineering Geol., 6, 42-428, 1971.

Kaila, K. L., and H. Narain, Evolution of the
Himalaya based on seismotectonics and deep
seismic soundings, New Delhi Seminar on Him.
Geol. (September), Sp. Publ. N. G. R. I.,
Hyderabad, 30p., 1976.

Krishnaswamy, V. S., S. P. Jalote and S. K.
Shome, Recent crustal movements in Northwest
Himalaya and the Gangetic foredeep and re-
lated pattern of seismicity, Proc. 4th Symp.
Earthquake Engg., Roorkee Univ., Roorkee,
419-439, 1970.

Kumar, G., Geology and sulphide mineralization
in the Pokhri area, Chamoli District, Uttar
Pradesh, In: Base Metals: Part I, Misc. Pub.
Geol. Surv. Ind., 16, Calcutta, 92-98, 1970.

Kumar, G., and N. C. Agarwal, Geology of the
Srinagar-Nandprayag area (Alaknanda valley),
Chamoli, Garhwal and Tehri Garhwal Districts,
Kumaun Himalaya, U.P., Him. Geol., 5, 29-59,
1975.

Kumar, G., S. R. Mehdi and G. Prakash, A review
of the stratigraphy of parts of Uttar Pradesh
Tethys Himalaya, Jour. Palaeont. Soc. Ind.,
15, 86-98, 1972.

Kumar, G., G. Prakash and B. Dayal, A note on
the cement-grade limestone bands in Calc Zone
of Tejam, Pithoragarh District, U.P., Indian
Minerals, 24, 123-130, 1970.

Kumar, G., G. Prakash and K. N. Singh, Geology
of the Deoprayag-Dwarahat area, Garhwal
Himalaya, U.P., Him. Geol., 4, 323-347,
1974.

LeFort, P., Himalayas: the collided range:
present knowledge of the continental arc,

Am. Jour. Sci., 275-A, 1-44, 1975.

Mehdi, H. S., G. Kumar and G. Prakash, Tectonic evolution of eastern Kumaun Himalaya: a new approach, Him. Geol., 2, 481-501, 1972.

Merh, S. S., Structural studies in the parts of Kumaun Himalaya, Him. Geol., 7, 26-42, 1977.

Misra, R. C. and R. P. Sharma, Structure of the Almora Crystallines, Lesser Himalaya: an interpretation, Him. Geol., 2, 330-341, 1972.

Misra, R. C. and K. S. Valdiya, The Calc Zone of Pithoragarh, with special reference to the occurrence of stromatolites, Jour. Geol. Soc. Ind., 2, 78-90, 1961.

Molnar, P., W. P. Chen, T.J. Fitch, P. Tapponnier, W.E.K. Warsi and F. T. Wu, Structure and tectonics of the Himalaya: brief summary of relevant geophysical observations, in Himalaya, Science de la Terra, C.N.R.S., Paris 7, 269-294, 1977.

Molnar, P., and P. Tapponnier, Cenozoic tectonics of Asia: effects of a continental collision, Science, 189, 419-426, 1975.

Nossin, J. J., Outline of the geomorphology of the Doon Valley, northern U.P., India, Zeit. Geomorph. N. F. 12, 18-50, 1971.

Pachauri, A. K., Stratigraphy correlation and tectonics of the area around Purola, Uttar-Kashi and Dehradum District U.P., Him. Geol., 2, 370-387, 1972.

Pande, I. C., A geological note on the Ramgarh area, District Nainital, U.P., Quart. Jour. Geol. Min. Metal. Soc. Ind., 27, 15-23, 1950.

Pecher, A., The M.C.T. of the Nepal Himalaya and the related metamorphism in the Modi Khola cross section, Annapurna Range, Him. Geol., 5, 115-131, 1975.

Pecher, A., Geology of the Nepal Himalaya: deformation and petrography in the M.C.T. zone, in: Himalaya, Science de la Terre, C.N.R.S., Paris, 301-318, 1977.

Pilgrim, C. E. and W. D. West, The structure and correlation of the Simla rocks, Mem. Geol. Surv. Ind., 53, 1-140, 1928.

Powar, K. D., Multiphased mesoscopic folding in the metasediments of Almora area, Kumaun Himalaya, Pub. Centre. Adv. Stud. Geol., Panjab Univ., Chandigarh, 61-67, 1970.

Raina, B. N., and B. D. Dungarkoti, Geology of the area between Nainital and Champawat, Kumaun Himalaya, U.P., Him. Geol., 5, 1-28, 1975.

Ranga Rao, A., On the Krol Nappe hypothesis, Jour. Geol. Soc. Ind., 9, 153-158, 1968.

Ranga Rao, A., Traverses in the Himalaya of Uttar Pradesh, Misc. Pub. Geol. Surv. Ind., 15, 31-44, 1972.

Rastogi, B. K., Earthquake mechanisms and plate tectonics in the Himalayan region, Tectonicphysics, 21, 47-56, 1974.

Rupke, J., Stratigraphic and structural evolution of the Kumaun Lesser Himalaya, Sedimentary Geology, 11, 81-265, 1974.

Saklani, P. S., Structure and tectonics of the Pratapnagar area, Garhwal Himalaya, Him. Geol., 1, 75-91, 1971.

Saxena, S. P., Geology of the Marchula-Bhikiasen area, District Almora, Uttar Pradesh, with special reference to the South Almora Thrust, Him. Geol., 4, 630-647, 1974.

Saxena, S. P. and P. N. Rao, Does Almora Nappe Exist? Him. Geol., 5, 169-184, 1975.

Shah, O. K., and S. S. Merh, Structural geology and stratigraphy of Bhimtal-Bhowali area in Kumaun Himalaya - a reinterpretation, Jour. Geol. Soc. Ind., 19, 91-105, 1978.

Shah, S. K., and A. K. Sinha, Stratigraphy and tectonics of the "Tethyan" Zone in a part of western Kumaun Himalaya, Him. Geol., 4, 1-27, 1974.

Sharma, Ram P., Elucidation of structure and stratigraphy of Deoban Belt, north of Tons River in Himachal Pradesh Himalaya, Appreciation Seminar Indian Photo-interpretation Institute Dehradun, (Abstract), 10-11, 1972.

Sinvhal, H., P. N. Agarwal, G. C. P. King and V. K. Gaur, Interpretation of measured movement at a Himalayan (Nahan) Thrust, Geophys. Jour. Royal Astr. Soc., 34, 203-210, 1973.

Srivastava, H. N., The crustal seismicity and the nature of faulting near India-Nepal and Tibet trijunction, Him. Geol., 3, 381-393, 1973.

Valdiya, K. S., An outline of the stratigraphy and structure of the southern part of the Pithoragarh District, U.P., Jour. Geol. Soc. Ind., 3, 27-48, 1962.

Valdiya, K. S., The stratigraphy and structures of the Lohaghat subdivision, District Almora U.P., Quart, Jour. Geol. Min. Metal. Soc. Ind., 35, 167-180, 1963.

Valdiya, K. S., The unfossiliferous formations of the Lesser Himalaya and correlation, Report 22nd Intn. Geol. Congr., 11, 15-36, 1964.

Valdiya, K. S., Petrography and sedimentation of the sedimentary zone of southern Pithoragarh, U.P., Himalaya, Min. Geol. Metal. Inst. Ind., Wadia Volume, 521-544, 1965.

Valdiya, K. S., Origin of the magnesite deposits of southern Pithoragarh, Kumaun Himalaya, Economic Geol., 63, 924-934, 1968.

Valdiya, K. S., Stromatolites of the Lesser Himalayan carbonate formations and the Vindhyan, Jour. Geol. Soc. Ind., 10, 1-25, 1969.

Valdiya, K. S., Origin of Phosphorite of the Late Precambrian Gangolihat Dolomite of Pithoragarh, Kumaun Himalaya, India, Sedimentology, 19, 115-128, 1972.

Valdiya, K. S., Tectonic framework of India: A review and interpretation of recent structural and tectonic studies, Geophys. Res. Bull., 11, 79-114, 1973.

Valdiya, K. S., Lithology and age of Tal Formation in Garhwal, and implication on stratigraphic scheme of Krol Belt in Kumaun Himalaya, Jour. Geol. Soc. Ind., 16, 119-134, 1975.

Valdiya, K. S., Himalayan transverse faults and folds and their parallelism with subsurface structures of north Indian plains, Tectonophysics, 32, 353-386, 1976.

Valdiya, K. S., Extension and analogues of the Chail Nappe in Kumaun Himalaya, Ind. Jour. Earth Sciences, 5, 1-19, 1978.

Valdiya, K. S., Structural set-up of the Kumaun Lesser Himalaya, in: Himalaya, Science de la Terre, C.N.R.S., Paris, 268, 449-462, 1977.

Valdiya, K. S., An outline of the structural set-up of the Kumaun Himalaya, Jour. Geol. Soc. Ind., 20, 145-157, 1979a.

Valdiya, K. S., Intracrustal boundary thrust of the Himalaya, Tectonophysics, 66, 323-348, 1980.

Valdiya, K. S., and V. J. Gupta, A contribution to the geology of the Tethys Himalaya in north-eastern Kumaun, with special reference to the Hercynian gap, Him. Geol., 2, 1-34, 1972.

Vashi, N. M. and S. S. Merh, Structural elements of the rocks in the vicinity of South Almora Thrust near Upradi (Almora District, U.P.), Jour. M. S. University, Baroda, 14, 27-32, 1965.

Vashi, N. M., and S. K. Laghate, Structural and metamorphic studies of the rocks to the west of Peora in Kumaun Himalaya, Him. Geol., 2, 515-526, 1972.

Verma, R. K., M. Mukhopadhyay and B. N. Roy, Seismotectonics of the Himalaya and continental plate convergence, Tectonophysics, 42, 319-335, 1977.

THE GEODYNAMIC HISTORY OF THE HIMALAYA

Augusto Gansser

Geologisches Institut der E.T.H., Zürich, Switzerland

Abstract. The geodynamic history of the Himalaya is discussed based on the well established subdivisions into Subhimalaya, Lesser Himalaya, High Himalaya, Tethys Himalaya and the Indus-Tsangpo Suture Zone.

From a fragmentary Gondwanic sedimentation in the south (Indian Shield) to the complete Tethyan sections in the north we note a gradual change, only disrupted by the Main Central Thrust (MCT), with over 100 km of displacement. No crystalline divide has ever existed. Northwards the Tethyan platform sediments change into flysch, leading to the ophiolitic suture zone which indicates the northern limit of peninsula India. This major structural element involves mantle rocks while southwards the main structures are younging and becoming intracrustal such as the MCT (Miocene), Main Boundary Thrust (MBT) (Pleistocene) and Main Frontal Thrust (MFT) (recent). The known structural and stratigraphical facts require that peninsular India was never far distant from a most complex southern front of Eurasia.

Orogenic and related metamorphic events are restricted to a late Precambrian polyphase overprinted by a much younger Himalayan phase, culminating in the Neogene. The Himalayan metamorphism is mostly restricted to already metamorphosed older sections and rarely involves rocks younger than lower Paleozoic.

The northern extension of peninsular India as well as the highly complicated southern Eurasian front are involved in the Himalayan orogen.

The Himalaya is not strictly a mountain range born from the Tethys since the greater part consists of the reworked northern edge of the Indian Shield (Indian plate), with Gondwana affinities. This fact dominates the structural evolution of the largest mountain range of our globe. The present paper is intended to be also an explanatory note to the schematic tabulation of the main Himalayan events (Fig. 1). Much of the information is still contradictory, but I have endeavoured to present the facts as much as possible based on my own field experience and less from the multitude of published information. (See also map Fig. 2).

We base our summary presentation of the geodynamic history of the Himalaya on the well-known and generally accepted subdivisions into five main structural units, since each of these units has its characteristic developments. From the south to the north we distinguish:

1. The Subhimalaya consisting mainly of the Siwalik molasse, which along the discontinuous Main Frontal Thrust (MFT) borders the north Indian plain in the south and is limited in the north by the Main Boundary Thrust (MBT).

2. The Lesser or Lower Himalaya thrust along the MBT onto 1 and limited to the north by the Main Central Thrust (MCT).

3. The High Himalaya, representing a huge crystalline thrust sheet thrust along the MCT onto 2.

4. The Tethys Himalaya (Tibetan Himalaya) following as a sedimentary cover on 3.

5. The Indus-Tsangpo Suture Zone, in the north, reflecting the collision between India and the complex Tibetan mass.

1. The Subhimalaya consists of the low foothills bordering the north Indian plains and exposes the southern molasse belt: the Siwaliks. The Siwalik sedimentation originated from the Himalaya in the north, began during the Upper Miocene and culminated in the Upper Pliocene-Lower Pleistocene. The preceding fine clastics, the Murrees, outcropping in the western Subhimalaya, were deposited between the Lower and Upper Miocene and had their origin most likely in the still pronounced hills of the northern Indian Shield, since the Himalaya did not yet exist as a potential source area. The northernmost outcrops of the Murrees in the western syntaxial bend consist of red, highly slickensided phyllites, with incipient metamorphism. Contrarily, the Siwaliks increase in thickness and grain size northwards towards the Himalayan front. Recent detailed investigations show a sedimentation rate in the Upper Siwaliks, where a flood plain and channel facies dominates, of up to 50 cm in 1000 y and an upper age limit, so far still disputed, of 0.5 to 0.4 m.y. (Johnson et al., 1979).

The structures of the Siwaliks in the Subhimalaya are exposed as simple faulted folds, with intensity increasing towards the north and with imbrications towards the overriding MBT. Usually

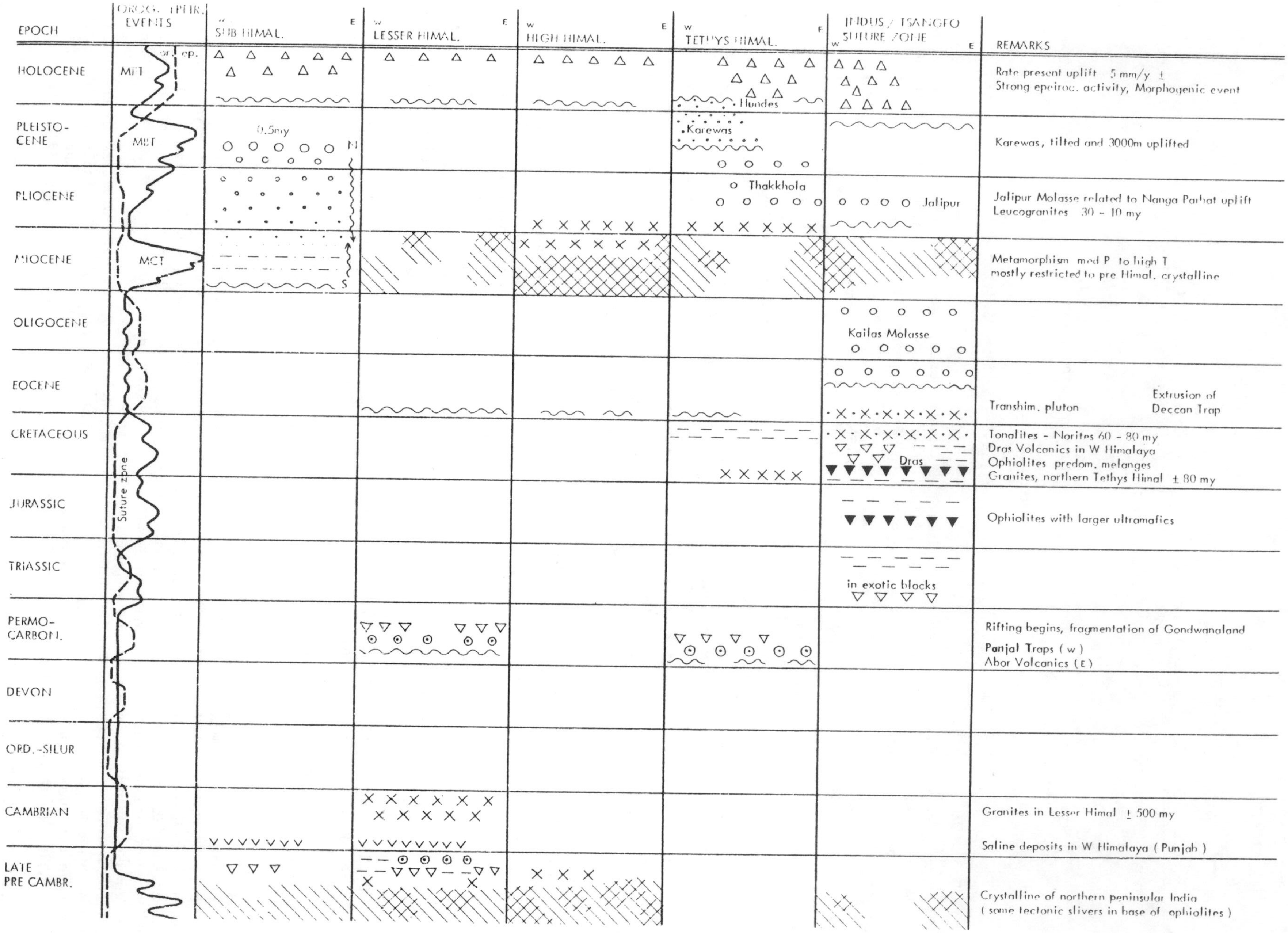

Fig. 1.   Tecto-genetic diagram of the Himalaya.

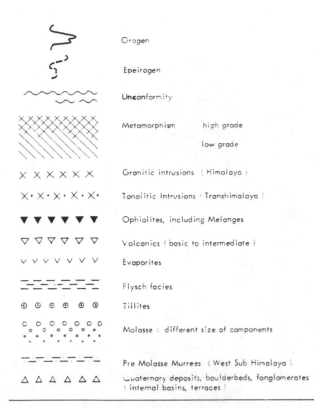

Orogen

Epeirogen

Unconformity

Metamorphism   high grade

low grade

× × × × × ×   Granitic intrusions (Himalaya)

X·X·X·X·X·   Tonalitic Intrusions (Transhimalaya)

▼ ▼ ▼ ▼ ▼ ▼   Ophiolites, including Melanges

▽ ▽ ▽ ▽ ▽ ▽   Volcanics (basic to intermediate)

ᵛ ᵛ ᵛ ᵛ ᵛ ᵛ   Evaporites

Flysch facies

⊙ ⊙ ⊙ ⊙ ⊙ ⊙   Tillites

Molasse : different size of components

Pre Molasse Murrees (West Sub Himalaya)

△ △ △ △ △ △   Quaternary deposits, boulderbeds, fanglomerates
(internal basins, terraces)

the upper conglomeratic Siwaliks are in contact with the thrusted northern boundary. Only in the eastern Himalayas (Arunachal Pradesh, formerly NEFA) are real shuppen zones exposed with tectonic slices of older Gondwana rocks, dipping at 30-40° below the MBT (Jhingran et al., 1976). In spite of the often severe tectonics, reversed sections within the Siwaliks are practically absent.

South of the Siwaliks, the bordering alluvial terraces are often gently folded and tilted and reflect the present site of the active Himalayan orogeny, underlined by a marked seismicity. Locally the terraces are overridden by the frontal Siwalik sediments, with a conspicuous steep thrust - the Main Frontal Thrust (MFT) - well exposed along the eastern foothills. Contrasting with the Siwalik sediments, the terraces along the foothills but also within the Himalaya are surprisingly coarse and often somewhat fanglomeratic. They reflect the recent morphogenic uplift of the Himalaya, in contrast to the frontal orogenic movements.

The north Indian plain, following to the south the Subhimalaya, is actually the Himalayan foreland and masks the northern continuation of the Indian shield. Sporadic drilling, magnetometer, gravity and seismic results confirm the continuation of the sharply northeast trending Indian shield below the Subhimalaya. Significantly, the up to 4000 m thick Siwaliks drilled in three wells south of the central foothills, transgress

directly on the late Precambrian shield sediments (Datta and Sastri, 1977). Considering the relatively shallow and narrow basin, particularly along the eastern Himalaya, the bulk of the Siwalik molasse comprises only a small fraction of the visible volume of the Himalayan range. This is in striking contrast to the Alps, where the volume of the Alpine molasse corresponds to the total volume of the exposed mountain range.

2. The Lesser Himalaya is clearly limited to the south by the MBT and to the north by the MCT, which forms the base of the High Himalaya. Neither thrust is everywhere clearly defined. In the south, various structural units can reach the MBT and actually take on the function of the thrust zone. Examples are the Krol thrust in the western and some crystalline thrusts in the eastern part of the Lesser Himalaya. The northern thrust boundary often begins with an intense schuppen zone of mostly crystalline imbrications which locally mask a clear-cut contact. Some of the larger crystalline thrust sheets, forming the highest structural elements in the Lesser Himalaya, seem to have their origin in these schuppen zones still below the actual MCT. Westwards, with the development of the Kashmir basin, a Tethyan Himalayan element, the Lesser Himalaya is strongly reduced. Similarly, in the eastern Himalaya crystalline thrusts, belonging to the Higher Himalaya, seem to encroach on the complicated Lesser Himalayan units and reach locally the MBT. Unfortunately this area is not yet sufficiently known.

Paleogeographically, the Lesser Himalaya belongs to the northern extension of the Indian shield, forming a complicated platform, which borders the shallow Tethyan sea. The gradual change from a southern shield/Gondwana facies to the northern Tethyan facies is now interrupted by the intracrustal MCT which a visible displacement of at least 100 km. Originally there was no divide, contrary to many different opinions (Saxena, 1971; Fuchs, 1967; Hashimoto et al., 1973, Ashgirei, 1970, and others), which speak of a Himalayan ridge or of a crystalline central axis. This fact is paramount in the understanding of the paleogeographic development of the Himalaya, and is becoming more widely accepted (Bassoullet et al., 1977; Colchen, 1977).

The evident lack of fossils in the widely exposed late Precambrian sediments, the marked facies changes in the shallow deposits and the complicated tectonics, coupled with relatively bad exposures, are the reasons why the Lesser Himalaya is still relatively little known, in spite of intense detailed work. This has resulted in a rather confusing amount of local formation and group names. Late Precambrian and early Cambrian detrital (base) and thick carbonate sediments (with Riphean stromatolites) can be correlated with sediments of the northern Indian shield (Srikantia, 1977). Similarly we note a large gap until the transgression of the Permo-Carboniferous Gondwana sediments with the char-

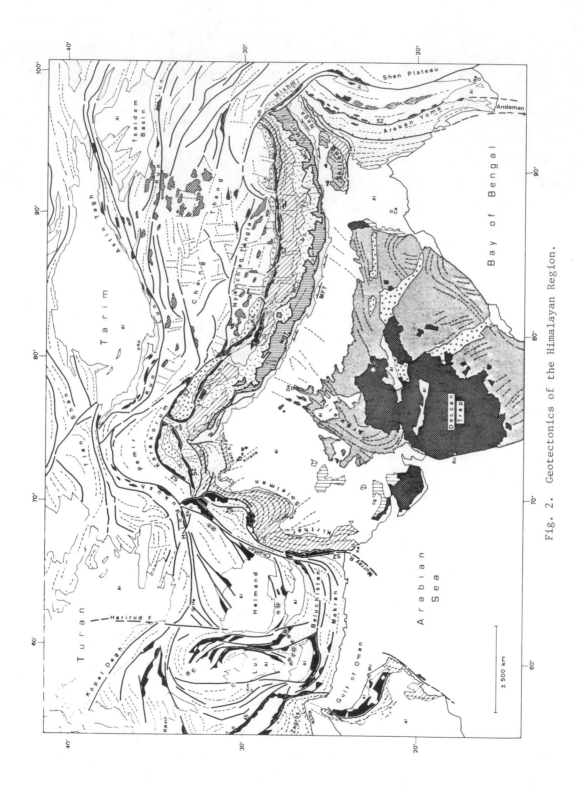

Fig. 2. Geotectonics of the Himalayan Region.

Al	Alluvial plain
	Pre-Gondwana basement, Indian shield
	Gondwana sediments on Indian shield
	Mesozoic Platform sediments on Indian shield
	Trap, Indian shield
	Molasse type sediments (Siwaliks of sub-Himalaya)
	Lesser Himalaya
	Crystalline of High Himalaya
	Tethyan sediments (platform)
	Flysch facies
	Ophiolites incl. ophiolitic melanges and related pelagic sediments
	Transhimalayan plutons
	Tertiary and Quaternary volcanics
	Quaternary volcanoes
	Main thrust and fault zones
	Lineaments (fold axis, sec. fault zones, fracture zones)

ABBREVIATIONS

Bo	Bombay	Ls	Lhasa
Ca	Calcutta	Mu	Muscat
De	Delhi	NB	Namche Barwa Mt.
He	Herat	NP	Nanga Parbat Mt.
Is	Islamabad	Qt	Quetta
Jl	Jolmo Lungma (Everest)	Ra	Rangoon
Ju	Jungbwa nappe	Sp	Sponglang Klippe
Kb	Kabul		
Ks	Kailas Mt.		main structural units:
Ka	Karachi	SZ	Suture Zone (Indus/Tsangpo)
Kt	Kathmandu	MCT	Main Central Thrust
Ko	Khotan	MBT	Main Boundary Thrust
La	Ladakh	MFT	Main Frontal Thrust

Outside wider Himalaya and Indian shield only structural trends, ophiolites and volcanoes are shown.

acteristic tilloid horizon (Blainis of the Himalaya, Talchirs on the Shield). The Blainis form
actually one of the very few marker beds, which
allow us to separate the older horizons from the
rudimentary Mesozoic section, known as the Krol
formation of Auden (1934, 1970). All Lesser
Himalaya sediments are capped by the Paleocene-
Eocene Subathus, exposing a marked unconformity
which, according to some authors, include the detrital Tal rocks normally above the Krol section.
Various authors have tried to close the large
stratigraphic gaps in the Lesser Himalayas, but
so far the fossil evidence is most doubtful - for
instance Middle Devonian brachiopods from a
quartzite pebble of unknown origin (Gupta, 1972).

In two places within the Lesser Himalaya exist
well preserved sediments with fossils from Ordovician to Devonian age, the Godavari-Pulchauki
basin of the Kathmandu area (Bordet, 1967;
Stoecklin, 1979) and the Tang-Chu basin in Central Bhutan (Gansser, 1964; Termier and Gansser,
1974). In both basins the sediments are underlain by crystalline rocks with granitic intrusions, and these crystallines form well outlined
nappes, lying on top of the Lesser Himalayan
sediments. In both areas these nappes seem to be
related to the MCT as klippen-like outlayers, the
Kathmandu nappe and the Tang Chu nappe respectively. We have to deal here with sediments of a
Tethyan facies, which, like the much larger Kashmir basin, have overriden the Lesser Himalaya.

Various granites have intruded the Lesser Himalayan Precambrian section, with the 500 m.y. old
Mandi granite as the best known example (Jaeger
et al., 1971). Intense folding and incipient low
grade metamorphism antedate the intrusions, while
both granites and sediments have undergone a marked Himalayan metamorphic overprint and give young
mica ages, etc. Regionally, metamorphism in the
Lesser Himalaya is only weak, and a real crystalline basement is unknown. On the other hand, the
highest structures of the Lesser Himalayas are
crystalline thrust sheets, such as the well-known
Almora nappe.

These crystalline sheets show a reversed metamorphism, a curious and still little understood
fact. This is in contrast to the thick section
above the MCT. Contrary to some authors (Powell
and Conaghan, 1973) the metamorphism of the nappes
forming the highest structural units of the Lesser
Himalaya is the result not of only one but of two
main metamorphic events at the end of the Precambrian followed by the Himalayan overprint. Recent
though still sporadic whole rock ages (Rb/Sr) show
this convincingly (see also discussion by Mehta,
1979). Furthermore, the rich content of granitic
and gneissose pebbles in the Gondwanas of the
eastern Lesser Himalayas disprove the dogma of a
unique Himalayan metamorphic phase (Jain and
Thakur, 1975). These facts support our views
(Gansser, 1964) that after strong late Precambrian
orogenic events no further orogeny affected the
Himalayan range (Lesser as well as High Himalaya
and Tethys Himalaya) until the Mesozoic and late

Tertiary Himalayan phases. Regional epirogenic
movements are possible for some gaps and marked
facies changes in the sedimentary sequence, including some of the granitic intrusions, but they
should not be correlated with Caledonian and Hercynian orogenies, terms better avoided within the
Himalayan domain.

From the south to the north, from the MBT to
the MCT, appear in the Lesser Himalaya succeed-
ingly older thrust masses. Where the Krol Nappes
are not developed, steep thrusts of Gondwana
rocks can be observed. They occur along the
Nepalese foothills and extend through Sikkim and
Bhutan into the eastern Himalaya (Arunachal
Pradesh). Inwards (to the north) follow further
thrusts with Precambrian and early Cambrian rocks,
well documented by the Daling thrust. Finally we
note the crystalline thrusts towards the MCT. The
relatively steep thrusts, particularly in the
Gondwanas, flatten northwards, a fact proven by
the Rangit window in Sikkim. A northwards flattening of major thrusts can be assumed for most
of the intracrustal events, while the more deep
seated suture zone thrusts may remain steep to a
considerable depth.

3. The High Himalaya is best expressed as the
"Dalle du Tibet" (Lombard, 1958), since it forms
the base of the thick pile of Tethyan or Tibetan
sediments. For this reason the upper limit is
rather arbitrarily chosen. It is placed where
the Lower Paleozoic sediments show rather independent tectonics, including the remarkable back-
folds, and where the imprinted Himalayan metamorphism is dying out. In contrast, the lower
border is sharply defined by the MCT. The High
Himalaya and its basal MCT is particularly well
outlined in the Central Himalaya. Towards the
west and particularly towards the east, the often
well-developed crystalline thrust sheets of the
Lesser Himalaya interfere. They seem to root in
the schuppen zones below the MCT but lack of
sufficient petrological and structural information, particularly in the eastern Himalayas,
leave many questions yet unsolved.

The best exposed sections were found in the
Kumaon Himalaya west of Nepal (Heim and Gansser,
1939; Gansser, 1964) as well as in Central Nepal
(Le Fort, 1975). They are surprisingly constant
in their regional composition. They invariably
begin with gneisses, coarse or banded, with two
micas, kyanite or sillimanite and garnet. Locally the gneisses can be migmatitic. They are
followed by coarse quartzites, becoming more cal-
cic towards the upper part, with sections of
silicate bands. Here begins the intrusion of
tourmaline granite dykes, which are related to
irregular bodies of massive tourmaline granites.
At the same time marbles and lime-silicate bands
increase. The top of the section is argillaceous
in the Kumaon and more calcareous in the Nepal
Himalaya. Here the metamorphism decreases and
the first fossils (Cambrian to Ordovician)
appear. The argillaceous and calcareous basal
sections can be over ten thousand meters thick,

and the crystalline slab below can measure 10 to 15 km. We note in this enormous pile of crystallines and sediments that the metamorphism of the High Himalaya decreases upwards, contrary to the reversed metamorphism in the crystalline nappes of the Lesser Himalaya (Almora nappe). Only locally has reversed metamorphism been noted near the MCT in central Nepal (Le Fort, 1975). Detailed investigations in the Kathmandu nappe have clearly established a normal metamorphic section (Stoecklin, 1979).

The metamorphism of the High Himalaya consists, as we have already noted for the Lesser Himalaya, of two general phases. One during the late Precambrian, overprinting still older events, which is documented by whole rock Rb/Sr analysis, which show ages of 1800 to 1500 m.y. (Bhanot et al, 1977, Mehta, 1979). The other are the Himalayan phases, active mostly during the main orogenies between Miocene and Pliocene. These late phases, to which many authors have given too much emphasis, are mostly restricted to the already metamorphosed sections and affect only rarely and locally the sedimentary pile in the western and easternmost Himalaya). This is a rather curious fact, typically Himalayan, and has not yet received a convincing explanation. In spite of a medium to high grade metamorphism of 20-10 m.y. age, the Cambrian sediments in the western Himalayas are not affected. In the Central Himalaya the regional metamorphism reaches into the Ordovician, but post-Ordovician sediments are still unmetamorphosed. In the middle part of the crystalline High Himalaya section, where marbles and lime silicate rocks dominate, we noted the tourmaline or leucogranites, leading from complicated dyke systems to irregular batholites, which can grow to larger leucogranite intrusions, such as the Badrinath granite (Heim and Gansser, 1939), the Mustang granite (Hagen, 1968), the Manaslu granite (Le Fort, 1973), the Makalu granite (Bordet, 1961) and the tourmaline granites in the Chomolhari and northeastern Bhutan region (Gansser, 1964). All these granites are surprisingly similar petrologically and are of the same age (Neogene). They display a weak to strong contact metamorphism (northeast Bhutan) within otherwise unmetamorphosed Mesozoic sediments.

4. The Tethys Himalaya (Tibetan Himalaya) follows to the north of the High Himalaya and begins with fossiliferous sections, where the metamorphism has ended and an independent structural style begins. Complicated folds and thrusts are disharmonic in relation to the underlying crystalline slab, often with larger back folding, beautifully displayed in the Nilgiri fold of the Thakkhola valley of central Nepal. Surprising is the practically unbroken sedimentary suite of the Tethys Himalaya from Cambrian to the Lower Eocene. Sedimentary gaps are minor (upper Carboniferous) and decrease with progressing detailed investigations. Furthermore, this thick pile of mostly platform-type sediments is conformable with a surprising facies continuity from east to west

all along the entire range, over 2000 km. Some noticeable changes can, however, be observed in the late Precambrian to Ordovician, which in the western Himalaya is mostly fine detrital to argillaceous (Martoli Garbyang section), while in the central part a calcareous facies is more dominant (Nilgiri facies of Central Nepal). This facies change has some influence on the general tectonics. Backfolding is more dominant in the calcareous sections of central Nepal. In contrast, the Triassic is dominantly calcareous and dolomitic in the west (Kashmir) and much more diversified towards the centre and the east (Bassoullet et al., 1977). On the other hand, some characteristic horizons are surprisingly constant, such as the Devonian Muth quartzite, the Gondwana pebble beds and the delicate pelagic Spiti horizons of late Jurassic/earliest Cretaceous age. Northwards, towards the Indus-Tsangpo suture zone the Upper Mesozoic changes into a flysch facies, which in the western Himalaya (Ladakh area) begins already with the Trias (Frank et al., 1977).

Today we recognize various basins where the Tethyan sediments are well preserved. These basins are differentiated structurally but not stratigraphically. Their edges are erosional remnants without any marked facies change. From the west to the east I have distinguished the following main basins (Gansser, 1974a): The Kashmir basin with a narrow steeply infolded extension to the east, the Chamba basin. It is separated through the Nun Kun crystalline high from the Spiti basin to the north east, the classical locality for the stratigraphy of the Tethys Himalaya. The Sutlej Cross High, connecting the central crystalline (Nun Kun continuation) with the Rupshu crystalline to the north forms the eastern limit of the Spiti basin. Further eastwards follows the northern Kumaon basin, a depression between the eastwards plunging Rupshu crystalline and the eastwards rising domelike Gurla Mandhata crystalline uplift. This depression exposes the largest Quaternary gravel basin in the whole Himalaya (the Hundes area of the upper Sutlej river), with gravels well over 1000 m thick, cut by most impressive deep canyons and in which the find of Rhinoceros in the early 19th century has led to animated discussions about several thousands of meter uplift in Pleistocene times of strata which are still horizontal (Lydekker, 1881). In the same basin we note also the ophiolitic Jungbwa nappe, the highest structural element of the Himalaya, lying flat on the Tethyan sediments. Compared to the Kashmir and Spiti basin, the Tethyan sediments of the northern Kumaon basin are more severely tectonized, with complicated folds and several important imbrications (Heim and Gansser, 1939).

The Gurla Mandhata uplift and its still very little-known eastern continuation separates the Kumaon basin from the west Nepal Basin, which includes the well known Thakkhola region. East of Manaslu high follows a very large basin within

Tibet, which I called the Kampa Dzong basin, after the type locality of the Upper Cretaceous-Eocene described already by Hayden (1907). This basin can be followed into the eastern Tethyan Himalaya and includes the region north of Jolmo Lungma (Everest) well known through the excellent investigations of the Chinese researchers, which culminated in the publication of five volumes in Chinese (1974-76) and in English review by Mu An-Tze et al., (1973).

In each of these basins, except Kashmir and Spiti, one recognizes northwards a facies change from platform marine sediments into deeper and more detrital deposits of a flysch facies, which leads into the Indus-Tsangpo suture zone. In the west as well as in the eastern part this flysch facies begins already in the Trias, while in the central part we note only a Middle and Upper Cretaceous flysch. This northern flysch zone forms an uninterrupted belt all along the suture zone and is not subdivided into the above-mentioned basins. Within the Kampa Dzong basin, from the Manaslu area to the famous Jamdrock Tso in the east occur equally spaced granite and granite gneiss uplifts, well outlined as domal features on the Landsat photos. They roughly follow the line where the platform sediments change into the flysch facies. Based on preliminary investigations (Chang and Chen, 1973; Chang et al., 1977), the age of the biotite granites is regarded as Upper Mesozoic but pre-Upper Cretaceous. They certainly cut the Jurassic sediments and, rich in potassium feldspars, are distinct from the Transhimalayan plutons (see below). The surprising alignment of these granites south of and parallel to the suture zone and limited to a belt between two north-south directed cross-features, which also break the suture zone (Gansser, 1977), is significant, but so far difficult to explain.

5. The Indus-Tsangpo Suture Zone. This most important structural feature of the whole Himalaya and of the adjoining north-south directed ranges in the west (Quetta belt) and in the east (Arakan Yoma belt) has received recently much attention as one of the best developed and exposed collision zone between greater India and various fragments of Gondwanic and Tethyan origin incorporated in the Eurasian plate. The wealth of often contradictory theoretical deductions greatly exceeds the amount of factual field observations from an area 5000 km long and mostly difficult of access (Fig. 2).

We have noted how all along the north side of the Himalaya the Tethyan platform sediments change into a flysch facies with all the indications of deeper and pelagic sediments. These facts suggest that we have actually reached the northern limit of the Indian shield with a rather steep continental slope towards a narrow but very pronounced oceanic basin. This was no simple feature but contained island arcs, complicated internal basins and slices of continental rocks. These deductions are based on the known struc-

tural configuration, lithology and petrology of the ophiolitic belt and its related rocks characterising the suture zone. Only rarely can we discover "normal" ophiolitic sections. Generally the steeply dipping belt is highly disrupted, except in some of the southwards thrust ophiolite nappes such as the Spongtang klippe in the west and the larger Jungbwa nappe in the center, where a conspicuous ophiolitic melange with flysch forms the base and the ultramafic bodies the top (Gansser, 1979). The recognition of the ophiolitic melanges as one of the most important elements of the ophiolitic belt in this area is most significant. In my opinion it reflects very steep and narrow oceanic basins, the sediments of which are now found as exotic blocks within the melanges. Basement-type crystalline rock from continental slices was involved during the compressive phase. These conditions, not readily recognizable, where the tectonisation was intense as in the Himalayan suture zone, are clearly displayed in the narrow ophiolitic belts of southeast Iran, which is one of the best examples of the intimate relations of continental slices and ophiolites with its melanges and the respective rapid facies changes (Stoecklin, 1977; Gansser, 1974b).

It has been now well established that the ophiolites characterize the Himalayan suture zone all along its west-east extension of 2500 km as well as its continuation in the Quetta ranges to Karachi in the west and along the Arakan Yoma ridges in the east (Fig. 2). However, this belt is by no means continuous. It is frequently broken by mostly north-south aligned structural anomalies, and in the western Himalayas we note even a clear doubling of this zone (Gansser, 1977).

From the Nanga Parbat Cross High to the east, the suture zone is bordered on its north side by a narrow belt of molasse (the Paleogene Ladakh/Kailas molasse). It is in tectonic contact with the ophiolites but transgresses clearly over the Transhimalayan plutons, which form again a continuous belt north of the suture from the Swat area in northern Pakistan to the Indus bend at Namche Barwa mountain in the east. They too are broken into segments, similar to the suture zone. They are mostly tonalitic, with acid and more basic differentiations and are genetically related to the suture zone. They are clearly distinct from the Himalayan plutons, a fact already stressed by Hayden (1907), while discussing the hornblende granites of the Lhasa region, which belong to the Transhimalayan plutons. Their average age is about 60 m.y. and their low Sr 87/Sr 86 ratios (0,704) suggest their oceanic origin (Frank, pers. com.). The volume of these batholites is considerable, and may be related to the size of the consumed oceanic crust.

The ophiolites of the suture zone and the related flysch sediments show a low grade green schist metamorphism. Towards the Nanga Parbat Cross High, and particularly further west, the metamorphism increases, with a dominating high pressure facies composed of glaucophane and eclogitic rocks of the Shang La, west of the middle Indus valley (Tahirkheli et al., 1977). Towards the east, already at Gyantse, southwest of Lhasa, staurolite garnet schists were mentioned for the first time by Hayden (1907). Further eastwards the metamorphism increases and towards Namche Barwa a high grade phase has been reported within the flysch belt (Chang et al., 1977).

Conclusions

The sedimentary history shows fragmentary sedimentation in the south, similar to peninsular India, which was involved in the greater part of the Lesser Himalaya. The present differences in facies from south to north are exaggerated by thrusting, which has brought more distant deposits, over 100 km apart, into juxtaposition. this is particularly well expressed on both sides of the MCT, where the change from a more Gondwanic facies with large sedimentary gaps towards a well developed and continuous section of Tethyan aspects has let to the erroneous concept of a dividing crystalline barrier. Paleogeographically we realize a gradual change into the deeper part of the Tethyan basin with recurrent Gondwana type deposits, probably far to the north, even within the Tibetan block. Similar conditions have prevailed on the west and less clearly on the east side of peninsular India, where fossiliferous Tethyan type Paleozoic and Mesozoic platform sediments have encroached onto the Indian Shield, well expressed in the Suleiman, Kirthar and Karachi ranges (Hunting Survey Corp., 1960; Crawford, 1979). All these facts demand a reappraisal for the drift history of greater India. The observations require that peninsular India was never far distant from the southern front of the Eurasian continent and that the Tethyan "ocean" now consumed along the northern suture zone was actually a narrow zone of complicated island arcs, internal basins and irregular slices of continental rocks (Crawford, 1979).

The structural history of the Himalayan range is unique (Fig. 1). We noted that the crystalline sections ("basement") of the Himalaya reflect late Precambrian orogenic phases known from the Indian shield, such as the Delhi and later Aravalli, and confirmed by the few, but significant age dates. A surprisingly long anorogenic period followed, broken by epeirogenic accidents only, for example the Lower Paleozoic with its 500 m.y. old granites and the earlier and later Gondwana phases with their related volcanism.

The following Himalayan orogeny is still little understood, but several important phases can be recognized, ultimately not very unlike the Alpine history. The earliest orogenic movements are noted along the suture zone with the recognition of the Triassic flysch. Further south in the Tethys Himalaya the Himalayan phase follows from the Middle to Upper Cretaceous onwards. Events culminated most likely during the Miocene, together with a pronounced medium pressure metamorphism containing all grades depending on the stratigraphic and structural position of the rocks. A major orogeny preceeded the main Siwalik sedimentation during the Miocene. In many cases the metamorphism outlasted the main deformation. Following the deposition of the Siwalik molasse we note the last major orogeny during the Middle Pleistocene. Still younger is the remarkable morphogenic phase, still active today. It may have rates of uplift up to 5 mm per year. This estimated amount, over five times the well-known Alpine value, is deduced from terrace elevations, the displaced Karewa beds and regional morphological spects.

The main Himalayan orogenic phases are well reflected in major accidents such as large nappes. Mesozoic events are restricted to the suture zone and involve the complete crust and the upper mantle. Further to the south we note the MCT, a remarkable intracrustal accident of Miocene age with late orogenic granitisation (tourmaline-bearing leuco-granites). The MBT along the foothills in the south reflects a Pleistocene event and involves only the upper part of the crust. Recent events are shown by the thrusting of the Siwalik front on alluvial terraces, the MFT, as well as the tilting and gentle folding of the latter.

We have here convincing evidence that the Himalayan orogenies are wandering and younging from the north to the south, while involving gradually more shallow structural levels.

In the western frontal Himalayas (Punjab) the late Precambrian to early Cambrian evaporite horizons (type Salt Range) must have played an important role in the detachment and lubrication of the MBT as well as the front hill structures in general. It is known that these saline beds extend further eastwards below the Lesser Himalaya as well as southwards along the Suleiman ranges. In both regions they may facilitate abnormal structures such as the western syntaxis and the Suleiman arc (Srikantia and Sharma, 1972).

Recent events should be reflected by the seismicity, though it is surprising that earthquakes are only rarely directly located along major structures. The main suture zone is at present seismically dead in spite of the spectacular hot spring activities in the Puga and Chumatang area. Microseismic investigations are planned here and may show interesting trends (Gupta and Singh, 1977). The MCT, with many hot springs along its trace is also seismically inactive. Microseismic investigations in relation with the Tarbela dam show alignments with hardly any relation to the visible young tectonics of this active region (Seeberg and Armbruster, 1979).

For a better understanding of the history of the Tethyan Himalaya and its suture zone we should know more about the geological development of Tibet north of the Transhimalayan plutons. Unfortunately this information is still rudimentary, and it is hoped that the great effort by the Chinese geologists, presently engaged in the exploration of Tibet, will produce the necessary results in order to arrive to a more comprehensive picture. The presence of older suture zones within Tibet has been reported (Chang et al., 1977; Chang and Chen, 1973) repeating the Himalayan history within older stratigraphy levels (Devonian flysch with ophiolitic melange belts). These facts have not yet been shown on any published map. Contrasting with the Himalaya is the widespread Quaternary volcanism, issuing from a most peculiar, exceptionally thick (70 km) but "soft and hot" continental crust (Molnar and Tapponier, 1977, 1978) (Fig. 2).

We certainly must realize that two completely different types of continental plates exist north and south of the Himalaya, this outstanding orogen, which has inherited certain features from both of them.

Acknowledgement. I thank William Lowrie for reading the manuscript and his valuable suggestions.

References

Ashgirei, G. D., A new approach for the understanding the geological structure of Himalaya and search of mineral deposit related with them, Himalayan Geol., 7, 1-21, 1978.

Auden, J. B., Geology of the Krol belt, Rec. Geol. Surv. India, 67, 357-454, 1934.

Auden, J. B., Discussion "On the Krol Nappe Hypothesis", Journ. Geol. Soc. India, 11/3, 288-302, 1970.

Bassoullet, J. P., M. Colchen and R. Mouterde, Esquisse paléogéographique et essai sur l' évolution géodynamique de l'Himalaya, Mém. h. ser. Soc. Géol. France, 8, 213-234, 1977.

Bhanot, V. B., V. P. Singh, A. K. Kansal and V. C. Thakur, Early Proterozoic Rb-Sr Whole-rock Age for Central Crystalline Gneiss of Higher Himalaya, Kumaon, Journ. Geol. Soc. India, 18/2, 90-91, 1977.

Bordet, P., Recherches géologiques dans l'Himalaya du Nepal, région du Makalu, Paris Centr. Natl. Recherche Sci., 275p., 1961.

Bordet, P., J. Cavet and J. Pillet, La faune de Phulchauki, près de Kathmandu (Himalaya du Nepal), Bull. Soc. Géol. France, 7/2, 3-14, 1960.

Chang, Cheng-Fa and Hsi-Ian Cheng, Some tectonic features of the Mt. Jolmo Lumgma area, southern Tibet, China, Sci. Sinica, 16, 257-265, 1973.

Chang, C., X. Zheng and Y. Pan, The Geological History, Tectonic Zonation and Origin of Uplifting of the Himalayas, Peking, Institute of Geology, Academic Sinica, 17p, 1977.

Colchen, M., Les caractères gondwaniens et téthysiens des séries himalayennes, Implications paléogéographiques, Soc. Géol. du Nord, T. XCVII, Lille, 279-286, 1977.

Crawford, A. R., The Indus suture line, the Himalaya, Tibet and Gondwanaland, Geol. Mag., 111, 369-383, 1974.

Crawford, A. R., Geodynamics of Pakistan (Farah, and De Jong, Ed.), Geological Surv. of Pak., Quetta, 103-110. 1979.

Datta, A. K. and V. V. Satri, Tectonic evolution of the Himalaya and the evaluation of the petroleum prospects of the Punjab and Ganga Basins and the Foot Hill Belt, Himalayan Geology, 7, 296-325, 1977.

Desio, A., Geological Tentative Map of the Western Karakorum, Institute of Geology, Univ. of Milan, Ist. ital. d'arti grafiche, Bergamo, 1964.

Desio, A., Geological Evolution of the Karakorum, in Geodynamics of Pakistan (Farah and De Jong, Ed.), Geol. Surv. of Pak., 111-124, 1979.

Frank, W. and G. Fuchs, Geological Investigations in West Nepal and their Significance for the Geology of the Himalayas, Geol. Rdsch., 59/2, 552-580, 1970.

Frank, W., A. Gansser and V. Trommsdorff, Geological observations in the Ladakh Area (Himalayas), a preliminary report, Schw. Min. Petr. Mitt., 57, 89-113, 1977.

Fuchs, G., Zum Bau des Himalaja, Oesterr. Akad. Wiss., Denkschr. Wien., 113, 211p., 1967.

Fuchs, G., Traverse of Zanskar from the Indus to the Valley of Kashmira - a preliminary note, Jahrb. Geol. B. A., 120/2 Wien, 219-229, 1977.

Fuchs, G., The Geology of the Karnali and Dolpo Regions, Western Nepal, Jahrb. Geol. B. A., 120/2 Wien, 165-217, 1977.

Fuchs, G., and A. K. Sinha, The Tectonics of the Garhwal-Kumaun Lesser Himalaya, Jahrb. Geol. B. A., 121/2 Wien, 219-241, 1978.

Gansser, A., Geology of the Himalayas, Wiley-Interscience, London, 289p., 1964.

Gansser, A., The Indian Ocean and the Himalayas-A geological interpretation, Eclogae Geol. Helv., 59, 831-848, 1966.

Gansser, A., Orogene Entwicklung in den Anden, in Himalaja und den Alpen, ein Vergleich, Eclogae Geol. Helv., 66/1, 23-40, 1973.

Gansser, A., The Himalayan Tethys, Riv. Ital. Paleont. Strat., Mem. XIV, 393-411, 1974a.

Gansser, A., The Ophiolitic Melange, a World-wide Problem on Tethyan Examples, Eclogae Geol. Helv. 67/3, 479-507, 1974b.

Gansser, A., The great suture zone between Himalaya and Tibet - a preliminary account. Colloq. intern. C. N. R. S., 268, Ecologie et Géologie de l'Himalaya, 181-191, 1977.

Gansser, A., The Ophiolitic Suture Zones of the Ladakh and the Kailas Region. A Comparison. Journ. Geol. Soc. India, 20, 277-281, 1979.

Gansser, A., Reconnaissance Visit to the Ophiolites in Baluchistan and the Himalaya. In Geodynamics of Pakistan, (Farah and De Jong, Ed.), Geol. Surv. of Pak., 193-213, 1979.

Gupta, V. J., A note on the stratigraphic position of the Sirdang quartzites of the type area, Kumaon Himalaya, Verh. Geol., B-A, H-2, Wien, 263-264, 1972.

Gupta, M. L. and S. B. Singh, The Himalayan Orogen and its geothermal resources, Himalayan Geol., 7, Wadia Inst. of Geol., Dehradun, India, 326-344, 1977.

Hagen, T., Geology of the Thakkhola, Denkschr. schweiz. naturf. Ges., 86/2, 160p., 1968.

Hashimoto, S., et al., Geology of the Nepal Himalayas, Saikong Pub. Co., Sapporo, Japan, 286p, 1973.

Hayden, H. H., The geology of the provinces of Tsang and U in Tibet, Geol. Surv. India Mem., 36/2, 122-201, 1907.

Heim, A. and A. Gansser, Central Himalaya, Geol. Observations of the Swiss Expedition 1936, Denkschr. schweiz. naturf. Ges., 73/1, 245p, 1939.

Hunting Survey Corp., Ltd., Reconnaissance Geology of part of west Pakistan, (with maps 1:253,440), Toronto, Canada, 550p, 1960.

Jaeger, E., A. K. Bhandari and V. B. Bhanot, Rb-Sr age determination of biotites and whole-rock samples from the Mandi and Chor Granites, Himachal Pradesh, India, Eclogae geol. Helv., 64, 521-527, 1971.

Jain, A. K., and V. C. Thakur, Stratigraphy and tectonic significance of the eastern Himalayan Gondwana belt with special reference to the Permo-Carboniferous Rangit pebble slate, Bull. Indian Geol. Assoc., 8/2, 50-70, 1975.

Jhingran, A. G., V. C. Thakur and S. K. Tandon, Structure and Tectonics of the Himalaya, Himalayan Geol. Seminar, Wadia Inst. of Himalayan Geol., New Delhi, 1-39, 1976.

Johnson, G. D., N. M. Johnson, N. D. Opdyke and R. A. K. Tahirheli, Magnetic Reversal Stratigraphy and Sedimentary Tectonic History of the Upper Siwalik Group Eastern Salt Range and Southwestern Kashmir, Geodynamics of Pakistan, (Farah and De Jong, Ed.), Geol. Surv. of Pak., 149-165. 1979.

Le Fort, P., Les leucogranites de l'Himalaya, sur l'exemple du granite du Manaslu (Népal central) B. S. G. F., 7/15, 555-561, 1973.

Le Fort, P., Himalayas: The collided range. Present knowledge of the continental arc., Am. J. Sci., 275-A, 1-44, 1975.

Lombard, A., Un itinéraire géologique dans l'Est du Népal (Massif du Mont Everest), Denkschr. schweiz. naturf. Ges., 82/1, 107p., 1958.

Lydekker, R., Observations on the ossiferous beds of Hundes in Tibet, Rec. Geol. Surv. India, 14, 178-184, 1881.

Mehta, P. K., Rb-Sr Geochronology of the Kulu-Mandi Belt: Its implications for the Himalayan Tectogenesis - A reply, Geol. Rundschau, 68/1, 383-392, 1979.

Meyerhoff, A. A., Petroleum in Tibet and the India-Asia Suture (?) Zone, Journ. of Pet.Geol. 1/2, 107-112, 1978.

Molnar, P., and P. Tapponnier, The collision be-

tween India and Eurasia, Sci. Am., 236/4, 30-42, 1977.

Molnar, P., and P. Tapponnier, Active tectonics of Tibet, J. Geophys. Res., 83, B-11, 5361-5375, 1978.

Mu, An-Tze, Shi-Hsuan Wen, Yi-Kang Wang, Ping-Kao Chang and Chi-Hsiang Yin, Stratigraphy of the Mount Jolmo Jungma region in Southern Tibet, China, Scientia Sinica, 16, 1, 96-111.

Powell, C. McA., and P. J. Conaghan, Plate tectonics and the Himalayas, Earth & Planetary Sci. Letters, 20, 1-12, 1973.

Saxena, M. N., The Crystalline Axis of the Himalaya: The Indian Shield and Continental Drift, Tectonophysics, 12, 433-447, 1971.

Seeber, L., and J. Armbruster, Seismicity of the Hazara Arc in Northern Pakistan: Decollement vs. Basement Faulting, Geodynamics of Pakistan, (Farah and De Jong, Ed.), Geol. Surv. of Pak., 131-142, 1979.

Shah, S. K., Facies pattern of Kashmir within the Tectonic frame work of the Himalaya, Tect. Geol. of the Himalaya (Saklani, Ed.), New Delhi, 63-78, 1978.

Srikantia, S. V., and R. P. Sharma, The Precambrian salt deposits of Himachal Pradesh Himalaya - its occurrence, tectonics and correlation, Himalayan Geology, 2, 222-238, 1972.

Srikantia, S. V., The Sundernager Group: Its geology, correlation and significance as stratigraphically the deepest sediment in the Peninsular of Lesser Himalaya, Journ. Geol. Soc. India, 18/1, 7-22, 1977.

Stoechlin, J., Structural correlation of the Alpine ranges between Iran and Central Asia, Mém. h. sér. Soc. géol. France, 8, 333-353, 1977.

Stoecklin, J., and K. D. Bhattarai, Geology of the Central Mahabharat Range, Tectonophysics, in press, 1979.

Tahirkheli, R. A. K., M. Mattauer, F. Proust and P. Tapponnier, Some new data on the India Eurasia convergence in the Pakistani Himalaya, Colloq. Inter. C. N. R. S., 268, Ecologie et Géologie de l'Himalaya, 209-212, 1977.

Termier, G., and A. Gansser, Les séries dévoniennes du Tang Chu (Himalaya du Bhoutan), Eclogae geol. Helv., 67/3, 587-596, 1974.

Thoni, M., Geology, structural evolution and metamorphic zoning in the Kulu Valley (Himachal Himalayas, India) with special reference to the reversed metamorphism, Mitt. Ges. Geol. Bergbaustud., Oesterr., 24, 124-187, 1977.

Valdiya, K. S., Outline of the structure of Kumaon Lesser Himalaya, Tect. Geol. of the Himalaya (Saklani, Ed.), New Delhi, 1-14, 1978.

Wadia, D., The Cretaceous volcanic series of Astor-Deosai, Kashmir, and its intrusions, Rec. Geol. Surv. India, 72/2, 151-1616, 1937.

York, J. E., R. Cardwell and J. Ni, Seismicity and Quaternary Faulting in China, Seism. Soc. Bull. Am., 66/6, 1983-2001, 1976.

GEOLOGY AND TECTONICS OF THE HIMALAYAN REGION OF LADAKH, HIMACHAL,
GARWHAL-KUMAUN AND ARUNACHAL PRADESH: A REVIEW

Anshu K. Sinha

Wadia Institute of Himalayan Geology, Dehradun 248001, India

Abstract. Geoscientists directly or indirectly associated with the Wadia Institute of Himalayan geology have worked between the extreme NW
end of the Himalaya in Ladakh and the eastern extremity in Arunachal Pradesh (Fig. 1); this paper
summarizes their results.

Throughout the north of the Himalayan arc, the
basement crystalline complex is mainly overlain
or overthrust by Precambrian to Tertiary sediments, whereas in the south Precambrian, Paleozoic, Mesozoic and Tertiary rocks have been overthrust from N to S by a succession of three or
four thrust sheets. The main axial lineament of
the Himalaya is characterized by vertically dipping crystalline rocks, with bivergence in the
disposition of the southern and northern limbs
intruded by younger granites. It has been possible to identify, besides the longitudinal lineaments along the Himalayan strike, transverse
folds and NW-SE lineaments cross cutting the
northwestern Karakorum as far as the Indian plain
in the SE. The syntaxis of the eastern Himalaya
arc over Burma is described. An uplift rate of
0.55 cm/year and the age of the main thrusting of
15 m.y. have been calculated.

1. Research in Ladakh (Northwestern Himalaya)

Since 1976 the Wadia Institute of Himalayan
Geology (WIHG) has organized multidisciplinary
expeditions to Ladakh, covering the basin of the
Indus, Zanskar, Shyok and Nubra rivers. Stoliczka (1865) and Lydekker (1880) first studied the
region. Early workers confined their observations to the western part of the Indus Suture Zone
(Fig. 2) and the information on the northern part,
beyond the Ladakh range, remained limited because
of the inaccessible and difficult nature of the
terrain.

WIHG geologists (Annual Report WIHG 1978-79)
proposed the litho-tectonic subdivisions given
in Table I. The present paper does not follow
these subdivisions. A geologic profile (Fig. 3)
across the Ladakh Himalaya illustrates vividly
the regional tectonics. Sharma and Kumar (1978a)
proposed a comparative litho-stratigraphic sequence of the Ladakh region, to which pertinent

data can be added from the works of Fuchs (1977),
Frank et al., (1977), Pal et al., (1978) to provide a composite picture (Fig. 4).

A. The Litho-Tectonic Units

a. Southern Crystalline("Central Crystalline")
The NW-SE Belt of the Central Crystallines passing through Kulu, Manali, Keylong and Nun-Kun extends as far as Sankoo in the Suru Valley, where
it plunges northward under the Permo-Triassic sequence of the Tethyan realm. The crystallines
contain quartzites, schists, gneisses and granites. This metamorphic belt forms the "root-
zones" of the crystalline nappe which was thrust
southwards; the Keylong-Badrinath-Malari axis of
the Himalaya (Sinha and Jhingran, 1977) lies
within the root zones. At the northern end a
transitional decrease of metamorphic grade from
staurolite-kyanite-bearing schists and gneisses
to quartz-mica schists is observed (Kumar, 1978).
South of Sankoo, a band of biotite-muscovite
granite gneiss occurs as a concordant body within the quartz-mica schist which further northward
grades into low grade greenish schistose phyllite. Schistose Panjal Trap with carbonaceous
phyllites overlie crystallines at Sankoo. The
rocks of the southern metamorphic belt belong to
the quartz-allocite-epidote-almandine subfacies
of the green-schist facies, kyanite-staurolite
subfacies of the epidote-amphibolite facies and
amphibolite facies (Kumar, 1978).

b. Northern (Tso-Morari) Crystallines: These
metamorphic rocks, comprising metasediments,
metabasics, gneisses and granites occupy a NW-SE
belt between the Tethyan Mesozoic complex in the
south and the ophiolitic melange and Indus Suture
Zone in the north. They are grouped in three
formations: Puga Formation, Sumdo (Taglang La)
Formation, and Polokong La and Rupshu granites.

The Puga Formation has been folded into a broad
antiform and synform between More plain and Rums-
te and is uplifted by faulting. (Overall, the
Tso-Morari crystallines form a very large doubly
plunging anticlinal fold, referred to as "dome"
by many workers.) It comprises mainly para-

Fig. 1. Location map of the Himalayan region showing the area of research conducted by Wadia Institute of Himalayan Geology.

gneisses, quartz-sericite schists with intrusive granite and amphibolite and quartz-sericite schists with subordinate marble and basic rocks. Kyanite and sillimanite metamorphism appears near the core and decreases progressively N and S to garnet and biotite zones.

The Taglang La Formation constitutes low to high grade metamorphic rocks derived from pelitic, calcareous and marly sediments. These sediments were intruded by basic rocks which now occur as concordant bands of amphibolites. The rocks show a southward increase in metamorphism from the chlorite zone near Gya, through a biotite zone near Taglang La, to garnet zones around Debring at the northern end of the Morari plain and to a kyanite zone on the western bank of Tsokar. Metamorphic grade decreases again southwards to reach low grade phyllite with bands of marble against the Tethyan Mesozoic sediments along the Morari-plain fault. Between the Spongchen and Pakja nalas - tributaries of the Zarra river - an interesting group of basic rocks form a 3 km wide zone of volcanics interbedded with fine-grained tuffaceous horizons and massive, unsorted, agglomerates and thin bands of limestone and gabbro. These basic rocks differ considerably from the surrounding meta-sediments. They resemble exotics in the ophiolitic melange exposed around Sumdoh and in the Yebat nala (a westerly flowing tributary of Riyul river).

On the northwest bank of Tso Morari the mica gneiss of the Puga Formation is overlain by 100 m of calcareous schist and a 50 m band of crystalline limestone. This band has yielded deformed, bivalve shells and crinoid stems and also contains thin layers of phosphatic limestone. Similar limestone have been also found in the Taglang La Formation at localities near Lato Prang, Sumdo, Ghaksa La, Kiamaru La and Shiul La. Micro fossils of conodonts viz. Gondolella rosenkrantzi Bender and Stoppel and foraminifera Hyperammina, sp., Thurramminoides sp., Hemidiscus sp., Lituotuba sp. and Ammobaculites sp., and ostracod By-

thocypris sp. have been recovered by Azmi from the Tanglang La limestone, and indicate a Permian age (Virdi et al., in press). This shows that the Puga Formation is also of late Paleozoic age. The Tso Morari Crystallines were considered to be Precambrian and were correlated by earlier workers with the central crystallines. The discovery of microfossils in Taglang La Formation indicates that the Tso Morari crystallines are metamorphosed and granitized sediments of Permian to Triassic age. Sharma and Kumar (1978) regard the patches of metamorphic Permo-Carboniferous rock as eroded outliers of the thick sediments deposited over the crystallines by the transgressive Tethys Sea.

Polokong La granite, Rupshu granite: The coarse grained, foliated and porphyritic granite with muscovite, biotite and amphibole of the Polokong La, has a fairly wide distribution and occurs as a stock intruding the Taglang La (Lower Permian?) and Puga Formations.

The Rupshu granite is a concordant body within the Taglang La Formation. It is coarse-grained, porphyritic, foliated to unfoliated and contains muscovite, biotite, hornblende and tourmaline. Xenoliths of phyllite, sandstone and gabbro occur. Surrounding metasediment have been hornfelsed, and intruded the coarse-grained granite.

c. Panjal Trap: In the Suru valley section rocks of the Indus Suture Zone are thrust against green basaltic flows and green schistose tuffs which are tightly folded and are overlain by carbonaceous and pyritiferous phyllites and slates which further eastward grade into grey shales near Mulbeckh, Bodhkharbu and Lamayuru and have yielded Triassic fossils.

In the Bodhkarbu-Sirwastu La section, near Yukimal (Fig. 2), a 10 m thick highly shattered basic body, intensely folded, has been observed. The body is equivalent to the Panjal Trap which occurs below the Triassic shale at the Zoji La. Its presence as far as Bodhkharbu and Padam (in

124 SINHA

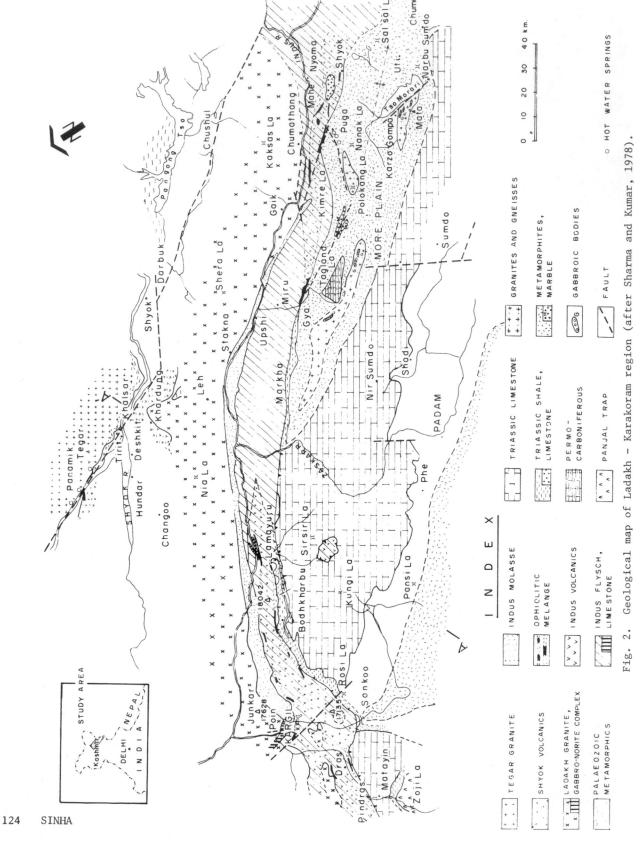

Fig. 2. Geological map of Ladakh – Karakoram region (after Sharma and Kumar, 1978).

I N D E X

INDUS MOLASSE

OPHIOLITIC MELANGE

INDUS VOLCANICS

INDUS FLYSCH, LIMESTONE

TRIASSIC LIMESTONE

TRIASSIC SHALE, LIMESTONE

PERMO-CARBONIFEROUS

PANJAL TRAP

GRANITES AND GNEISSES

METAMORPHITES, MARBLE

GABBROIC BODIES

FAULT

○ HOT WATER SPRINGS

TEGAR GRANITE

SHYOK VOLCANICS

LADAKH GRANITE, GABBRO-NORITE COMPLEX

PALAEOZOIC METAMORPHICS

STUDY AREA

Kashmir

DELHI NEPAL

I N D I A

TABLE I. Litho-tectonic division of Ladakh region (from N to S)

Karakoram batholith	Mio- Pliocene (N_1-N_2)
_____Shyok Thrust_____	
Shyok Fmn	Permo-Trias (P-T)
_____Khalsar Thrust_____	
Khardung Volcanics and Volcano-sedimentaries	Cret- Eocene $(K_2-\underline{P})$
_____Intrusive Contact_____	
Ladakh Batholith	Cret. (K)
_____Transgressive Contact_____	
Karu Fmn (Kargil Molasse)	Mio- Plio (N_1-N_2)
_____Upshi Thrust_____	
Indus Fmn	Cret + Eoc $(K_2-\underline{P})$
_____Indus Thrust_____	
Nidar Ophiolites	Cret. (K)
_____Thrust_____	
Sumdo Ophiolitic Melange	Cret. (K)
_____Zildat Thrust_____	
Tso Morari Crystalline (Thaglang La Fmn (Puga Fmn (	Up. Paleo. (PZ)
_____Morey Plains Fault_____	
Tethyan Mesozoic Sediments	Mesoz. (MZ)

the Zanskar region), indicates that Panjal vol-
canicity was not only restricted to the Kashmir
basin, but extended up the western margin of the
Spiti basin.

d. **Tethyan Mesozoic Sediments: Permian Limestone and Trias Shale:**
Permian Limestone: A small lensoid body of Permian limestone is exposed between Triassic phyllites and slates and the Indus Suture Zone rocks.
This white, pink and grey limestone is brecciated;
it has yielded Permian fossils, Paradoxiella sp.,
Pachypholia sp., Colaniella sp., Hertschia, Leon-
ardophyllum, Waagenophyllum, Frondicularia, Sul-
coretepora Liosphaera, Ethmosphaera, Lithomitra,
Dictyomitra (Tewari and Pande, 1970; Sharma and
Shah, 1977). Smaller lensoid bodies of the Permian limestone north of Fatu La in contact with
the Indus Suture Zone are truncated westward
where the Triassic phyllites come in contact with
the Indus Suture Zone.
Triassic Shale: The shales above the Panjal Trap
and the Permian limestone have been variously
folded and metamorphosed. In general they consist of carbonaceous and calcareous shales with
interbedded limestone, slates, phyllites, schistose phyllites and crystalline limestone. Near
Bodhkharbu, the limestone has yielded Triassic
bryozoa, crinoid stems and corals (Pal and Mathur,
1977). The contact between the limestone and
the enclosing incompetent Triassic slates and
phyllites is highly tectonized.
As observed near Lamayuru, Bodhkharbu, and
Sankoo, the basal Triassic shales, slates and
phyllites are carbonaceous and pyritiferous. The
carbonaceous shales and slates above the Panjal
Trap near Yukimal yielded a broken ammonite. The
unit contains also dark grey limestone bands (5

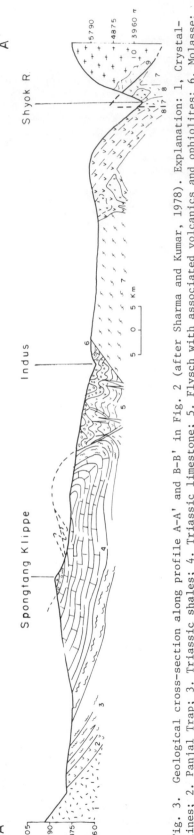

Fig. 3. Geological cross-section along profile A-A' and B-B' in Fig. 2 (after Sharma and Kumar, 1978). Explanation: 1, Crystallines; 2. Panjal Trap; 3. Triassic shales; 4. Triassic limestone; 5. Flysch with associated volcanics and ophiolites; 6. Molasse; 7. Ladakh granite; 8. Shyok volcanics; 9. Paleozoic host rocks; 10. Tegar granite.

to 20 m thick), calcareous shales and siltstones. Above the carbonaceous shales and slates, dark grey phyllites, slates, calcareous shales and siltstones near Fatu La contained flattened ammonites.

e. Kioto Limestone: The Triassic shales and phyllites are in contact to the south with a wide zone of limestone, partly dolomitic, which outcrops from Matayin to Rosi La, Kingi La, Zang La to the More plain and further east, except in the Suru valley where it is pinched out. Near Matayin a thick sequence of Triassic limestone shows tight folds with a low angle SE plunge (Fig. 2). Near Pindras, this limestone is thrust over the Indus Suture Zone rocks. The sequence below this limestone, towards the northwestern closure of synclinal though, has been truncated by the thrust. The appearance of underlying Triassic shales, southeast of Dras, is also as a result of an oblique cut of the NW-SE trending syncline by the thrust.

f. Indus Suture Zone: The structural highs (southern and northern Crystallines) and the low separating them (Tethyan basin) is obliquely cut by the WNW-ESE trending 'Indus Suture Zone'. This zone is intensively and complexly folded. However at places, it has been possible to differentiate pelagic and neritic sediments, volcanics and ophiolitic melange as follows:
    4. Indus molasse
    3. Indus flysch
    2. Ophiolitic melange and deep sea sediments
    1. Indus volcanics and associated sediments

1. Indus Volcanics (Dras Volcanics): Between the Indus flysch formation in the north and Tso Morari crystallines in the south lie the ophiolites and ophiolitic melange sequence of the Sumdo Formation which is considered to be the lower part of the flysch as a whole. The belt can be traced SE of Rumtsey through Kiameru La, Thachang La, Shiul, Kidmang, Zildat La, Sumdo Chunglung La and Nidar; southeast of Nidar the belt continues through Hanley and extends into Tibet. The belt which is only few tens of metres wide SE of Rumtsey widens to a maximum of some 20 km in Nidar-Kyun Tso section. A number of isolated klippen of ophiolitic melange and serpentinite rest on the Tso Morari crystallines indicating the wide areal extent of the Sumdo Formation prior to erosion. The ophiolitic melange forms the southernmost zone and consists of agglomerate, volcanic tuffs, volcanic serpentinites and limestone. Isolated lenticular blocks of limestone and serpentinite occur in a volcanic groundmass. This belt contains a band of glaucophane schist NW of Sumdo (Virdi et al., 1977), SE of Sumdo, in Ribal Phu and NW of the northern Kyun La on the eastern slopes of Nenacle peak (6,370 m). The glaucophane schist is overlain by highly deformed and metamorphosed phyllitic rocks

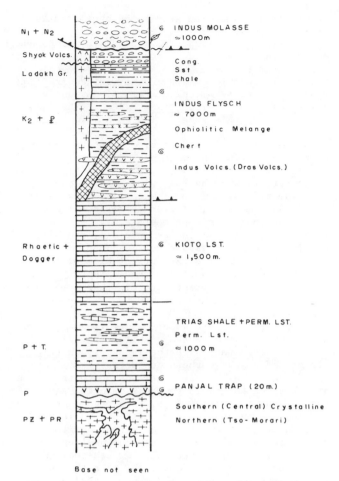

$N_1 + N_2$	INDUS MOLASSE ≈1000m
Shyok Volcs.	Cong. Sst Shale
Ladakh Gr.	
$K_2 + P$	INDUS FLYSCH ≈ 7000m
	Ophiolitic Melange
	Chert
	Indus Volcs. (Dras Volcs.)
Rhaetic + Dogger	KIOTO LST. ≈ 1,500 m.
P + T.	TRIAS SHALE +PERM. LST. Perm. Lst. ≈ 1000 m
P	PANJAL TRAP (20 m.)
PZ + PR	Southern (Central) Crystalline Northern (Tso- Morari)

Base not seen

Fig. 4. Tectono-stratigraphic column of the Ladakh-Karakoram region (based on the data of Fuchs, 1977; Sharma and Kumar, 1978; Pal, Srivastava and Mathur, 1978).

derived from pelites, basic volcanics and agglomerates. Chromite occurs in serpentinite near Kiameri La. The serpentinite-peridotite zone occurs N of the melange zone from which it is separated by a tectonic contact: it consists of ultrabasic rocks serpentinized to varying degrees. The zone is about 2 km wide N of Tso Kar, and some 8 - 10 km wide in the Kyun Tso Nidar section. The serpentinites contain podiform chromite bodies (near Chaksa La on the way to Kidmang); magnesite is associated with peridotites N of Tso Kar and chromite with serpentinite in a klippe at Karzok. The gabbro zone N of the serpentinite zone, consists of layered gabbro and other basic rocks which intrude the serpentinite-peridotite, as clearly seen at Zildat La, and in Chunglung nala. In the Nidar river section, at Churak Sumdo, the contact between gabbro and serpentinite is characterized by an agglomerate with fragments of serpentinite in a gabbroid groundmass. Leucocratic intrusions in the gabbro may be late stage differentiates of basic magma. Very coarse-

grained gabbro dykes cut the pillow lavas which constitute the uppermost zone of the ophiolitic sequence: pillows vary in size from 0.5 to 1 m across. They are well exposed in the Kalra nala in the Nidar section where they are interbedded with blue cherts and jaspers. The serpentinite-peridotite zone lies in tectonic contact with the Indus Formation.

Shah and Gergan (1978) studied the western extension of the Indus Suture Zone in the Kargil-Dras area. Here the Dras Formation constitutes the major unit and is composed of volcanics and interbedded sediments, red and olive green shales interbedded with radiolarian cherts). The cherts have yielded well preserved fossils of Orbitolina lenticulata var. trochus, O. lent. var. chitralensis radiolarica Lisosphaera sp., ellipsodium sp., stavrostylus sp., chitonastrum sp., lithostrobus sp., Pelecepod Hippurites sp., Bryozoa Reptomuthcava ladakhensis sp. nov. The assemblage is indicative of Middle Cretaceous.

The Dras Formation also includes two zones of ophiolitic melange consisting of serpentinized peridotite, dunite, radiolarites, olistostrome limestone associated with shales and basic intrusives.

2. Ophiolitic Melange: The ophiolitic melange is emplaced along the fracture zones in the Indus volcanics and the associated sedimentary rocks. Two distinct zones occur in the northern and in the southern margins of the Indus Suture Zone, varying in thickness from a few meters to tens of meters. North of Dras, this zone is marked by the emplacement of a chromite- and platinum-bearing serpentinite body. The eastwards extension of this melange zone in the Suru valley lies south of Trizpen, where it is quite wide and contains related limestone pebbles and shows intense shearing. Sharma et al. (1978) have reported a K-Ar age of 77.5 ± 1 m.y. for the volcanic rocks associated with the ophiolitic melange zone near the thrust contact with the Triassic shale. Shah and Sharma (1977) describe Globotoruncana sp., Heterohelix sp., Rugoglobigerina sp., Liosphaera sp., Cenosphaera sp., Dictyomitra sp. and Sethocyrtia sp., of Middle Cretaceous age from chert and jasperite beds of the ophiolitic melange around Mulbekh. This zone is very thick and ultrabasics of various composition, e.g., dunites, peridotites, pyroxenites and serpentinite, occur. Near Sumdo these ultramafics associated with green volcano-sediments (green phyllites) lie in contact with the Puga schists and gneisses along the Zildet fault along which serpentinite bodies of varying dimensions have been emplaced. The sequence observed is:

Ophiolite melange	Jasperoid shales, sandstones and conglomerates
	Limestone (cherty)
Indus Volcanics	Radiolarian chert, jasperoid shales and tuffs

3. Indus Flysch: The Indus flysch constitutes a NE-SW trending belt about 5-10 km wide. The belt is separated from the Ladakh granite and Karu molasse in the north by the south dipping Upshi Thrust. Towards the south, the belt is in tectonic contact with the Zaskar sediments, the Tso Morari crystalline, the Nidar ophiolites and Sumdo ophiolitic melange. This tectonic line is the Indus 'suture', also referred to as the Indus Thrust.

Pal and Mathur (1977) have divided the Indus flysch into five members. E. Interbedded slates and conglomerates (259 m), D. Red shales (200 m), C. Grey shales (250 m), B. Sandstones (150 m), A. Red and green slates with basics (400 m) (bottom).

The Indus Formation has yielded a rich marine fauna of bivalves, gastropods and foraminifera besides corals, ostracodes and algal remains. On the basis of this, Pal and Mathur (1977) have proposed the following biostratigraphic zones which range from Cretaceous to Eocene.

(I) Orbitolina zone (zone I) (Aptian): The zone contains Orbitolina parma Fossa Mancini, O. discoidea Gras, O. sp. cf. O. morelensis Ayala-Castraneres, 'Hippurites' sp. I, 'Hippurites' sp. II, besides unidentifiable oysters, cephalopods, and foraminifera. The occurrence of Orbitolina parma and O. discoidea suggests an Aptian age.
(II) Plant fossils zone (zone II) (Middle to Upper Cretaceous): This zone contains plant impressions, trace fossils, and worm burrows. Although this zone has yielded no characteristic fossil, it has tentatively been assigned a Middle to Upper Cretaceous age.
(III) Pitar-Callista zone (zone III) (Paleocene-Recent): Grey gritty sandstones and shales have Callista semilunaris and Pitar (Calipitaria) carteri. Although none of these taxa is a guide fossil, the stratigraphic position of the zone strongly suggests a Paleocene age. The bivalve genera Callista and Pitar range in age from Paleocene to recent.
(IV) Clio zone (zone IV) Paleocene - Recent: In the Miru section in addition to Clio sp., also occur: Trachycardium halaense, Venericardia mutabilis, Potamides pascoei, Melania marginata, and "Turritellids?"
(V) Assilina sp. ex. gr. A. daviesi var. nammalensis zone (zone V) (Lower Eocene): This zone is separated from the underlying zone by the first appearance of characteristic Lower Eocene species, namely Assilina granulosa. In addition Assilina sp. ex. gr. Adavesi nammalensis, Nummulites atacicus, and N. manilla occur. The zone is represented by unfossiliferous conglomerates in the Miru section.
(VI) Ostrea zone (zone VI): Besides oysters, the zone contains "Turritellids"? Cerithium hookeri, Venericardi mutabilis, and Pelecyora (Cordiopsis), subathuensis.
(IX) Barren zone IX: This zone is exposed in the Kunda La-Miru sections where it is represented by unfossiliferous red to grey sandstones, shales and conglomerates.
(X) Assilina granulosa zone (zone X)(Lower Eocene): A 100 m thick sequence of conglomerates and red to greenish-grey shales. The zone is developed only in the Miru section where it has yielded Assilina granulosa Form A, A. sp. ex. gr. A. daviesi var. nammalensis, and Nummulites atacicus. The faunal assemblage of this zone indicates a Lower Eocene age.

Sedimentological work carried by Srivastava as discussed in WIHG Annual Report (1978-79), show that the Indus flysch presents typical flysch characteristics. The following megacycles have been identified: I. Conglomerate-sandstone-shale, II. Sandstone-shale, III. Conglomerate-sandstone, IV. Sandstone-Shale Conglomerate, V. Shale-sandstone-limestone, VI. Shale-sandstone, and VII. conglomerate-shale sandstone.

In most of the area the current direction is from SSE to NNW. Though in the northern part of the area a NW-SE current direction has been noted, thus indicating two source areas.

4. Indus (Kargil) Molasse: The molasse lies unconformably on the Ladakh granite (Fig. 2) and is in tectonic contact with the Indus flysch (Fig.3). Its northern and southern contacts are well exposed near Karu. It is confined to a narrow linear belt from Kargil in the west through Hagnis, Dumker, Saspol, Nimmu, Karu, Upshi, to south of Kiari. In the eastern sector this unit is well exposed, resting on the Indus flysch and the associated volcanics and ophiolitic melange in the Chunglung-Liyan area. The molasse near Kargil occurs widely but tapers out in a thin belt towards the east where it forms detached outcrops unconformably overlying the Samdo Formation. Locally the underlying contact between the Ladakh granite and the Indus flysch has been exposed and is generally unconformable as well seen in the Mahe-Nyoma section. At Kargil, the molasse sequence consists of alternating red shales, grey sandstones and conglomerates. The conglomerate pebbles are of gabbro, norite, basics, red shales, chert, limestone, etc., of variable sizes, cemented in a shaly arkosic matrix. Near Nimmu and Likir the grey shales and sandstones showing current bedding, ripple marks, etc., overlie the granite and the conglomerate beds here contain more granitic pebbles than near Kargil. Near the southern limit of the molasse, pebbles of Indus flysch have been found embedded in the red shale. Towards the end of the period of flysch deposition, the orogen appears to have migrated from the south towards the exterior, or north, displacing the area of marine sedimentation to the north where miogeosynclinal conditions reigned.

Fresh water molluscs (unio), vertebrates (Hypboops, femur) and plant fossils (palm-leaf, Sabal sp.) suggest a Miocene-Pliocene age for the Indus molasse (Dixit et al., 1971; Shanker et al., in press).

g. _Ladakh Granite_: The Ladakh batholith is exposed along the NW-SE trending Ladakh range which runs parallel to the Indus Suture Zone. The composition varies from tonalite and granodiorite to granite. Three zones in general have been recognized, viz., aureole, border and core zones.

The border zone is mostly exposed along the southern flank of the Ladakh range between Upshi and Khalsi. It is mainly tonalitic and granodioritic with occasional tongues of hornblende-poor pink granite as noticed near Khardung La. This zone is also characterized by the emplacement of acid (aplitic and pegmatitic) and basic (doleritic) dykes as well as gabbroic-noritic intrusives as noticed near Kargil, Likir and Karu. One such igneous complex has been found intruding the contact zone of the ophiolitic melange and volcanics with granites near Kargil. Near Kargil a younger leucogranite phase intrudes the volcanics and younger basic (doleritic) veins intrude these granites.

The core consists of a pink porphyritic granite and leucogranite which are more or less separated in space and time. Pink porphyritic granite from the core zone of the batholith exposed near Hemiya has given a K/Ar age of 27.8 ± 0.6 Ma (Oligocene) Sharma et al. (1978a). The leucogranite which intrudes the pink porphyritic granite east of Giak and also occurs at a number of places such as south of Kargil, is therefore, still younger and may be Lower Miocene in age. It is suggested that the crystallization of the Ladakh batholith (pink porphyritic granite and leucogranite) continued up to 27.8 ± 0.6 m.y. and possibly beyond into the Lower Miocene.

h. Shyok Volcanics: As mentioned above a suite of volcanic rocks overlies unconformably the Ladakh granite near Khardung and continues northwards to the Shyok valley (Fig. 2); it has been named the Shyok volcanics (Sharma and Gupta, 1978). The suite comprises volcano-sedimentary and volcanic rocks, including basalts, spilites (?), andesites, trachytes, keratophyres, rhyolite, serpentinite, epidosite, chert, tuff, volcanic breccia, jasperoidal shales, etc.: it is some 8 km thick. K/Ar dating of the basal volcanics near Khardung has given an age of 28 ± 2 m.y. (Upper Eocene), (Sharma et al., 1978a). The overlying volcano-sedimentary rocks might be of Upper Eocene or Lower Oligocene age. From the point of view of magmatism associated with tectonism, it is significant that the Shyok volcanics represent a much younger phase of eruption and are a separate unit in space and time compared with the Dras volcanics (77.8 ± 1 m.y.).

Towards the base of the Karakorum range, the Ladakh granite and the overlying Shyok volcanics are overthrust by the Karakorum batholith and its Paleozoic (?) host metamorphic rocks.

B. Tectonics of the Ladakh Region

The impetus to interpret the tectonics of the region gathered momentum after the pioneer work of Wadia (1931) on the syntaxis of the northwest Himalaya, its tectonics and orogeny. The peripheral fractures of the heterogenous batholithic Ladakh granite are due to structural deformation and doming of the basement ridges (Sharma and Kumar, 1978). The Shyok volcanics, lying along the northern fault, are fractured and mylonitized near the fault zone. Sharma and Gupta (1978) traced this fault through Tirit to the Nubra valley in the NW (Khalsar-Tirit fault).

The Karakorum fault (Molnar and Tapponier, 1975), Nubra-Gartang lineament (Gupta and Sharma, 1978) or Karakorum-Gandak lineament (Kumar and Sharma, pers. com.) can be clearly seen in Landsat and Met-satellite imageries. The lineament, marked by a prominent geomorphic pattern cutting the Himalayan arc at an oblique angle, is, according to geophysical and depth contour data, a recent fault with dextral movement. Along the Nubra-Shyok thrust, the Karakorum granite along with its metamorphic (? Paleozoic) envelope has been thrust over the Shyok volcanics. Serpentine-bearing ophiolites have been noticed along this thrust (Sharma et al., 1978a). The thrust has been offset by transverse faults such as the one near Panamik (Fig. 2); it is probable that these faults are active and of a comparatively young age.

The attention of geologists world-wide has been attracted to the complicated ophiolites of the Indus zone in relation to its interpretation as "Indus-suture" by Gansser (1964), as perhaps the relict of Eurasian and Indian plate boundaries. Subsequently Sinha and Jhingran (1977) interpreted this lineament as the Gilgit-Dras-Darchen-deep seated mantle fault associated with ophiolites and indeed the system of faults occurs throughout this suture zone parallel to the main trend as well as _en echolon_. This zone has a faulted north boundary whereas the southern boundary is overthrust. Cross-cutting N-S faults occur in the Kargil area.

According to Pal et al. (1978), the Indus Formation has undergone three fold events dated as Upper Eocene (F_1) Middle Miocene (F_2) and Upper Pliocene (F_3). The authors also describe that the flysch is thrust over the Ladakh granite in the north along one Bazgo-Upshi thrust. They inferred that the Indus and Dras furrows came into existence in mid-Mesozoic time, representing mio- and eugeosynclinal troughs respectively. Tectonic effects along northern and southern margins are seen in the Tethyan sediments. Faults trend NW-SE. The Tso Morari crystallines are faulted to the south against the Tethyan sediments, along the More plain - Skio fault. Another fault follows the axis of anticline of the Tso Morari crystallines. The Zildet fault brings the crystalline in juxtaposition with flysch in the Puga valley. Some N-S or NNE-SSW faults oblique to the basement lineaments observed in Padam-Zang La section, Zara Mala-Morang La section, Tsokar-Sumdo Puga section Rusi La - Soman section, the western margin of Tso Morari lake, etc., indicate a younger tectonic phase.

The anticlinal folds of the northern crystalline plunge both in SE and NW and seems to have a direct bearing on the Tethyan sedimentation of Kashmir and Spiti region.

II.   Research in the Himachal Himalayan Region

The geology of Simla region in the Himachal Pradesh was studied at the end of last century. It was in the Simla Hills where the revolutionary nappe hypothesis (Pilgrim and West, 1928) was introduced in the Himalayan mobile belt. Since then continuous efforts have been made to solve the stratigraphic problems in order to explain the tectonics of the area. From south to north the region (Fig. 1) may be divided into the following structural-facies zones:
1. Molasse zone of the Siwalik,
2. Tertiary sedimentary zone,
3. Allochthonous Krol carbonate-terrigenous tectonic unit,
4. Parautochthonous flyschoidal Simla and carbonate-terrigenous Shali Formations,
5. Chail tectonic unit with early geosynclinal sedimentary facies,
6. Rampur Formation in a tectonic window,
7. Allochthonous metamorphosed crystalline Jutogh unit forming the uppermost thrust-sheet,
8. Vaikrita Group of high grade crystallines.

Since the beginning of geologic studies tectonic interpretation has been based on the inversion of metamorphism of the exposed rocks around the Simla Hills. Except the well established Tertiary fauna nowhere have any authentic megafossils been discovered to provide conclusive evidence as to the age of the various stratigraphic horizons. In the absence of age data neither could the correct configuration of the sedimentary basins be reconstructed, nor could a consensus be reached on the stratigraphic succession. Consequently, the tectonic evolution remains obscure. Recently WIHG scientists have attempted to determine the age of Shali carbonate, Krol, Simla, Chail and Jutogh Formations by searching for microfossils, fossil algae, and by isotope dating of igneous and metamorphosed suites of acid and basic rocks.

A.   Geological Setting: The geology of the area (Sinha, 1978, in press) as shown in the geologic map of the Simla hills area (Fig. 5), in the tectono-stratigraphic column (Fig. 6) and in the geologic profiles (Fig. 7) is based on the author's decade of field work in the area and on published material as cited in the text and references.

B.   Litho-Tectonic Units
a.   Vaikrita Group (Inner Crystalline Unit): This lithotectonic unit consists of high grade metamorphic rocks (kyanite-sillimanite grade) and is thrust over the Jutogh Formation (Outer Crystalline Unit) along the Vaikrita thrust as

shown by a break in metamorphism (i.e., from sericite-muscovite schist with tiny garnets to kyanite-sillimanite-bearing gneisses) and a zone of mylonitized gneisses, quartzites and calc-silicate rocks. The term Vaikrita thrust was originally suggested by Valdiya (1973) as demarcating the base of his Joshimath Formation of the Vaikrita Group of high grade metamorphic rocks. A thick pile of these metamorphics consists of garnetiferous staurolite-kyanite schists and gneisses, sillimanite gneisses, migmatites, quartzites, marble, calc-silicate rocks and amphibolites. These rocks are intruded by medium- to coarse-grained biotite, granite, tourmaline-bearing leucogranite and an associated pegmatite phase.

b.   Chor Granite: The Chor granite forms a prominent feature of the Chor mountain southeast of Simla town and is the last granitic outcrop of the outer granite belt. It is a well foliated porphyritic biotite granite, circular in outcrop (Dixit, 1973), and is concordant with the surrounding metasediments of the Jutogh Formation. The heterogenous granite contains numerous inclusions of schist and quartzite. The central part (7.5 km^2) is composed of massive magmatic granites, surrounded by metasomatic granite gneiss (114 km^2). K/Ar dating of biotite from granites and granite gneisses has given ages of 1000 m.y. and 246 $\pm$ 9 m.y. respectively. Micaschists surrounding the granite gneisses have been proved to be youngest, being 56 $\pm$ m.y.

c.   Jutogh Formation: The name Jutogh Formation has been assigned to a lithological association comprising schists, phyllites, slates, meta-quartzites, limestone and gneisses which Pilgrim and West (1928) had described as the Jutogh Series. The following litho-units can be recognized:
3. Mica schist member,
2. Carbonaceous schist-limestone member,
1. Metaquartzite and quartz schists (Boileaugunge member).

Geochronology (Ashigirei et al., 1976, 1977): K/Ar dating on Jutogh rocks has given the following ages:

Sample No.
17-64   Biotite-chlorite schists, (Simla)(whole rock) 23 $\pm$ m.y.
35/19   Garnetiferous schist, (Chaupal)(biotite) 30 $\pm$ m.y.
92-52   Crystalline schist, (Chor-Didar-Naura) (muscovite) 37 $\pm$ m.y.
80-64   Chlorite graphitic schist, (Simla)(whole rock) 42 $\pm$ m.y.
129-44   Garnetiferous schist, (Kulu)(biotite) 85 $\pm$ m.y.
13-64   Biotitic Boileaugunge Qzt. (Simla)(whole rock) 86 $\pm$ m.y.
0-64   Muscovite Boileaugunge Qzt. (Simla) (whole rock) 120 $\pm$ m.y.
16-64   Biotite Tourmaline Boileaugunge (Simla) (whole rock) 123 $\pm$ m.y.

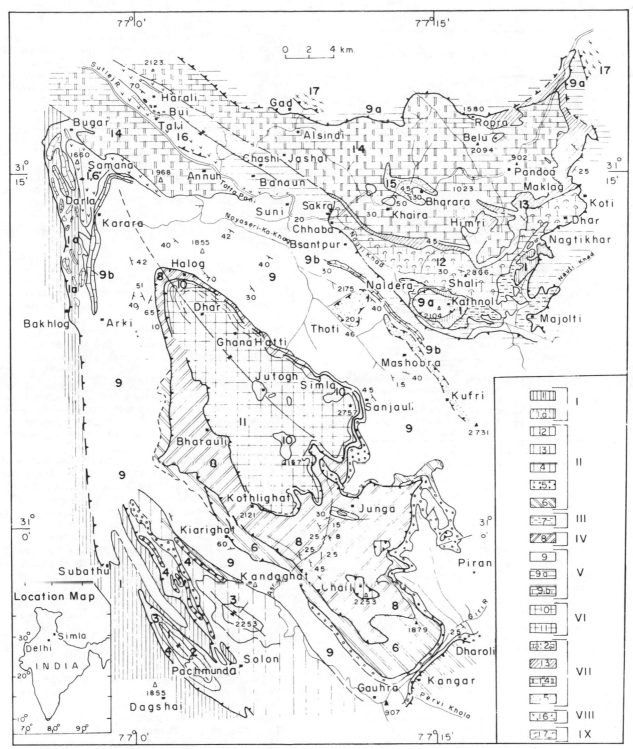

Fig. 5. Geological map of Simla Himalaya, Himachal Pradesh (after Sinha, 1978, compiled from author's observation and other data as mentioned in referred paper). I. Tertiary; 1. Undifferentiated; 1a. Paleocene; II. Krol thrust sheet; 2 & 3 Krol A, B, C, D; 4. Infra Krol; 5. Blaini; 6. Jaunsar; III. Madhan slates; IV. Chail; V. Simla unit; 9. Simla flyschoidal type in general; 9a. Basantpur; 9b. Naldera and Kakarhatti Stromatolitic limestone; VI. Jutogh; 10. Graphic schist and limestone; 11. Boileaugunge quartzites, mica schist; VII. Shali unit; 12. Upper Shali stromatolitic limestone; 13. Shali slates; 14. Lower Shali stromatolitic limestone; 15. Khaira quartzite; VIII. Basic rock suite; IX. Granito-gneiss.

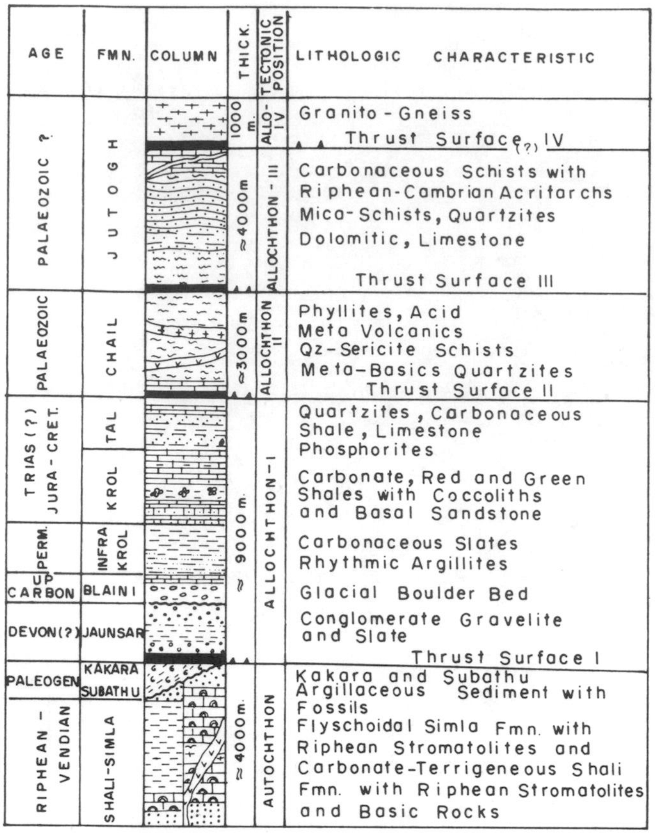

AGE	FMN.	COLUMN	THICK.	TECTONIC POSITION	LITHOLOGIC CHARACTERISTIC
PALAEOZOIC ?	JUTOGH (H)		1000	ALLO-IV	Granito-Gneiss / Thrust Surface(?) IV
PALAEOZOIC	JUTOGH		~4000m	ALLOCHTHON-III	Carbonaceous Schists with Riphean-Cambrian Acritarchs, Mica-Schists, Quartzites, Dolomitic, Limestone / Thrust Surface III
PALAEOZOIC	CHAIL		~3000m	ALLOCHTHON-II	Phyllites, Acid Meta Volcanics, Qz-Sericite Schists, Meta-Basics Quartzites / Thrust Surface II
TRIAS (?) JURA-CRET.	TAL			ALLOCHTHON-I	Quartzites, Carbonaceous Shale, Limestone Phosphorites
	KROL		9000m		Carbonate, Red and Green Shales with Coccoliths and Basal Sandstone
UP PERM	INFRA KROL				Carbonaceous Slates Rhythmic Argillites
UP CARBON	BLAINI				Glacial Boulder Bed
DEVON(?)	JAUNSAR				Conglomerate Gravelite and Slate / Thrust Surface I
PALEOGEN	KAKARA SUBATHU			AUTOCHTHON	Kakara and Subathu Argillaceous Sediment with Fossils
RIPHEAN - VENDIAN	SHALI-SIMLA		~4000m		Flyschoidal Simla Fmn. with Riphean Stromatolites and Carbonate-Terrigeneous Shali Fmn. with Riphean Stromatolites and Basic Rocks

Fig. 6. Tectono- stratigraphic column of the Simla-Himalaya (after Sinha, 1978).

The uppermost thrust sheet consisting of crystalline schists and garnet gneisses (The Upper "Garhwal Nappe") has shown a wide variation of radiometric dates:

70X Biotite gneiss, Sarahan, north of Rampur (biotite) 13 ± 2 m.y.

M-15, Migm. quartz-biotite crystalline schist, Manali (Beas River)(biotite) 37 ± 3 m.y.

116-41, Biotite schist (Manali Beas River) (biotite) 49 + 3 m.y.

72/40, Biotite gneiss migmatized, Nirth (biotite) 50 ± 3 m. y.

3/3, Kyanite-staurolite migmatized cryst. schist, Manali (biotite) 59 ± 3 m.y. (Pande and Kumar, 1974).

B-1, Biotite-porphyroblast cryst. schist (Manali) (biotite) 73 ± 5 m.y. (Pande and Kumar, 1974).

20-B, Gneiss (collection of R. N. Chatterjee) Chor (whole rock) 148 ± 5 m. y.

33, Garnet-micaschist (collection of R.N.Chatterjee) Chor (whole rock) 160 ± 14 m.y.

Garnet gneiss (collection of R. N. Chatterjee) (Chor) 250 m.y.

Ko-20, Garnet- mica schist, Manali (biotite) 730 ± 20 m.y. (Pande and Kumar, 1974).

The rocks of the Jutogh Formation differ from those of the Vaikrita Group in their grade of metamorphism and the nature of the granitic bodies.

They are predominantly of garnet grade and staurolite-kyanite may be rarely developed. Sharma (1976) studied the Jutogh Formation of Nirth and Rampur area and described the migmatization of meta-pelites and meta-semipelites that have been migmatized into banded and augen migmatites.

d. Rampur Formation: This formation composed mainly of quartzites with chloritic phyllite, metavolcanics at appreciable thickness and in a number of intrusive layers and paragneisses. It is exposed in the Rampur window of the Sutlej gorge at Rampur-Bushair township. Rb/Sr dating of 1840 ± 150 m.y. (Frank, 1975) of the associated porphyroid gives a Precambrian age equivalent to the Berinag quartzite of the Kumaun Himalaya.

e. Chail Unit: The name "Chail Series" was given by Pilgrim and West (1928) to a series of talcose quartzite, quartz schist and phyllites and the carbonaceous slate (the Jutogh Series) of Rajgarh.

Different types of black, carbonaceous (graphitic), dark-grey and greenish-grey shales converted into phyllites and black slates constitute the major part of Chail complex. They have been regionally metamorphosed upto the grade of green schist facies. In the Simla and Kulu area the slate-phyllite component constitutes up to 60-70%.

The sandstones are a subordinate component in the Chail. They are oligomictic, greywacke, and tuffaceous.

The volcano-sedimentary origin of the above rocks is proved by their extreme irregular granularity, the presence of well rounded, angular grains of quartz and plagioclase; fragments of microgranular rocks of volcanic appearance; and a wide and common occurrence of albitization, characteristic of eugeosynclinal volcanic origin. According to chemical analysis the sodium content in the tuffaceous sandstones in conspicuously higher than that of potassium (Ashgirei et al., 1977). Moreover, in the greywacke the fragments of disseminated quartz with corroded embayment inlets and inclusion of glass, typical of very shallowly extruded volcanic rock, have been identified.

From the Nirth area further north of Simla, Virdi (1976) describes the Chail Formation as comprising phyllite, mica schist and mylonitic gneiss overlain along a thrust contact by the Jutogh Formation containing schistose quartzite, granet mica schist and graphitic schist with marble. The Rampur Formation underlies the Chail Formation with a thrust contact. Structurally the Chail and Jutogh are allochthonous in this area and the Rampur is parautochthonous. The rocks exhibit three phases of deformation, the first of which affected both the allochthon and autochthon, but is particularly well marked in the former. This phase of deformation produced major isoclinal to recumbent folds associated with an axial planar schistosity. Towards the end stages of this tectonic phase, the crystalline rocks were thrust over the sedimentaries.

Geochronology: K/Ar dating on rock of the Chail tectonic unit has given the following results:
45/53 Quartz-chlorite schist (Pandoh)(muscovite) 28 $\pm$ 2 m.y.
144/62 Quartz-chlorite schist (Pandoh) (muscovite) 32 $\pm$ 2 m.y.
32/14 Quartz sericite schist (Simla)(sericite) 45 $\pm$ 2 m.y.
24-10 Porphyroid (tuffo-argillite)(whole rock) 63 $\pm$ 2 m.y.
20-26 Muscovite-chlorite schist (whole rock) 63 $\pm$ 2 m.y.
70-28 Sericite schist (Pandoh) (sericite) 89 $\pm$ 6 m.y.
26/66 Quartz-sericite schist (Kathlighat) (sericite) 124 $\pm$ 9 m.y.

The analyses on the minerals were done in the laboratory of the Institute of Geology of the Ore Deposits, Petrography, Mineralogy and Geochemistry; Academy of Sciences, USSR, Moscow, and analyses on the whole rocks in the laboratories of Institute of Geology and Geophysics, Siberian Branch of Academy of Sciences, USSR, Novosibirsk.

f. Simla Turbiditic Flyschoidal Formation:
Certain well bedded grey slates (without sandstone lenses or carbonaceous matter), underlying the limestone and boulder bed of Blaini, were first referred to as "Simla Slates" or Infra-Blaini by Medlicott (1864). Pilgrim and West (1928) included dark unaltered slates and micaceous sandstones, limestone with pseudo-organic structures and also possibly the Deoban limestone under the "Simla Series".

Auden (1934) described separately the "Simla Slates" between Subathu and Arki, as comprising puckered leafy phyllites, massive quartzitic bands, finely ripple marked shaly quartzite and green nodular micaceous sandstones. According to him an important feature of these slates is the intercalation of Kakarhatti limestone.

Litho-petrographically the rocks of the Simla turbiditic-flyschoidal formation can be divided into the following four litho-facies group (Sinha, 1978; in press).
Group I. Terrigenous rocks
    a. Arenaceous and silty arenaceous rocks,
    b. Silty and argillaceous-silty rocks,
    c. Argillaceous and silty argillaceous rocks.
Group II. Terrigenous-carbonate rocks,
Group III. Carbonate rocks,
Group IV. Mixed rocks (rhythmites).

g. Shali Formation: Pilgrim and West (1928) record a gradual transition from the grey slates (Madhan Slates), through slates which contain thin bands or lenticles of limestone, often impure, up in to typical Shali limestone with its black chert bands. The limestone and slates,
with some quartzites at the top and bottom have therefore been grouped as the Shali series and undoubtedly underlie the Madhans. The Shali series has been sub-divided by West (1939) and Srikantia and Sharma (1976).

The Shali dolomitic limestones are profusely dotted with Ripean stromatolites (Sinha, 1977b).

h. Krol Tectonic Unit:
1. Junsar Formation: The name "Jaunsar Series" was first proposed by Oldham (1883) who divided the series into three:
    3. Traps and volcanic ashes (Upper),
    2. Red quartzite and slates (Middle)
    1. Grey slates and blue limestone (Lower).
The "Jaunsars" according to Pilgrim and West (1928) are older than the "Simla slate series" and younger than the "Chail Series". Later West (1939) considered the "Jaunsars" to be younger than the "Simla slates", and still later he reported an apparent lithological gradation from "Simla slates" upwards in the "Jaunsars" in a normal stratigraphic sequence.

Moreover, the conspicuous thrust plane over the Blaini in the eastern slope of Pervi Khala clearly suggests the overthrusting of the Jaunsars over the Simla-Blaini complex. To assume it a part of Simla would be a preposterous denial of the laws of nature. Moreover, Jaunsar quartzites always form the lower part of Krol-thrust sheet in the area of the Krol Mountains.
2. Blaini Formations: The Blaini Formation comprises three important lithological units, viz., the boulder bed, the intervening arenaceous and argillaceous horizons and dolomitic limestone (Sinha, 1975b). Auden (1934) divided this formation in two units a) Boulder bed and associated bleached shale, b) Pink dolomite limestone, and has been followed by Bhargava (1976).

More than one horizon of the Blaini boulder bed, which is closely associated with the Infra-Krol Formation, exists in the Krol belt, as in the similar zones of Kumaun and Garhwal (Fuchs and Sinha, 1978).
3. Infra-Krol Formation: The Infra-Krol rocks were described by Auden (1934) as being chiefly composed of dark shales and slates, interbedded in a varve-like manner with buff-weathering bands of impure slaty quartzite from 1/4 to four inches in thickness. Auden suggested that much of the Blaini and Infra-Krol may represent varved sediments associated with the Blaini glacial beds.

The Infra-Krol Formation in the area between Chandpur and Nao is predominantly quartzitic.
4. Krol Formation: Auden (1934) subdivided the Krol Formation.

C. Tectonic Implication of Simla Himalayan Region

The carbonate suites of the Shali and Larji window-zone have yielded Lower to Upper Riphean stromatolitic forms: Tungussia Semikh.; Conophyton cylindricus (Grabau), Colonella, Newlandia,

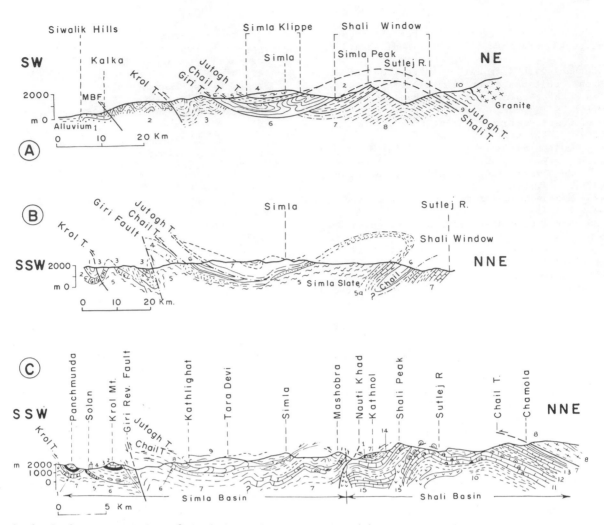

Fig. 7. Geological cross-section of Simla, Himalayan region. (a) After West (1939); (b) After Berthelsen (1951); (c) Author's interpretation (after Sinha, 1978). Explanation: (A) 1. Siwalik; 2. Tertiary;
3. Krol; 4. Jutogh; 5. Chail; 6. Simla; 7. Shali; 8. Khaira quartzite; 9. Genissose and migmatized rocks;
10. Granite. (B) 1. Younger Tertiary; 2. Eocene with nummulites; 3. Krol; 4. Jaunsar; 5. Simla slates;
5a. Naldera limestone; 6. Chail; 7. Jutogh. (C) 1. Subathu with nummulitics; 2. Krol C, D; 3. Krol B
with Mesozoic coccolith; 4. Infra Krol; 5. Blaini; 6. Jaunsar; 7. Simla; 8. Chail-phyllite, metavolcanics
and gneiss; 9. Jutogh; 10. Khaira quartzite; 11. Lower Shali limestone; 12. Shali slate; 13. Upper Shali
limestone; 14. Madhan slate of West (1939); 15. Basic volcanics.

Irregularia, Mound-like stromatolites (Ashgirei
et al., 1975; Sinha, 1977b) and Kussiella. Other
evidence of the age of the Shali carbonate complex is either not authentic or lacks direct
bearing on the series. The Simla flyschoidal
formation has also yielded typical Jurusania and
a new form Jurusania himalayica from Naldera and
Kakarhatti limestones which are Upper Riphean to
Vendian upwards in age (Sinha, 1977b). Recent
investigations discovered some typical acritarchs
varying in age in the broad spectrum of Upper
Precambrian. Work is in progress and the new
data should be published soon.
    The Chail and Chandpur rocks in their typical
area as well as in their extension from Mandi to

Srinagar (Garhwal) have yielded a rich assemblage
of Middle to Upper Paleozoic palynomorphs (Ashgirei, Sinha and Naumova, unpublished data, 1980).
Interesting forms of definite Paleozoic age have
been discovered from the Jutogh Formation. Moreover, very baffling Upper Cretaceous to Paleogene forms are being discovered from the Chail
schuppen zone in the Mandi and Nautikhad areas.
This identification of a new stratigraphic horizon in this part of the Lesser Himalaya may require a fundamental review of existing geological
maps. The above mentioned results have been com-
municated by G. D. Ashgirei (Friendship University, Moscow) after the discovery of forms in
our joint collection by different laboratories

of the Academy of Sciences of USSR, Moscow and Leningrad.

Member B of the Krols yielded a large assemblage of Oxfordian to Danian (Sinha, 1975a) coccoliths discovered under the electron microscope. The following forms were identified:

Tergestiella margereli (Noël) Rain, Microrhabdulus orbitosus Shumeko, Micula Staurophora (Vekshina) Gardet, Stradner, Zygolithus concinnus Martini, Tetralithus cf. gothicus Deflandre, Lithraphidites carniolensis Deflandre, Discorhabdus sp. Deflandrius sp. and ex. gr. Nannoconus. Most of these forms which also occur in other parts of the world indicate an Upper Jurassic to Upper Cretaceous age for the Krol B horizon of Simla area.

The Nummulitics from Subathu are well established besides the Ostrea flemengi d' Arch, Turritella subathuensis d' Arch, Cardita sp., Assilina etc., showing its Lower Eocene affinity.

A paleogologic reconstruction could be envisaged dividing the Lesser Himalayan zone of the Simla area into parallel structural- facies zones from south to north, viz., Siwalik Molasse; Tertiary zone with the base of Krol and Simla; the Krol-Blaini-Nagthat zone with Simla in the base, the flyschoidal turbiditic graben zone of Simla with the transitionally interfingering carbonate suite of Shali-overlapping in the upper part, the Chail/Chandpur Rampur Quartzite zone and the northermost Jutogh zone.

Subsequently, orogenic upheaval and tremendous horizontal compression with appreciable crustal shortening, caused the Krol basin to move southwards on its Simla base, carrying its transgressive capping of nummulitics. The tectonic style of deformation in the Simla silty slate in the Subathu area can be seen in Fig. 10 (Sinha, in press) where the wedge of Simla has been elevated against the down-thrown and squeezed Tertiaries on either side. The thrust tectonics involving the basal part of the Krol thrust along with the Jaunsar quartzites is very conspicuous in the Giri River valley, over Blaini boulder bed transgressing over Simla slates. The subsequent fault line, called Giri Reverse Fault, has complicated the contact between the Simla and Krol units near Kandaghat. The prominent Chail Thrust incorporating the early geosynclinal formation, is thrust over the Simla klippe and is in turn, succeeded by the Jutogh Thrust. The tectonic position of Simla-Blaini floor with respect to Chail and Jutogh nappe in the northern slope of Simla Klippe at Summer hill is proved by a well measured section which shows the presence of squeezed Chail (Sinha, 1978).

The floor of Simla flyschoidal formation extends as far as, and is sliced into, the schuppen zone at Sataun and near Kalka, cropping out along the Main Boundary Fault (MBF). The schuppen contain typical Riphean stromatolite-bearing dolomitic limestones, slates with typical Simla character and subsidiary quartzites along the exposed MBF alignment. MBF has acted as dividing line between southern Himalayan geosyncline and Indian platform (Sinha and Jhingran, 1977). As the lower part of the Simla is dominantly carbonate and slate the correlation of the Sataun rocks with the Simla is beyond doubt.

The basic magmatic episodes in the Darlaghat-Tattapani area are described by Sinha (1977a), and by Sinha and Bagdasarian (1977). K-Ar dating of the whole rock gives three sets of figures, viz., 42 ± 2, 410 ± 10, 338 ± 12 and 1190 ± 35 m.y., which suggest three different magmatic cycles.

The newly discovered palynomorphs as mentioned above hold the clue of further revolutionizing the paleogeographic reconstruction and its tectonic implications. The tectonic position of thrust sheets and their style of disposition can be seen in the map, structural section and tectono-stratigraphic column of the area (Fig.5,6,7).

III. Garhwal- Kumaun Himalayan Region

The contribution of the WIGH's scientists in the Garhwal and Kumaun Himalayan region started with the researches of Valdiya, who subsequently was nominated to the chair of the Department of Geology at Kumaun University, Nainital (Uttar Pradesh). In recent years notable contributions, in the field of tectonics are due to Valdiya and Gupta (1972), Valdiya (1971, 1976a, 1976b), Shah and Sinha (1974), Fuchs and Sinha (1974, 1978), Pal and Merh (1974), Sinha and Bagdasarian (1976), Bhanot et al.(1977), Gaur et al.(1977), Sinha and Nanda (in press), Srivastava (in press), Mehrotra and Sinha (1978, in press), Sinha (1979, unpublished data), Thakur (1976) and other works.

A. Higher Tethys Himalaya and Central Crystalline.

The entire Garhwal-Kumaun Himalayan region could be broadly divided into three major tectonic units from south to north, the Lesser Himalayan Zone, the Central Crystallines of the Great Himalayan range, and the Higher Tethyan Higher Himalaya.

1. Geology: The Tethyan Higher Himalaya comprises mainly a thick sequence of sediments with Tertiary granites cutting the base. The sequence extends from the Precambrian to the Eocene (Sinha, 1979).

The stratigraphic column of the Tethyan Zone of Garhwal and the Kumaun Higher Himalaya was given by Shah and Sinha (1974) and Sinha and Nanda (in press) and it has been summarized in Table II, and the Figures (8, 9, Map 1).

2. Tectonics: The NW-SE regional fold axis of the Paleozoic sedimentary zone changes to N-S in the Mesozoic sequence. The Martoli Formation shows cross-folds plunging NW-SE, as mentioned earlier, and this folding is absent in all the younger rocks. Characteristic secondary folds

Table II. Lithostratigraphic framework of the Tethyan Garhwal-Kumaun higher Himalaya.

Time Unit	Litho-units	Lithology	Assemblage Zones	Fossils
(1)	(2)	(3)	(4)	(5)
	Sangcha Malla Formation	Greenish shales with bands of radiolarian cherts; Greenish-grey wackes and dark shales		Radiolaria *Odontochitina*, *Otiogosphaeridium* Systematophora, *Diphyes*, *Cordosphaeridium*, *Aerosphaeridium*, *Hystrichokolpoma*, etc.
		Purple marly shale; with forminifera ooze; Dark greenish shale with greywacke bands;	Globotruncana-Heterohelix zone	Radiolaria Fucoid markings, *Globotruncana*, *Heterohelix*, *Plummerita*, *Shackoina* *Eouvigerina*.
CRETACEOUS Eocene	Giumal Sandstone	Greenish-grey sandstone and sandy shale with thick bands of massive radiolarian cherts;		No determinable fossils except radiolaria
		Thick bedded glauconitic sandy shales and sandstones.		
Portlandian	Spiti Shale: Saligram Member	Black shales with phosphatic, ferruginous and calcareous concretions;	Perisphinctes Hoplites zone	*Macrocephalites*, *Inoceramus*, *Belemnopsis gerardi*, *Perisphinctes* (*Virgatosphinctes*) *frequens*, *Lytoceras* sp., *Hoplites*, etc.
Callovian	Sulcacutus Member	Ferruginous oolite with coquina	Belemnopsis Sulcautus zone	*Belemnopsis* sp., *Reineckites* *Macrocephalites*, etc.
JURASSIC	Lias Lapthal Formation	Dark-blue to grey limestone with band of coquina	Pecten-Astarte zone	*Pecten*, sp. *Astarte* *Cardium* sp. *Trigonia* sp., *Arca* sp., *Avicula* sp., *Belemnites* *Rhynchonella*, *Lima* *Ostrea*. *Pleurotomaria*, *Lima*
Rhaetic	Kioto Limestone	Grey limestone with numerous bands of coquina;		
		Nodular and oolitic limestone;		
TRIASSIC		Cross-bedded calcarenite and arenaceous limestone.	Megaldon zone	Megaldon, Spiriferina, Pecten etc.
		Grey and blue dolomitic limestone.		

Table 11. (Cont.)

Time Unit (1)	Litho-units (2)	Lithology (3)	Assemblage Zones (4)	Fossils (5)
Noric	Kuti Shale	Alternating bands of black shale and limestone		Few pelecypod shells Juvavites, Lilangina, Pecten, Lima.
Carnic Ladinic	Kalapani Limestone	Nodular limestone	Ptychites zone	Ptychites; Gymnites, Halobia, Arcestes, etc.
		Grey massive limestone		
PERMIAN	Kuling Shale	Black crumbly shale with thin bands of limestone with concretions towards the top	Cyclolobus Oldhami zone	Paramarginifera himalayensis, Spiriferella rajah, Linoproductus cancrini, Chonetes vishnu, Dielaema, sp. Waagenoconcha purdoni, Coeiothyridnim roysii, etc.
			Paramarginifera himalayensis zone	
DEVONIAN	Muth Quartzite	White sugary orthoquartzite with bands of dirty white quartzite;	Schellwienella williami zone	S. williami, Leptaena rhomboidalis, Strophomena.
		Chocolate brown quartzite and dolomitic limestone	Pentamerus zone	Pentamerus sp., Camarophoria sp. Leptaena rhomboidalis, etc.
SILURIAN	"Variegated" Formation	Purple limestone and shale with bands of quartzite	Strophonella zone	Atrypa reticularis, Strophonella sp.? Orthis (Dalmanella) basalis, Favosites, Leptaena rhomboidalis.
	Young Limestone	Green nodular biohermal and biostromal limestone	Calostylis zone	Calostylis? dravidiana? Streptelasma sp. and massive stromatoporids
ORDOVICIAN	Shiala Formation	Grey to pinkish sandstone and quartzite with bands of limestone to the top.	Rafinesquina alternate zone	R. alternate, Leptaena halo, Favosites sp. Saffordia sp., Monotrypa sp. Strophomena
		Alternating bands of sandstone and shale.	Monotrypa zone	Chaenaerops, Laptaena? tracheali Rhinidictya sp., Asaphus sp.,? Lioclema sp.
		Alternating bands of greenish shale and biostromal limestone. Green splintery shale with thin bands of arenite.	Refinesquina aranea zone, Orthis testudinaria zone	Rafinesquina aranea, R. muthensis Triplecia sp., Skendium sp. et.

139

Table II. (Cont.)

Time Unit (1)	Litho-units (2)	Lithology (3)	Assemblage Zones (4)	Fossils (5)
	Garbyang Formation	Green needle shales with occasional bands of limestone;		_Eccliopteris kushanensis_
		Alternating bands of sandstone and shale with graded bedding;	Eccliopteris Zone	
		Cross-bedded calcareous sandstone;		
CAMBRIAN		Greyish green graded bedded sandy shales;		
		Crinoidal and oolitic limestone	Horizon bearing indeterminate trilobite fragments and lingulids	
		Brown marl;		
		Brown dolomitic limestone with alternating bands of shale;		
PRECAMBRIAN	Ralam Formation	Arenaceous shale		
		Dark coloured quartzite		
		Conglomerate alternating with quartzite		
	Martoli Formation	Granded bedded black shales, slates and phylites.		
	Vaikrita Group	Quartzite, quartz-schist, calcsilicates, kyanite and sillimanite gneisses, migmatites etc.		

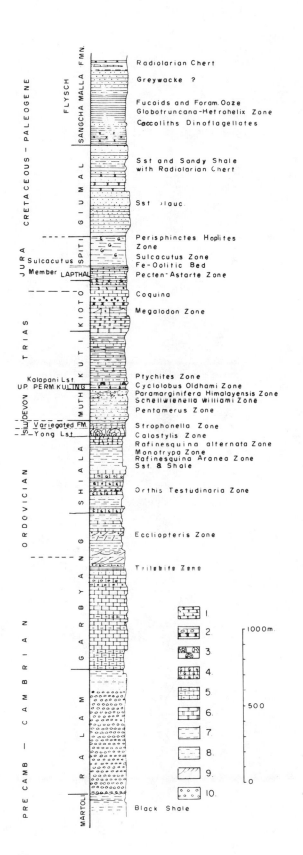

Fig. 8. Biostratigraphic column of higher and Tethyan zone of Garhwal Kumaun Himalaya (Modified after Shah and Sinha, 1974). Explanation: 1. Cherty horizon; 2. Oolitic and nodular limestone; 3. Algal-biohermal limestone; 4. Fossiliferous limestone; 5. Dolomitic limestone; 6. Dolomite; 7. Shales and argillites; 8. Sandy argillite; 9. Sandstone/quartzite with cross bedding and other sedimentary structure; 10. Conglomerates.

---

generally reflect the competence of the various formations. The Garbyang Formation, especially at its base has been involved in a series of north and north-west directed gravity structures producing a feature that may be termed "toothpaste" folding, including flat or low angle-dislocations (Sinha, 1979). A continuous zone of such structures characterizes the entire Lower Garbyang belt along the Girthi Ganga valley. Here the normal regional folding prevails. The Kuti Shales are involved in a series of disharmonic folds and local thrusts. The pattern of folding undergoes a definite change in the Mesozoic rocks, especially within the Kioto Limestone and Lapthal Formation. Not only does the fold axis change gradually to N-S, but these rocks have also suffered a high degree of shallow seated tectonic deformation producing a series of en echelon structures within which the less competent Spiti shale has developed disharmonic folding. E-W crosspuckers have also developed in the Lapthal Formation, leading to a fluted appearance of outcrops of the Sulcacutus Member, the younger unit having suffered this folding.

Faults: No large scale thrusting is reported in Kumaun and eastern Kumaun (Gansser, 1964). The boundary between the metamorphosed Vaikrita group and the pile of sedimentaries differs from that in the eastern region by being faulted and truncating most of the Martoli Formation. The major faults of the area can be described: (i) essentially strike faults which however become occasionally strike slip faults and trend NW-SE and NNW-SSE, (ii) NE-SW to E-W transverse faults. An important fault of the first category is the site of hydrothermal mineralization of barytes and copper ore near Barmatiya (Sinha, 1977c).

Structure of the Kiogad Exotics: An exotic thrust is thought to be responsible for the large translation of the rocks from their original place of deposition. According to this concept a sheet of exotics comprising the limestone of the Permian to Lias Chitichun facies and serpentines and lavas were thrust on a sole of flysch rocks of the Sangcha Malla Formation. Von Kraft (1902) had earlier reported the presence of slickenside structures near the contact between the exotic rocks and the flysch, and Heim and Gansser (1939) regarded this contact as the thrust plane. However, the occurrence of a large number of smaller blocks of limestone, serpentine, etc., within the flysch rocks of the

GEOLOGY AND TECTONICS 141

Sangcha Malla Formation below the above mentioned'contact, cannot be accounted for by the supposed thrust. To overcome this objection Gansser
(1964) subsequently placed this thrust within the
flysch sequence in his structural section. While
the contact of the Sangcha Malla Formation with
the exotics is highly disturbed and generally
marked by basic to ultrabasic extrusives and intrusives, the sequence below is uniformly undisturbed in the Sangcha Malla region with little
strike faulting and there is no evidence whatsoever to postulate a large scale thrust within it.

B. Lesser Himalayan Zone.

The pioneer contribution in the region of
Lesser Kumaun Himalaya from W.I.H.G. has come
from the works of Valdiya who recognized four
tectonic units (Valdiya, 1976b).
1. In the north the pronouncedly mylonitized
and pervasively retrograde metamorphic rocks
and Precambrian augen gneisses and synkinematic
granodiorites and post-kinematic granites of the
Munsiari Formation, are sandwiched between the
Vaikrita and Main Central Thrusts at the base of
the Great Himalaya, they represent a zone close
to the root of the Almora Nappe and of its many
klippen that cover vast areas of the Lesser Himalaya in the eastern half of Kumaun. The Munsiari
Formation extends northwest to join up with the
Jutogh of Himachal Pradesh.
2. Imbricately underlying the Almora nappe is
the second thrust sheet of little metamorphosed
early Riphean flysch penetrated by voluminous
granitic porphyroids of the Ramgarh group, the
northwestern prolongation of which is involved,
together with the underlying quartzites, in imbricate structures. One of the sheets of the
Purola schuppen zone extends northwest to join
up with the Chail Nappe of Himachal.
3. In the inner Lesser Himalaya the crystalline
klippen and scales of the schuppen zones overlie
a vast thrust sheet of Berinag quartzites and
penecontemporaneous basic volcanics.
4. The synclinorial Krol nappe comprises, in
addition to the Berinag nappe, several other
formations of possibly Paleozoic age. Complexly folded and faulted autochthonous early Riphean flysch (Damtha) and Middle Riphean to
Vendian Tejam group (Deoban + Mandhali Formations)
outcrop in vast windows in the inner Lesser
Himalaya.
The stratigraphy at Garhwal and Kumaun has been
satisfactorily correlated with the well established succession in the adjacent Himachal Pradesh
area, as shown on the map and section (Fig. 10).
Fuchs and Sinha (1978) published a monograph giving geologic maps and sections with an account of
Garhwal and Kumaun Himalaya and proposed the
following units:

1. Parautochthonous Unit (including Krol Belt
    and Simla Slates).
2. Chail nappes.

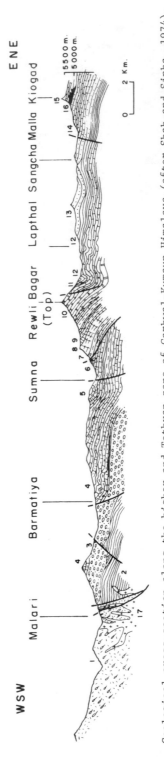

Fig. 9. Geological cross-section along the higher and Tethyan zone of Garhwal-Kumaun Himalaya (after Shah and Sinha, 1974).
Explanation: 1. Vaikrita group; 2. Martoli Fmn.; 3. Ralam Fmn.; 4. Garbyang Fmn.; 5. Shiala Fmn.; 6. Yong limestone; 7. Varie-
gated Fmn.; 8. Muth qzt; 9. Kuling shale; 10. Trias and Kioto lst; 11. Lapthal Fmn.; 12. Spiti shale; 13. Giumal sst.; 14.
Flysch; 15. Exotics; 16. Ophiolite; 17. Granitic intrusive.

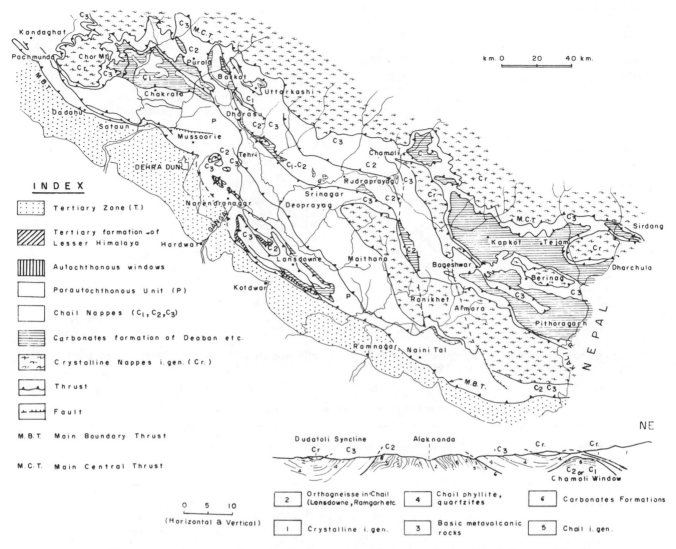

Fig. 10. Tectonic map and geological section of Lesser Garhwal and Kumaun Himalaya (modified after Fuchs and Sinha, 1978).

3. Carbonate terrigenous zone with Precambrian stromatolites.
4. Crystalline nappe and klippen.

The Tertiary zone, consisting mainly of Siwaliks, is overridden by the various units of the Lesser Himalaya along the Main Boundary Thrust (MBT). At depth this structural plane is not thought to merge into a low angle thrust. Sinha and Jhingran (1977) consider the MBT as a steep angle tectonic lineament. The southern portions of the Lesser Himalaya are regarded as sheared off from their original base but still in a parautochthonous position. Though existence of some windows is evidence that the frontal portions of the Krol belt are allochthonous, this does not imply that a "Krol nappe" is derived from somewhere far north. Furthermore there are no traces of "Krol nappe" over the Deobans, or roots of such a nappe

north of the Deoban-Tejam belt. The lowest unit we also consider to be parautochthonous, being derived from immediately north of the Krol belt and having marginally overridden the latter. The higher Chail units, however, are true nappes. The repetition of clastic and carbonate formations is not stratigraphic but tectonic. Particularly the phyllites of the quartzite-metavolcanic complexes overlying the carbonates have gained their position by thrusting. Up to three subsidiary units are discerned, the upper two definitely being nappes. The uppermost, Chail nappe 3, overlaps the lower units and several formations of the parauthchthonous unit (Krol Belt), thus showing tectonic unconformity. The Chail nappe system is overlain by the crystalline nappes and the "reversed metamorphism" was thus created. These folds are crossed by N-S zones of axial depression or culmination in which windows and outliers of higher nappes occur. The Chor area is a typical zone of axial depression and is followed to the east by the Tons zone of culmination. Further SE there is slight SE axial plunge and outliers of higher nappes in that direction are larger and more frequent. The lowest unit of the Lesser Himalaya-the parautochthonous unit - also becomes more and more reduced and finally disappears SSE of Naini Tal.

Faults also influence the structure of Garhwal-Kumaun, but to a lesser degree. Faults probably traverse the fold structures in the sectors Bhowali-Dwarahat and Ladhiya Valley-Someshwar-Baijnath and also the southwestern limb of the Lansdowne syncline. The Aglar fault in the northern limb of the Mussoorie syncline is approximately parallel to the strike of the rocks.

Rawal (in press) and Rawat and Varadrajan (1979) have recently tried to analyze the importance of subsequent fault structures at Kaliasor on the Rishikesh-Badrinath Highway and describe the "Alaknanda-thrust" as a reverse fault with varying angles of dip.

IV. Eastern Himalayan Region of Arunachal Pradesh

The Arunachal region (Fig. 1) of eastern Himalaya has been investigated since 1972 by a team of geoscientists including Tandon, Jain, Thakur, Verma, Kumar and Singh. In subsequent years Tandon, Jain and Verma shifted to other institutions and Thakur opted to work in other regions of Himalaya. Thus the work is being continued by Kumar and Singh. The information has been compiled from Jain et al.(1974), Thakur and Jain (1974, 1975), Verma and Tandon (1976), Jain and Thakur (1978), Annual Report W.I.H.G., 1977-78, and 1978-79.

The state of Arunachal Pradesh is divided into five districts from west to east Kameng, Subansiri, Siang, Lohit and Tirap.

a. Geology: Kumar and Singh (WIHG Annual Report, 1977-78, 1978), on the basis of field and laboratory investigations propose lithotectonic units of the region as shown in Table III.

b. Tectonics: Structural analysis of the Lohit district reveals four different phases of ductile folds in the Mishmi and Tiding Formations, whereas only the foliation (S_2) and crenulation (F_4) has been developed in certain zones of the Lohit meta-granodiorite and metadiorite (Thakur and Jain, 1975). The Mishmi Formation, representing the Lesser Himalaya zone, now lies in close juxtaposition against the Lohit igneous complex of probable trans-Himalayan province. The NW-SE trends in the Mishmi metamorphics may continue uninterruptedly to the Dihang Valley, where the Miri and

Table III. Litho-tectonic units of eastern Himalaya of Arunachal Pradesh

Se Le Group	- High grade crystalline schist and gneisses, migmatites and granites with basic intrusions.

————————————————————————Thrust————————————————————————

Bomdila Group	- Sericitic quartzite, limestone, gypsum, phyllitic schist bands in quartzite at the contact between quartzite and gneisses. Fine and coarse grained gneisses with amphibolite veins and lenses.

————————————————————————Thrust————————————————————————

Miri Quartzite	- Boulder quartzite, pink and white quartzite which shows iron ore mineralization along the hinge region of fold, cherty limestone, thin bands of slate with pebbles of limestone, green and purple slate and calcareous shale.

————————————————————Contact not clear————————————————————

Rangit Pebble Slate	- Pebble or boulder slate and thin bands of siliceous tuffs (greenish) and thin bands of compact sandstones.

————————————————————Contact not clear————————————————————

Gondwana Group	- Black shale, black slate with coal seam, micaceous sandstone and gritty sandstone, pebbly slate, mudstones and thin bands of white quartzite. The black shale shows the presence of plant fossils.

————————————————————————Thrust————————————————————————

Tertiaries (Siwaliks)	- Salt and pepper sandstone with grey silt and mudstone, soft sandstone, sand, clay, shale and conglomerates, etc.

Siang group rocks also trend NW-SE and N-S. These trends are clearly cut and overlapped by the ENE-WSE Himalayan trends of the Gondwana and Siwalik belts (Jain et al., 1974). The Mishmi Thrust has been observed to rest on the recent alluvial sediments of the Lohit River. It is generally thought that the "Eastern Himalayan syntaxis" was formed due to bending the Himalayan orogenic belt around the north easterly projecting foreland of the Shillong plateau which forms a promontory of peninsular shield mass. Several arguments do not support this over simplified version; firstly the principal lithologic units in the west of the syntaxis region cannot be compared with their counterparts in the east, and secondly the regional ENE-WSE structural trends in the Siang district and NW-SE trends in the Lohit district represent two different episodes of tectonic movements of Himalaya age (Thakur and Jain, 1974).

Some of the salient features from the recent work by Kumar and Singh (Annual Report, WIHG for 1978-79, 1979) in the Kameng and Subansiri districts can be summarized as follows: ENE-WSW to NE-SW trends of the Siwaliks, Gondwana, Miri, Bomdila and Se La groups of rocks as shown in the Landsat imagery (Fig. 11). A 50-100 m thick mylonitized zone within the Bomdila Group near hergaon, along the Rupa-Kalaktang road section

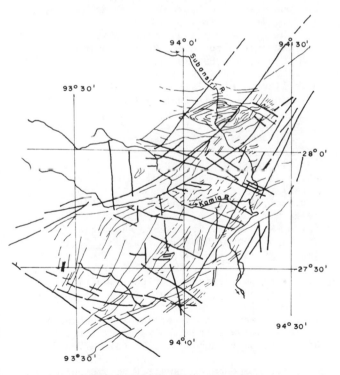

MAIN TRENDS AND LINEAMENTS FROM LANDSAT-IMAGERY-145-041

Fig. 11. Landsat imagery lineament map of eastern Himalaya (after Kumar, Kumar and Singh, in press).

was recognized. The thrust along which the Se La Group overlies the Bomdila Group may be equivalent to the Main Central Thrust of the Western Himalaya. In this region the tourmaline granite is Tertiary.

V. Some Observations on Regional Himalayan Tectonics

The recent papers by W.I.H.G. scientists can be reviewed as follows:

Valdiya (1976a) described a large number of fractures, faults and folds trending normal and oblique to the Himalayan tectonic trend. The tear faults of Kumaun and Nepal show predominant right lateral shear movements. A colossal transverse NW-SE structure across the Himalayan terrain has been defined as the Karakoram lineament on the basis of features on Landsat imagery and combined geological and geophysical data. Gupta (1974) substantiating and elaborating the findings of Santo (1969, 1970) concluded that the very localized V-shaped pocket of intermediate depth earthquakes beneath Hindu Kush is indicative of the existence of the underthrust relict plate, whereas between Hindu Kush and Burma, foci are restricted to shallow focal depths. Compiling the foci distribution of upper mantle earthquakes, Kumar (1975) inferred that the V-shaped pockets are related to dip-slip or normal faulting at the borders of the continental plates. Many double and multiple events are closely related in both time and space. Thus almost all took place around the nominal depth of 33-60 km in the Himalaya whilst at 170-200 km depths in Hindu Kush region.

Jhingran et al. (1976) and Thakur (in press) attempted to review Himalayan structure and tectonics, deal with the principal features of five structural zones of Himalaya, viz., Siwalik, Krol nappe, Simla-Shali unit, crystalline unit and Tethyan Sedimentary unit, from south to north and give an overview of thrust and nappes of Western Himalaya.

Sinha and Jhingran (1977) after a detailed study of geologic, seismic, stratigraphic, magmatic and metallogenic features and comparing them with development of the Caucasus mobile zone, established the presence of deep-seated lineaments, mainly longitudinal to the strike of Himalaya and grouped them as follows: 1) Gilgit-Dras-Darchan deep seated lineament with ophiolitic melange of Cretaceous age with K/Ar dates of 107.5, 75 and 73 m.y. (Sinha and Bagdasarian, 1976); 2) Keylang-Badrinath main central axial structural line of Himalaya incorporating the "root-zones" of upper crystalline nappes with younger granites of 42, 52 and 73 m.y. (Ashgirei et al., 1976); 3) Pir Panjal-Darla deep seated lineament associated with three distinct phases of magmatism viz., Precambrian 1190 $\pm$ 35 and 710 m.y., Paleozoic-410 $\pm$ 10, 338 $\pm$ 12, 228 $\pm$ 10 m.y. and Tertiary-69 $\pm$ 5.51 and 42 $\pm$ 2 m.y. (Sinha, 1977a). The last phase obviously coincides with the final phase Himalayan tectonic movements and, 4) the Main Boundary lineament which is also called Main Boundary Fault separating Siwalik molasse from the pre-Tertiary sequence.

Evidence obtained from metamorphism, deformation and radiometric dating indicate that the Central Crystalline unit of the Himalaya represents an old Precambrian basement which has been reactivated during Caledonian (?) and Alpine orogenic movement (Thakur, 1976, 1980). Fission track annealing of apatite from the Mandi granite in the Himachal Pradesh intruding into the Chail tectonic unit shows an average rate of cooling and uplift rate of $3.4^{O}C/m.y.$ and 0.55 mm/year respectively (Sharma et al., 1978b). The same method has also been applied to date the thrusts and it has been suggested that the Vaikrita and other allied thrusts in the Sutlej valley of the Himachal Pradesh were formed during Middle Miocene (15 m.y.) (Sharma and Nagpal, in press).

Acknowledgement. In course of compilation of the paper the structure and tectonics group of WIHG has taken keen interest, and valuable help in material and fruitful discussion have been provided from Drs. V. C. Thakur, S. Kumar, K. K. Sharma, N. S. Virdi, D. Pal, R.A.K. Srivastava, K. R. Gupta, R. S. Rawat and other scientists of the WIHG.

President of WIGH Shri S. P. Nautiyal's appreciation of creative research along new lines of thought and the encouragement of the Director, WIHG, are gratefully acknowledged.

References

Ashgirei, G. D., A. K. Sinha, M. E. Raaben and O. B. Dmitrenko, New findings on the geology of Lower Himalaya, Himachal Pradesh, India, Chayanica Geologica, Delhi, 1(2), 143-151, 1975.

Ashgirei, G. D., I. C. Pande, A. K. Sinha and B. C. Mallik, About the history of metamorphism of western Himalaya (in Russian), Geol. Min. Deposit Asia, Africa, Latin Amer. Countries, Moscow, 1, 19-36, 1976.

Ashgirei, G. D., A. K. Sinha, I. C. Pande and B. C. Mallik, A contribution to the geology, geochronology and history of regional metamorphism of Himachal Himalaya, Him. Geol., 7, 102-117, 1977.

Auden, J. B., The geology of the Krol belt, Rec. Geol. Surv. India, 67(4), 357-454, 1934.

Bhanot, V. B., B. K. Pande, V. P. Singh and W. C. Thakur, Rb-Sr whole-rock age of the granitic gneiss from Askote area, eastern Kumaun and its implication on tectonic interpretation of the area, Him. Geol., 7, 118-122, 1977.

Bhargava, O. N., Geology of the Krol Belt and associated formations: A reappraisal, Mem. Geol. Surv. India, 106(1), 168-227, 1976.

Dixit, P. C., R. K. Kachroo, H. Rai and N. L. Sharma, Discovery of vertebrate fossils from the Kargil basin, Ladakh (Jammu and Kashmir), Curr. Sci., 40, 633-634, 1971.

Dixit, A. K., Petrology and geology of Chor Massif area, H. P. (in Russian), Ph. D. Thesis, Friendship Univ., Moscow, 1973.

Frank, W., Daten and Gedanken Zur Entwicklungs-geschichte des Himalaya, Mitteil. der. Geol. Gesellschaft in Wien, 66-67, Band, 1973/74, 1-7, 1975.

Frank, W., A. Gansser and V. Trommsdorff, Geological observations in the Ladakh area (Himalayas) - a preliminary report, Schweiz. Mineral. Petrogr. Mitt., 57, 89-113, 1977.

Fuchs, G., Traverse of Zanskar from the Indus to the valley of Kashmir - a preliminary note, Jahrb. Geol. B.-A., Wien, Bd. 120, Heft 1, 219-229, 1977.

Fuchs, G. and A. K. Sinha, On the geology of Naini Tal area (Kumaun Himalaya), Him. Geol., 4, 563-579, 1974.

Fuchs, G., and A. K. Sinha, The Tectonics of the Garhwal-Kumaun Lesser Himalaya, (with 3 plates), Jahrb. Geol., B.-A., Bd. 121, Heft 2, Wien, 219-241, 1978.

Gansser, A., Geology of Himalayas, Interscience, London, p. 289, 1964.

Gaur, G. C. S., V. K. S. Dave and R. S. Mithal, Stratigraphy, structure and tectonics of the carbonate suite of Chamoli, Garhwal Himalaya, Him. Geol., 7, 416-455, 1977.

Gupta, H. K., Some seismological observations and tectonics from Hindu Kush to Burma Region, Him. Geol., 4, 465-479, 1974.

Gupta, K. R., and K. K. Sharma, Nubra-Gartang Superlineament, Ladakh - some geological observations and their implications, Photonir-vachak, (Jour. Indian Soc. Photo-interpretation) Dehra Dun, 6, 15-22, 1978.

Heim, A., and A. Gansser, Central Himalaya, Mem. Soc. Helvetique Sciences Naturelles, 73 (1), pp. 246, Zurich, 1939.

Jain, A. K., V. C. Thakur and S. K. Tandon, Stratigraphy and structure of the Siang District, Arunachal Pradesh (NEFA) Himalaya, Him. Geol., 4, 28-60, 1974.

Jain, A. K., and V. C. Thakur, Abor volcanics of the Arunachal Himalaya, J. Geol. Soc. India, 19 (8), 335-349, 1978.

Jhingran, A. G., V. C. Thakur and S. K. Tandon, Structure and tectonics of the Himalaya, Preprint, Hima. Geol. Seminar, New Delhi, pp.39, in press, 1976.

Kumar, S., Tectonics and earthquake mechanism of the shallow earthquake seismic belt, the Himalaya, Geol. Rundschau, 64 (3), 977-992, 1975.

Kumar, S., Presence of parallel metamorphic belts in the northwest Himalaya, Tectono-physics, 46, 117-133, 1978.

Kumar, S., and T. Singh, Tectono-stratigraphy of Subansiri District, Arunachal Pradesh, in Proceedings Workshop on Stratigraphy and

Correlation of Lesser Himalayan Fmns.,
Kumaun Univ., Naini Tal, India, edited by
K. S. Valdiya (in press).

Lydekker, R., Geology of the Ladakh and neighbouring districts, etc., Rec. Geol. Surv. Ind., 13
(1), 26-58, 1880.

Medlicott, H. B., On the geological structure
and relations of the southern portion of the
Himalayan ranges between the rivers Ganges
and the Ravee, Mem. Geol. Surv. India, 3 (2),
1-212, 1864.

Mehrotra, N. C., and A. K. Sinha, Discovery of
microplanktons and the evidences of the Younger Age of Sangcha Malla Formation (Upper Flysch of Malla Johar area in the Tethyan Zone of
Kumaun Himalaya, Him. Geol., 8 (II), 1978.

Mehrotra, N. C., and A. K. Sinha, Further studies
of microplanktons from the Sangcha Malla Formation (Upper Flysch) of Malla Johar, Higher
Himalaya, (in press).

Molnar, P., and P. Tapponier, Cenozoic tectonics
of Asia: Effects of a continental collision,
Science, 189, 419-426, 1975.

Oldham, R. B., Note on the geology of Jaunsar and
the Lower Himalaya, Rec. Geol. Surv. India, 16
(4), 193-198, 1883.

Pal, D., and S. S. Merh, Stratigraphy and structure of the Naini Tal Area in Kumaun Himalaya,
Him. Geol., 4, 547-562, 1974.

Pal, D., and N. S. Mathur, Some observations on
stratigraphy and structure of Indus Flysch,
Ladakh Region, Him. Geol., 7, 464-478, 1977.

Pal, D., R. A. K. Srivastava and N. S. Mathur,
Tectonic framework of the miogeosynclinal
sedimentation in Ladakh Himalaya: A critical
analysis, Him. Geol., 8 (1), 500-523, 1978.

Pande, I. C., and S. Kumar, Absolute age determination of crystalline of Manali-Jaspa Region,
Northwestern Himalaya, Geol. Rundschau, 63 (2),
539-548, 1974.

Pilgrim, G. E., and W. D. West, The structure
and correlation of Simla rocks, Mem. Geol.
Surv. India, 140, 53 pp., 1928.

Rawat, R. S., and S. Varadarajan, The Alaknanda
Thrust, Curr. Sci., 48 (19), 1979.

Rawat, R. S., Road problems in Rudraprayag area:
Geological Considerations, Causes and Suggestions, Jour. Indian Soc. Engg. Geol., (in press).

Santo, T., On the characteristic seismicity in
South Asia from Hindu Kush to Burma, Bull. Int.
Inst. Seism. Earthq. Eng., 6, 1969.

Santo, T., Regional variation of the passive
detectability of earthquakes in the world,
Bull. Earthq. Res. Inst., 48, 1107-1119,
1970.

Shah, S. K., and M. L. Sharma, A preliminary
report on the fauna in radiolarites of Ophiolite-Melange Zone around Mulbekh, Ladakh,
Curr. Sci., 46, 817, 1977.

Shah, S. K. and A. K. Sinha, Stratigraphy and
tectonics of the "Tethyan" Zone in a part of
Western Kumaun Himalaya, Him. Geol., 4, 1-27,
1974.

Shah, S. K. and J. T. Gergan, Fauna from the in-

terbedded sedimentaries of Dras volcanics,
Abst. 9th Seminar on Himalayan Geology, Dehra
Dun, W.I.H.G., 104, 1978.

Shankar, R., R. N. Padhi, G. Prakash, J.L.Thusu
and R. N. Das, The evolution of the Indus
Basin, Ladakh, India, Preprint Himalayan
Geol. Seminar, New Delhi, in press, 1976.

Sharma, K. K., A contribution to the geology of
the Satlej Valley, Kinnaur, Himachal Pradesh,
India, Sci. de la Terre, C. N. R. S., Paris,
369-378, 1976.

Sharma, K. K., and K. R. Gupta, Some observations
on the geology of the Indus and Shyok valleys
between Leh and Panamik, Districts, Ladakh,
Jammu and Kashmir, India, Recent Researches
in Geology, Delhi, 7, 193-143, 1978.

Sharma, K. K., and S. Kumar, Contribution to the
geology of Ladakh, Northwestern Himalaya, Him.
Geol., 8 (1), 252-287, 1978.

Sharma, K. K. and K. K. Nagpaul, Dating of
thrusts in Himalaya by fission track method,
Earth Plan. Sci. Letter, in press.

Sharma, K. K. and A. K. Sinha, G. P. Bagdasarian
and R. Cn. Gukasian, Potassium-Argon dating of
Dras Volcanics, Shyok Volcanics and Ladakh
Granite, Ladakh, Northwest Himalaya, Him. Geol.,
8, 288-295, 1978a.

Sharma, K. K., H. S. Saini and K. K. Nagpaul,
Fission track annealing, ages of apatite from
Mandi Granite and their application to tectonic
problems, Him. Geol., 8 (1), 298-312, 1978b.

Sharma, M. L., and S. K. Shah, New genera from
"Exotic Block" at Lamayuru, Ladakh, Curr. Sci.,
46, 790, 1977.

Sinha, A. K., Calcareous nannofossils from Simla
Hills, (Himalaya, India) with a discussion on
their age in the Tectono-stratigraphic column,
Jour. Geol. Soc. India, 16 (1), 69-77, 1975a.

Sinha, A. K., The tectonic-stratigraphic significance of the Blaini Formation of the Simla
Hills, Bull. Indian Geol. Assoc., Chandigarh,
8 (2), 151-161, 1975b.

Sinha, A. K., Geochronology, petrography-petrochemistry and tectonic significance of basic
rock suites of north-western Himalaya- with
special reference to Himachal Himalaya, India,
Recent Researches in Geology, Delhi, 3, 478-
494, 1977a.

Sinha, A. K., Riphean stromatolites from western
Lower Himalaya, Himachal Pradesh, India, Fossil
Algae, edited by E. Flugel, Springer Verlag,
Berlin, 86-100, 1977b.

Sinha, A. K., A discovery of barite and assoc-
iated polymetallic mineralized zone in Tethyan
Zone of Higher Garhwal and Kumaun Himalaya,
Him. Geol., 7, 456-463, 1977c.

Sinha, A. K., Para-autochthonous turbiditic-
flyschoidal simla and terrigenous carbonate
Shali Formations of Himachal Himalaya, their
litho-petrography and tectonic setting, Him.
Geol., 8 (1), 425-455, 1978.

Sinha, A. K., Geology and tectonics of the part
of Higher Central Himalayan Tethyan Zone,
paper presented at the International Committee

on Geodynamics (working group 6) Meeting, Peshawar (Pakistan), Nov. 23, 1979.

Sinha, A. K., Tectono-stratigraphic problem in the Lesser Himalayan Zone of Simla Region, Himachal Pradesh, in Proceedings Workshop on Stratigraphy and Correlation of Lesser Him. Fmns., Kumaun Univ., Naini Tal, India, edited by K. S. Valdiya, in press.

Sinha, A. K., and G. P. Bagdasarian, Potassium-Argon dating of some magmatic and metamorphic rocks from Tethyan and Lesser Zones of Kumaun and Garhwal Indian Himalaya: And its implication in Himalayan Tectogenesis, Sci. de la Terre, C. N. R. S., Paris, 387-394, 1976.

Sinha, A. K., and A. G. Jhingran, Deep seated lineament structures in Himalaya and Caucasus: Their role in the history of geological development and metallogeny, Him. Geol., 7, 46-64, 1977.

Sinha, A. K. and R. A. K. Srivastava, In the occurrence of glauconite with radiolarites in the flysch sediments of Malla Johar, in Higher Himalaya; and its significance in tectonics and sedimentation, Him. Geol., 8 (II), in press.

Sinha, A. K. and A. C. Nanda, Biostratigraphic zonation of Tethyan Zone of Unta-Dhura Lapthal region of Malla Johar, Higher Kumaun Himalaya, Him. Geol., 9, in press.

Srikantia, S. V., and R. P. Sharma, Geology of the Shali belt and the adjoining areas, Mem. Geol. Surv. India, 106 (1), 31-66, 1976.

Stoliczka, F., Geological sections across the Himalayan mountains, from Wangtu Bridge on the River Sutlej to Sungo on the Indus, with an account of the formations in Spiti, accompanied by a revision of all known fossils from that district, Mem. Geol. Surv. India, 5 (1), 1-154, 1865.

Tewari, B. S., and I. C. Pande, Permian fossiliferous limestone from Lamayuru, Ladakh, Pub. Cent. Adv. Study, Geol. Punjab Univ., 7, 188-190, 1970.

Thakur, V. C., and A. K. Jain, Tectonics of the region of eastern Himalayan syntaxis, Curr. Sci., 43 (24), 783-785, 1974.

Thakur, V. C. and A. K. Jain, Some observations of deformation, metamorphism and tectonic significance of the rocks of some parts of the Mishmi Hills, Lohit District (NEFA), Arunachal Pradesh, Him. Geol., 5, 339-364, 1975.

Thakur, V. C., Divergent isogrades of metamorphism in some part of Higher Himalaya Zone, Sci. de la Terre, CNRS, Paris, 433-441, 1976.

Thakur, V. C., Tectonics of the central crystallines of western Himalaya, Tectonophysics, 62, 1980.

Thakur, V. C., An overview of thrusts and nappes of western Himalaya, in Thrust and Nappe Tectonics, edited by K. McClay and N. J. Price, Blackwell Scientific Publications Ltd., Oxford, in press.

Valdiya, K. S., Origin of phosphorite of the Late Precambrian Gangolihat dolomites of Pithoragarh, Kumaun Himalaya, India, Sedimentology, Amsterdam, 19, 115-128, 1971.

Valdiya, K. S., Lithological subdivision and tectonics of the "Central Crystalline Zone" of Kumaun (abst.), Proc. Sem. Geodyn. Him. Region, N.G.R.I., Hyderabad, 204-205, 1973.

Valdiya, K. S., Himalayan transverse and oblique faults and folds and their parallelism with sub-surface structures of the North Indian Plains, Tectonophysics, 32, 353-386, 1976a.

Valdiya, K. S., Structural set-up of the Kumaun Lesser Himalaya, Sci. de la Terre, CNRS, Paris, 449-462, 1976b.

Valdiya, K. S., and V. J. Gupta, A contribution to the geology of northwestern Kumaun, with special reference to the Hercynian Gap in Tethys Himalaya, Him. Geol., 2, 1-33, 1972.

Verma, P. K., and S. K. Tandon, Geologic observations in a part of the Kameng district, Arunachal Pradesh (NEFA), Him. Geol., 6, 259-286, 1976.

Virdi, N. S., Stratigraphy and structure of the area around Nirath, Dist. Simla, Himachal Pradesh, Him. Geol., 6, 163-175, 1976.

Virdi, N. S., V. C. Thakur and S. Kumar, Blueschist facies metamorphism from the Indus Suture Zone of Ladakh and its significance, Him. Geol., 7, 479-482, 1977.

Virdi, N. S., V. C. Thakur and R. J. Azmi, Discovery and significance of Permian microfossils in the Tso Morari crystallines of Ladakh, J & K, India, Him. Geol., 8 (II), in press.

Von Kraft, A., Notes on the "Exotic Blocks" of Malla Johar, Men. Geol. Surv. Ind., 32 (3), 127-183, 1902.

Wadia, D. N., The syntaxis of the north-west Himalaya, tectonics and orogeny, Rec. Geol. Surv. India, 65, 189-220, 1931.

Wadia Institute of Himalayan Geology, Annual Report 1977-78, Dehra Dun India, pp. 3-19, 1978.

Wadia Institute of Himalayan Geology, Annual Report 1978-79, Dehra Dun, India, pp. 2-45, 1979.

West, W. D., Structure of the Shali Window near Simla, Rec. Geol. Surv. India, 65, 125-132, 1939.

THE GEODYNAMIC EVOLUTION OF THE HIMALAYA-
TEN YEARS OF RESEARCH IN CENTRAL NEPAL HIMALAYA AND SOME OTHER REGIONS

P. Bordet

Facultés Catholiques - 21, rue d'Assas, F-75006 Paris

M. Colchen

Université de Poitiers - Laboratoire de Pétrologie de la Surface
40, avenue du Recteur Pineau, F-86022 Poitiers Cedex

P. Le Fort

Centre de Recherches Pétrographiques et Géochimiques
Case Officielle n° 1, F-54500 Vandoeuvre-lès-Nancy

A. Pêcher

Ecole des Mines, Parc de Saurupt, F-54042 Nancy Cedex

Abstract. Ten years of field and laboratory
studies by French teams have yielded interesting
and novel views on some of the long out-standing
problems of Himalayan geology. Following original
observations by Heim, Gansser and others, detailed investigations of the Ladakh ophiolite zone
and the Main Central Thrust (M.C.T.) in Nepal,
followed by analysis of fluid inclusions, petro-
fabrics and metamorphism have led to descriptions
of the evolution of the M.C.T., to an explanation
of the famous reverse metamorphism associated
with the M.C.T. and of the leucogranite intrusions. A calender of the geodynamic evolution of
the Himalaya is given in conclusion.

INTRODUCTION [P.B.]

French geological research in the Himalaya
started in 1950 with the climbing of Annapurna I,
during which the first Spiti ammonites of Thakkhola were collected (Ichac & Pruvost, 1951). Since
then, climbing expeditions with geologists as
party members and entirely geologic expeditions
took place regularly in Nepal. Efforts were soon
concentrated on :
- Eastern Nepal : the Arun valley; the Everest,
Makalu and Kangchenjunga (Janu) massifs;
- Central Nepal : the Kali Gandaki - Thakkhola,
Marsyandi and Burhi Gandaki valleys; the Annapurnas, Manaslu and Ganesh Himal massifs.
The results were published mainly in memoires

and as maps (Bordet, 1961; Bordet *et al.* 1971;
Bordet, Colchen & Le Fort, 1975) and showed in
particular :
- that there is a rather good lateral correlation along the Nepal Himalaya. However, differences exist between regions that necessitate careful field study;
- that structural units are rather easy to characterize in the fossiliferous Tibetan sedimentary zone. However, the lack of chronostratigraphic
markers makes it a much more difficult task in
the Midlands and the metamorphic areas and hin-
ders correlations of Higher and Lesser Himalayan
events across the M.C.T.
- that the lack of important unconformities in
all the Himalayan units does not help to elucidate the tectonic, metamorphic and magmatic events.
The "Alps model", developed by the first geologists working in Nepal (e.g. T. Hagen), was soon
abandoned. The main aim of research since 1970
became the finding of another model, better fitted to field observations. It was linked with the
clear idea that the interpretation of the range
could not be dissociated from the global tectonics theory.
A research programme was developed, which was
further integrated with the ICG WG6. This programme had the following three main objectives :
1 - to delineate, as precisely as possible, the
pre-orogenic history of the Himalaya :
- by determining the chronologic position of

the formations engaged in the mountain building (paleontological or geochronological data),
- by analysing the old deformations, starting from the sedimentological and paleogeographic reconstructions,
- by investigating plutonic and volcanic episodes that characterize these periods;
2 - to analyse the orogenic deformation, metamorphism and magmatism, studying :
- the individual, successive phases,
- the system of constraints, the intensive parameters of thermodynamic conditions and their evolution,
- the links between these deformations, metamorphism and magmatism;
3 - to look for recent and present indications of orogenic activity in the Himalaya :
- by geomorphologic and neotectonic analyses,
- by geophysical, in particular seismic surveys presently going on.

This programme has been mainly executed in Central Nepal, but, over recent years, investigations have been extended to the innermost parts of the Himalaya in the Ladakh region.

The following lines will, after a short summary of the geological divisions adopted by our team, deal with the main results of the last ten years of our research, results that bear on the geodynamic modelling of the Himalaya.

During the course of these studies, the new information obtained, along with the maps and results of earlier authors, were compiled in map form. The "Carte géologique du haut Himalaya" (1:200 000) thus drafted was published in 1980 by the CNRS, along with an explanatory text; the map is included in the present volume (Map 2).

GEOLOGICAL DIVISIONS OF THE HIMALAYA [P.L.F.]

The Indus-Tsangpo suture zone lies north of the Himalaya and north of Nepal.

In the Nepal Himalaya, the following main lithostratigraphic and structural units are differentiated, from north to south :
- the Tibetan sedimentary series, which forms many of the highest summits, and overlies the gneisses of the Tibetan Slab,
- the Lesser Himalaya formations, also called in Nepal the Midland formations,
- the Siwalik Series, the youngest and the more external part of the belt, which have been very little studied by our team.
Two major thrusts divide the previous units :
- the Main Central Thrust (M.C.T.), between the Tibetan Slab and the Midland formations;
- the Main Boundary Thrust (M.B.T.), between the Midland formations and the Siwaliks : this thrust, still active, resembles more a large listric fault than a thrust; therefore, it is better called the Main Boundary Fault (M.B.F.).
A very brief lithostratigraphic and structural summary of these units is given below.
1) The Indus-Tsangpo suture zone of Ladakh

(Andrieux *et al.* 1977b; Bassoullet *et al.* 1978a, b,c; Colchen, 1977b)
This zone is underlined by a belt of discontinuous outcrops of ophiolites, approximately aligned in NW-SE direction with the exception of a few massifs lying to the south (fig.1). The ophiolites are tectonically associated with Cretaceous to Eocene sedimentary units. In a cross section, the suture appears as an asymetric fan-shaped structure over-thrusting both India and Eurasia to the south and to the north.
2) The Tibetan sedimentary series (Colchen, 1971, 1975; Le Fort, 1975a; Bassoullet, Colchen & Mouterde, 1977)
The Tibetan sedimentary series is made of mainly marine epicontinental fossiliferous deposits from below the Lower Ordovician to Lower Cretaceous without any major break or unconformity. It is intruded by leucogranite massifs (Manaslu in Central Nepal, Makalu in Eastern Nepal). At its base, the Tibetan sedimentary series becomes more and more metamorphosed and progressively passes into the Tibetan Slab, which constitutes the infrastructure.
3) The Tibetan Slab (Le Fort, 1971b, 1975a-c; Bordet, 1977; Pêcher, 1978) (see fig.10)
The Tibetan Slab comprises various gneisses and marbles forming in Central Nepal a well-identified morphological unit at the base of the Tibetan sedimentary series. This pile of gneisses is very massive, northward dipping, without any megascopic visible folds other than large flexures. Consisting of three different formations (from bottom to top : quartz pelitic micaschists and gneisses, metamorphic limestones, augen gneisses), its thickness increases from 5 km in west to 10 km in east.
4) The Midland formations (Pêcher, 1978; Mascle, 1979) (see fig.10)
The Midland volcano-sedimentary formations can be divided into two main groups : the lower, with strong volcanic (mainly felsic) affinities includes the Ulleri augen gneisses; the upper one is much more diversified (shales, carbonaceous shales, dolomitic limestones, quartzites, mafic tufs and volcanics...). This upper group of formations outcrops on both sides (north and south) of the large Pokhra-Gorkha anticlinorium, and is variously deformed and metamorphosed. The less metamorphosed carbonaceous schists of the southern side, contain impressions of plant remains (including leaves) (Pêcher, 1978, p. 75; Mascle, 1979); they are so far the only fossils found in the Nepalese Midland formations.

THE INDUS ZONE : POSSIBLE EVIDENCE FOR
EARLY TRIASSIC OCEANIZATION
(Bassoullet *et al.*, 1978a, b and c) [M.C.]

Recent studies in Ladakh bring new pertinent information on the suture :
- the ophiolites are characterized by the pre-

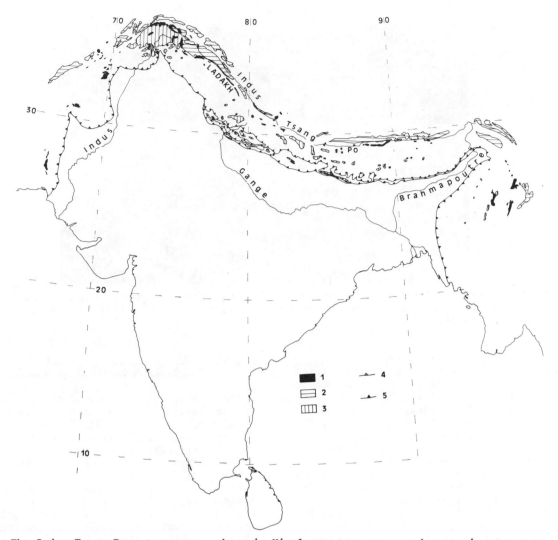

Fig. 1 - The Indus-Tsang Po suture zone, the main Himalayan structures and magmatic outcrops :
(1) "ophiolites", (2) granitic bodies north of the suture zone, (3) granitic bodies south of the
suture zone (Higher Himalaya leucogranites and Lesser Himalaya cordierite granites), (4) Main Central
Thrust, (5) Main Boundary Thrust and equivalents.

sence of several facies typical of an ophiolite
assemblage : ultramafic complex (particularly
harzburgites with a metamorphic tectonic fabric :
blastomylonites), gabbroic complex containing cu-
mulates and pegmatites, volcanic flows (pillow-
lavas and agglomerates), tectonically associated
manganiferous red radiolarites and pink limestones (Hallstadt facies);

- the ophiolites constitute in fact an ophiolitic nappe thrust southwards onto the Himalayan
series. Near Photaksar, about 25 km south of the
main suture line, they outcrop particularly well
in a large klippe (Gansser, 1964), very similar
to that of the Amlang La area (Heim & Gansser,
1939);

- within the area, two other structural units
also exist. They are characterized either by a

Middle Triassic-Jurassic calcareo-pelitic flysch
(the Lamayuru unit) or by a Jurassic (?) to Cretaceous greywacke-pelitic flysch with some vul-
canites (the Dras-Nindam unit);

- structural and microstructural studies indicate that the pseudo-fan pattern aspect of the
structure is the result of several phases of tectonic evolution in which two major events can be
distinguished. A first episode, occurring before
the sedimentation of the Indus detritic series
(Aptian-Albian to Eocene) characterizes essentially the ophiolitic and flysch nappes which
were displaced southwards. The second post-Middle
Eocene event resulted in the refolding and northward thrusting of the nappes onto the Indus detritic series and even locally onto the Ladakh
granodiorites.

Fig. 2 – Photograph of the top of the Lamayuru exotic block (Colchen unpublished). (1) neritic Permian limestones, (2) polymetallic crust and sedimentary dykes, (3) infilling of pelagic Triassic limestones, (4) volcanic tuffs.

– a few exotic limestone blocks are associated with the ophiolites, either within the suture or within the klippes, such as the Kiogar block (Heim & Gansser, 1939; Gansser, 1964). One of them, near the monastery of Lamayuru, displays a particular succession of facies (fig.2) : a late Permian (late Djulfian) neritic limestone rich in Algae, Foraminifera, Brachiopoda and Crinoida (*Colaniella* and *Palaeofusulina* biozone according to M. Lys, Orsay), appears eroded and locally coated with polymetallic crusts.

The surface is locally infilled by pelagic limestone (rich in Ammonites of Scythian age, several species of Meekoceras similar to the fauna of the "Meekoceras beds" of the Lilang section in the Spiti area according to J. Gueix, Lausanne) mixed with a few volcanic fragments; the series ends with a succession of tuffs, volcanic agglomerates, pillow lavas and radiolarites. A similar succession, associated with Triassic red limestones, also exists within the tectonized base of the Photaksar ophiolite klippe.

In conclusion, in the internal part of the Himalayan orogen, there are evidences of :
– the presence of oceanic crust : the ophiolitic suite, regionally associated with Triassic pelagic facies;
– the occurrence of a zone, possibly related to a continental crust, with the succession in time of : neritic limestones, polymetallic crusts, pelagic sediments and volcanics. In our opinion, this succession characterizes the typical evolution of continental break-up and the associated creation of a passive margin regime.

In this case such an evolution, occurring just between the Permian and the Trias, could emphasize the activity of the Tethys Ocean.

THE TIBETAN SEDIMENTARY SERIES [M.C.]

The general stratigraphy of the Tibetan sedimentary series has been unravelled in Central Nepal (Colchen, 1971, 1975; Bordet *et al.*, 1972; Bassoullet *et al.*, 1977). Two recent discoveries may be mentioned here :

1) The Permian-Triassic boundary (Bassoullet & Colchen, 1977).
Paleontological data show the Permian-Triassic boundary to lie at the lower part of *Otoceras* sp. aff. *woodwardi* limestones of Lower Scythian. The uppermost Permian is not in evidence and is thought to be missing.

Lithostratigraphic and paleontological data strengthen this hypothesis, showing that a change of sedimentation occurred as early as Lower Triassic : low energy carbonate sedimentation (biomicrite with thin-shelled Molluscs) succeeded by heterogeneous detrital deposits was followed by further carbonate sedimentation of higher energy level. This succession is known elsewhere in the Himalaya and indicates palaeogeographical changes of sedimentation, connected to the incursion of the sea in a domain that was particularly unstable during the Carboniferous and the Permian.

2) The problem of Gondwanan and Tethyan characters of the Himalayan series (Colchen, 1977b).
Analogies of lithofacies and fauna have been described between the Carboniferous and Permian series of the Lesser and Higher Himalaya. These analogies indicate similar palaeogeographic conditions with an imbrication of the Gondwanan and Tethyan characters. Although the Higher Himalaya is a part of the epicontinental Tethyan domain, Gondwana influences appeared from time to time

during the entire Mesozoic so that it can be described as a peri-Gondwana area. The opening of an oceanic Tethys (intra or extra Gondwana ?) is again suggested by Colchen (1977) who addresses the problem of the limit of the Gondwana in north India.

THE MAIN CENTRAL THRUST ZONE (M.C.T. ZONE) [A.P.]

1) Introduction

The M.C.T. Zone has been studied mainly by Pêcher since 1972 in Central Nepal (Annapurna-Manaslu and Arun valley). Several publications deal with the results of the study of deformation and metamorphism in these areas (Pêcher, 1974, 1975, 1977, 1978, 1979; Brunel, 1975; Brunel & Andrieux, 1977; Brunel *et al.*, 1979). The main metamorphic and deformational events in Central and Eastern Nepal are shown to be very similar.

The methods of study used by these authors include :

- field investigations and mapping at all scales,
- microscope study of thin sections,
- whole rock chemical analysis and microprobe mineral analysis,
- X Ray diffractometry,
- quartz C axis determinations and statistical analysis in oriented thin sections of quartzites,
- study of quartz fluid inclusions and interpretation on a heating and freezing stage.

The main results concerning the deformation and the metamorphism are summarized below.

2) The deformation

The deformation is typically un-coaxial, the rotational character increases progressively in the proximity of the thrust plane (i.e. the plane of highest deformation between the Tibetan Slab and the Midland Formations). This zone appears to be a very large shear zone, the thickness of which may reach 10 km; it gradually passes to the more superficially deformed and more intensely folded areas (the Tibetan sedimentary series, above, and the southern part of the Midland formations, below).

a) The mesoscopic geometry of the M.C.T. Zone

The mesostructure has three main aspects :

i   - a flat cleavage,
ii   - a conspicuous "type a" line,
iii - a scarcity of folds.

i - At the top of the Tibetan Slab, the fracture cleavage S2, axial plane of the south-vergent B2 folds soon becomes a metamorphic cleavage, and farther down a foliation plane (main structural surface of the Tibetan Slab). No structural discontinuity is observed at the base of the pile of gneisses, across the thrust : the same S2 metamorphic cleavage can be found in the Midland formations, usually parallel to the S0 boundaries, or, if not, slightly more northward dipping than S0.

ii - The existence of an omnipresent mineralogical or stretching lineation, NNE-SSW, is probably the most conspicuous feature of the M.C.T.

Zone. In the higher part of the Tibetan Slab and in the Tibetan sedimentary series, the main line is an intersection line. Lower, as the metamorphic recrystallization increases, a mineralogical line (elongation of minerals) is substituted for the intersection line. Together with it, the orientation turns away from the general E-W trend of the belt : the lines scatter in the cleavage plane towards a N-S direction.

In the underlying Midland formations, the line is a stretching line, marked by the elongation of pebbles in detritic layers, by mullions in more homogeneous and competent ones, or by metamorphic striae in the schistose layers. This line, very well expressed near the M.C.T., vanishes away from it in the lower Midland formation, or in the southerly folded area; but in all the areas studied, its direction remains remarkably constant - NNE-SSW - as in the lower part of the Tibetan Slab : this direction, perpendicular to the general cartographic trace of the thrust, can be considered as the *transport direction*, and the line must be regarded as an "a" line.

The finite strain ellipsoid had been measured on the pebbles of several samples. In the Arun area (Brunel, 1975), it is typically constrictive; in Central Nepal (Pêcher, 1978), it is usually of flattening type, except in narrow (hundred of meters) bands parallel to the line, where mullion structures are particularly well expressed. So, despite the great homogeneities in orientation, these bands (kinds of "transform faults") and the differences in the shape of the strain ellipsoid, indicate strong flow heterogeneities.

iii - The scarcity of the folds : in the transition zone between the Tibetan sedimentary Series and the Tibetan Slab, the large south-vergent B2 folds flatten drastically; lower in the pile of gneisses, no large synmetamorphic folds can be clearly seen. In the part of the Midland formations within the shear zone, the still-recognizable stratigraphic polarities of the gently northward dipping metasediments exclude the presence of large recumbent folds; the only large folds are post-metamorphic concentric folds and are related to the Pokhra-Gorkha anticlinorium.

Nevertheless, in outcrop, or in thin sections, some syn- to late- metamorphic folds are visible; their axial directions follow the same variations as the line : roughly E-W at the top of the Tibetan Slab, they scatter in the cleavage plane when nearing the M.C.T.; in the highly sheared zone close to the M.C.T., the axial B direction is usually parallel to the NNE-SSW flow direction.

Brunel and Andrieux (1977) are of the opinion that these folds in Eastern Nepal were probably initiated parallel to the "a" line, due to the constrictive character of the finite strain. In Central Nepal, this explanation does not fit very well with the mainly flattening type of the finite strain; but some rare outcrops, where folds are numerous enough to be statistically informative, show a dispersion of the axis directions throughout the cleavage plane, with a strong maximum near the "a" direction (sheath folds) : here, one may propose that the NNE-SSW oriented folds are due to reorientation of earlier folds by sliding in the cleavage plane, (see Pêcher, 1978, fig.91).

b) Criteria of un-coaxial (rotational) deformation

In the M.C.T. Zone, several features show the rotational character of the deformation, as well as its progressivity :

i - The rotational character is fossilized in the symmetry of the microstructures : in XZ sections (perpendicular to the cleavage, parallel to the line), they are monoclinic (i.e. "dissymmetric"); the sense of dissymmetry is then in accordance with the direction of shear, as inferred from regional considerations (the Tibetan Slab is thrust southward over the Midland formations); in YZ sections (perpendicular to both lineation and cleavage), the apparent symmetry is statistically orthorhombic;

ii - Progressive increase of shearing (increase of γ value) is revealed by the progressive changes in microstructures, and by the evolution of the dissymmetries, which become increasingly acute closer to the thrust plane.

In these aspects, the Himalaya appears as a model, where deformation on a global tectonic scale can be followed down in scale to the internal structure of the minerals. The inventory of the shear structures has been made. They concern either the geometric dissymmetries, or the preferred orientation of minerals.

i - Dissymmetries :

- almonds S-S' : in detail, the metamorphic cleavage S2 appears as a microscopic juxtaposition of almond-shaped bodies, giving a characteristic cupular aspect to the rocks. These almonds result from the association of the main cleavage plane S with an S', which must not be attributed to another deformation phase; visible in the same metamorphic assemblages, S and S' are penecontemporaneous, S' revealing sliding movements on S as soon as S was formed; the obliquity of the relation S-S' is very constant (S' more northward dipping than S) and gives the direction of sliding;

- rotation and fracturation of the porphyroclasts;

- intrafolial drag folds;

- rotated internal schistosity of porphyroblasts, associated with dissymmetric pressure shadows.

ii - Quartz microstructures and preferred orientation :

Based on microstructures in quartz-rich formations (quartzites, sandstones, quartzo-feldspathic gneisses), several main microstructural zones have been recognized of which the cartographic patterns are roughly the same as those of the metamorphic zones (see below). From bottom to top (i.e. from the Lower Midland formations to the M.C.T., and up into the Tibetan Slab), they are:

- zone of sedimentary microstructures;
- zone of porphyroclastic microstructures : the shapes of the old, flattened, detrital grains are still recognizable, surrounded by the phyllites; but most of the grains are polygonized partially recrystallized, and extended by pressure-shadows with a mosaic structure;
- mosaic microstructures are very characteristic of the upper part of the Midland formations, close to the M.C.T.; here, some samples also show characteristic mylonitic ribbon microstructure;
- exaggerated grain growth microstructures, always found in the gneisses of the Tibetan Slab, are also observed in the higher part of the Midland formations in the Burhi River area.

These various microstructures reflect the combined increase of plastic deformation and temperature when crossing the M.C.T. The influence of temperature predominates in the Tibetan Slab where the microstructures reveal a strong thermic overprint younger than the plastic deformation. Increase of the rate of deformation has been mainly demonstrated by the study of the preferred orientations of quartz C-axes (Pêcher & Bouchez, 1976; Bouchez & Pêcher, 1976; Brunel, 1979).

Using the metamorphic cleavage plane and the stretching lineation as the external geometric reference, quartz C-axes are distributed along two cross-girdles (diverging at an angle of = 2 Θ), roughly symmetric in respect to the XY or YZ plane; but one of the girdles is more densely populated than the other, attesting the rotational component of the deformation. Towards the M.C.T., the dissymmetry of distribution increases, and one of the girdles even vanish completely (as in the ribbons fabrics)(see Pêcher, 1978; fig.121)

Multiplication of measurements has shown that the value of Θ (which can appear "a priori" as a good shearing rate criterion) seems to rapidly reach a minimum value (15 to 20°), and then remains constant whatever the value of the shear ratio; thus the main criterion for the γ value seems to be the rate of population density of one girdle compared to the other.

In one or two samples from the Arun area, Brunel (1979) has shown that the plastic deformation of the quartz was accompanied by a new shape fabric and the elongation of the quartz grains are oblique to the cleavage plane as defined by the phyllites. This obliquity poses the problem of the definition of the schistosity (the plane of flattening) and of the choice of the cinematic reference point, as well as the problem of the relative chronology between metamorphism (i.e. recrystallizations) and the acquisition of preferred orientations in the quartz.

3) Metamorphism

The following data on the metamorphism are based mainly on detailed petrographic investigations of Pêcher (1975-1978) in the Annapurna - Manaslu area. Brunel's observations in the Arun Valley (1975) are in good agreement with the features described here.

a) Age of the main metamorphic events in the
M.C.T. Zone

The metamorphic assemblages define the S2 plane, associated to the evolution of the M.C.T.;
in regard to shearing, metamorphism starts rather
early (the S2 metamorphic cleavage acted as the
sliding plane, chanelling the main flow deformation), and its thermal print continues throughout
the movement (late-deformation figures such as in
-S pegmatitic lenses, or filling of open cracks,
show metamorphic assemblages similar to those of
their host rock) : the main metamorphic print is
therefore contemporaneous with the M.C.T. thrusting (i.e. Miocene).

Radiochronometric data on cooling ages, lead to
similar conclusions : the intrusion of the Manaslu granite, related to the migmatisation of the
Tibetan Slab (Le Fort, 1973), is dated as 16 to
23 m.y. by Rb/Sr measurements on muscovites (Vidal, 1978; Hamet & Allègre, 1978). Similarly, two
muscovites from the front part of the M.C.T. nappe, south of Kathmandu, gave ages of 22 and 26 m.
y. (Andrieux, Brunel & Hamet, 1977). A series of
K-Ar ages in the Tibetan Slab of the Everest area
(Krummenacher *et al.*, 1978) varies from 20.5 m.
y., 9 km above the M.C.T., to 9 m.y. close to it,
the decrease in age being related to the processes of erosion and cooling throughout the Miocene.

b) The possibility of an earlier metamorphism

The slight metamorphism of the Tibetan sedimentary series is not yet well correlated with the
phases of deformation, and might be older than D2
– i.e. contemporaneous with D1, or, very hypothe-
tically, a distant replica of the high pressure
metamorphism of Kashmir and Pakistan (Thakur &
Virdi, 1978; Bard *et al.*, 1979), dated as Upper
Cretaceous (Bard, oral comm. 1979).

Within the Tibetan Slab, the syn-D2 metamorphism is strong enough to obliterate most of the
previous events; nevertheless, throughout the M.
C.T. Zone, structural superposition of metamorphic minerals can be seen locally : but rather
than a succession of metamorphic phases (i.e.
thermobarometric events separated by large lapses
of time), such superpositions reflect minor discontinuities in the recrystallization-deformation
processes synchronous with deformation of the M.
C.T., from the initial cleavage up to the late
retromorphic events, some million years later.

Thus if pre-Alpine metamorphic events had taken
place in Central Nepal (some radiochronometric
data might be slight indications of them), they
are today masked by the Alpine, mainly Miocene
metamorphism, in the structural domain of the Himalayan belt described here.

c) Main characteristics of the metamorphism in
the M.C.T. Zone

Several metamorphic zones have been recognized
and mapped based on the mineralogical equilibria
in the KFMASH chemical system (K_2O, FeO, MgO,
Al_2O_3, SiO_2, H_2O). They refer to the paragenesis
of the meta-sandstones and the meta-pelites, the
most abundant types of rocks in the thrust zone
(fig.3 and 4).

The following minerals of the system, KFMASH,
have been identified : quartz and muscovite, both
very frequent; chlorite (chl; only prograde chlo-
rites, and not retromorphic ones, are taken into
account here), chloritoid (ctd), biotite (bio),
garnet (grt; almandine : 60-75%, pyrope : 3-30%,
spessartite : 1-25%), staurolite (sta), K-feldspar (KF), kyanite (ky) and sillimanite (sill).

Those index minerals define the following zonal
succession, from the south to the north of the M.
C.T. (when in brackets the mineral is scarce) :

   i - chl, (ctd) zone,
  ii - chl, bio, (ctd) zone,
 iii - chl, bio, grt, (ctd) zone,
  iv - (chl), bio, grt, sta, ky zone,
   v - bio, grt, ky, (KF) zone, and
  vi - bio, grt, sill, KF, (ky) zone.

In CaO rich rocks, the other main index minerals are plagioclase (scarce or detrital before
zone iv), zoisite (restricted to close to the M.
C.T. - that is to the limit iv-v), actinolite
(zones i-iii), hornblende and diopside (zone v-
vi).

The average dip of the M.C.T. Zone being northward, the succession above corresponds to an apparent increase in metamorphism from the structurally lower part of the zone (lower Midland formation outcrops in the innermost part of the
Pokhra-Gorkha anticlinorium) to the higher one
(upper part of the Tibetan Slab) : this succession illustrates typically the famous Himalayan
"reverse metamorphism", previously described, by
many authors, in the Central and Eastern Himalaya
(see Le Fort, 1975a).

Higher in the Tibetan sedimentary series, where
observations are too scattered to delineate precisely such mineralogical zones, the metamorphism
decreases "normally", the main index minerals
being sillimanite, garnet, biotite and chloritoid
(plus diopside in limestone); unpublished data on
illite crystallinities show that the higher epizo-
nal or anchizonal metamorphism reaches up into
the Lower Mesozoic (Dunoyer de Segonzac, Colchen
& Le Fort, unpublished data).

- *Geometry of the isograd surfaces*

The surfaces bounding the metamorphic zones can
be regarded as isograd surfaces; their geometric
relation to the thrust plane throws light on the
relationship between metamorphism and deformation
with due regard to the uncertainty on their position - which may be precise to several 10 m or
1 km - according to the number of samples studied, (some 800 for the Annapurna-Manaslu area),
and on the rock chemistry restrictions.

The salient point of the geometry of the isograds, clearly illustrated on the map or along
cross-sections (cf. Pêcher, 1975, 1978, 1979;
Brunel, 1975), is the conspicuous regional parallelism between the M.C.T. and the isograds (fig.
3, 4) : such a constant geometrical relationship
also implies a genetic relationship, and the metamorphic distribution observed must result from
the thrust-shearing events.

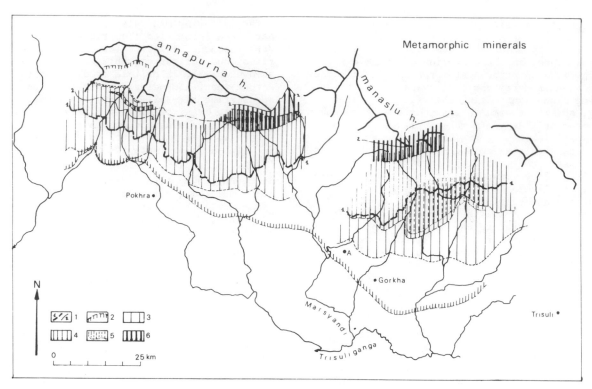

Fig. 3 - A map of the distribution of metamorphic minerals in Central Nepal (Pêcher, 1977, fig. 12).
(1) lower limit of the Tibetan Slab -1- and upper limit of formation I -2-, (2) biotite appearance,
(3) garnet (also present in the Tibetan Slab), (4) kyanite, (5) staurolite ; (6) sillimanite (fibro-
lite). A : nepheline syenite and alkaline gneisses of Ampipal.

The other following geometrical aspects must also be emphasized :

• the spacing of the isograds is closer near the front of the thrust (in the Arun Valley, where the thrust can be followed for more than 50 km from north to south, Brunel, 1975) than in areas farther back from it (in the Burhi Gandaki area, Pêcher, 1978);

• the isograds may cut lithostratigraphic boundaries: the reverse metamorphism is not due to the pile of variously metamorphozed scales or sheets, but reflects an abnormal thermobarometric distribution in the shear zone;

• no metamorphic hiatus can be clearly observed across the M.C.T. plane : the post-metamorphic displacements must be slight compared to the syn-metamorphic ones. From this point of view, the western part of the area studied (Kali Gandaki-Annapurna) might be an exception : here the ky-sta zone seems to disappear against the thrust plane; it may be due to the existence of a somewhat different P.T. pattern in the frontal part of the

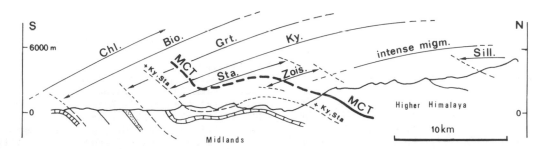

Fig. 4 - The Burhi Gandaki section and distribution of metamorphic minerals (Pêcher, 1979a).
Abbreviations for minerals as in the text.

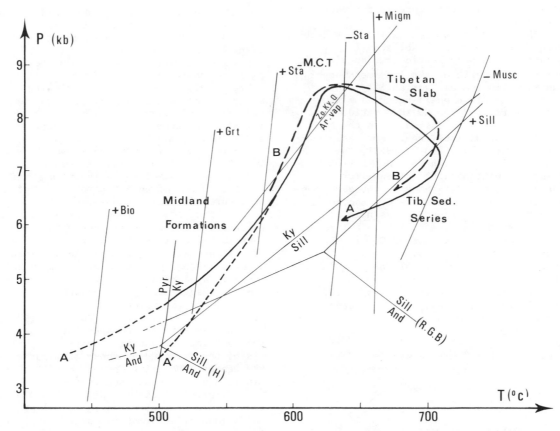

Fig. 5 - The Himalayan reverse metamorphism : pressure-temperature distribution around the M.C.T. The fluid phase composition is supposed to be either pure water (curve A) or a mixture of water and carbon dioxyde (curve B) ; curve A' takes into account the hypothetical presense of andalusite (Pêcher, 1978).

Abbreviations : see the text ; + bio..., - sta... means apparition of bio..., disappearance of sta... on the high-temperature side of the curve ; Migm = migmatization (liquidus curve), Pyr = pyrophyllite; the univariant curves for the $SiAl_2O_5$ polymorphs are from Richardson *et al.*, 1968 (R.G.B.) or Holdaway, 1971 (H.).

thrust, or it may be a precursor of the nappe system found farther west (Western Nepal or Kumaon).

- The pressure-temperature distribution in the shear zone

This distribution can be deduced from the chemico-mineralogical reactions at the boundaries of the metamorphic zones. Such observations will give the P-T conditions during the main recrystallization of the rock; but it does not imply that all the equilibria were reached at the same time everywhere in the shear-zone : for example, there can be still strong recrystallization synchronous with deformation in the central part of the zone, coeval with colder, "late-metamorphic", deformations in the more external parts.

With this rectriction in mind, one can deduce the pressure-temperature evolution through the shear-zone from the following equilibria (fig.5) :

- appearance of biotite, soon followed by apparition of garnet;

- disappearance of chloritoid, apparition of kyanite and staurolite;

- disappearance of staurolite soon after its apparition (due to the chemical composition of the rocks, sta forms from grt and chl, at approximately 590°C, and reacts with chl to give bio + ky at approximately 635°C : this rather narrow stability field could possibly explain the absence of the ky-sta zone in the Annapurna area for kinematic considerations, supposing that we have here a more quickly cooled frontal part of the thrust);

- stability field of coexisting kyanite and zoisite (near the M.C.T.);

- liquidus curves of ky-pl-bio-musc and sill-FK-bio-musc assemblages (migmatisation of the Tibetan Slab), and

- absence of H.P. paragenesis under the M.C.T. : no evidence of H.P. paragenesis, even as relicts, have been found here; moreover some authors (Hashimoto *et al.*, 1973; Rémy, 1974) have mentio-

ned andalusite, but its existence is not very reliably established.

Remark : The P-T distribution in fig.5 is dedu-
ced from equilibria curves established for $P_f = P_{H_2O}$. In fact, microthermometric studies of fluid
inclusions in the quartz of the late-metamorphic
lenses has revealed a great variability in the
composition of the fluid phase being, a mixture
of CO_2, H_2O and salts : the CO_2/H_2O ratio, very
low away from the M.C.T., increases in the central part of the shear zone, where some samples
contain nearly pure CO_2 (Pêcher, 1979). The fluid
trapped in these late lenses does not reflect the
fluid acting during mineralogical equilibration
of the host-rock (considerations on the paragenesis stability field show that the CO_2 content in
the host rock could not have been as high as in
the lenses); nevertheless, a tentative P-T curve,
taking into account some amount of CO_2 in the
fluid phase, is given in fig.6.

These two curves differ only slightly, and both
show that :

- the highest temperature values (approx. 710°)
are reached in the upper part of the Tibetan
Slab, 5 to 8 km above the M.C.T.;

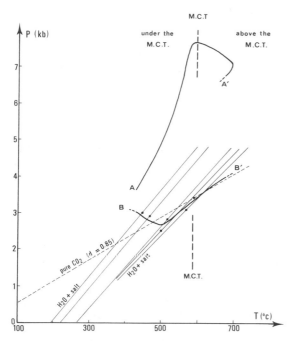

Fig. 6 - P.T. distribution around the M.C.T.,as
inferred from the main metamorphic assemblages
(curve AA', see fig. 5) and from the fluid
inclusions in the late-metamorphic exsudation
lenses (curve BB'). For the curve BB' : data from
Potter and Brown (1977) (full lines, isochors for
a filling by a brine) and from Kennedy (1954)
(dashed line, isochor for a filling by pure CO_2).
True formation temperature in the inclusion is
estimated from the mineralogical paragenesis of
the considered lens (Pêcher, 1978).

- the pressures increase downwards to attain a
maximum near the M.C.T., (approximately 8,5 kb)
and then apparently decrease lower in the underlying Midland formations.

- *Thermobarometric decrease and post-metamorphic events in the M.C.T. Zone*
The previous observations deal with the defor-
mation-pressure-temperature pattern during the
main stage of petrographic evolution of the
thrust zone. Some field or laboratory results
describe its variations in time :

i - the thermal influence is still evident after the major deformation. Indeed :

- the position of the veins of migmatitic mobi-
lizates, which often cut across the gneissic
structures (relaxation phenomena);

- the presence of Riedel's extension fractures
(with orientations consistent with the overall
shearing directions) filled with "hot" minerals :
for example sillimanite, in the sillimanite zone,
or even in the upper part of the kyanite zone,
where it appears as a "retromorphic" mineral;

- the exaggerated grain growth microstructure,
which can be interpreted as reflecting post-plas-
tic-deformation annealing (Bouchez, 1977);

- the similarities between the paragenesis of
the in-S late exsudation lenses and the paragenesis of the surrounding rocks (the paragenesis of
the lenses, when rich enough, show the same index
minerals as the host rock).

Thus, particularly in the Tibetan Slab, the
temperature decreased little or not at all whilst
conditions of deformation varied considerably
(conditions of ruptural deformation, probably
corresponding to a decrease of the prevailing
stress).

ii - The fluid pressure (Pf) decreased faster
than the temperature : a strong decrease in the
fluid pressure between the main period of metamorphism and the late exsudation lenses has been
demonstrated by study of the fluid inclusions in
the lenses.

The Pf in these lenses has been deduced from
the fluid density (in the inclusions for which it
was possible to draw the isochore, i.e. inclusions of brine or pure CO_2) combined with the
trapping temperature as estimated from the mineral associations.

The Pf-T curve across the M.C.T. Zone based on
these data (see fig.6 : a cross-section in the
Annapurna area) shows a drastic decrease of fluid
pressure, of up to several kb (3 to 4 kb at the
level of the M.C.T. However these values are only
a first approximation, as the exact role of fluid
pressure during the metamorphic processes has not
as yet been clarified).

iii - Some deformation however continued in the
shear-zone after the decrease in temperature :

- according to Brunel (1979), the preferred
orientations of quartz C-axis would reflect mainly a late, colder, stage of the deformation;

- low grade retromorphic equilibrium are observed (for instance destabilisation of garnet and
biotite to chlorite), the newly formed minerals

being then often located in Riedel's fractures
("cold" Riedels) or in open extension cracks perpendicular to the "a" rock fabric;

- slickenslide-striae occur on the older structural surfaces (the striae are usually best expressed on the steep reverse-side of the S-S' almonds, or at the surface of the exsudation lenses).

Thus there has been some cold sliding in the M.
C.T. Zone, at first restricted to the old main
structural discontinuity, the S2 plane. As cooling proceeded, some blocking must have occurred;
new discontinuities and folding are then needed
to absorb the continuing shortening. This phase
may be associated with the irregular apparition
of the new S3 strain slip cleavage, which is
steeper than the older S2, and the formation of
large B3 folds (for instance the Pokhra-Gorkha
anticlinorium) in the front (south) of the previously strain-hardened (metamorphosed) wide
shear-zone.

*4) Origin of the reverse metamorphism, some
geodynamic implications*

The reverse metamorphic zonality expresses an
abnormal distribution of both the pressure and
the temperature, at the time of the stabilization
of mineral equilibria, equilibria which could
have been tempered by a rapid decrease of either
temperature or pressure.

a) The pressure distribution

A very astonishing feature of the pressure distribution is its downward decrease under the M.C.
T. : the total thickness of the Tibetan sedimentary series plus the Tibetan Slab is more than
20 km, and corresponds to a lithostratigraphic
load which fits rather well with the 8 kb pressure proposed near the M.C.T.; in the underlying
terranes, pressure should be still higher.

No really good answer to this paradox has yet
been put forward. Meanwhile, one must notice that
the pressures given are deduced from equilibria
in rocks spaced out along several south-north
cross sections following the topography, i.e. not
perpendicular to the thrust plane, but quite
oblique to it; the low pressure metamorphic zone
of the Midland formations, although structurally
the lowest, lies in fact in front of the M.C.T.
trace, and not under the thrust plane; this zone
might not have been covered by all the thrust
pile at the time of metamorphism. Such an asser-
tion would imply that :

- the erosion commenced very early (as shown by
the strong pressure decrease printed in the late-
metamorphic lenses);

- the present frontal trace of the thrust is
probably not far from its maximum extension;

- the mineralogical assemblages above and below
the thrust plane were not coeval.

b) The temperature distribution

Whatever the true pressure pattern, the temperature pattern can be rather easily explained
(Le Fort, 1975a) by subduction type thermal models. Toksöz and Bird (1977) calculated the crustal thermic distribution in a collision belt such
as the Himalaya, taking into account crustal
shortening in large M.C.T. type cleavages.

These models require transitional S-shaped
forms of the isotherms on both sides of large
thrusts (true subduction, or "intracontinental
subductions"), with the superposition of zones of
normal thermal gradients above zones of reverse
ones. If recrystallisation is sufficiently rapid
to fossilize this particular thermal pattern, a
zone of apparent reverse metamorphism (in respect
to temperature) will exist.

According to the models, the zone of maximum
temperature depends mainly on the relative amount
of shear friction heating versus heating due to
the more or less high initial temperature in the
trust area. As the maximum temperature lies
clearly a few km above the M.C.T. plane, the
Himalayan reverse metamorphism in the lower part
of the Tibetan Slab and in the Midland Formations
must be essentially due to the high temperature
of the Tibetan Slab prior to thrusting.

THE SOUTHERLY FOLDED AREA
(Mascle & Pêcher, 1977; Mascle, 1979;
Brunel *et al.*, 1979) [A.P.]

1) The southern Lesser Himalaya

Although there is no important structural break
between the northern and the southern parts of
the Midlands, changes in the type and intensity
of deformation give a different aspect to the
southern area : tectonically, this area appears
much more complex than the monoclinal and unfolded M.C.T. Zone; several phases of deformation
(D2 to D5), can be recognized, on the basis of
successive refolding of the cleavage; most of
them were not described by previous authors
(Fuchs & Frank, 1970; Rémy, 1972; Hashimoto *et
al.*, 1973).

a) Structures associated to the main S1-2 cleavage :

The oldest structures are characterized by a
regional cleavage which clearly affects the Midland formations up to the Eocene Tansen Formation. This cleavage, marked by very low grade metamorphic recrystallizations (sericites), appears
as the prolongation of the cleavage synchronous
with the shearing in the M.C.T. Zone. In relation
to the tectonics of the Tibetan Slab, it can be
called S1-2, although no D1 deformation has been
observed.

S2 is related to isoclinal recumbent folds
(nappes), of several kilometers, with large overturned limbs visible with polarity criteria such
as upside-down stromatolithes and cross-bedding;
the overthrusting may exceed 15 km. The D2 features visible in the field are mainly lineations;
their pattern shows, in a shallower environment
than in the M.C.T. Zone, that the same north-
south "a" sliding is an important process of internal deformation of the rock; the sliding decreases southwards, i.e. away from the shear-
zone. These lineations are :

- intersections S0 - S1-2 and microfolds axis, showing a dispersion pattern with two maxima roughly north-south and east-west, the latter becoming sharper to the South;
- mineral streching lineations, about north-south and better expressed in the northern area.

   b) Post S1-2 structures

After the flow deformation, the preexisting structures of the Lesser Himalaya, shears or nappes, were refolded by concentric folds, which correspond to a deformation at higher structural level in the frontal part of the belt. These folds give the southernmost part of the Himalaya its characteristic "Jurassian" aspect, and are :
- B3 folds, frequently dissymetric to the north, and rarely so in the south, associated to a S3 cleavage of kink-band or crenulation type,
- large sub-meridian B4 folds, parallel to the Thakkhola fault zone, only exceptionally accompanied by a N-S fracture cleavage (as in the Arun valley);
- B5 large open E-W antiforms and synforms, sometimes accompanied by a S5 strain slip cleavage.

   2) *The Main Boundary Fault zone (M.B.F. zone)*

All the previous deformations are directly related to movement along the M.C.T., and are cut by the Main Boundary Thrust, which appears in the field as a fault, plunging 60 to 70° to the north; it is accompanied by minor reverse and strike-slip faults, or by minor folds, in agreement with the direction of shortening about north-south and perfectly horizontal.

In front of the M.B.T., the Churia Hills (the Siwaliks) show very regular parallel anticlines (B5-6), with short southern and long northern limbs, which dip regularly north at some 70° near the M.B.T. to 35° a few km south of it.

QUATERNARY GEOLOGY AND GEOMORPHOLOGY [M.C.]

The team has not yet undertaken any study in the Siwaliks, but has worked on one of the major intramontane basins of the Himalaya, the Plio-Quaternary Thakkhola graben (Colchen, Fort & Freytet, 1979). This study was associated with research on the recent evolution of the Himalayan range, in particular on the Quaternary formations and the geomorphology of Ladakh and Higher Himalaya (Fort, 1977, 1978a and b).

   1) *Sedimentation and Plio-Quaternary tectonics in the Thakkhola basin* (Colchen *et al.*, 1979)

This N-S intramontane graben lies north of the high range and cuts the Himalayan structures. The sediments, deposited on folded and faulted Triassic to Cretaceous series, show the following sequence :
- deposition of the Tetang formations (probably Pliocene) in a small basin. Sediments include polygenic conglomerates, lenses of stromatolitic limestones and an Ostracode bearing argillaceous horizon;
- phase of deformation (F2 faults);
- partial erosion of the Tetang formations and deposition of the Thakkhola formations (Plio-Qua-

ternary) in a larger basin limited to the west by a system of F3 faults. These sediments include polygenic conglomerates, argillaceous beds and oncolithes with lenses of limestones;
- phase of deformation (F4 faults);
- erosion followed by the deposition of fluvioglacial formations (recent Quaternary).

   2) *Geomorphology of the Ladakh region* (Fort, 1978a)

The geomorphology of the whole region, which is a good example of arid, continental and subtropical high mountains, is controlled by the different natures of the various formations : the heterogeneous Ladakh batholith and the various Zanskar sedimentaries.

The cold Quaternary and present times gave rise to a typical morphology : regular valley-slopes and periglacial glacis (pediment), which are mainly due to the processes of gelifraction and snow-solifluxion. The Indus valley lies along a prominent structural axis of the Himalaya. Along the valley, the great extension of the Quaternary deposits (moraines and mainly fluvio-glacial deposits) and their relative positions are likely a result of both colder climates, probably wetter than at present times, and uplift, which still persists today throughout the area.

   3) *Quaternary deposits and periglacial aspects in the upper Burhi Gandaki*

Observations of great interest have been made in the upper Burhi Gandaki valley (Fort, 1977) :
- the present glaciation is characterized by the coexistence of Himalayan and transitional Tibetan types of glaciers and by a great variety of marginal dynamics;
- the existence of a rather large unglaciated area, where the periglacial forms and the different aspects of frost-and-snow morphodynamics enable us to distinguish active and inherited cold morphologies;
- numerous remains of older glacial stages, show the relative chronology of the late Pleistocene and Holocene events;
- travertine deposits, the development of which is greatly influenced by the environment conditions.

In particular, the most original periglacial forms are the inherited and inactive ones (Fort, 1978b). The rock glaciers and the soli(geli)-fluxion lobes are indications of a colder and wetter climate than the present one. The increasing dryness of the present climate stops the formation of such typical forms and deposits, to the nearly exclusive advantage of the thermo (cryo)clastic process.

Thus, the upper Burhi Gandaki valley can be considered as an exemplary illustration of the geomorphological and sedimentological evolution suffered during the last thousands of years by the high and northern valleys of the Himalaya. It provides good observations for further comparisons with the other parts of the Himalaya and other dry, high and continental mountains of the world.

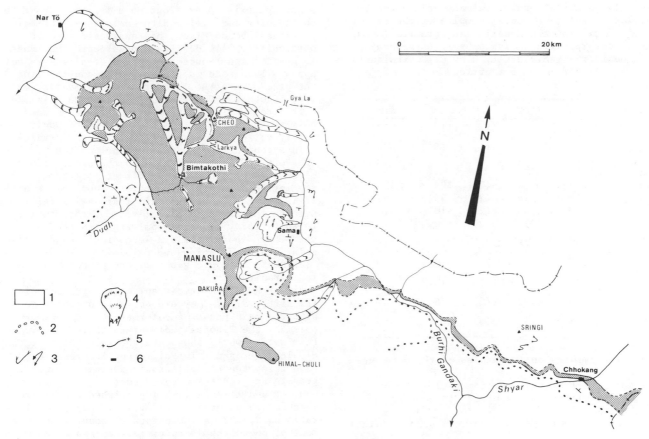

Fig. 7 - Geological sketch map of the Manaslu leucogranite (Le Fort).
(1) Manaslu leucogranite, (2) augen gneisses of the Tibetan Slab (Formation II), (3) major synclines and anticlines, (4) glaciers, (5) Nepal border, (6) main villages.

MAGMATISM. PETROLOGY AND GEOCHEMISTRY [P.L.F.]

1) The Manaslu leucogranite (Higher Himalaya)

The Higher Himalaya contains a dozen massifs of very typical leucogranites covering in all some 10.000 sq km. Difficult to reach and study, a few only have been given some attention in the past as by Misch (1949, Nanga Parbat), Bordet (1961, Makalu) and Gansser (1964, Bhutan).

In Central Nepal, a detailed study has been made of the Manaslu leucogranite. Previous results and unpublished data are summarized below.

a) Petrologic characteristics of the Manaslu leucogranite

Several preliminary reports have been published on the Manaslu granite (Le Fort, 1973, 1974b, 1975a and c). Geological mapping of the eastern part (fig.7) enables a more thorough description of this typical Higher Himalaya leucogranite. The massif covers an area of some 450 sq km in the shape of a lenticular slab, and shows three main parts :

i - *the core* is made of a very homogeneous two mica, medium grained granite, completely devoid of enclaves and thoroughly foliated. Dykes and

patches of tourmaline aplopegmatites criss-cross the granite with variable abundance. The granite is composed of quartz (30 %), euhedral zoned plagioclase (33 %, rim An 3 to 8, center An 12 to 20), K feldspar (27 %, mixing of orthoclase and microcline), muscovite (8 %) sometimes associated with a little fibrolitic sillimanite, biotite (2 %), rare accessories (apatite, zircon, opaques). Table 1 gives the average chemical analysis of this very homogeneous granite.

ii - *the lower* part of the slab of granite shows an increasing abundance of the tourmaline aplopegmatites. They invade the thinly banded gneisses on which rests the slab, leaving metasedimentary enclaves of various size (micaschists and marbles). A few hundred meters thick, this part is concordant with the upper part of the Tibetan Slab. Towards *the roof* of the granite, masses of tourmaline, muscovite, amazonite and beryl-bearing pegmatites appear.

iii - the eastern part of the massif narrows abruptly, east of Manaslu, into a very long sheet some 300 m thick, concordant with the country rocks at the top of the Tibetan Slab. This *eastern "arm"* has been followed for over 50 km until

TABLE 1. Chemical analysis, mean values for 36 samples of the core and 23 samples of the Eastern sheet ("arm") of the Manaslu leucogranite (CRPG, Nancy, quantometric analysis by K. Govindaraju, wet chemistry by M. Vernet, U by C. Kosztolanyi). P.F. stands for loss of ignition.

	Core	Arm		Core	Arm
SiO_2	74.04	73.94	F(1)	1110	940
Al_2O_3	14.67	14.72	Cl(1)	15	22
$Fe_2O_3t.$	0.84	0.81	B(1)	from < 50 to	820
MnO	0.02	0.02	Ba(2)	192	128
MgO	0.19	0.14	Co(2)	49	50
CaO	0.50	0.46	Cr(2)	<10	<10
Na_2O	4.06	4.14	Cu(2)	<10	<10
K_2O	4.49	4.48	Ni(2)	<10	<10
TiO_2	0.07	0.07	V(2)	<10	<10
P_2O_5(1)	0.14	0.13	U(3)	10.8	7.1
P.F.	0.76	0.72	Sr(2)	76	41
Total	99.78	99.63	Rb(4)	385	-

(1) wet chemistry on 10 samples of the core and 2 of the arm
(2) mass spectrometry (quantometer)
(3) wet chemistry on 7 samples of the core and 2 of the arm
(4) isotope-dilution analysis on 4 samples (Vidal, 1978)

it enters China, north of the Ganesh Himal. The granite is similar to the granite of the core (table 1) although with a more pronounced foliation. This "arm" does not exist in the west. Here a huge network of aplopegmatitic dykes with two mica or two mica and tourmaline invade the lower Tibetan sedimentary series for several kilometers beyond the granite.

The bottom of the slab of Manaslu granite shows no contact metamorphism with the surrounding country rocks of amphibolite facies regional metamorphism, other than a few occurrences of wollastonite within a few meters of the contact. However, as the Manaslu granite intrudes higher and higher levels of the Tibetan sedimentary series, a more and more pronounced contact halo appears. Near the top, where the granite intrudes the thin interbedded Upper Triassic shales, sandstones and lumachelles, the contact metamorphism has converted these rocks into staurolite garnet mica-schists, muscovite quartzites and pyroxene marbles over some 50 m.

The general foliation of the granite corresponds to the main schistosity of the country rocks (S2). The granite is syn- to late metamorphic.

A few kilometers below the Manaslu lenticular

slab, anatexis spreads out on a regional scale in the kyanite and sillimanite gneisses of Formation I of the Tibetan Slab. This anatectic zone is particularly well developed underneath the Manaslu granite and reduced where the granite does not appear. This zone is considered as constituting the place where the leucogranitic magma was formed, in other words the *roots* of the Manaslu leucogranite. The aplopegmatitic network between this zone and the bottom of the granite would partly correspond to the feeders of the granite.

In a general thermodynamic model of the M.C.T. (Le Fort, 1975a), anatexis and the production of close to eutectic magma is directly linked with the intracontinental subduction. Thus this granite is thought to be typical of crustal origin with no interference of activity of the mantle. Le Fort (1973, 1974b, 1975a) suggests that all the other Higher Himalaya leucogranites have similar characteristics and origin.

b) Rb-Sr isotopic geochemistry of the Manaslu leucogranite

Higher Himalaya leucogranites are syntectonic (Le Fort, 1973); they are nearly contemporaneous with the main phase of deformation and metamorphism of the Tibetan sedimentary series. Radiometric dating would thus date the main phase of the Himalayan orogeny. Two different laboratories have tried the Rb/Sr isotopic dating on Manaslu granite samples taken by Le Fort.

Hamet and Allègre (1976) published a whole rock isochron (9 points) of 28 m.y. with an initial ratio of 0,7408. A whole rock – muscovite measurement of one of the samples gave an age of 23 m.y. Vidal (Rennes) measured anew three of the same samples analyzed by Hamet and Allègre as well as four new samples. The previous isochron was not met again (Vidal in the Himalaya International Colloquium of the C.N.R.S., 1977, pp. 539-540; Vidal, 1978), and a scattering of points was found; the new muscovite – whole rock data gave an age of 16 m.y.

Although Hamet and Allègre (1978) acknowledged their faulty dating but tried to maintain their isochron, actually this isochron does not exist (fig.8). The high value of the initial Sr ratio is confirmed within the possible range of age (Eocene to Pliocene).

Several explanations have been put forward for the scattering of the points :

- non representative sampling : however the major elements geochemistry shows the homogeneity of the different samples collected by only one geologist accustomed to such problems;

- the smallness of the samples : due to the difficulties of sampling and transport, the samples barely exceed 1 kg. However an "adequate" weight is not known;

- isotopic homogeneity has not been attained during melting of the sialic parent rocks.

This last explanation seems to be the most realistic one. In fact, leucogranites quite frequently present such a scattering, as shown for the Lesser Himalayan leucogranite of Palung by

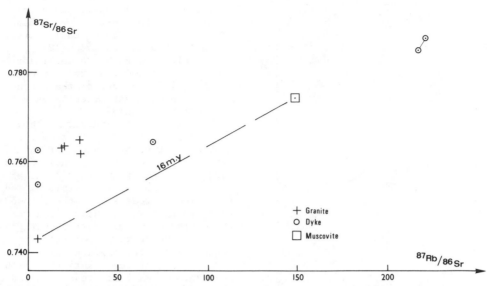

Fig. 8 - Rb-Sr diagram for the Manaslu leucogranite : samples from the core (D14, D22, D37, D65, U464, U476), the border zones (D8, D45, U303), the arm (U277) and aplitic dykes in the country rocks (B6, B7, N67). The important scattering of the whole set as within each category far exceeds analytical error.

Andrieux *et al.* (1977a). The role of intense fluid circulation remains unknown.

c) Geochemistry of Rare Earth Elements

The geochemistry of Rare Earth Elements has been studied (Cocherie, 1977 and 1978; Cocherie *et al.*, 1977) in order to evaluate of its compatibility with the proposed origin of the Higher Himalaya leucogranites through partial melting of the crustal material of the Tibetan Slab.

In this respect, seven samples of leucogranite, two of gneisses and two of anatexites from the Tibetan Slab have been analysed by radiochemical neutron activation (granites) or mass spectrometry (gneisses and anatexites).

The results for the leucogranites (fig.9) show :

- a general low REE content (10 to 25 chondrites) compared to granitic rocks;
- a regular enrichment of the light REE against the heavy REE (La/Yb > 10), in other words a regular fractionation from La to Lu;
- a conspicuous negative anomaly in Eu;
- a REE content even lower than that of the gneisses and anatexites from which they would have originated.

This last point raises a problem, as in partial melting, the incompatible elements (the light REE and to a lesser degree the heavy REE) should concentrate in the melt. However to produce granite melts with the observed REE pattern, one would need a parent gneiss with very low REE content. Such gneisses are not known. In explanation, Cocherie (1978) suggests either that these elements may not have an incompatible behaviour and/ or that the fluid phases that were certainly very abundant may have driven out the REE by forming carbonate complexes.

A satisfactory explanation has not yet been reached, though the same problems appear to characterize all the leucogranites (Cocherie, 1978).

2) *The Midland augen gneiss and Lesser Himalayan granites*

a) Midland augen gneiss (Ulleri formation) (Le Fort & Pêcher, 1974; Le Fort, 1975a; Pêcher & Le Fort, 1977; Pêcher, 1978).

Characteristic horizons of feldspathic augen gneiss have been recognized and mapped within the Midland Nepal formations (fig.10). They appear at the same lithostratigraphic level, towards the top of the Lower Midland group, with a highly variable thickness ranging from zero to 1500 m in the region of Ulleri; they constitute the *Ulleri formation* of which the main characteristics can be summarized as follows :

- lithostratigraphic control;
- tectonic and metamorphic characters similar to those of the surrounding rocks;
- intercalations of schists and quartzites similar to those of the surrounding rocks;
- transitional contacts when visible;
- lateral variations with conglomeratic beds in the same lithostratigraphic position;
- heterogeneity of the augen gneisses;
- variability of the alkaline ratio;
- occurrence of bluish rounded quartz;
- presence in their vicinity of metamorphosed tholeiĩtic volcanics (amphibolites), (Lasserre, 1977).

These field and geochemical characteristics all fit in well with a felsic volcano-sedimentary origin, with subordinate mafic layers.

No fossil or radiometric dating evidence is available to assert the age of the Ulleri formation. However, it took part in the Himalayan oro-

BORDET ET AL.

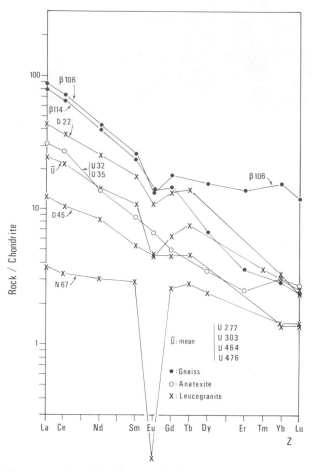

Fig. 9 - Chondrite normalized rare-earth patterns
(logarithmic) for the Manaslu leucogranite, two
samples of anatexites and two samples of non
migmatized schists from Formation I of the
Tibetan Slab (Cocherie, 1978, fig. 11).

geny (K/Ar ages on minerals) and a Lower, or Middle, Palaeozoic age is suggested.

Similar augen gneisses occur widely in the
Himalaya, always in the same lithostratigraphic
context, although they have been described under
various names as a result of the different hypotheses concerning their origin.

The wide occurrence of this volcano-sedimentary
episode leads to several important geodynamic implications :

- it should eliminate a certain number of tectonic nappes and injected scales which were inferred when the augen gneiss were supposed to be
part of an old crystalline basement;

- it eliminates the need of a Palaeozoic orogeny, as claimed by several workers, on the basis
of pre-Himalayan granites and radiometric dating;

- it implies a peculiar pre-Himalayan geodynamic pattern, little of which is known at present.

b) Lesser Himalayan granites (Le Fort, Debon &
Stebbins, 1978)

In the south of Nepal, eight to ten granite
massifs form the core of the Mahabharat range.
Poorly studied, they recently have been partly
mapped by Stöcklin *et al.* (1977) and one of them
was sampled for Rb/Sr dating (Andrieux *et al.*,
1977a). Debon and Le Fort started their field
study in 1977 on four of them, south of Kathmandu
and only preliminary results have been published
(Le Fort *et al.*, 1978).

They can be described as follows :

- the elongated massifs show sharp contacts;
generally structurally concordant with the country rocks, which are regionally metamorphosed micaschists with two micas and garnet. Hardly any
contact metamorphism is visible. The concordant
foliation increases near their boundary. In certain cases, this foliation is a metamorphic one,
supporting a two mica-garnet paragenesis;

- the granite is in general a porphyritic monzogranite of medium grain, with two micas and
cordierite (altered). Varieties with varyingly
abundant tourmaline, are associated with it. Garnet and andalusite may occur. Chemical analyses
of the two mica-cordierite monzogranite show its
hypofeldspathic, hypo-alcaline (hyposodic) aluminous and relatively quartz-rich character;

- in addition to the metasedimentary enclaves,
these granites contain numerous closely spaced
microgranular mafic enclaves, that are more common in granodiorites than in granites;

- they always intrude the formations of the
M.C.T. overthrust just north of the M.B.T. (cf.
Mahabharat nappe of Brunel, 1976), and remain as
elongated klippes. They were emplaced before
overthrusting took place and no trace of feeders
is visible beneath the klippes in the Midland
formations. They were followed, not only in
Nepal, but also along two thirds of the Himalaya
(1600 km from 73 to 87° E) and with the same characteristics.

The fact that the Lesser Himalayan granites
have very similar petrological and geochemical
characteristics, and a very constant structural
position along most of the entire length of the
Himalaya, speak in favour of their major geodynamic significance.

Highly foliated and transformed to orthogneiss
near their contacts, and at their ends (the "Outer Band" of the Dalhousie granite; Mac Mahon,
1882) they were tectonized and metamorphosed during their Himalayan nappe transport.

*3) Geochemistry of amphibolites and alkaline
gneisses from the Midland formations of Nepal*
(J.L. Lasserre, 1977)

Amphibolites occur at several levels in the
Midland formations of Nepal. They are mainly developed in the upper part of these formations,
where four main horizons have been studied (Lasserre, 1977). Their mineralogy is constant : actinolite, quartz, biotite, clinozoisite, albite,
ilmenite, apatite and sphene. Results of chemical
analysis (17 elements) of some fifty samples form
a fairly homogeneous group showing that they have
been derived from more or less differentiated

tholeiĩtic basalts, of island-arc tholeiĩte type. Spilitisation does not seem to occur.

Lasserre also suggests that a part of the Midland formations, including the Ulleri augen gneisses (Le Fort & Pêcher, 1974; Pêcher & Le Fort, 1977), are result of associated felsic and mafic volcanism, that could be the consequence of the interaction of an old island-arc with a continental margin.

Independently, a massif of nepheline syenite and alkaline gneisses has been discovered for the first time in the Himalaya, at Ampipal, north of Gorkha (Lasserre, Pêcher & Le Fort, 1976). Of moderate extension (8 x 2 km) (fig.3), this massif comports miaskitic nepheline syenites cut by a few mafic and ultramafic alkaline dykes (melteigite, jacupirangite), both more or less foliated. Chemical analyses (19 elements) show that these rocks are of igneous origin and that they are probably a result of fractional crystallization of an initial olivine alkaline basalt magma. Lasserre (1977) also related their genesis to the rifting of the Indian continent at the time of the Himalayan collision. However, Rb/Sr dating (Le Fort & Sonet unpublished) seems to give a much older age to these rocks.

CONCLUSION [M.C., P.L.F.]

This paper has summarized facts of geodynamic significance observed by our team. In conclusion, we would give the main ten successive steps of the geodynamic evolution of the Himalaya (see also Le Fort, 1971a, 1975a; Bassoullet et al., 1977) :

1 - Upper Precambrian to Devonian : epicontinental sedimentation in the Higher (and Lesser ?) Himalaya. Widespread felsic volcanic activity in the Lesser Himalaya;

2 - Carboniferous to Permian : epicontinental sedimentation, basaltic volcanism and epirogenesis throughout the Himalaya; overlap of Gondwan and Tethyan realms;

3 - Triassic to Middle Jurassic (Dogger) : platform sedimentation in the (Lesser and) Higher Himalaya; fragmentation of the Northern "edge" of the Indian craton, oceanization : opening of the Tethys Ocean (within Gondwana ?);

4 - Upper Jurassic (Malm) to Neocomian : platform sedimentation in the Higher Himalaya, northward subduction of oceanic crust with possible birth of an island arc, first emplacement of ophiolitic nappes towards the south;

5 - Aptian to Upper Cretaceous : mafic volcanism and platform sedimentation at the beginning in the Higher Himalaya, subduction continues, deposition of the Indus molasse following the erosion of the first Transhimalayan reliefs;

6 - Eocene : important slowing of the spreading rate of the Indian ocean, marine transgression, closure of the oceanic crust and collision of Indian and Eurasian plates, second emplacement of ophiolitic nappes towards the south;

7 - Oligocene : no marine deposit, M.C.T.

starts functioning as a intracontinental zone of subduction, the first Himalayan reliefs rise;

8 - Miocene : metamorphism (normal and reverse) and deformation linked to the M.C.T. reach their maximum, emplacement of the Higher Himalaya leucogranites, erosion of the Himalayan range and consequent sedimentation in the Siwalik basin;

9 - Pliocene : movement on M.C.T. stops and is relayed by M.B.T., sedimentation continues in Siwaliks and starts in intramontane basins;

10 - Pleistocene : M.B.T. continues and overthrusts the Siwaliks, erosion fills the Ganga Basin, widespread epirogenic movements.

Acknowledgments. We would like to thank Frances Delany for thoroughly reviewing this manuscript and making many useful suggestions for improving the English. This work was supported in the field by the Centre National de la Recherche Scientifique, Greco n° 12 : "Himalaya-Karakorum". Laboratory work was carried out at the Centre de Recherches Pétrographiques et Géochimiques, Nancy. We also thank J. Gerbaut for her careful work.

References

The list gives not only those mentioned in the text but also all the teams publications. Main references are shown with an asterisk.

*Andrieux, J., M. Brunel & J. Hamet, Metamorphism and relations with the Main Central Thrust in Central Nepal. 87Rb/87Sr age determinations and discussion, *Colloque int. n° 268 Himalaya, Paris 1976, Centre Natl. Rech. Sci., vol. Sci. de la Terre*, 31-40, 1977a.

Andrieux, J., M. Brunel, J. Hamet & P. Le Fort, La suture de l'Indus au Ladakh (Inde), *5ème Réunion annuelle Sci. de la Terre, Rennes, Soc. Géol. France éd.*, 11, 1977b.

Andrieux, J., M. Brunel & S.K. Shah, La suture de l'Indus au Ladakh (Inde), *C.R. Acad. Sci., Paris, t. 284*, 2327-2330, 1977c.

Andrieux, J., F. Arthaud, M. Brunel & S. Sauniac, Données nouvelles sur la géométrie et les mécanismes de formation des chevauchements dans la partie occidentale du Bas-Himalaya (Koumaon et Cachemire), *7ème Réunion annuelle Sci. de la Terre, Lyon, Soc. Géol. France éd.*, 12, 1979.

Bard, J.P., Q. Jan, H. Maluski, Ph. Matte & F. Proust, Position et extension de la "ceinture" métamorphique à faciès schistes bleus dans l'Himalaya du Pakistan Nord, *7ème Réunion annuelle Sci. de la Terre, Lyon, Soc. Géol. France éd.*, 29, 1979.

Bassoullet, J.P. & M. Colchen, Les caractères gondwaniens des formations tibétaines de l'Himalaya, *4ème Réunion annuelle Sci. de la Terre, Paris, Soc. Géol. France éd.*, 35, 1974.

Bassoullet, J.P. & M. Colchen, La limite Permien-Trias dans le domaine tibétain de l'Himalaya du Népal (Annapurnas - Ganesh Himal), *Colloque int. n° 268 Himalaya, Paris 1976, Centre Natl. Rech. Sci., vol. Sci. de la Terre*, 41-60, 1977.

*Bassoullet, J.P., M. Colchen & R. Mouterde, Esquisse paléogéographique et essai sur l'évolution géodynamique de l'Himalaya, *In : Livre à la mémoire de A.F. de Lapparent : recherches géologiques dans les chaînes alpines d'Asie du Sud-Ouest, Mém. h. sér. Soc. Géol. France*, n° 8, 213-234, 1977.

Bassoullet, J.P., M. Colchen, J. Marcoux & G. Mascle, Une transversale de la zone de l'Indus de Khalsi à Phothaksar, Himalaya du Ladakh, *C.R. Acad. Sci., Paris, t. 286*, 563-566, 1978a.

Bassoullet, J.P., M. Colchen, J. Guex, M. Lys, J. Marcoux & G. Mascle, Permien terminal néritique, Scythien pélagique et volcanisme sous-marin, indices de processus tectono-sédimentaires distensifs à la limite Permien-Trias dans un bloc exotique de la suture de l'Indus (Himalaya du Ladakh), *C.R. Acad. Sci., Paris, t. 287*, 675-678, 1978b.

Bassoullet, J.P., M. Colchen, J. Marcoux & G. Mascle, The Indus suture zone : possible evidence for Early Triassic oceanization, *XXXVI congrès, assemblée pleinière de la C.I.E.S.M., Antalya (Turquie), nov.-déc. 1978*, 2 p., 1978c.

*Bordet, P., Recherches géologiques dans l'Himalaya du Népal, région du Makalu, *Centre Natl. Rech. Sci., Paris*, 275 p., 1961.

Bordet, P., On the position of the Himalayan Main Central Thrust within Nepal Himalaya, *Seminar on Geodynamics Himal. région, Natl. Geophy. Res. Inst., Hyderabad*, 48-55, 1973.

Bordet, P., Géologie de la Dalle du Tibet (Himalaya central), *Mém. h. sér. Soc. géol. France*, n° 8, 235-250, 1977.

Bordet, P., M. Colchen, D. Krummenacher, P. Le Fort, R. Mouterde & J.M. Rémy, Recherches géologiques dans l'Himalaya du Népal, région de la Thakkhola, *Centre Natl. Rech. Sci., Paris*, 279 p., 1 map 75.000, 1971.

Bordet, P., M. Colchen & P. Le Fort, Some features of the geology of the Annapurna range, Nepal, *Himal. Geol., vol. 2*, 537-563, 1972.

Bordet, P., M. Colchen & P. Le Fort, Recherches géologiques dans l'Himalaya du Népal, région du Nyi-Shang, *Centre Natl. Rech. Sci., Paris*, 138 p., 1 map 75.000, 1975.

Bouchez, J.L., Le quartz et la cinématique des zones ductiles, *Thèse d'Etat, Nantes*, 176 p., 1977.

Bouchez, J.L. & A. Pêcher, Textures et orientations préférentielles du quartz en relation avec le Grand Chevauchement Central himalayen, *4ème Réunion annuelle Sci. de la Terre, Paris, Soc. géol. France éd.*, 67, 1976a.

*Bouchez, J.L. & A. Pêcher, Plasticité du quartz et sens du cisaillement dans les quartzites du Grand Chevauchement Central himalayen (M.C.T.), *Bull. Soc. géol. France, vol. 18, n° 6*, 1375-1383, 1976b.

Brunel, M., La nappe du Mahabharat, Himalaya du Népal central, *C.R. Acad. Sci., Paris, t. 280*, 551-554, 1975.

Brunel, M., Quartz (C-axis) fabrics in shearzones mylonite : evidence for a late incremental deformation major imprint, *In press, Tectonophysics*, 1979.

*Brunel, M. & J. Andrieux, Déformations superposées et mécanismes associés au chevauchement central himalayen "M.C.T." : Népal oriental, *Colloque int. n° 268 Himalaya, Paris 1976, Centre Natl. Rech. Sci., vol. Sci. de la Terre*, 69-83, 1977.

Brunel, M., M. Colchen, P. Le Fort, G. Mascle & A. Pêcher, Structural analysis and tectonic evolution of the Central Himalaya of Nepal, *in Structural Geology of the Himalaya. Today and Tomorrow's printers and publishers, New-Delhi*, 247-264, 1979.

Cocherie, A., Données préliminaires sur la géochimie des terres rares dans le massif leucogranitique du Manaslu (Népal Central), *Colloque int. n° 268 Himalaya, Paris 1976, Centre Natl. Rech. Sci., vol. Sci. de la Terre*, 93-110, 1977.

*Cocherie, A., Géochimie des terres rares dans les granitoïdes, *Thèse 3ème cycle, Rennes*, 116 p., 1978.

Cocherie, A., Ph. Vidal, R. Capdevila & J. Hameurt, Géochimie des terres rares et des isotopes du Sr dans le leucogranite du Manaslu et les gneiss de la Dalle du Tibet (Népal Central), *5ème Réunion annuelle Sci. de la Terre, Rennes, Soc. géol. France éd.*, p. 160, 1977a.

*Colchen, M., Les formations paléozoïques de la Thakkhola, *In : Recherches géologiques dans l'Himalaya du Népal, région de la Thakkhola, Centre Natl. Rech. Sci., Paris*, 83-117, 1971.

*Colchen, M., Les séries tibétaines, *In : Recherches géologiques dans l'Himalaya du Népal, région de Nyi-Shang, Chap. IV, Centre Natl. Rech. Sci., Paris*, 67-97, 1975.

Colchen, M., Sur le flysch et la molasse de l'Indus, Himalaya du Ladakh, *5ème Réunion annuelle Sci. de la Terre, Rennes, Soc. géol. France éd.*, p. 160, 1977a.

*Colchen, M., Gondwanian and Tethysian characters of the Himalayan series, *Annales Soc. géol. Nord, France, t. 47*, 279-286, 1977b.

Colchen, M. & D. Vachard, Nouvelles données sur la stratigraphie des terrains carbonifères et permiens du domaine tibétain de l'Himalaya du Népal, *C.R. Acad. Sci., Paris, t. 281*, 1963-1966, 1975.

Colchen, M. & P. Le Fort, Some remarks and questions concerning the geology of the Himalaya, *Colloque int. n° 268 Himalaya, Paris 1976, Centre Natl. Rech. Sci., vol. Sci. de la Terre*, 131-137, 1977.

Colchen, M., M. Fort & P. Freytet, Sédimentation et tectonique plio-quaternaire dans le Haut-Himalaya : l'exemple du fossé de la Thakkhola (Himalaya du Népal), *7ème Réunion annuelle Sci. de la Terre, Lyon, Soc. géol. France éd.*, p. 121, 1979.

*Fort, M., Les formations quaternaires de la haute vallée de la Buri Gandaki, Himalaya du Népal, *Thèse 3ème cycle, Paris 6*, 1977a, *Centre Natl. Rech. Sci., Paris*, 209 p. et 21 pl., 1979.

Fort, M., Contribution à l'étude de la sédimentation quaternaire de la Haute-Chaîne himalayenne : le bassin de Sama (haute vallée de la Buri Gandaki, Népal Central), *Colloque int. n° 268 Himalaya, Paris 1976, Centre Natl. Rech. Sci., vol. Sci. de la Terre*, 139-146, 1977b.

Fort, M., Geomorphology of the Ladakh, *Bull. Assoc. Géogr. Franç., Paris*, n° 242, 159-175, 1978a.

Fort, M., About some periglacial aspects of the Tibetan side of the Himalayan high range, with the upper Gya Valley as example (West-Central Nepal), *Actes colloque "Le périglaciaire d'altitude du domaine méditerranéen et abords", Assoc. géogr. d'Alsace éd.*, 311-320, 1978b.

Fourcade, S., J. Hamet & C.J. Allègre, Données de la méthode rubidium-strontium et détermination des terres rares dans le leucogranite du Manaslu : implications pour l'orogenèse himalayenne, *C.R. Acad. Sci., Paris, t. 284*, 717-720, 1977.

Fuchs, G. & W. Frank, The geology of West Nepal between the rivers Kali Gandaki and Thulo Bheri, *Jahrb. Geol. Bund. Anst., S. 18*, 103 p., 1970.

Gansser, A., Geology of Himalayas, *Interscience, John Wiley and S., London*, 289 p., 1964.

Hamet, J. & C.J. Allègre, Rb-Sr systematics in granite from central Nepal (Manaslu) : significance of the oligocene age and high 87 Sr/86 Sr ratio in Himalayan orogeny, *Geology, v. 4*, 470-472, 1976.

Hamet, J. & C.J. Allègre, Reply, *Geology, v. 6*, p. 197, 1978.

Hashimoto, S. & al., Geology of the Nepal Himalayas, *Saikon Publ. Co., Sapporo, Japan*, 286 p., 1973.

Heim, A. & A. Gansser, Central Himalaya. Geological observations of the Swiss expedition 1936, *Mém. Soc. Helv. Sci. Nat., 73*, 1-247, 1939.

Holdaway, M.J., Stability of andalusite and the aluminium silicate phase diagram, *Amer. J. Sci., vol. 271*, n° 2, 97-131, 1971.

Ichac, M. & P. Pruvost, Résultats géologiques de l'expédition française de 1950 à l'Himalaya, *C.R. Acad. Sci., Paris, t. 232*, 1617-1620, 1951.

Kennedy, G.C., Pressure - volume - temperature relations in CO_2 at elevated temperatures and pressures, *Amer. J. Sci., 252*, 225-241, 1954.

Krummenacher, D., A.M. Bassett, F.A. Kingery & H.F. Layne, Petrology, metamorphism and K/Ar age determinations in Eastern Nepal, *In : Tectonic geology of the Himalaya, Today and Tomorrow's print. and publ., New Delhi*, 151-166, 1979.

*Lasserre, J.L., Amphibolites and alkaline gneisses in the Midlands Formations of Nepal. Petrography, Geochemistry, Geodynamic involments, *Colloque int. n° 268 Himalaya, Paris 1976, Centre Natl. Rech. Sci., vol. Sci. de la Terre*, 213-236, 1977.

Lasserre, J.L., A. Pêcher & P. Le Fort, An occurrence of nepheline syenite and alcaline gneiss at Ampipal, Lesser Himalaya of Central Nepal, *Preliminary note, Chayanica Geologica, 2/1*, 71-78, 1976.

*Le Fort, P., La chaîne himalayenne et la dérive des continents, *Rev. Géogr. phys. Géol. dyn., (2), t. 13*, 5-12, 1971a.

Le Fort, P., Les formations cristallophylliennes de la Thakkhola, *In : Recherches géologiques dans l'Himalaya du Népal, région de la Thakkhola, Paris, Centre Natl. Rech. Sci.*, 42-81, 1971b.

Le Fort, P., Les leucogranites à tourmaline de l'Himalaya sur l'exemple du granite du Manaslu (Népal central), *Bull. Soc. géol. France, V. 15, n° 5-6*, 555-561, 1973.

Le Fort, P., Modèle thermique de la subduction intracontinentale himalayenne, *2ème Réunion annuelle Sci. Terre, Nancy, Soc. géol. France éd.*, p. 253, 1974a.

*Le Fort, P., The anatectic Himalayan leucogranites with emphasis on the Manaslu tourmaline granite, *In Recent Res. in Geol. Hindustan Pub., Delhi, V. 2*, 76-90, 1974b.

*Le Fort, P., Himalaya : the collided range. Present knowledge of the continental arc, *Amer. J. Sci., 75*, 1-44, 1975a.

*Le Fort, P., Les formations cristallophylliennes de la "Dalle du Tibet" en Marsyandi, *In "Recherches géologiques dans l'Himalaya du Népal, région du Nyi-Shang, Centre Natl. Rech. Sci., Paris*, 21-42, 1975b.

*Le Fort, P., Le granite du Manaslu, *In "Recherches géologiques dans l'Himalaya du Népal, région du Nyi-Shang, Centre Natl. Rech. Sci., Paris*, 49-66, 1975c.

Le Fort, P., A spilitic episode in the Tibetan Upper Paleozoic Series of Central Nepal, *Bull. ind. Geol. Assoc., 8 (2)*, 100-105, 1975d.

Le Fort, P. & C. Jest, Les sources thermales, *In Objets et Mondes : "L'Homme et la haute montagne : l'Himalaya", t. XIV, fasc. 4*, 213-218, 1974.

Le Fort, P. & A. Pêcher, Les gneiss oeillés du Moyen Pays népalais; un ensemble volcano-sédimentaire acide d'âge paléozoïque ou plus ancien en Himalaya, *C.R. Acad. Sci., Paris, t. 278*, 3283-3286, 1974.

*Le Fort, P., F. Debon & J. Stebbins, Mise en évidence d'une ceinture de granites à cordiérite et enclaves "microgrenues" en Bas-Himalaya (Népal-Indes-Pakistan), *6ème Réunion annuelle Sci. Terre, Orsay, Soc. géol. France éd.*, p. 243, 1978.

Mac-Mahon, C.A., The geology of Dalhousie, *Rec. Geol. Surv. India, V. 15*, 36-51, 1882.

Mascle, G.H., Structure du Mahabharat au méridien de Lumbini (Himalaya du Népal central),

C.R. Som. Soc. géol. France, 6, 279-281, 1976.

Mascle, G.H. & A. Pêcher, Tectonic "polyphasée" des séries du Moyen Pays himalayen (transversale de Lumbini, Himalaya du Népal), *C.R. Som. Soc. géol. France*, 4, 231-234, 1977.

*Mascle, G.H., Tentative stratigraphical reconstruction of the Midlands in Central Nepal; interest of the structural analysis, *In press, Himal. Geology*, 1979.

Misch, P., Metasomatic granitization of batholitic dimensions, *Amer. J. Sci., V. 247, Part I :* 209-245; *Part III : Relationships of synkinematic and static granitization*, 673-705, 1949.

Pêcher, A., Métamorphisme et tectonique en régime de "Subduction intracontinentale". Exemple du massif des Annapurnas (Himalaya du Népal Central), *2ème Réunion annuelle Sci. Terre, Nancy, Soc. géol. France éd.*, p. 311, 1974.

Pêcher, A., The main central thrust of the Nepal Himalaya and the related metamorphism in the Modi-Khola cross section (Annapurna range), *Himal. Geol. 5*, 115-132, 1975.

Pêcher, A., Geology of the Nepal Himalaya : deformation and petrography in the Main Central Thrust zone, *Colloque int. n° 268 Himalaya, Paris 1976, Centre Natl. Rech. Sci., vol. Sci. de la Terre*, 301-318, 1977.

Pêcher, A., Les fluides dans la zone du Grand Chevauchement Central himalayen : la distribution du CO_2 au Népal Central, *6ème réunion annuelle Sci. Terre, Orsay, Soc. géol. France éd.*, p. 303, 1978.

*Pêcher, A., Déformations et métamorphisme associés à une zone de cisaillement. Exemple du grand chevauchement central himalayen (M.C.T.), transversale des Annapurnas et du Manaslu, Népal, *Thèse de Doctorat ès-Sciences, Grenoble*, 354 p., 1978.

Pêcher, A., Remarques sur le métamorphisme inverse himalayen, *7ème Réunion annuelle Sci. Terre, Lyon, Soc. géol. France éd.*, p. 358, 1979a.

*Pêcher, A., Les inclusions fluides des "quartz d'exsudation" de la zone du chevauchement central himalayen (Népal central) : données sur la phase fluide dans une grande zone de cisaillement intracrustale, *In "Minéraux et mi-*

nerais" *special issue Bull. Minéralogie*, 102, 537-554, 1979b.

*Pêcher, A. & J.L. Bouchez, Microstructures and quartz preferred orientations in quartzites of the Annapurna area (Annapurna Range), *Himalayan Geology*, 6, 118-132, 1976.

*Pêcher, A. & P. Le Fort, Origin and significance of the Lesser Himalaya augen gneisses, *Colloque int. n° 268 Himalaya, Paris 1976, Centre Natl. Rech. Sci., vol. Sci. de la Terre*, 319-329, 1977.

Potter, R.W. & D.L. Brown, The volumetric properties of aqueous sodium chloride solutions from 0° to 500°C at pressures up to 2000 bars, based on a regression of available data in literature, *U.S. Geol. Surv. Bull., n° 1421c*, 36 p., 1977.

Rémy, M., Résultats de l'étude géologique de l'Ouest du Népal, les séries népalaises, *C.R. Acad. Sci., Paris*, 275, 2299-2302, 1972.

Rémy, M., Le métamorphisme et ses divers types dans l'Ouest du Népal, les séries népalaises, *C.R. Acad. Sci., Paris*, 279, 461-464, 1974.

Richardson, S.W., P.M. Bell & M.C. Gilbert, Kyanite-sillimanite equilibrium between 700° and 1500°C, *Amer. J. Sci.*, 266, 513-541, 1968.

Stöcklin, J. & K.D. Bhattarai, Geological map of Kathmandu area and Central Mahabharat range, *United Nations Dev. Project, Kathmandu*, 1977.

Thakur, V.C. & N.S. Virdi, Lithostratigraphy, structure and tectono-metamorphic history of the South-Eastern part of Ladakh, Kashmir Himalaya, *Abst. of 9th seminar on Himalayan Geology, Dehra-Dun*, 96-98, 1978.

Toksöz, M.N. & P. Bird, Tectonophysics of the continuing Himalayan orogeny, *Colloque int. n° 268 Himalaya, Paris 1976, Centre Natl. Rech. Sci., vol. Sci. de la Terre*, 443-448, 1977.

*Vidal, Ph., Rb-Sr systematics in granite from central Nepal (Manaslu) : significance of the oligocene age and high 87 Sr/86 Sr ratio in Himalayan orogeny. Comment, *Geology, V. 6*, p. 196, 1978.

(Received June 9, 1979;
revised September 13, 1979.)

STATUS REPORT OF THE WORK CARRIED OUT BY GEOLOGICAL SURVEY OF INDIA
IN THE FRAMEWORK OF THE INTERNATIONAL GEODYNAMICS PROJECT

V. S. Krishnaswamy

Director General, Geological Survey of India
General Overview

The geoscientific activities under the geodynamics project in Himalaya were coordinated
with the Annual Programmes under the Five Year
Plans of the Geological Survey of India. These
multidisciplinary programmes included field and
laboratory studies, systematic geological mapping, Quaternary geology, geotechnical investigations, mineral exploration, special geophysical and geochemical, geochronological and remote
sensing/photogeological investigations.

Out of the 384,000 sq. km. area of Indian
Himalaya, the regional geology in all the sectors
is now known though detailed information through
geological mapping on various scales is available
in a little over 56% of the territory. Besides
preparation of geological maps to different
scales, the data so collected were used to syn-
thesize tectonic, metamorphic, metallogenic,
lineament, seismo-tectonic and geomorphic thematic maps either for the whole region or for
some sectors only, depending on the objectives.

Since geodynamic processes are best understood in global or regional perspectives,
opportunities were created and availed of for
periodical synthesis and review of the composite data and interpretation at national and
international levels. Symposia/workshops on
geological, structural, natural resources, seismotectonics, basin studies and other topics were
organized by the Geological Survey of India between 1974 and 1979, and Survey staff also participated in other symposia held elsewhere. Quite
a few of these geoscientists had intimate knowledge of Himalayan Geology and geodynamic problems and substantial contributions have either
been published or are in press. Much of the data
on geodynamics was presented at the International Seminar on Himalayan Geology in September 1976.

Concurrently the international geoscientific
collaboration programmes of the Commission for
the Geological Map of the World (CGMW), International Geological Corelation Programme (IGCP),
Economic and Social Commission for Asia and
Pacific (ESCAP), United Nations Development Programme (UNDP), etc., which to some extent had a
geodynamic content, were pursued by the Geologi-

cal Survey of India. Indeed, the subject of geodynamics is such that progressive understanding,
synthesis of new data and refinement of earlier
interpretations are necessary. This concurrent
synthesis is attempted and is expected to be presented in new publications on Himalaya.

Salient Features of Achievements

In the earlier status report (1973), certain
geodynamic programmes under broad topics were
identified by the Geological Survey of India for
further studies. A brief review of the important activities reflecting progress on the understanding of various aspects of geodynamic processes is accordingly presented here. In addition, fairly comprehensive synthesised summaries
of significant progress made on important components of the programme are also included, in
later chapters.

1. Studies of recent dynamics.

(a) Geomorphological and Quaternary geological
studies for deciphering geometry of neotectonic
movements in the Sutlej, Beas, Giri, Bhagirathi-
Ganga valleys; Tista-Brahmaputra valleys, etc.

Neotectonic features in the sub-Himalayan and
Lesser Himalayan parts of Jammu and Kashmir,
Himachal Pradesh and Uttar Pradesh Himalaya and
particularly along the major river valleys were
studied during Quaternary geological investigations and also in connection with the geotechnical evaluation of stability conditions of
the river-valley project investigations.

In general, five to six terraces were recognized in the Sutlej, Jamuna Bhagirathi, Alaknanda
valleys. Terrace displacements, opposing disposition in the same terrace and faulting, indicate neotectonic movements. Observations on en-
trenched gorges, breaks in river gradients when
studied with terrace characteristics and existing
faults in the rock formations, reflected neotectonic movements along the vertical to sub-
vertical faults. The systematic terrace studies,
in terms of composition, their disposition and

correlation along the profiles have helped to re-construct the changing palaeogradients and char-acteristics of the river valleys which have been responding to the rather imperceptible neotec-tonics, secular retreat of glaciers (in higher to medium parts) and isostatic adjustments. It was interesting to note the Permo-Carboniferous fossiliferous boulders in the Alaknanda T_3 to T_5 terraces derived from about 200 km upstream while the recent or younger terraces are devoid of them although the basinal provenance has not changed. The relict flats of the elevated ridges, etc., and their slopes have also been studied in a preliminary way to reflect upon the geomorphic response to the endogenic and exogenic forces.

The geotechnical studies were carried out in the form of detailed study of terraces and geo-morphic features across and along the faults (thrusts, normal faults, tears); computation of seismological data and observations (isoseismal maps), assistance from the instrumental data (precision levelling, triangulation, data from tiltmeter, strainmeter and micrometer); study of creep, and study of compressions and release of strain along the tunnels in Salal, Beas-Sutlej link, Giri project, Giri-Bata project, Jamuna project, Tehri Maneri-kali projects (Bhagirathi valley), etc. Similar studies were also carried out in parts of Tista-Brahmaputra-Siang, Suban-siri valleys of Eastern Himalaya to some extent.

It has been generally found that the thrusts and faults in the sub-Himalayan belts and also those close to the Main Boundary Fault (MBF) and those at the junction of frontal hills with al-luvial plain are active. Present-day creep type movements of the order of 1 cm/year have been re-corded in the Jamuna Hydel Project area by tilt-meter measurements; many tectonic dislocations like Krol and other thrusts are active. The Kinnaur Earthquake (1975) and Dharamsala Earth-quake (1978) have been genetically related to the release of strain along N-S tear faults. Active tear faults near Chamera Hydroelectric Project (Himachal Pradesh) affect the Tertiary rocks but stop at pre-Tertiary rocks. This feature has been explained by some workers as evidence of underthrusting of a Tertiary block in which case the latter alone would show the tear effects.

The enormous pressures/compressions observed in the tunnel cavities in the Giri valley is a classic example of neotectonics. This study enabled a suitable construction strategy to be suggested.

The neotectonics of the Tista valley was de-ciphered on the basis of terraces and geological studies. The seismological features, terrace faulting, tears, sinking and subsiding topo-graphy near the gorge of Subansiri, and the active faults close to the emergence of the Siang river onto the Brahmaputra (Assam) plain have been subjects of constant study during the in-vestigations for hydro-electric projects and flood control and in the course of detailed basin studies in parts of the Brahmaputra valley. Some of the results of these investigations have been published particularly in relation to the Brahma-putra basin.

(b) Study of Thermal Springs.

During the last six years systematic study of the known thermal springs (their present tally is 114) in the NW Himalaya has been in progress with a view to assessing their potential for harnessing geothermal energy. So far about 75% of them have been reconnoitered and as a result considerable geological, geochemical, geophysical, isotopic and seismotectonic information is now available.

Based on the available information the NW Hima-laya could be subdivided in three district geo-thermal sub-provinces or belts. The region north of the Main Central Thrust (MCT) and along the Indus-Shyok tectonic zones could be grouped as belt I, which one school of thought describes as a collision junction of two crustal plates. This belt has late Mesozoic-Cenozoic magmatism, moderately high seismicity and high microseis-mic activity, indicating shallow disturbances, possibly related to the ongoing deformation, very high heat flow and a high temperature grad-ient (100-150°C/km), high temperature of the cir-culating hydrothermal convection system, as indi-cated by the discharging geothermal waters and the high base temperatures (>200°C) of the deep geothermal waters.

Belt II encompases the region between the Krol Thrust (or its equivalents) and the Main Central Thrust (MCT) and is characterized by a folded mountain belt with Tertiary metamorphism, deep intrabasinal reversed fault/thrust, moderately strong intermediate depth focus earthquakes; large tear faults with a chain of thermal springs having base temperatures generally between 150-200°C and with only moderate temperatures; slight-ly elevated temperature gradient (up to 50-70°C/Km) and heat flow. Recent movement along many faults in this belt is evidenced by tilted river terraces; fairly active micro-tremor activity is observed in some areas; strong deformation obser-ved in many river valley hydel projects during construction.

Belt III lies south of the Krol Thrust and covers the foothills of the Himalaya. The tem-perature data obtained from deep oil-well drill-ing indicates subnormal temperature gradients (20-26°C/Km), low base temperatures of the ther-mal waters (90-120°C) but strong neotectonic movements along the major faults, tilting of terraces and some strong earthquakes. The low temperature gradients observed could be possibly due to the measurements being taken in the older formations and consequently they could not be taken as an operating gradient for the belts.

II. Studies of Palaeodynamics

(a) Faults, dislocations, fractures and lineaments.

Field mapping, photogeologic and Landsat data were utilized to study the nature of some of the major faults, thrusts and lineaments of the Himalaya. In certain specialized investigations geophysical studies were also taken up to delineate the fault or fracture zones. The major activities were as follows: (i) determination of the fault or thrust parameters in the Indus tectonic zone; (ii) demarcation of fault and fracture systems by geological and geophysical methods in Puga and Manikaran geothermal areas and in the Alaknanda valley; (iii) geological studies on Main Central Thrust; North and South Almora thrusts; (iv) studies on thrusts in Eastern Himalaya in relation to minor structures, stratigraphy and nappe as autochthonous concepts; (v) demarcation of important faults/thrusts in Siang districts, etc., of Arunachal Pradesh Himalaya on the basis of new fluids including the recent Palaeogene rocks in the Lesser Himalayan (vi) studies of faults and fractures in relation to base metal mineralization in Kashmir, Pithoragarh district (Utter Pradesh) and Darjeeling district (West Bengal) and (vii) interpretative analysis of sealed thrusts in Eastern Himalaya due to post-thrusting granitic activity.

The studies on various aspects of the Main Central Thrust and Indus Tectonic Zone are synthesised in fairly comprehensive summaries in subsequent chapters of this review.

The South Almora Thrust has been particularly studied with a view to defining it in terms of litho-stratigraphic associations, degree of cataclasis, metamorphism, and its relation with the North Almora Thrust. There is considerable difference of opinion on the South Almora Thrust being a folded counterpart of the North Almora Thrust. It has also been suggested that there may be more than one tectonic plane in the zone of this thrust.

The North Almora Thrust has also been defined as a sharp plane with a variable southwards dip of moderate to high angle and which according to some workers merges with the Srinagar fault in the Ganga valley.

In the Eastern Himalaya, the recent structural studies and the concept of a pile of thrust sheets in the Sikkim-Darjeeling Lesser Himalaya, emanating from Tibetan hinterland, has added new thinking on the geodynamic processes in the Lesser Himalayan geology. The discovery of nummulite bearing rocks, just north of the Siwalik (Neogene) belt in Dihang valley, appears to indicate that the Main Boundary Thrust (MBT) separates the older Tertiary from the Neogene sediments, in Arumachal Himalaya also.

(b) In recent years the Phanerozoic sedimentary basins of the Himalaya have been studied, resulting in the compilation of maps and stratigraphic columns depicting lithostratigraphic, biostratigraphic and geochronological features, volcanism, discontinuities, etc. The delineation of basin boundaries and the interrelation with other basins, as also the picture of basin configurations in the total framework of SE Asia on a smaller scale has been evolved. This study of sedimentary basins has been accelerated as a result of the project IGCP 32 by a W. G. within the Economic and Social Commission for Asia and the Pacific.

The nature of the contact between the Central Crystalline metamorphites and the Tethyan basins of Uttar Pradesh and Himachal Pradesh Himalaya as also in parts of Sikkim-Himalaya was studied by various expeditions during the seventies. It has been suggested that in the Spiti-Zanskar basin this contact is welded, whereas in the Uttar Pradesh Tethyan basin, the Dhar-Martoli fault at the base has been found to extend into the lowermost lithologic units of the sedimentary column in certain sectors, thereby implying, at best, the reactivation of a possible earlier marginal fault bounding the basin. In the Sikkim Himalaya, the tourmaline granitic sheets at the base of the Tethyan basin are regarded as having sealed the tectonic plane between metamorphites and the sedimentaries.

(c) During the seventies, studies on flysch-ophiolite/melange sequences of the mobile belts of the Andaman-Nicobar Island arc, Disang flysch and ultramafics of Nagaland and Manipur, have been published in a number of papers (Ray. 1974; Acharya et al., 1976; Sinha et al., 1976; Varadrajan et al., 1976). Under the programme of International Geological Correlation Programme Project No. 39, the ophiolites of the Himalaya have been studied for geological setting, tectonic framework, metallogenic aspects, application of plate-tectonic and other concepts, and definition of the ophiolitic constituents and its parameters in Indian Mesozoic-Tertiary mobile belts. A consolidated report on the various aspects and a bibliography is expected to be prepared at the end of the project.

The project Indus Flysch Ophiolite Project (INFLO) of the Geological Survey of India has been launched to study the ophiolite and flyschoids of the Indus tectonic zone. A comprehensive summary of work to date and current ideas on this ophiolite belt are given in a separate chapter in this review.

The classical concept of geosynclinal evolution and orogeny has been applied by some workers to the whole of North West Himalayan-Eastern Himalayan, Indo-Burma-Andaman-Nicobar Island arc into a common regional late Mesozoic-Tertiary geotectonic feature with similarities in basinal evolution and orogeny and deformation of styles.

(d) Studies on the origin, structural relations and mode of emplacement of selected plutonic and volcanic complexes vis-a-vis their tectonic setting:

(i) The acid plutons of Lesser Himalaya which occur mainly in the crystalline thrust sheets of Lesser Himalaya, as well as in the Central Crystalline group, and in the Ladakh granitic complex, have been regionally mapped and some laboratory studies done (Bhandari et al., 1976; Das Gupta et al., 1976; Chatterjee, 1976). The Dudatolic granite of the Lesser Garhwal Himalaya has been systematically mapped and studied; the intrusive and migmatic aspects, deformational phases with the surrounding thermal and regional metamorphic setting, have been defined; evidence of lead mineralization (with possibly zinc, copper, and traces of pneumatolytic mineralization) has been recorded in parts of the surrounding schistose rocks.

The granites of Gangotri have been studied, and the relations between biotite and tourmaline-muscovite-granitoids and gneisses described. The host rocks have been regarded as parts of the Haimanta succession in Bhagirathi and adjoining western valleys of Himachal Pradesh.

The granitoids of Bara Shigri have been the subject of field and some laboratory studies, in the course of regional mapping and in connection with the exploration of polymetallic mineralization, including stibnite. The granites of Zojila pass and Shingo pass have been studied in the course of regional mapping and geothermal exploration. Most of the younger granitoids have been regarded as high level emplacements without significant thermal aureoles, disposed as linear or tabular sheets, veins and injections in the older metamorphites.

(ii) The basic and ultrabasic rocks of the Himalayan mobile belt have been under investigation for their distribution in relation to various sedimentary basins and tectonic settings. They have been classified in respect to their genetic and crust-mantle relationships, and their significance in the evolution of the mobile belt. Significant progress has been made in understanding the problems of palaeodynamics, and a comprehensive summary of the subject also forms a part of this review.

Studies on structural evolution and metamorphic history (including isotopic age determinations)

The structural evolution of the Panjal volcanics and their metamorphis history is typical of the tectonic-volcanic evolution of the Tethyan basins. Based on the field relations, petrochemistry, and structural setting, the evolution of these basalt andesites from tholeiitic magma and their eruption in sub-aerial conditions, has been reflected in the comprehensive summary referred to above. The spilitisation has been considered as a secondary change during alteration or metamorphism in green schist facies.

The Precambrian of Banital-Ramban-Bhadarwah is a part of the Salkhala thrust sheet which has undergone multiphase deformation (Himalayan &

pre-Himalayan), with pre-Himalayan regional metamorphism and intrusives (probably early to mid-Palaeozoic), and dislocation metamorphism during Tertiary orogeny. The Precambrian rocks of the Paddar area, which form part of the Central Crystalline Group, occur in a higher tectonic unit (above the Main Central Thrust). These Precambrian formations show intense migmatisation and metamorphism up to the amphibolite facies; local development or relicts of eclogite facies has also been reported in the Paddar area. Two of three generations of deformation are discernible in these rocks; some workers (unpublished progress report of GSI, Janpangi et al., 1977; Varadrajan et al., 1976; personal communication from A. P. Tewari, GSI) have considered the ultramafites of Paddar as allochthonous and metamorphosed equivalents of the Indus ophiolite zone.

The Precambrian metamorphic formations of the Lesser Kumaon Himalaya, Uttar Pradesh have been systematically mapped. These rocks mostly occur in the crystalline thrust sheets and show three to four phases of deformation, with associated polymetamorphism and, in general, a reversed metamorphism from chlorite zone to biotite-garnet zone; local development of kyanite or sillimanite occurs in the vicinity of syntectonic acid intrusions. The thermal aureole of biotite granites, or tourmaline granites is insignificant. Retrograde metamorphism is usually present in the dislocation zones. Geochronological data on the Lesser Himalayan acid to intermediate intrusives in the Precambrian formations indicate their Paleozoic age; Proterozoic ages are reflected only in a few cases, individual mineral ages denote an Oligocene metamorphic cycle which seems to coincide with the development of the Lesser Himalayan thrust sheets.

The geochronological data, on the whole, indicate the following three phases of Himalayan orogeny which have been superposed on the Precambrian-Palaeozoic events in the Precambrian belts:

Himalayan Orogeny I -Late Cretaceous/early
                      Eocene (Fission track
                      dating)
Himalayan Orogeny II -Oligocene (K/Ar dating)
Himalayan Orogeny III-Pliocene (Fission track
                      dating - K/Ar also tried)

A major period of Miocene activity at about 16-18 Ma has also been indicated in certain cases by K/Ar and fission tract dating; whole rock Rb/Sr isochrons indicate Paleozoic and Proterozoic ages in certain cases and data sent to a pronounced migmatitic phase related to Caledonian activity.

(iv) The tectonic and metamorphic evolution of the Sikkim-Darjeeling Himalaya has received considerable attention and a comprehensive summary follows in the later chapters of this review.

(v) The distribution of metamorphic facies in the Himalayan belt has been compiled on a regional scale, as part of the metamorphic facies map of South and East Asia, under the thematic map projects of the Commission for the Geological Map of the World.

Studies on the structural and tectonic framework of the Himalaya have been supplemented by the use of Landsat imageries and aerial photographs to prepare the lineament map of the Himalaya which will be published in due course on a suitable scale.

The Main Central Thrust of Himalaya - A Review

The Main Central Thrust (MCT) is one of the prominent geotectonic features along the entire length of the Himalaya. It borders to its north the Tethyan sedimentary belt together with its basement of Central Crystalline rocks. To the South of Main Central Thrust (MCT) lies the Lesser Himalayan geotectonic zone.
Dispositional and Structural Characteristics.
Recent work in different sectors of Uttar Pradesh, Himachal Pradesh and Jammu Himalaya shows that the Main Central Thrust includes two or more tectonic planes. The two major tectonic planes, the lower (southern) one and the upper (northern) one, show overlapping relationship. Locally the upper thrust is directly in contact with the sedimentaries of the Lesser Himalaya tectonically overlapping and thus obscuring the lower thrust block.

The upper, northern, thrust is well identified in the Himachal Pradesh Himalaya as the Jutogh Thrust, whereas the lower, southern, thrust is designated by different names e.g., Panjal thrust (Salkala), Chail thrust, etc. The presence of the Panjal Thrust is disputed by some, while others consider it as a dislocation of fundamental importance (Sharma, 1977; Raina and Dutta, 1979). Two or more thrust components of the Main Central Thrust have been recognized in the Jamuna, Bhagirathi, Bhilangana, Alaknanda and Kali valleys of the Uttar Pradesh, Himalaya, where the grades of metamorphism, dislocations and intrusive phases in and around the tectonic planes have been studied in detail.

In Sikkim and Bhutan Himalaya, the plane separating the overlying Darjeeling Formation from the Daling Formation is considered to be the equivalent of the Main Central Thrust (MCT). However, some workers are of the view that the Darjeeling, Daling and Buxa Formations belong to one group of rocks, occurring in reversed order and thrust southwards over the Gondwana rocks and that the Chuntang Formation comprising sillimanite gneiss, quartzite, calc-granulite and calc-gneisses is thrust southwards over the Darjeeling gneiss; and this thrust would thus represent the Main Central Thrust sensu stricto. Others opine that in the Darjeeling-Sikkim Himalaya a belt of Paro Group separates the Darjeeling gneiss and the Daling Formation. A narrow but persistent belt of sheared and lineated granitic gneiss, mylonite and diaphthoritic phyllonite, separating the Paro and Daling, possibly represents the Main Central Thrust.

In the Bhutan Himalaya, the Shumar, Samchi and Chekha Formations (Nautiyal et al., 1964; Jangpangi, 1974, 1978) are thrust southwards over the Buxa Formation and the tectonic planes have elements comparable to Main Central Thrust. These formations are overlain by rocks of the Thimpu Formation along another thrust plane, which has been locally considered as equivalent to the Main Central Thrust.

Some of the Workers in Bhutan (Janpangi, 1974) have given greater weight to the process of granitisation and granitic emplacement and have related them to metamorphic grades; similarly facies variation in the sedimentary columns have been stressed, with the result that quite a few thrust planes, according to them, may not exist or be only of local significance. However, in the overall geological set up, regional thrust components in the Thimpu Formation have been recognized (cf. MCT) which is supposed to have been sealed by post-tectonic acid igneous activity.

In Arunachal Pradesh, the Bomdila Group of epi- to meso-grade metamorphism in the Kameng district, is thrust over the Tenga Formation (Buxa Formation) and is itself overlain by higher grade rocks of the Sela Group (Thimpu Group of Bhutan) along another tectonic contact. The contact of the Bomdila Group and the Sela Group is regarded as the (upper thrust of) Main Central Thrust (MCT). In the Lohit district of Arunachal Pradesh, the Lohit Thrust separates the subjacent low grade metamorphic rocks and serpentinite and the overthrust diorite-granodiorite crystalline complex of the inner Lohit Himalaya and has been mapped for about 200 km with a 50^0-70^0 NE dip and may represent or be equivalent to the Main Central Thrust (MCT). The outer (western) metamorphic belt in the Lohit sector has almost completely overridden the sedimentary belt of the Lesser Himalaya. Whether or not it represents the lower thrust in relation to the Main Central Thrust (MCT) Zone is a matter for further inquiry.

The recent discovery of nummulite-bearing Eocene rocks in the frontal zone of the Lesser Himalaya of Dihang Valley and other areas, and the older Tertiary outcrops within the tectonic frame of Abor Volcanic zone in Dahang Valley, lends credence to post-Eocene thrust movements within the Palaeozoic volcano-sedimentary sequences. The crystalline or metamorphites which may be related to southward translation along the Main Central Thrust (MCT), are thrust over these sedimentary volcanic sequences.

Metamorphism and Plutonism: Regionally, there is always a contrasting metamorphism and plutonism on either side of the Main Central Thrust (MCT). Above the thrust zone the rocks are highly metamorphosed with poly-phase deformation and intense acidic igneous activity.

The complexities of the Main Central Thrust (MCT), therefore, require imperatively that extensive studies on the structural fabric, plutonism, metamorphism and geochronological studies be undertaken to establish the sequence of the rock formations above the Main Central Thrust Zone, which are designated as the Central gneiss or Central Himalayan Crystalline Group. This Group consists of metasediments, migmatites and intrusive granitoids. In selected areas, polyphase deformation and metamorphism has been investigated. These features have obliterated the stratigraphic relations to the extent that it is yet to be established whether sequences younger than the Precambrian are involved. This aspect has obviously a great bearing on the geotectonic analysis and age of the activity along the Main Central Thrust (MCT).

Geochronological investigations on a number of granitic and basic elements on either side of the Main Central Thrust (MCT), indicate that the majority of the granitoids are of Paleozoic age. However Tertiary ages for some of the tourmaline-muscovite granites and phases of metamorphism have also been indicated in the North West Himalaya. The outcrops provide little or no evidence of the intrusion into the Main Central Thrust (MCT) by Tertiary or Mesozoic granitoids although this has been inferred in the supposed higher tectonic planes. On the other hand intense dislocation metamorphism is apparent in many sectors.

Recently, considerable data on the nature of metamorphism in the blocks beneath the Main Central Thrust (MCT), as well as in the crystalline klippen of Lesser Himalaya and from the Central Crystallines north of the Main Central Thrust (MCT) have been obtained. In the autochthonous sub-thrust mass, phyllite, quartz-schist, quartz-chlorite schist and chlorite-sericite schist occur. The overall mineral assemblage of these rocks suggests that they are of green schist facies.

The schuppen zone immediately north of the sub-thrust block comprises various crystalline metamorphites. All rocks, augen and banded gneisses, mesograde-schists, phyllonites-phyllites, etc., of this zone show a pervasive cataclasis and widespread diaphthoresis. In Himachal Pradesh in general low grade Chail (Salkhala) and mesograde Jutogh rocks are associated with this zone.

The banded gneiss are the main constituents of the lower part of the next higher zone of the Central Crystalline Group. They show upper and middle amphibolite facies in most of the areas. Development of coarse grained textures is seen further up along the axial zone of the refolded banded gneiss. Higher up and the upper granitic migmatites show an upper amphibolite facies with cordierite and sillimanite.

The apparent increase of the so-called reversed progressive regional metamorphism thus observed north of the Main Central Thrust (MCT) from the green schist to amphibolite facies has been explained by some workers (Ghosh et al., 1974) as the effect of polyphase metamorphism and deformation. The juxtaposition of the rocks of contrasting metamorphic recrystallization, partly

obliterating the effect of thrusting could contribute to such a situation. However, the metamorphism again decreases upwards towards the contact with the Tethyan basins, thus following the normal trends.

In the Sikkim Himalaya, some workers (Sinha Roy, 1974) feel that the Daling rocks show polyphase metamorphism, as evidenced by the relationship of mineral paragenesis and deformation phases. Of the three phases of deformation present in the rocks, the second deformational phase is responsible for the development of Barrovian zones, and the inversion of metamorphism has been caused by the overlapping of the Daling-Darjeeling thrust block by the infrastructural migmatic complex in the axial block of Central Crystalines.

In Bhutan, it has been suggested that the deformation along Thimpu Thrust has retrograded high grade rocks to blastomylonites and phyllonites. It is believed by some workers (Janpang, 1974) that the inversion of metamorphic grades in Bhutan, as well as in the Darjeeling area, can be explained by the juxtaposition of high grade rocks within the diaphthoric zones with the less metamorphosed rocks near the thrust.

The normal isograde pattern of the Central Crystallines and the inverted isograde pattern of the Lesser Himalaya on regional or local scales, has also been explained by various workers (Thakur, 1977) as being due to (i) connate metamorphism due to granitic magma under stress, (ii) presence of recumbent folds, (iii) presence of thrust slices, (iv) large scale underthrusting of two continental slabs and the resultant thermodynamic changes.

Acid Plutons. Widespread granitic activity (anatectic, magmatic, residuals, etc.) is seen in the Central Crystallines in the form of gneisses and granites. The gneissic sheets are highly foliated and show extensive and pronounced mylonitization. Foliated and non-foliated types occur as concordant sheets, and exhibit networks of aplitic and pegmatitic veins. The 11-17 Ma (Miocene) age recorded from some of the Himalayan granites by K/Ar dating perhaps represents the reactivation of rocks during the Tertiary orogeny; however some younger intrusives as well, are present. On the other hand the Lower Tertiary granites (hornblende granite of Burzil Valley, Kashmir, granodiorite of Lower Dras Valley, etc.) which are intrusive into the Cretaceous basic volcanics, show Himalayan deformation. According to some French workers (Andrieux et al., 1977) (mostly working in Nepal) the anatectic orgin is related to crustal shear along the Main Central Thrust (MCT) at deeper levels or in selected lithologic levels and their distribution may be related to subduction or other geotectonic phenomena associated with the Main Central Thrust (MCT)

Basic rocks. Several bodies of basic igneous rocks are also associated with the Central Crystalline rocks. In Kumaun and Nepal, these are represented by epidiorites, amphibolites, hornblende and chlorite schists, dolerites and pyroxenites. Little geochronological data is available on these rocks. Most of the dates are on minerals and record ages of 56 and 39 Ma. Although the precise data sequences are complex, it is possible that these basic bodies represent basic activity associated with different phases of deformation. Some workers (Mishra et al., 1976; Bordet, 1973; and Srikantia et al., 1976) regard the Central Crystallines as an integral part of the Sino-Tibetan plate and have suggested that the evidence of basic rocks being present along the Main Central Thrust (MCT) may represent simatic material emplaced during the collision of the continents of India and Asia.

Deformation related to Tertiary and other orogenies. Four phases of deformation are generally observed in the Central Crystallines and the klippen of metamorphites of the NW Lesser Himalaya. The first phase shows isoclinally inclined, rootless folding with development of axial plane schistosity which does not seem to be related to Himalayan orogeny. This episode is also marked by a first episode of progressive metamorphism. The second phase of folding normally resulted in folds coaxial to the first phase and developed thrust sheets. Second generation biotite is related to this phase. This was followed by gentle folds and warps developed during the third and fourth phases. However these deformations seem to be related to the Tertiary orogeny only, and clear imprints of earlier orogenies could not be recognized in the western sector of the Himalaya. Nevertheless workers (Srikantia, 1977) have suggested a number of pre-Himalayan orogenic phases such as the Sundernagar orogeny, Shali orogeny, Kurgiakh orogeny and Hercynian orogeny, etc., in the metasedimentary and sedimentary formations adjacent to the MCT in the sub-thrust and in some places in the supra-thrust blocks.

In Arunachal Pradesh Himalaya, pre-Hercynian and Hercynian and Himalayan trends on both sides of MCT have been broadly defined.

It has also been reported from the Sikkim-Darjeeling Himalaya (Sinha, 1973; Raina et al., 1976) that the first phase of deformation seen in the Crystallines is not imprinted on the Gondwana rocks seen in the Ranjit valley. The deformational phases in eastern Himalaya are described elsewhere in this review.

The fact that metasediments forming the base of the Tethys sediments and occurring to the north of Great Himalaya are not found as inliers in the southern part in NW Himalaya indicates the possibility that the crystalline nappes of Lesser Himalaya could be rooted in the region of Central Crystallines which are themselves regarded by some workers (Heim et al., 1939; Gansser, 1964; Brunel et al., 1977; Molnar et al, 1977) as allochthonous. This would mean significant crustal shortening. No direct correlation between

the thrust sheets of the Lesser Himalaya with the zone of Central Crystalline is possibly because of differences of metamorphism and stratigraphy in Uttar Pradesh and parts of NW Himalaya. The present trace of Main Central Thrust (MCT) is in the nature of a scar with thrust lobes extending far south at some places, and the actual site of the root zone of high angled faults is perhaps buried under the southward thrust mass of the Central Crystallines. Differential southward advances of the individual crystalline thrust sheets during the Himalayan orogeny and subsequent vertical uplift and erosion, accompanied by isostatic adjustments has provided the setting of large scale crystalline nappe and klippe structures and the present geomorphic configurations of the MCT Zone.

Significance in Relation to Sedimentation. The Main Central Thrust (MCT) has also been viewed in the context of sedimentation and sedimentary basins to its north and south. There is meagre evidences to suggest a high crystalline arch in the region of the Central Crystallines or MCT, contributing to the sedimentation of the Paleozoic-Mesozoic basins, excepting perhaps the granite pebbles in Upper Paleozoic diamictites. Sedimentary basin analysis, though in a preliminary stage, suggests that the MCT and Central Crystallines became areas of positive provenance in Late Mesozoic-Tertiary times.

Age of MCT. In consideration of the deformational history of old Tertiary-late Mesozoic granites, undeformed Miocene or younger granitoids, as also from the stratigraphic relations of the frontal lobes of MCT components in Lesser Himalaya it appears that the thrust might have developed in Paleogene times, but maximum movement occurred in mid-Miocene.

Main Central Thrust in the light of Global Tectonics. Various geotectonic models for the Himalaya describe the MCT in different ways, e.g.
    (i) a suture zone representing collision of Indian and Asian plate.
    (ii) a crustal shear responsible for underthrusting of Indian shield below the Himalaya.
    (iii) a deep rooted thrust, from within the sial, perhaps extending to 30 km depth (Peive, 1960).
Besides the above concepts, the concept of an abyssal fundamental fault sometimes extending over to 700 km and accompanied by later warping and bending of basement blocks, has been under scrutiny and application in Himalaya.

The field data so far collected by the GSI (unpublished progress report) and other workers in recent years has not yielded convincing data to fit any one of these models without reservations. The MCT is essentially a schuppen belt and not a single tectonic plane; no basic or ultrabasic activity of significance has been noticed along it, nor are the old basement blocks upwedged or exposed along it, the paired metamorphic belts of type collision areas, or characteristics of subduction zones have not been found along it; there are no ophiolites, melange or blue schist metamorphism along the MCT; nor is this a zone of deep seismic activity; nor is there convincing evidence of a concealed root zone below the MCT cover. As such the above mentioned tectonic models cannot be well applied to the MCT as it is exposed or known todate. It is undoubtedly a deep rooted thrust zone with strong imprints.

Problems Requiring Attention in Future.
(i) The delineation of the MCT with uniform definition throughout the length of Himalaya by systematic mapping is essential to attribute the same geotectonic significance to this great thrust plane.
(ii) Detailed stratigraphic, structural, magmatic, metamorphic, petrofabric and geochronological studies of the rocks loosely termed "Central Crystallines" are suggested, to determine if metasediments, etc., other than of Precambrian age are involved in this group.
(iii) Deeper geophysical probes along the MCT, may enhance geological or geotectonic evaluations.
(iv) Evaluation of resource possibilities including geothermal energy resources may be investigated by multidisciplinary approach.

Study of the Basic Rocks of Himalaya

I. Lithologic structural association. The study of basic rocks of the Himalaya provides a means to understand the palaeogeodynamic processes particularly in the context of crust-mantle interrelations in space and time. They are separately discussed according to the regime to which they belong.
A.1. Basic rocks in higher tectonic levels of Northwestern Lesser Himalaya-(with metasediments, acid rocks, etc.)

Basics within Crystalline Nappes. The Crystalline nappes of the Jutogh and Almora in the Lesser Himalaya forming the highest tectonic level contain numerous sills and dykes of basic and metabasic rocks. The Salkhala nappe, at the lower level, also includes a variety of basic rocks as intrusives in the metasediments and granitoids.
Basics within Mandi granite. The Mandi granitic massif has been intruded by basic rocks at many places. The suite comprises metagabbro, olivine gabbro, gabbro, diorite, and granodiorite, the last two being hybrid varieties in contact with granite. Olivine is restricted to the rocks in the core, where the texture is intergranular.
Basics within the Almora crystallines. Massive to schistose amphibolites, epidiorites and dolerites occur as dykes and sills within the Almora

metamorphites at a few places. Dolerites are moderately altered, recrystallized and from subophitic to blasto-ophitic texture.
An olivine gabbro body has been studied from Sitlakhet area; the differentiated body is intrusive into the Almora granite.
Amphibolites occurring in the zone of the North Almora Thrust show effects of retrogressive metamorphism on the rocks. Epidioritic rocks and amphibolites are reported from Askot and Baijnath Klippen also. Some of these basic bodies are likely to be penecontemporaneous flows.
Bafliaz-Thanamandi Volcanics (Proterozoic). Metamorphosed lavas and associated hypabyssal rocks occur in the metamorphic sequence of Salkhala group of rocks (Proterozoic) and Dogra Slates of Kashmir nappe on its western border (Suran Valley of Poonch district and Rajauri district). Panjal Thrust separates these rocks from the Basantgarh metavolcanics of the Kashmir parautochthonous folded belt. These have also been recognized as distinctly different from the younger Panjal Volcanics, associated with overlying Tethyan sequence of the Kashmir nappe.
The assemblage comprises flows and sills of spilites (sometimes glomeroporphyrites) sills and dykes of 'keratophyres', tuffs, agglomerates and ash beds. Limburgitic lava has also been reported from a few places. Rocks are generally massive to crudely foliated; ellipsoidal amygdales of chlorite, epidote, calcite and chalcedony are common.

A.2. Volcanics associated with sedimentary sequences of Lesser Himalaya.

(i) Mandi-Darla Volcanics (Proterozoic - early Paleozoic): The Mandi-Darla Volcanics occur stratigraphically associated with the rock types of the Shali structural belt of the Himachal Himalaya which is essentially regarded as consisting of Proterozoic to Lower Palaeozoic rocks; structurally the belt has been interpreted as parautochthonous to allochthonous by different workers.
The Mandi-Darla volcanics comprise epimetamorphosed basic lavas, ash beds, few intermediate lavas, and agglomerates interbedded with slate-quartzites of the Sundernagar Group. Intertrap beds of slate-phyllite and quartzites are common in thick lava sequences. Numerous sills, dykes and bosses of gabbro-norites, dolerites and epidiorites invade the Mandi Volcanic formation and the associated sedimentaries.
(ii) Banjar-Rampur Volcanics (Proterozoic-Paleozoic): In the Larji-Rampur Window (framed by Salkhala Thrust Sheet) basic lava beds are associated with sediments of the Rampur Group.
The volcanics comprising epimetamorphosed basic flows, sills and tuffaceous slates are generally interstratified with quartzites and phyllites, from the Parvati valley on the north to the Sutlej valley on the south. A number of basic sills and dykes invade Larji-Banjar-Manikaran Formation.

Near Larji, a few thin flows of basic lavas are exposed below the Larji Carbonate belt.

(iii) Rudraprayag Volcanic formation (Protero-zoic-early Palaeozoic): associated with a sedimentary sequence overlying the Pithoragarh-Deoban Carbonate belt or Garhwal Group of rocks (Deoban group), a zone of volcanic rocks extends from near Gangulihat (Sarju valley) on the SE upto Kuthnaur (Jammu valley) on the NW spanning the entire Inner Sedimentary belt of Kumaon and the Garhwal Lesser Himalaya. These volcanics occur in all the sectors, almost in physical continuity and can be grouped broadly into two horizons as seen in the type section of Rudraprayag (Uttar Pradesh, Himalaya) hence are referred to here as the Rudraprayag Volcanic Formation. The metasedimentary-metavolcanic sequence is bounded by the North Almora Thrust on the S and Main Central Thrust on its N. The volcanic formations comprise mostly basic flows, tuffites, sub-volcanites and basic intrusives.

Petrographically, the rocks are of basaltic and spilitic composition, the latter being more dominant. Primary fabric is seen only in less altered basalts and never in the spilites. Chlorite, epidote, calcite, prehnite, pumpelly-ite, quartz, chalcedony, albite and zeolite (rare) are common minerals filling the vesicles, fractures and veins.

The hypabyssal varieties are dolerites, epi-diorites and a locally differentiated suite of pyroxenite-gabbro-norites and diorites.

(iv) Bhowali Volcanic Formation (early - mid-Paleozoic): In the southeastern extremity of the Chandpur-Krol belt near Nainital, a group of metasedimentary-metavolcanic rocks are exposed in an anticline bounded by the Main Boundary Fault/Thrust on the south and South Almora Thrust on the north. This group has been considered as equivalent to the Nagthats by many investigators (Valdiya, 1964; Gansser, 1964; Tewari and Mehdi, 1964; Raina and Dungrakoti, 1975) while a different school prefers to correlate it with the Deoban Group. The contact of this group with Ramgarh porphyry is, however, disputable. The Krol group of rocks of Nainital syncline are considered by a few (Raina and Dungarkoti, 1975; Taron, 1969) to rest tectonically over the Bhowali Volcanic Formation and associated sedimentaries.

In field relations and petrography characteristics, these epimetamorphosed basic lavas, tuffs and the associated intrusives are very similar to the rocks of the Rudraprayag Volcanic Formation. Different views have been expressed on their spilitic nature. Emplacement environs apart, the relative abundance of different secondary minerals gave rise to spilitic composition. Some workers have even called the volcanic suite spilite-keratophyres; while the porphyrites of Ramgarh area are also identified as keratophyres.

(v) Basantgarh Volcanics (Permo-Carboniferous); In the "autochthonous/parautochthonous folded belt" of Kashmir Himalaya, "Panjal Volcanics" and "Agglomeratic Slates" with outliers of Subathu (Eocene) and discontinuous 'faulted strips' of Triassic limestone are exposed in a zone bounded by the Murree Thrust and the Panjal Thrust as a tightly compressed package of strata. This Volcanic belt is considered to be the extension of the Mandi Volcanics of the Shali Structural belt by some while according to others it is equivalent of the Panjal Volcanics of the Kashmir basin.

(vi) Basic rocks within the Krol-Tal (Permian-? Jurassic): These are only a few and far between and have not been systematically studied. A few outcrops in the Nainital area, around Ayarpatha peak, are intrusive into Krol Limestone and comprise doleritic sills and dykes.

B. Tethyan Sedimentary Regime.

(i) Panjal Volcanics (Mainly Lower Permian): The Panjal Volcanics occupy a large tract of the Kashmir synclinorium in the Pir Panjal range, the Jhelum, Lidar, Sin and Marbal Valleys, Banihal area, the Wular Lake area, Mansbal Lake area, Basnai-Kolahoi, Nagmarg and Zoji la. Strati-graphically, the Panjal volcanics are well defined both at the base and at the top and they mainly belong to the Lower Permian though they may range from Mid Carboniferous to Upper Triassic. Mostly they occupy a position between the Carboniferous and Middle Permian sediments. However in the NW part of the Kashmir basin they show marked overlap and lie over sediments ranging in age from Devonian to Precambrian. They are 1500 to 2400 m thick. Individual flows are generally lenticular, though same flows extend for miles. Bosses of coarse gabbro and norite and numerous dykes and sills of dolerites invade the Panjal Volcanics and smaller sills and dykes of lamprophyres occur. Pillow structures have been identified at a few places.

The Panjal Volcanics predominantly comprise basalts. Other rock types described are andesites, dacites, rhyolites, trachytes and spilites. Most common amygdules are chlorite, quartz, chalcedony and epidote. Limburgite, ankaramite and picrite are also reported.

(ii) Chamba Volcanics (Permian): In the Chamba synolinorium thin basic lava flows have been reported within the Salooni Formation (Permian).

(iii) Phe Volcanics (Upper Carboniferous-Lower Permian): In the Zanskar Tethyan basin, basic volcanic rocks designated as the Phe or Ralakung Volcanics have been recently studied over a linear stretch of 170 km. In the SE the volcanics occur over the marine, fossiliferous Upper Carboniferous rocks, but in the NW, they show unconformable relationships with the underlying successively older formations. The volcanics are about 500 m thick and are overlain at places by Permian and elsewhere by Triassic sediments. They consist of basalts and diabases whick have undergone extensive epidotisation. The Phe volcanics are comparable in age and geological setting to the Panjal Volcanics.

C. Basic/ultrabasics of Indus Tectonic Zone
Indus Tectonic Zone. (i) Dras Volcanic Formation
(late Mesozoic): Considerable data is now available on the general geologic and tectonic aspects
of the late Mesozoic - early Cenozoic Dras Volcanics of the Indus Tectonic Zone which is now
considered as distinct from the ophiolite suite
of the same zone.

The Dras Volcanic Formation forms part of the
continuous sequence of the Sangeluma Group in
which according to recent work, the ultramafics
occur as tectonic emplacements. The ophiolite
bodies in the Dras Volcanic Formation include
dunite, diopsidite and serpentinite. These
ophiolite bodies, sometimes rimmed with a melange
of serpentine and magnetite, have a tectonic
fabric and there is a lack of any contact metamorphic effect on the surrounding lavas and sediments. (For details see the accompanying description of the Indus Tectonic Zone).

Amphibolites occurring in the zone of the North
Almora Thrust show effects of retrogressive metamorphism on the rocks. Epidioritic rocks and
amphibolites are reported from Askot and Baijnath
Klippen also. Some of these basic bodies are
likely to be penecontemporaneous flows.

D. Shyok Volcanics (associated with Trans-Himalayan Tethyan Zone). In the Shyok - Nubra
valley of Ladakh (bounded by Ladakh batholith on
the south and Axial Karakorum batholith on the
north) calc-alkaline volcanics (Khardung Volcanics) occur along with a basalt-limestone-slate
sequence (Tegar Formation). Recent field studies
(Srimal, 1979; Rodcliffe, 1979) reveal that volcanism and sedimentation is of Upper Cretaceous
to Lower Tertiary age. The Shyok Volcanics are
however considered to be different from the Indus
zone volcanics, the former may be a part of the
trans-Himalayan calc-alkaline volcanic suites.

Petrographically, the calc-alkaline suite comprises andesite, dacite, rhyodacite and rhyolite
along with agglomerates, tuffites and ignimbrites.
Epidotisation is common.

E. Basic rocks of Higher Himalayan metamorphic
regime.
a) Ladakh granite: A mafic complex comprising
hornblendite, gabbro, gabbro-norite and gabbroic
anorthosite intrudes the autochthonous Ladakh
granite near Kargil and other places along with
diorite and tonalite. They are locally cut by
felsic and mafic dykes.
b) Basics along Main Central Thrust: Widespread
occurrences of ampbibolite sills all along the
Main Central Thrust (MCT) and a suggestion of
their ophiolitic nature was followed up by some
workers (Gansser, 1964; Valdiya and Gupta, 1972;
Mishra and Bhattacharya, 1976) by describing a
universal basic fringe at the base of the Central
Crystallines along the Main Central Thrust (MCT),
probably indicating the "Sima" beneath the sialic
layer thus supporting a locale of place junction.

The reported or conjectured universal occurence

of basic rocks all along the MCT could not be
substantiated by field mapping in different sectors; nor are these of ultrabasic composition.
c) Basics within Central Crystalline Metamorphites: Central Crystalline Metamorphics occurring all along the domain of Great Himalaya are
associated with numerous bodies of basic rocks.
From petrographic studies it is concluded that
some of the basic schists are derived from ultramafic rocks, while others may represent contemporaneous flows. The metabasites of Sumcham
sapphire mines, Doda district (J & K) are generally regarded as metamorphosed ultramafics, a
correlation of these tectonized masses with the
ophiolites has been proposed (? late Mesozoic),
(Jangpangi et al., 1977).

The metabasites associated with Daling or
equivalent rocks of Eastern Himalaya, are highly sheared and of deformed vesicular nature and
are suggestive of contemporaneous lava flows
(cf. basic flows in Crystalline thrust sheets of
NW Himalaya). The basics in the Darjeelings are
more or less comparable to those seen in granitoids and higher level tectonic units of NW
Himalaya.

Lohit Ultramafite-mafite Rocks (Pre-Permian-
Paleozoic): In the Lohit Himalaya, the Tidding
ultramafics/mafics associated with green schist
facies metamorphism are much older than the
Cretaceous serpentinites of Manipur-Nagaland-
Andaman belt and are likely to be pre-Permian in
age. Their tectonic significance has yet to be
fully understood.

Bands of hornblende schist, chloritic basic
schists, metanorite, metadolerite and lamprophyre
are seen in the granodiorite-diorite complex and
associated metasediments which are thrust over
the Tidding serpentinites, etc.
Abor Volcanics (Upper Palaeozoic): The geological setting of Abor volcanics and those associated with the Himalayan Gondwana formation is described elsewhere in this review. Prevalence of
the basalts, andesite, rhyolite, lava breccia,
volcanic conglomerates, and presence of basaltic
dykes and sills are common.

II. Petrochemistry Petrogenesis and tectonic
environment:
i) Petrochemical studies have been carried out
to some extent on the various volcanic series:
Panjal, Phe, Bafling, Mandi, Rudraprayag and
Bhowali, etc.

Since most of these Paleozoic volcanics are
considerably altered, the petrochemical interpretations will inevitably have to be more dependent on immobile or significantly 'immobile'
elements, e.c., Ti, Y, P, Zr, Nb. Trace element
data are available only on the Bhowali Volcanics
and in view the paucity of available data, minor
element data have to be utilized judiciously.

It has been proposed, on the basis of reinterpretation of data on various elemental/molecular
ratios, that most of the suites fall within the

field of tholeiitic basalt as against the earlier interpretations favouring the genetic relation with alkali basalts. The spilitization is considered to be due to later alkali remobilization phenomenon in most of the suites. It is however true that all basaltic rocks, as analyzed, are not of uniform composition and spilitic character.

While using 'mobile' elements only the rocks showing basaltic mineralogy and undergoing little alteration have been considered. In a conventional AFM diagram, the Panjal lavas occupy mainly the tholeiitic field with little iron enrichment. Intermediates are abundant, while rhyolites are rarer. The Phe lavas are comparable with the Panjal. The Bafliaz suite varies from basic to acid. Rudraprayag and Mandi lavas are prominently tholeiitic and basic, with few intermediate and acid differentiates. Much scattering is noticeable in all the suites, which might be due to alteration which might have contributed to the calc-alkaline nature of many rocks.

(ii) The Panjal and Phe lavas show more anorthite, while the Rudraprayag and Mandi lavas have more albite and fall in the field of spilitic rocks. The Bafliaz lavas are albite rich. Many of the altered rocks are nepheline-normative, although nepheline never appears in the model norms. Similarly although most of the less altered rocks are quartz-normative, a few are hypersthene-normative or have olivine in norms which never appear in the mode. The interpretations also clearly indicate that alteration of basalt to spilite brings forth increasing normative albite and finally nepheline-normative character.

The chemistry of metamorphic alteration is further illustrated in ACF and $H_2O - Fe_2O_3/FeO$ diagrams, which bring out the variability in oxidation ratio and antiperthitic oxidation/hydration relationship. More hydrated rocks are chlorite-rich with higher MgO while more oxidized rocks are epidote-rich with higher CaO. Prehnite-pumpellyite rich spilites are rich in Na_2O, CaO and Al_2O_3 but contain less H_2O (in respect of chlorite spilites).

(iii) Petrogenesis: From field, petrographical and petrochemical data cited above, it is concluded that spilitic rocks in these volcanic suites are genetically related to the associated tholeiitic basalts. Chemical redistribution during regional metamorphism/burial metamorphism contributed to the formation of spilitic rocks. Reviewing the Lesser Himalayan spilites in the light of primary spilites, autometasomatic spilites and metamorphic spilites, it may be reasonable to classify them as metamorphic spilites. The degree of magmatic differentiation is indicated by the fact that in all the suites there are intermediate rocks as well as acid rocks but all the suites are predominantly derivations of basic magma. The trend of magmatic differentiation is obscured by secondary processes.

While analyzing the metamorphic facies, the mineralogical and chemical changes in the light of the temperature - pressure, depth and fluid pressure considerations, it has been tentatively inferred that fluid pressure from channel ways to massive parts was instrumental in causing metamorphic differentiation during low grade regional metamorphism. It also explains the fact that permeable parts are most spilitized.

On an extensive analysis of suitable geochemical parameters viz., Ti, Zr, Y, P_2O_5, K_2O, some workers attempted (Taron, 1978) to show that metabasalts of UP, and HP Himalaya have transitional characteristics between continental basalts ("Within-place") to oceanic basalts ("Diverging plate margins"); whereas the Mandi and Phe lavas exhibit more oceanic affinity than Rudraprayag lavas; Panjal lavas are mostly of ocean island type. On geochemical considerations it has also been tentatively indicated that the Panjal and Bafliaz lavas are of deeper derivation than the Rudraprayag and Mandi lavas while the Phe lavas perhaps came from shallowest levels. Darla lavas, (if considered separately from Mandi lavas), are similar to the Phe lavas, in this respect. These variations also explain the transitional tectonic nature of Lesser Himalayan suites, between typical continental regime and oceanic regime. But none came from crustal levels and therefore they are distinctly different from orogenic magmatic suites. The Panjal lavas are more akin to plateau basalts while other suites have intermediate characteristics. Field data show that while the Panjal and Phe Volcanics are mostly subaerial, other suites were emplaced mainly in aqueous environment.

Thus on the basis of geological, petrochemical and tectonic studies some workers propose that the Lesser Himalayan mobile belt evolved from epicontinental basins; the tectonic environment was continental and continental to oceanic; magmatism was neither eugeosynclinal nor orogenic, but reflects intracratonic, epeirogenic facies. Divakara Rao (this volume) throws further light on this problem.

Indus Tectonic Zone of Ladakh Himalaya - Its Geology and Geodynamic Significance

In the Central Ladakh Himalaya, from Nanga Parbat in the NW to Hanle in the SE, there is a long, linear and sharply defined tectonic zone which separates the Tethyan Phanerozoic belts of Kashmir and Spiti-Zanskar from the granitic batholith of the Ladakh Range; the zone is referred to hereafter as the "Indus Tectonic Zone".

The quest for new mineral and energy resources and the need for understanding of geodynamic processes operative in the Ladakh Region led to intensive geoscientific investigations in the Upper Indus Valley by the Geological Survey of India (GSI) during 1973-1975. This included, besides other investigations, intensive geological mapping in the Chumathang - Puga - Nyoma area of the Upper Indus Valley.

Despite the considerable attention the Indus

Tectonic Zone has received, there is a lack of an overall geological picture of this fascinating belt. In order to make up this deficiency a multidisciplinary investigation under the name "project INFLO" was launched in 1977. This project envisaged the preparation of a geological map for the entire Indus Tectonic Zone based on regional lithostratigraphic classification in addition to sedimentologic, petrographic, geo-chemical and paleontological studies. An area of about 4000 sq. km. has already been mapped on 1:50,000 scale.

Geologic Setting

In the Central Ladakh Himalaya, there are three distinct geotectonic zones respectively: 1. Ladakh Crystalline Zone of granitoids and metamorphites on the north; 2. Indus Tectonic Zone comprising the Indus Group and the Sangeluma Group; and 3. Spiti-Zanskar and Kashmir Phanerozoic Zone with Precambrian Crystalline basement and Shillakong ophiolite nappe on the south. All the three belts are disposed parallel to each other along a WNW - ESE to E-W direction.

Ladakh Crystalline Zone: The Ladakh Crystalline Zone comprises a heterogeneous association of granitoids, gabbroids, basic intrusive, metavol-canics and metasediments. The granitoids have an intrusive relationship with a newly recognized metamorphic rock association designated as the Kharbu Group, in the Dras Valley (or Giambal Group; or the Puga Group). The Kharbu Group comprises garnetiferous marble, calc-phyllite, slate, metavolcanics and amphibolite. The Kharbu Group is intruded by the granitoids of the Ladakh Complex.

Ladakh Granitic Complex: This represents a complex association of granite, granodiorite, dio-rite, tonalite, gneisses, gabbroids, together with trachytes, trachyandesites and diabases. Granitoids predominate over other rocks.

The Ladakh Granitic Complex borders the Indus Tectonic Zone all along its northern border from Nanga Parbat area to Hanle.

The isolated granitoid and gabbroid bodies occurring in the Batambas-Trizpan area of the Suru Valley have now been considered as part of the Ladakh Granitic Complex, now outcropping in a tectonic window which is framed by the Dras Volcanics.

The age of the Ladakh Granitic Complex is determined by the oldest sediment that overlies it. At Nyoma, this is the Indus Group with basal fossiliferous, Cenomanian beds. Therefore, the upper age limit of the complex is pre-Cenomanian; its lower age limit is not known.

The Ladakh Granitic Complex together with the Kharbu Group forms the foundation for the Indus Group sediments.

Indus Tectonic Zone: The Indus Tectonic Zone comprises two independent, parallel and almost homotaxial sedimentary belts. The northern one is designated as the Indus Group and the south-ern belt as the Sangeluma Group after the Sange-luma Valley in central Ladakh. The Indus Group and the Sangeluma Group are totally dissimilar to each other in lithology, sedimentation and rock association, with the former being non-ophiolitic and the latter ophiolitic.

Indus Group: The sediments of the Indus Group, variously referred as the "Indus Flysch", "Ladakh Molasse", "Kargil Formation", "Indus Molasse" and "Hemis Conglomerate" unconformably overlie the Ladakh Granitic Complex along its southern slope and are traceable from Kargil in the NW to Hanle in the SE. They further extend toward Kailas in southern Tibet. Isolated outcrops of the Indus Group are known in the area SE of Kargil where they occur within a window zone framed by the Dras Volcanics.

The Indus group is divisible, on the basis of lithostratigraphy, into four formations respect-ively (from older to younger). The Skinding Formation; (diamictites, sandstones, grits); the Kuksho Formation; (alternating siltstones, shales, fine grained sandstone, turbidites, etc.), the Maklishun Formation (carbonaceous shale, grey sandstone, grits, etc.) and the Karit Formation (purple greyish shales, silt, sandstone, grits, shales, diamictites with clasts of Eocene to Triassic rocks). Its contact with the Ladakh Granitic Complex is an angular unconformity though at places it is tectonic in nature.

The recent survey has confirmed that the Indus Group comprises different stages of sedimentation from residual-fluvial to flyschoid to molassic type during different stages of geodynamic de-velopment of the basin.

Chumathang granitoid intrusion: In the Chuma-thang area of the Upper Indus Valley, a late orogenic, granitoid intrudes the Indus Group and is dated at 7 M.y.

Sangeluma Group: The name Sangelum Group is adopted for a sequence of formations compris-ing sediments and volcanics in which ophiolites, are emplaced in a tectonic belt to the south of the Indus Group. The Sangeluma Group broadly incorporates formations designated as the Sumdo Formation, Dras Complex, Dras Formation and the Dras Volcanics and the ophiolite melange.

The Sangeluma belt is tectonically bounded in the N by the Indus Group from Hanle to Kargil and by the Ladakh Granitic Complex from Kargil to the Nanga Parbat area. In the S, it is tectonic-ally delimited by the Proterozoic Phanerozoic Tethyan belts of Kashmir and Spiti-Zanskar. The lower part of the Sangeluma Group is everywhere delimited by the Thrust. It can be divided into four formations; from older to respectively younger the Khalsi Formation (volcanogenic; Orbitolina limestone, cherts, etc.) the Dras Volcanics (basalt, andesite flows, cherts, etc., later emplacement of ultramafics, etc.; the Nindam Formation shales, sandstones, limestones, graywacke; nummulitic limestones, etc.) and the Shergol Formation (shales, sandstone, diamictites, etc., with ophiolitic emplacement). The Khalsi Formation is Albian to Cenomanian in age; the Dras Volcanics, represent the development of an island arc system. The Shergol formation shows tectonic contacts with ophiolite ultramafic rocks within it.

The Dras Volcanic Formation, perhaps represents a stage of development of island-arc system in the evolution of the Sangeluma Group. It is characterized by abundant basalt-andesite lava flows with chert beds. The lavas are marine, vesicular and amygdaloidal. Pillow lavas are seen. Explosive volcanism is evident as indicated by volcanic breccia.

Sedimentation of the Sangeluma Group: The sedimentation of the Sangeluma Group was probably along the continental margin. The Khalsi Formation represents a shelf type deposit. Subsequently unstable conditions resulted in submarine volcanic activity and the development of an island arc. With the deepening of the sea, continental slope deposits of the Nindam were formed. The Shergol represents essentially a shallow water deposit. The ophiolite emplacement in the Sangeluma Group is independent of sedimentation. The Sangeluma sedimentation has a distinct geosynclinal character which is rather unique to the belt.

Spiti-Zanskar and Kashmir Proterozoic-Phanerozoic Zone

The Indus Tectonic Zone along its southern margin, is bounded by the Spiti-Zanskar and Kashmir Proterozoic-Phanerozoic belt. The Proterozoic crystalline basement is represented by the Salkhala Group along the Kashmir belt, by the Giambal Group - Batal Formation in the Suru Valley and the Puga Group in the eastern Ladakh along the Spiti-Zanskar belt. The crystalline basement rocks show amphibolite facies and in Puga area they even contain eclogitic rocks.

The Phanerozoic sediments along the Indus Tectonic Zone mainly belong to Permian - Kuling Formation, Triassic - Lower Jurassic - Lilang Group, Jurassic - Spiti Formation, Cretaceous-Giamal Formation and the Eocene - Kanji Group.

In the Prinkiti La area south of Lamayuru, there is a lenticular body of serpentinite-serpentinized harzburgite emplaced within the Kuling Formation.

Shillakong Ophiolite: Along the crest of the Zanskar Mountains, in the structurally depressed part of the Zanskar synclinorium, large ophiolite blocks are seen resting over the Eocene Kanji Group and the Jurassic - Cretaceous Spiti and Guumal Formations (Srikantia and Razdan, in press). This represents the Shillakong ophiolite nappe the lower part of which comprises basalt, pillow lava, volcanogenic sediments, tuffite, purple chert and serpentinite and in the upper part harzburgite, lherzolite, dunite and serpentinite.

Tectonics

The tectonic sequence of the various structural units in the Central Ladakh Himalaya is presented in the Table below. The various structural belts have a NW-SE to WNE-ESE trend.

The autochthonous zone of the Ladakh Granitic Complex has acted as a buttress for the N directed forces which brought various parauthochthonous units to lie upon each other. Though this unit has not suffered any significant horizontal translation it appears to have undergone vertical uplift.

The autochthonous Indus Group forms a cover for the Ladakh Granite and represents the next higher tectonic unit. This largely autochtonous structural unit has borne the brunt of movement of the overlying parautochthonous units. This is reflected in the northward overturning of folds.

Batambas - Trizpon - Tasgam window zone is the first record of a tectonic window in the Trans-Himalayan zone. This window falls within the Indus Tectonic Zone and it largely consists of Ladakh granitic rocks with a cover of Indus Group sediments framed by the thrust sheet of the Dras Volcanic Formation of the Sangeluma parautochtonous.

Sangeluma parautochthonous belt is defined by two distinct tectonic discontinuities respectively the Pashkyum Thrust in the basal part and the Wakha - Sanko Thrusts in the upper part.

Pashkyum Thrust forms the most remarkable tectonic feature in the Central Ladakh Himalaya. Along this thrust plane the Sangeluma belt has moved over the Indus Group and the Ladakh Granitic Complex. This thrust has been earlier variously referred as the "Great Counter Thrust" "Great Ladakh Thrust", "Mahe Thrust" and "Kargil Thrust" (Berthelsen, 1952; Tewari, 1964; Shankar et al., 1976; Raiverman and Mishra, 1975). The Pashkyum Thrust trends WNW - ESE with a southerly hade of 30°-70°.

Khalsi-Nurla schuppen zone along the Pashkyum Thrust, locally, several schuppen zones have developed involving the Indus Group sediments. The Khalsi - Nurla schuppen zone is a prominent feature along the sole of the Pashkyum Thrust. This shows intense imbrication and tectonic enclosure of Orbitolina limestone within the Karit Formation of the Indus Group.

Spiti-Zanskar parautochtonous is a major tectonic unit comprising the Kuling - Lilang schuppen zone and Mesozoic-Cenozoic parautochthonous. The Precambrian crystallines of Puga and Suru Valley are also part of this parautochthonous.

Kuling-Lilang schuppen zone: This zone is sandwiched between the Sangeluma belt and the

Table I. Tectonic Sequence in Central Ladakh Himalaya

Shillakong ophiolite nappe	Spontang sub-nappe: ultramafics	
	————————Thrust————————	
	Photang sub-nappe: basalt, pillow lava, volcanic-sediment, purple chert, Serpentinite.	
	————————Thrust————————	
	Mesozoic-Cainozoic parautochthonous	
Spiti – Zanskar	————————Sanko Thrust————————	
Phanerozoic Proterozoic Main parautochthonous Zone	Kuling – Lilang schuppen belt	
	————————Wakha Thrust————————	
Sangeluma Group	Shergol	
Parautochthonous belt	Nindam	
	Dras Volcanics	
	Khalsi	
	————————Pashkyum Thrust————————	
Indus Group	Karit	Window Zone
	Maklishun	Karit
Autochthonous	Kuksho	Maklishun
	Skinding	
	————————Unconformity————————	
	Ladakh Granitic Complex and Kharbu Group- Autochthonous	Ladakh Granitic Complex

Mesozoic-Cenozoic parautochthonous and is prominently seen in the area between Mulbekh and Lamayuru and extends further east toward the Zanskar river. The zone contains numerous imbricated scales of Lilang limestone and Kuling shales and locally even of the Shergol Formation and also ophiolite bodies. The schuppen zone is bounded by the Wakha and Sanko Thrusts.

Mesozoic-Cainozoic parauthochton: This is a highly compressed and tightly folded zone of carbonate rocks and shales. It exhibits many spectacular décollement folds. This parautochthonous zone is interrupted towards the south by steep to vertical faults along the contact between the Lilang and the Spiti shale.

Shillakong Ophiolite Nappe: This represents the most remarkable tectonic unit in the Zanskar Mountains and occupies the highest allochthon in the Ladakh Himalaya. The nappe is attributed to obduction from N to S and is independent of movement of parautochthonous units from south to north.

Suru Crystalline Zone: This forms a major zone of structural culmination along NNE-SSW direction along the Suru valley. It separates the Spiti-Zanskar Tethyan basin from the Kashmir Tethyan basin. It also appears to have controlled the Kashmir parautochthonous.

The Precambrian – Permo-Triassic belt of the Kashmir parautochthonous is thrust over the Indus

Tectonic zone as seen in Dras – Pindras section. Towards the Nanga Parbat – Harmosh sector, the Nanga Parbat granites appear to have been thrust over the Indus Tectonic Zone. Along this sector, all the major structural elements turn from NW to N and trend parallel to the Astor river.

Metamorphism

Surprisingly, metamorphic changes are the least within the formations of the Indus Tectonic Zone. Metamorphic rocks are prevalent in the Giambal and the Puga Groups which intermittently border the Indus Tectonic Zone in the southern part. These Groups contain high grade amphibolite facies rocks and the Puga Group contains even eclogitic rocks, and apparently were produced by regional metamorphism.

Metamorphic rock types occur within the Ladakh Granitic Complex. The Kharbu Group contains high grade green schist facies to low grade amphibolite facies rocks.

The singular absence of any high pressure or even high temperature metamorphic rocks in the Indus Tectonic Zone is significant. The Indus Group sediments show only local development of cleavages in shales of the Kuksho Formation. In the Sangeluma Group has developed secondary planar structures without any distinct mineralogical changes. The reported occurrence of glaucophane schist slices in the Dras Volcanics along the Pashkyum thrust near Pashkyum, has yet to be confirmed. In the Shergol Formation however, a slice of amphibolite is seen emplaced within the sediments and the Formation contains local olistromes of marble. The Dras Volcanics are also similarly unmetamorphosed. It is significant that the ophiolites of the Sangeluma belt, Kuling-Lilang schuppen zone and the Shillakong nappe are all unmetamorphosed excepting the effect of dislocation metamorphism (Frank et al., 1977; Casnedi, 1976). However, in the NW part of the Indus Tectonic Zone, in the Astor Valley, regional and dislocation metamorphism of sediments and basic rocks has been reported (Casnedi, 1976).

Tectonic Evolution

The geotectonic importance of the Indus Tectonic Zone has been a subject of various opinions ranging from typical eugeosyncline, foundered geosyncline, to the great Suture Zone developed on the collision of the Indian with the Eurasian plate.

Recent research in the Indus Tectonic Zone have provided considerable data on the stratigraphy, sedimentation and structure of Central Ladakh Himalaya.

Srikantia and Razdon (in press) consider that this zone existed as a mega-lineament from early times and represents a structural feature of the greatest importance for studying the successive phenomenon of tension, distension, ophiolite development along a spreading centre, subduction phenomenon, compression, obduction, closing of sea and development of nappes. A careful understanding of stratigraphy tends to suggest that the Sangeluma Group represents a sequence of formations which include both sediments and volcanics in a normal stratigraphic relationship. The ophiolites are unrelated to sedimentation. Their emplacement is by tectonic processes and obduction at the time of closing of the sea. It, thus, represents a totally dismembered ophiolite belt of allocthonous emplacement with no sequential order. According to this view there is no evidence to prove that the Indus Tectonic Zone is a convergent plate margin.

Similarly, the Shyok - Nubra - Chushul zone is also considered as representing a region of intra-crustal dislocation, resulting in the development of deep fractures which became the conduit for magmatic intrusions. As the fractures widened the zone became a site for magmatic effusions and with further widening, a site for sediment deposition. Thus there is a remarkable inter-relationship between the Shyok tectonic zone and the Indus Tectonic Zone.

The Central Ladakh Himalaya, in which the Indus Tectonic Zone is located, deserves a detailed geophysical and geochemical study to understand the crust-mantle relationship, along with continued geologic/tectonic and sedimentologic studies.

Tectonic Evolution of the Eastern Himalaya[1] (Sikkim - Darjeeling - Bhutan - Himachal Pradesh)

The Eastern Himalaya forms a large segment of the range with complex geologic and tectonic patterns.

Stratigraphy of the sub-Himalayan range: The Upper Tertiary formations of the sub-Himalayan range represent the eastern continuation of the Siwalik fore-deep basin. Its lower stratigraphic succession may correspond to parts of the Dharmasala-Lower Siwalik formations of the Western Himalaya. A major part of the remaining section may be broadly correlated with the Upper and Middle Siwaliks. Rock fragments, heavy minerals and paleo-currents in the Siwalik point to a Lesser Himalayan provenance. Higher up in the sequence, the existence of high-grade metamorphic clastics is significant. This observation possibly indicates concomitant sedimentation with the advance of gneiss nappes in the lesser Himalayan zone, during Siwalik sedimentation.

The north Brahmaputra-sub Himalayan Siwalik Basin was separated from the Upper Assam Tertiary shelf by the subsurface high extending from the Garo-Mikir massif. Within the Siwalik belt of Kameng foothills (Arunachal Pradesh) kaolinised sandstones, basic rocks and clays, occurring as sheared basement wedges, may be correlated with Rajmahal-Sylhet Trap (Mesozoic) and associated sediments. However, in view of the latest find of nummulitic rocks in the Dihang valley, there is a possibility of these rocks extending upto Paleogene age.

Stratigraphy of the Lesser, Great and Tibetan Himalaya: The Tibetan Himalayan sedimentary belt, characterized by monotonous facies and well-documented marine fauna from Kashmir to Bhutan, is distinct from the Lesser Himalayan formations which are largely unfossiliferous and structurally complex.

Paleogene formations: Recently nummulitic rocks of Eocene age have been discovered N of the Siwalik belt in the Dihang valley; Tertiary rocks with plant fossils are also known in the Dihang valley within the tectonic frame of the Abor volcanics.

Mesozoic formations: These are present only in the Tibetan zone. Intercalated limestone and shaly limestone constitute the Triassic. The Jurassic-Lower Cretaceous in Bhutan comprise calcareous shale. Well preserved Upper Gondwana flora has been recently recorded from Bhutan within the upper part of the paralic succession. Late Paleozoic formations: A remarkably homogeneous late Paleozoic tilloid facies is develop-

---

[1] Compiled by S. K. Acharya, N. S. Krisnaswamy, K. K. Ray and S. Sinha Roy.

ed in the Lesser Himalaya and Tethyan basins. This is sporadically associated with similar and related marine fauna, acid-intermediate pyroclastics and basic-intermediate volcanics. Part of this faunal assemblage is akin to the fauna known from the agglomeratic slate of Kashmir, and the Talchir Formation of Central India. *Glossopteris* flora within the succeeding paralic facies with carbonaceous and coaly intercalations are also broadly similar.

The volcanic intercalations within the late Paleozoic sequence have been correlated with the Abor Volcanics associated with the Miri quartzite in the Siang district. Lower Gondwana sporomorphs have been recorded from some of the intratrappean beds of the Siang type section. The Abor Volcanic succession essentially comprises basaltic-andesitic flows and subaerial to water deposited rhyolitic-dacitic volcanoclastics.

The basal Gondwana tilloid containing profuse clasts of sedimentary, volcanic, metamorphic and granitoid rocks from the older rocks is usually believed to overlie the Buxa unconformably. However conformable transitional contact of the Gondwana tilloid with the Buxa dolomite and Thungsing/Miri quartzite has also been recorded from Sikkim window and eastern Bhutan.

Infra-Gondwana formations of Lesser and Great Himalaya: In the Lesser Himalaya, broadly three litholo-tectono-stratigraphic associations have been recognized within the feeble to moderately metamorphosed Daling-Buxa/Miri sequence (Acharya, 1971a, 1978a, b; Ray, 1971, 1976; Acharya and Ray, 1977; Acharya et al., 1976). There is a lack of general agreement on the reconstructed stratigraphy. Interfingering of the various litho-tectonic units are also regarded by some (Sinha, 1974b; Sengupta and Raina, 1978) as sedimentary facies variations and/or structural closures of northerly plunging reclined folds.

Based on metamorphic transitions, structural consanguinity and broad lithological similarity, some authors (Sinha, 1973; Verma and Tandon, 1976) have preferred to include low and high grade metamorphic rocks within one composite stratigraphic-metamorphic unit, e.g., Daling-Darjeeling Group and Bomdila Group. On the other hand, some authors (Acharya, 1971, 1977, 1978; Ray 1971, 1976) differentiate the metamorphites into two structural belts viz., the lower comprising Daling-Buxa/Miri sequence and the Gondwana and the upper unit made up of higher grade gneissic formations which also constitute the Central Crystallines. The association of quartzite, calcsilicate, marble, graphite schist, meta-volcanics and varied gneisses constitute the most widespread lithotectonic association of the upper unit, e.g., Paro group (sensu-lato), Chungthang Formation, Dirang Schist, Sela Group, etc.

Biotite augen gneiss, porphyroblastic granitic or quartzose gneisses are intimately associated at various levels of the Daling-Buxa/Miri sequence and the gneissic formations. These are mylonitic, strongly lineated and diaphthorazed within the Daling-Buxa sequence or at the base of the gneissic formations. These augen gneiss are regarded either as basement wedges or partly recrystallized tuffaceous Daling rocks and/or syntectonic metasomatic or intrusive granites within the Daling-Buxa and Paro assemblage.

The Tidding Formation (=? Paro Group of Bhutan) well developed in between the frontal zone of the Lohit Himalaya and the orographic bend in the upper reaches of the Siang gorge, predominantly is comprised of actinolite-albite-epidote-bearing green schist, crystalline limestone, serpentinites, micaceous quartzite, etc. Some authors (Ray, 1976a, b) consider that the serpentinite and serpentinized limestone may not represent a ophiolitic suite, as usually believed.

The base of the Tibetan sediments is variably truncated by overlapping thrusts and decollements and in many cases obliterated by the intrusion of sheet-like bodies of tourmaline-bearing leucogranites. According to one view, the infra-Gondwana Everest limestone and Everest pelite of north Sikkim-Everest area can be correlated with the Buxa and Daling formations of the frontal zone, whilst the granite injected and migmatised Everest pelite (in the lower part of the sequence) and the Lower Limestones, etc., are lithologically comparable with the Darjeeling and the Paro Groups. Infra-Gondwana (=Infra Permian) formations of Bhutan Higher Himalaya are broadly comparable with those of North Sikkim.

Structural Aspects: Field observations have provided new dimensions and interpretations to the structural disposition of the Siwalik belt, Palaeogene and Main Boundary Thrust (here after MBT) in the Eastern Himalaya. Essentially homoclinal, the Siwalik-Dharamsala sequence of the sub-Himalaya is demarcated to the south by a frontal fault against the subhorizontal Quaternary sediments of the Ganga-North Brahmaputra basin. The northern beds of the Siwalik belt usually appear to be more deformed and limited by the MBT against the Himalayan Gondwana or structurally higher rocks of the Lesser Himalaya. Structurally the MBT is conformable with the overriding rocks and Lesser Himalayan thrusts but oblique to the subjacent Upper Tertiary rocks. A narrow, open to closely folded belt, essentially constituted of Lower Siwalik-Dharamsala rocks, occurs to the north. Thin slices of Damuda rocks occur within this parautochthonous belt in Darjeeling and Kameng sub-Himalaya foothills and western Duars; slices of Upper Tertiary rocks, occasionally with angiosperm fossils occur within the Gondwana and pre-Gondwana formations of the Lesser Himalaya. Klippen of the Domuda and Daling formations also occur 3 km within the Siwalik range in Dumai area, Darjeeling foothills. These indicate that the MBT, akin to the lower thrust surfaces of the Lesser Himalaya, is also likely to flatten out northwards at depth.

Newly located fossiliferous Eocene rocks just north of the Siwalik frontal zone in the Siang valley occur in a tectonic disposition which may be comparable to the similar tectonic relations of Paleogene and Neogene in NW Himalaya. Search for extension of the Paleogene sequence west of the Siang Valley is continuing. Lower Tertiary plant-bearing rocks, possibly with volcanic rock associations, seem to be tectonically surrounded by older rocks. This find may throw more light on the palaeodynamics and basin configuration in Arunachal, Himalaya.

Structure and Metamorphism of the Lesser, Great and Tibetan Himalaya: Within the tectonic-stratigraphic units in the Lesser and the Great Himalaya, several thrusts normally limit the structural belts of nappes but there is no consensus about the number and location of these thrusts. The tectonic base of the Central Crystallines is regarded as the Main Central Thrust but, as indicated elsewhere in this review, there is lack of agreement about its location. However, recent interpretations of Bhutan geology tend to reduce the regional significance of the thrusts.

The tectonic base of the Tibetan sedimentary sequence above the Central Crystallines, which also often coincides with sheet-like intrusion of tourmaline-bearing leucogranites, is designated as the Trans-axial Thrust. Several formational and intra-formational thrusts and decollements have been recognized within the Tibetan zone sediments of North Sikkim and by some workers in Bhutan.

In the Lesser Himalayan region, there is general consensus regarding broadly E-W to WNW-ESE trending longitudinal F_3 folds, L_3 puckers and S_3 fracture cleavage and contemporaneous to subsequent N-S trending F_4 warps which affect all the lithotectonic units, the bounding dislocation zone, and regional metamorphic isogrades in Darjeeling-Sikkim-Bhutan Himalaya. Almost all the major river valleys of the Eastern Himalaya are parallel to the axial trace of regional F_4 antiforms. Almost orthogonal and often diagonal to this trend, are the synformal gneiss hills of Darjeeling, Kalimpong, etc. These are thus clearly post-thrust, post-metamorphic, late Cenozoic structures. But within the medium-to high-grade rocks and migmatites, prograde syn-to-post-kinematic crystallization occurs, as also the outgrowths of mica and garnet, which are also synkinematic with respect to late stage migmatitic material.

There is lack of agreement (Mukhopadhyay and Gangopadhyay, 1971; Acharya, 1971, 1977; Ray, 1971, 1976a; Sinha Roy, 1973a; Verma and Tandon, 1978) regarding earlier structures, metamorphism and thrust movement. The earliest D_1 deformation with highly impressed isoclinal folds (F_1) with axial plane cleavage/schistosity (S_1) and intersection type striping lineation (L_1) is usually considered as Hercynian or even older.

This is followed by co-axial to slightly oblique refolded isoclinal F_2 folds during D_2 deformation. However the break or protracted continuity between $D_1 - D_2$, the significance of a pronounced flattening fabric parallel to foliation and axial stretching and mineral lineation are subjects of continued enquiry as there is no unanimity regarding the correlation of D_1 and D_2 deformations.

Kinematic analysis of metamorphism in the Lesser Himalayan belt is interesting since the rocks are polymetamorphic and the zonal sequences of most significant Barrovian metamorphism are inverted. This has been a subject of rethinking in recent years. The main phase of regional metamorphism and large scale nappe movement have been linked with the L-S fabric which is dated late Cenozoic by some workers (Sinha Roy, 1974b), but considered broadly contemporaneous with Himalayan Gondwana sedimentation by others (Acharya and Ray, 1976). Those advocating a number of thrust sheets in Darjeeling-Sikkim Himalaya feel that the earlier and lower nappe system has escaped high grade metamorphism and granitization in contrast to later and higher gneissic nappes; later phase of Himalayan nappe movement has caused lithostratigraphic and metamorphic discontinuity with well marked diaphthoresis. But a prograde Cenozoic neometamorphism and migmatisation is also reported, which largely healed the effects of these dislocations. Such scheme of tectonics is interpreted as having produced a regionally inverted Barrovian metamorphism in the Lesser Himalaya. Some authors (Ray, 1947; Ghosh, 1956) consider that the high grade Central Crystalline, with 'infra-structural' character, overlaps the 'supra-structural' Daling rocks with its high level granite injections to produce the reversed regional Barrovian metamorphism.

Locally blue schist type stilpnomelane-pumpeyllite-lawsonite assemblage have been recorded from the Daling greywacke overriding the Gondwana rocks in east Sikkim.

Available K/Ar whole rock and mineral dates of the Lesser Himalayan low and high grade metamorphic rocks and Higher Himalayan leuco-granites cluster around late Oliogocene-Neogene (35-10 M.y.) age. These ages signify late Cenozoic reworking of much older rocks and time the main Himalayan metamorphism or late phase of Himalayan neometamorphism.

Four broadly consistent K/Ar whole-rock dates (178-213 M.y.) have been obtained from biotite augen gneisses within the Paro Group of North and East Sikkim. Comparable K/Ar whole-rock ages (170-190 M.y.) from biotite granite/augen gneiss or gneissose quartz porphyry/metavolcanics have also been recorded from certain sections of Nepal and Western Himalaya. These may correspond to a regional thermal or other disturbance outlasting late Palaeozoic magmatic-metamorphic event.

Eastern Orographic Bend of the Himalaya: There is no consensus regarding the nature and origin of the eastern orographic bend of the Himalaya.

The sharp bend of regional strike occurring at the Siang gorge is possibly an effect of large scale F_4 warp. On the contrary the rocks of Siang and Lohit Himalayas are regarded as distinct from each other by some workers.

Geotectonic Models: Various hypotheses of recumbent nappe, high level granitic injection, crystalline nappe, etc., and morpho-tectonic developments have been a subject of lively discussions (Auden, 1935; Gansser, 1964; Fuchs, 1968; Le Fort, 1975; Sinha, 1973a, 1974b and others). Besides these, the plate tectonic model based on continents, or the presence of a microcontinent between the two have been put forward. In the Eastern Himalaya plate tectonic models have been applied in terms of the MCT and MBT and crystalline nappes, whereas the Indus-Tspango suture falls far to the north (Dewey and Bird, 1970; Powell et al., 1973; Le Fort, 1975; Sinha, 1974b, 1976). The collision of the Indian plate with the eastern plate has been postulated to explain the geological pecularities of the Lohit Himalaya and the eastern orographic bend or 'syntaxis'.

In addition, the concept of overthrust nappes from the Tibetan hinterland has been conceived, which have covered fully or partly the Mesozoic-Cenozoic master basin (geosyncline) from Indus Zone to Andaman area.

References

Acharya, S. K., Structure and stratigraphy of the Darjeeling frontal zone, Eastern Himalaya, in Recent Geological Studies in the Himalaya Seminar, Calcutta, Geol. Surv. India Misc. Pub., 24(1), 71-90, 1971.

Acharya, S. K., Palaeogeography and Orogenic evolution of the Eastern Himalayas, in Himalaya, Sci. Terre., Colloq. Int., No. 268 CNRS, Paris, 24-30, 1977.

Acharya, S. K., Stratigraphy and tectonic features of the Eastern Himalaya, in Tectonic Geology of the Himalaya, P. S. Saklani (ed.), 243-268, 1978a.

Acharya, S. K., Mobile belts of the Burma-Malaya and the Himalaya and their implications on Gondwana and Cathaysia/Laurasia continent configurations, P. Nutalaya (ed)., Proc. 3rd. Regional Conference Geology & Min. Resources SE Asia, Asian Inst. Tech. Pub. Bangkok, 121-127, 1978b.

Acharya, S. K., S. C. Ghosh and R. N. Ghosh, Geological frame work of the Eastern Himalayas in parts of Kameng, Subansiri, Siang districts, Arunachal Pradesh, Sem. Geology and Mineral resources of NE Himalaya, Shillong, Geol. Surv. India. Abs., 39-40, 1976.

Acharya, S. K. and K. K. Ray, Geotectonic evolution of the Himalaya - a model. Himalayan Geol. Sem., New Delhi, Section III, 1976.

Acharya, S. K. and K. K. Ray, Geology of the Darjeeling-Sikkim-Himalaya, Guide to excursion, 4th Int. Gondwana Symp., Calcutta, 25p., 1977.

Andrieux, J., M. Brunel and J. Hamet, Metamorphism, granitisation and relations with the Main Central Thrust in Central Nepal: 87 Rb/87 Sr age determinations and discussions, Himalaya, Sci. Terre, Colloq. Int. No. 268, CNRS, 1977.

Auden, J. B., Traverses in the Himalaya, Rec. Geol. Surv. Ind., 69, 123-167, 1935.

Berthelsen, A., On the Geology of the Rupshu district NW Himalaya, Sertryk of Meddleleser fra Dansk Geologisk Forening, 12, 350-414, 1953.

Bhandari, A. K., and K. N. Singh, Granitic rocks of northwestern Himalaya and their geochronological status, Himalaya Geol. Sem., New Delhi, Section-I, 1976.

Bordet, P., On the position of the Himalayan Main Central Thrust within the Nepal Himalaya, Seminar on Geodynamics of the Himalayan region, Nat. Geophy. Res. Inst., Hyderabad, 148-155, 1973.

Brunel, M., and J. Andrieux, Déformations Superposés et Mécanisms associés a chevachement Central Himalaya "MCT" Nepal Oriental Himalaya, Sciences de la terre, Colloq. Int. No. 268, CNRS, 69-84, 1977.

Casnedi, R., The ophiolites of the Indus Suture line (Dras and Astor Zones-Kashmir) Ofiolitic Bell Gruppo Lav. Sulle, Ofiolite Mediterranee, 1, pt. 3, 365-371, 1976.

Chatterjee, Barin, A note on occurrences of microgranite and porphyry along the main boundary fault in Amritpur area, Nainital district, Himalayan Geol. Sem., New Delhi, Section-I, 1976.

Das Gupta, S. P., D. P. Bhattacharya, B. K. Chakrabarti and P. R. Sen Gupta, Metamorphism and Igneous Activity in the Himalaya, Himalayan Geol. Sem., New Delhi, Section - I, 1976.

Dewey, J. F., and J. M. Bird, Mountain belts and the new global tectonics, Journ. Geophys., 75, 2625-2647, 1970.

Frank, W., M. Thoni and F. Purtscheller, Geology and Petrography of Kulu-south Lahul area, Himalayas, Sci. Terre. Colloq. Int. n° 268, CNRS Paris, 147-172, 1977.

Fuchs, G., The geological history of the Himalayas, 23rd Int. Geol. Cong. Canada, Proc. Sec. 3, 161-174, 1968.

Gansser, A., Geology of the Himalayas, Inter Sc. Pub., London, 269p., 1964.

Ghosh, A. M. N., Recent advances on geology and structure of eastern Himalayas, 43rd Ind. Sci. Cong. Proc., pt. 2, 85-99, 1956.

Ghosh, A., B. Chakraborti and R. K. Singh, Structural and metamorphic history of the Almora Group, Kumaun Himalaya, U.P., Him. Geol., 4, 171-194, 1974.

Heim, A., and A. Gansser, Central Himalaya, Geological observations of the Swiss Expedition, Soc. Helv. Sci. Nat., 1939.

Jangpangi, B. S., Stratigraphy and tectonics of parts of Eastern Bhutan, Him. Geol., 4, 117-136, 1974.

Jangpangi, B. S., Stratigraphy and structure of Bhutan Himalaya Himalaya, P. S. Saklani (ed.), Today and Tomorrow's Printers and Publishers, New Delhi, 221-240, 1978.

Jangpangi, B. S., D. P. Dhoundial, G. Kumar and J. N. Dhoundial, On the geology of Sumbham Sapphire Mines area, Doda district, Jammu and Kashmir State, (abst), Him. Geol. 9th Sem., 1977.

Le Fort, P., Himalaya: The Collided range. Present knowledge of the continental area, Am. Jour. Sci., 275(A), 1-44, 1975.

Mishra, R. C. and A. R. Bhattacharya, The Central Crystalline Zone of North Kumaon Himalaya; its lithostratigraphy, structure and tectonics with special reference to plate tectonics, Him. Geol., 6, 133-154.

Molnar, P., W. P. Chew, T. J. Fitch, P.Tapponier, W. E. F. Warsi and F. T. Wu, Structure and tectonics of the Himalaya: A brief summary of relevant geophysical observations, Himalaya, Sci. Terre, Colloq. Int., No. 268, CNRS, 269-294, 1971.

Mukhopadhyay, M. K. and P. K. Gangopadhyay, Structural characteristics of rocks around Kalimpong, West Bengal, Him. Geol. 1, 214-230, 1971.

Nautiyal, S. P., B. S. Jangpangi, P. Singh, T.K. Guha Sarkar, V. D. Bhate, R. M. Raghavan and T. N. Sahai, A preliminary note on the geology of Bhutan Himalaya, Int. Geol. Cong. 22nd Sess. 11, 1-14, 1964.

Peive, A. V., Fractures and their role in the structures and development of the earth's crust, Int. Geol. Cong. 21 Sess., Copenhagen, 18, 280-286, 1960.

Powell, C. McA and P. J. Conaghan, Plate tectonics and the Himalayas, Earth Planet. Sci. Letters, 20, 1-2, 1973.

Raina, V. K. and U. Bhattacharya, Sedimentaries of north Sikkim. Rec. Geol. Surv. Ind., 106(2), 75-85, 1976.

Raina, B. N. and B. D. Dungrakoti, Geology of the area between Nainital and Champawat, Kumaon Himalaya U.P., Him. Geol., 5, 1-25, 1975.

Raina and Dutta, Unpublished GSI Progress Report, 1979.

Raiverman, V. and V. N. Mishra, Suru tectonic axis, Kargil area, Ladakh, Bull. Geol. Min. and Met. Soc. Ind., N. 48, 1-16, 1975.

Ray, S., Zonal metamorphism in the Eastern Himalaya and some aspects of local geology, Quart. Geol. Min. Met. Soc. Ind., 19, 117-140.

Ray, K. K., Some problems on stratigrahy and tectonics of the Darjeeling and Sikkim Hiamalayas, in Recent Geological Studies in the Himalayas, Seminar, Calcutta, Geol.

Surv. Ind. Misc. Pub., 24(2), 379-394, 1971.

Ray, K. K., Geotectonics of the circum Indian Meso-Cainozoic Mobile belt with special reference to oil and natural gas possibilities. C.G.N.W. Sem./Tectonics and Metallogeny, SE Asia and Far East, Calcutta, India, Geol. Surv. Ind. Misc. Pub., 34, 85-97, 1974.

Ray, K. K., A review of the geology of Darjeeling-Sikkim Himalayas, in Himalayan Geology Seminar, New Delhi, Geol. Surv. Ind. Abs., 58, (paper in press), 1976a.

Ray, K. K., Dalingian metallogeny in Darjeeling-Sikkim Himalaya, in Himalayan Geology Seminar, New Delhi, Geol. Surv. Ind. Abs., 157-158 (paper in press), 1976b.

Rodcliffe, R. P., Unpublished GSI Progress Report, 1979.

Sengupta, S., and P. L. Raina, Geology of parts of Bhutan foothills adjacent to Darjeeling dist., Ind. Jour. Earth Sci., 5, 20-33, 1978.

Shankar, R., R. N. Padhi, G. Prakash, J. L. Thussu and C. Wangdus, Recent Geological studies in Upper Indus Valley and the plate tectonics, Misc. Pub. Geol. Surv. Ind., V. 35, Pt. I, 41., 1976.

Sharma, V. P., Stratigraphy and structure of Hammu Himalaya, J & K State, India, Sci. Terre, Colloq. Int., No. 268, 379-386, 1977.

Sinha Roy, S., Kinematic significance of conjugate folds in the Daling metamorphites from Kalimpong Hills, Sikkim Himalaya, Him. Geol., 3, 176-184, 1973.

Sinha Roy, S., Tectonic belts in Sikkim-Darjeeling Himalaya and their geodynamic significance in Geodynamics of the Himalayan region Seminar, Hyderabad, Nat. Geophys. Res. Inst., 156-166, 1973a.

Sinha Roy, S., Polymetamorphism in Daling rocks from a part of Eastern Himalaya and some problems of Himalayan metamorphism, Him. Geol., 4, 74-101, 1974a.

Sinha Roy, S., Tectonic elements in the Eastern Himalaya and geodynamic model of evolution of the Himalaya, C.G.N.W. Sem./ Tectonics and Metallogeny, SE Asia and Far Asia, Calcutta, S. Ray (ed.), 1974b.

Sinha Roy, S., Structure of the allochthonous Proterozoic rocks in a part of the Tista Valley, Eastern Himalayas, Ind. Jour. Earth Sci., 4, 20-38, 1977.

Sinha Roy, S., and R. Stoneley, On the origin of ophiolite complexes in the Southern Tethys region, Tectonophysics, 34 (3-4), 257-265, 1976.

Srikantia, S. V., Sedimentary cycles in the Himalaya and their significance on the orogenic evolution of the mountain belt, Himalaya Sciences de la terre, editions du CNRS, 1977.

Srikantia, S. V., and O. N. Bhargave, Tectonic evolution of the Himachal Himalaya, Tectonics and Metallogeny of the SE Asia, Calcutta, S. Ray (ed.), <u>Geol. Surv. Ind. Misc. Pub.</u>, <u>34</u>, 217-236, 1974.

Srikantia, S. V., and M. L. Razdon, Geology of part of Central Ladakh Himalaya with particular reference to Indus Tectonic Zone - A late Mesozoic-Cenozoic ophiolite sediment association (in press).

Srikantia, S. V. and M. L. Razdon, The Ophiolite - Sedimentary Belt of the Indus Tectonic Zone of Ladakh Himalaya- its stratigraphic and tectonic significance, Intl. Symposium on Ophiolites, April, 1979, Nicosia, Cyprus, (in press).

Srimal, N., Unpublished GSI Progress Report, 1979.

Taron, P. B., Unpublished GSI Progress Report, 1969.

Taron, P. B., Tectonic environment of the Basic Volcanic rocks of U.P. and H.P. Lesser Himalaya determined using chemical discriminant, (in press), <u>Him. Geol.</u>, <u>9</u>, 1978.

Tewari, A. P., On the Upper Tertiary deposits of Ladakh Himalaya and correlation to the various geotectonic units of Ladakh with those of Kumaun - Tibet region, <u>Proc. 22nd Inter. Geol. Congr.</u>, <u>11</u>, 37-58, 1964.

Thakur, V. C., Divergent isogrades of metamorphism in some parts of Higher Himalaya Zone, Himalaya, <u>Sciences de la terre</u>, editions du CNRS, 433- , 1977.

Valdiya, K. S., A note on the tectonic history and the evolution of the Himalaya, <u>Int. Geol. Cong. 22nd Sess.</u>, Delhi, <u>11</u>, 269-278, 1964.

Valdiya, K. S. and V. J. Gupta, A contribution to the geology of northeastern Kumaon with special reference to the Hercynian gap in Tethys Himalaya, <u>Him. Geol.</u>, <u>2</u>, 1-33, 1972.

Varadrajan, S. and A. G. Jhingran, Ophiolite along Indus-Brahmaputra Valley, Himalaya and their tectonic significance (abst.), <u>Int. Geol. Cong., 25th Sess.</u>, <u>3</u>, 702, 1976.

Verma, P. K. and S. K. Tandon, Geologic observations in a part of the Kamend District, Arunachal Pradesh (NEFA), <u>Hima. Geol.</u>, <u>6</u>, 259-286, 1976.

ISOTOPIC AGE DATA FOR THE EASTERN HALF OF THE ALPINE-HIMALAYAN BELT

A.R. Crawford

Department of Geology, University of Canterbury, Christchurch, New Zealand

Abstract. A review of isotopic age data for the Working Group 6 region (eastern half of the Alpine-Himalayan Belt) is given, with particular reference to the Himalaya. Data for the Himalaya are still few and of variable quality. Reliable Precambrian ages (so far, up to 1620±90 Ma, and perhaps 2000 Ma) are known. Controversy continues about the meaning of these ages in respect of Himalayan metamorphism and evolution. Late Cenozoic ages relating to uplift need refinement. Late Palaeozoic and Mesozoic ages seem mostly spurious but indications exist of events which need confirmation as they would greatly clarify concepts of Himalayan origin.

Ages for the remainder of the belt are still very much at the reconnaissance stage. In Iran attempts to date the known Precambrian have been largely frustrated by the complexity of the orogen. In Afghanistan evidence of Precambrian ages is accumulating.

Introduction

At the beginning of the Geodynamics Project, age determinations were correctly recognised to be likely to help greatly in the understanding of the tectonic evolution of the eastern part of the Alpine-Himalayan belt, and, in particular, of the Himalaya. This report attempts to review the state of knowledge at mid-1979.

Age data are given as reported, except that Rb-Sr ages have been recalculated, where necessary assuming the now generally agreed decay constant for ^{87}Rb of 1.42×10^{-11} yr^{-1}.

It will be immediately apparent that data are very few for so vast an area, extending from the Irano-Turkish border to Burma. No systematic study has been practicable during the period of the project. Most data have been obtained by geochronologists making analyses on samples obtained by field geologists, not usually collected specifically for age work.

Geological Documentation

Maps

Of maps on a scale larger than 1:10M and smaller than 1:1M there is none which is reasonably up-to-date for the entire region. *The International Geological Map of Asia and the Far East* 1971, which is unfortunately named as it excludes Soviet territory, is badly out of date. The corresponding *International Tectonic Map of South and East Asia* is still in preparation. The map in Gansser (1964) remains very useful.

For sections of the belt, usually national territories, geological and tectonic maps exist of varying scale and of different periods. For Iran the Tectonic Map 1:2.5M (Stöcklin & Nabavi, compilers) is invaluable and for general purposes supercedes the Geological Map on the same scale produced in 1957 by the National Iranian Oil Corporation. For Afghanistan a 1:2.5M map of the whole territory exists. A geological map, scale 1:500,000 exists for central and southern Afghanistan (Wittekindt & Weippert, compilers, 1973), and Desio (1964) produced a 'geological tentative map' of the western Karakorum. Pakistan is completely covered by a 1:2M geological map (Pakistan Geological Survey, 1964) and Kazmi has produced, but not published, a tectonic map on that scale. The Indian Geological Survey 1:2M geological (1962) and tectonic (1963) maps are now rather out of date, as is the 1:5M geological map. Maps of Nepal at present are mostly those of individual workers of teams; the 1:500,000 map of the Hokkaido group (Ohta & Akiba, eds, 1973) is useful. No satisfactory map of northern Burma exists. The 1:2M map of the Himalaya (Gansser 1964), if somewhat dated, is still extremely useful, and complemented by his 1:3.5M structural map (Gansser 1977).

Maps of a scale larger than 1:500,000 are not plentiful. There is no satisfactory coverage of 1:250,000 maps though substantial areas of Iran have geological maps on that scale and some exist for parts of the Himalaya.

Texts

No text exists for the whole area which gives comparably detailed coverage for each section. Useful succinct summary accounts covering much of it are in the volume *Mesozoic-Cenozoic Orogenic Belts/Data for Orogenic Studies* (Spencer, ed., 1974), comprising chapters by Falcon

(S. Iran), Stöcklin (N. Iran), Auden (West Pakistan-Afghanistan), Desio (Karakorum), Gansser (Himalaya) and Brunnschweiler (Indo-Burman Ranges).

Of more conventional studies, Gansser (1964) remains outstanding for a readable account of the Himalaya, including the Karakorum in their regional setting and in detail, with full references and superb illustrations. No comparably elegant work of this type exists for any part of the remainder of the belt, though de Lapparent (1972) is very useful. A complementary and invaluable account by Stöcklin (1977), which deals with all the region except the central and eastern Himalaya, presents an up-to-date analysis in terms of structural zones, and has a useful detailed map scale about 1:7.7M.

Subdivision: Problem Areas

Subdivision of Belt

Most of the region was discussed in an earlier contribution (Crawford 1977) dealing with Iran, Pakistan and India. Since then the only significant addition to the geochronological data has been in respect of the Himalaya. The 1977 contribution made the simplest of divisions of the region as a whole and, because it did not deal with Afghanistan, essentially summarised data for Iran and for the Himalaya.

Stöcklin (1977) has provided a much more sophisticated tectonic subdivision, which covers more than the region in that areas in Soviet Central Asia, and Chinese territory in Tibet are also mentioned, as well as the Arabian platform; on the other hand he extends his study only into the western part of the Himalaya. His divisions are so useful that they are used here as the major units. This division of the first order is:

(a) A Southern Domain,
i the forelands, or as I prefer to call them, the platforms, of Arabia and India;
ii the southern marginal fold belts, which include the Zagros, Kirthar-Soleiman Range, Potwar Plateau-Salt Range and Siwalik Range;
iii the Himalayan thrust sheets. By this Stöcklin means the entire Himalaya south of his Axial Ophiolite Belt, but excluding the Salt Range. Stöcklin regards the Himalaya as unique, with no westward or eastward continuation.

(b) The Axial Ophiolite Belt. This Stöcklin regards as dual in Iran, and connects with a comparable belt on the northern side of the Himalaya proper (which is, he thinks, following Stoneley (1973) and Gansser (1964, 1966) also probably dual).

(c) A Central Domain. This is the largest part of the whole orogenic belt Stöcklin considers; it is markedly heterogeneous, but shows sufficient unity to be regarded as distinctive. It includes fragments of Palaeozoic platforms in central Iran, central Afghanistan, Nuristan and the Pamirs, passing through the Karakorum Black Slate zone to its eastern continuation in Tibet.

(d) A Northern Domain. Stöcklin defines this as that part of the whole orogen in which Tertiary orogenic processes were superimposed on a Hercynian basement or an 'epi-Hercynian platform'. It is not easy to define its boundaries, but the domain includes the Alborz, the South Caspian Depression (a unique element), the Kopet Dagh, a northern fold belt including the West Hindu Kush, North Pamir and Kuen Lun; and a northern foreland which is necessarily divided into the Turan 'plate', the Tien Shan, the Tarim Block. The northern foldbelt overlaps the Palaeozoic foldbelt of Central Asia, preventing the drawing of a sharp boundary.

Problem Areas

Foldbelts such as the Zagros and the Siwaliks do not need isotopic age determinations. We are now adequately served by the data for the Indian platform, if rather less so still for the Arabian. The principal areas of interest in the Southern Domain are in the Himalaya. The Axial Ophiolite Belt is of the greatest importance, but much can be learned from the presence of fossiliferous sediments. Real problems exist in the Central Domain in Iran and Afghanistan and the entire area to the east; though access is extremely difficult to much of this because of physical and administrative difficulties. There are also major problems in the Northern Domain, particularly in the Alborz and in the complex area of northeastern Iran and western Afghanistan, and also in the Hindu Kush and eastwards.

The area between the exposed rocks of the Indian platform and the Himalaya, buried by the sediments of the Plains, is of great interest. Attempts by the writer to obtain samples suitable for age determination from basement material in drill cores were unsuccessful, as the limited quantity of material was too much altered. There are useful comments in Auden (in Spencer, 1974, p.251).

Definition of Himalaya

The Himalaya can now be defined as extending eastwards for 2500 km from the major NW-SE fault zone between Kabul and Jalalabad to a system of parallel faults, of the same trend, which lie immediately beyond the great bend of the Dihang-Brahmaputra, and which include the Mishmi Thrust.

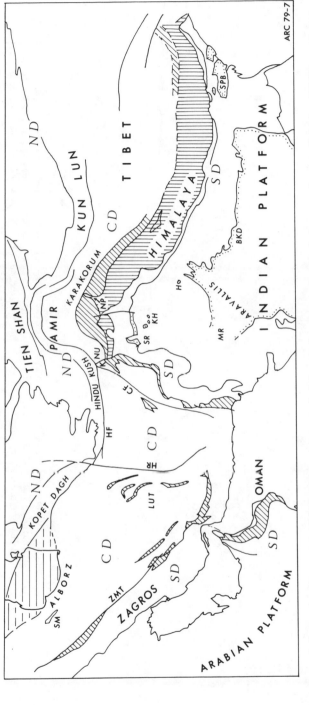

Fig. 1. Alpine-Himalayan Mountain Belt, eastern half. Main tectonic divisions, after Stöcklin (1977). ND, Northern Domain. CD, Central Domain. SD, Southern Domain. Axial Ophiolite Zone diagonally hatched (associated flysch of SE Iran and Pakistan not shown). Himalaya, vertically hatched. Indian Shield boundary, lined with dots. BKD, Bundelkhand. CF, Chaman-Arghandeh Fault system. H, Hissar hills. HF, Herat Fault. HR, Harirud Line. K, Kabul. KH, Kirana Hills. MR, Malani Volcanics region. NP, Nanga Parbat. NU, Nuristan. SPB, Shillong Plateau Block. SM, Soltanieh Mountains. ZMT, Zagros Main Thrust.

Fig. 2. Locality Map for Himalaya, mainly after Gansser (1964). Sub-Himalaya (Siwaliks),
diagonal hatching. Tethyan Himalaya, horizontal hatching. Lower and Higher Himalaya not
separately distinguished; black line indicates approximately the division. Am, Amlang La.
Al, Almora. As, Askot. Bh, Banihal Pass. Ch, Chandra Valley. Hz, Hazara. Ka, Kathmandu.
Kg, Kalimpong. Ki, Kiogar. Mg, Mustang. La, Lansdowne. Rh, Rhotang Pass. Pi, Pithoragarh.
Th, Thakkola.

It is conventional to regard the Siwalik
Ranges as Sub-Himalaya and the first of
the great zones which extend along
the length of the Himalaya.
To the north of this narrow marginal fold belt
(equivalent to the Alpine molasse), and beyond a
'Main Boundary Thrust', it is common practice to
distinguish a Lower (Lesser) Himalayan zone
structurally separated from a Higher Himalaya
by a 'Main Central Thrust'; and beyond that,
but as a stratigraphic zone with sequences the
base of which often cap the high peaks of the
zone to the south, a Tethyan Himalaya. These
vary in width along the length of the mountain
belt, and in the northwestern section the
division is rather less clear. Elsewhere it is
prominent; and in comparison with other great
mountain belts the division is remarkably
definite.

A sectional division is also usual, for con-
venience: Punjab Himalaya (which includes
sections in Afghanistan, Pakistan and India);
Kumaon Himalaya, with the Sutlej as the boundary;
the Nepal Himalaya; Darjeeling-Sikkim Himalaya;
Bhutan Himalaya; and the easternmost section
which used to be called Assam, then NEFA, and
now Arunachal Himalaya, from changes in the
names and boundaries of the eastern provinces
of India.

Main Problems of Himalaya

As Gansser (1964) emphasised, the main
problem in the Himalaya is the age of the
metamorphism. This is made very difficult by
repetition of thrusting in monotonously similar
rocks apparently almost completely unfossiliferous, though stromatolites have been recognised fairly widely in recent years. As time
markers they are not agreed to be completely
reliable.

Most of the difficulty is in the very complex
and still enigmatic Lower Himalaya. The Tethyan
Himalaya, if remote to workers coming from the
south, and in part in Chinese territory, lend
themselves to conventional, if arduous, stratigraphic mapping. The marine sequence is
complete and superb. Problems arise in
connection with granitoid and other intrusions,
often sill-like, and with what has commonly
been mapped as high-grade metamorphic rock in
the very thick (20-30 km according to Gansser)
basal 'slab' of crystalline rocks, the 'Tibetan
Slab' of the High Himalaya of such authors as
Bordet (1977). These rocks are sometimes
sandwiched between others of much lower grade
and which are apparently Upper Precambrian
sediments.

But in the Lower Himalaya, unlike the
Higher, it is rarely possible to pass down
through Phanerozoic sequences conformably
into Precambrian without structural breaks.
The extraordinary frequency with which thrust
planes coincide with stratification makes recogn-

ition often very uncertain. For many years it
was generally accepted that there was one
reliable marker, at least: the Blaini Boulder
Bed, which was regarded as the Himalayan equivalent of the Peninsular Talchir glacial boulder
beds of the Upper Paleozoic. This is now suspected by several workers to be misidentified
and thereby unreliable (Rupke 1973). The Lower
Himalayan problem is not just one of metamorphism
in a repeatedly-thrusted, folded sequence of
monotonously similar rocks of low grade which
probably span from well down in the Upper
Precambrian to well up into the Palaeozoic.
There is also the problem of reversed metamorphism. While in the High Himalaya the grade
usually decreases upwards, in the Lower there
are many areas where high grade rocks overlie
lower. Some can be shown to be overthrust
remnants. Others appear to lack structural
breaks. Yet the lower sequence is not itself
reversed. The Darjeeling area presents the
classic example (Mallet 1875; von Loczy 1907;
Gansser 1964).

This problem remains difficult. There is a
useful discussion by Le Fort (1975), who pro-
poses thermal metamorphic effects from frictional
heating of a descending slab (also this volume).

Since Gansser's publication of 1964, several
workers have discovered evidence of multiple
metamorphism in the Lower Himalaya. This has
made matters even more complicated, although
structural geologists appear to agree that four
phases are characteristic of the entire Himalaya,
perhaps excluding the unique Nanga Parbat area.

It is also desirable to find the oldest rocks
in the Himalayan sequences and relate them to
the metamorphic history; and to date the
intrusives. Curiously, the Lower Himalaya show
rather few intrusive granitoid rocks, and these
are not well documented. There is a marked
absence of batholiths; bodies are local. In
the Higher Himalaya there are much larger areas
of granite, mainly tourmaline-rich, all of
which appears to be Cenozoic. Volcanic rocks
are also uncommon in the Lower Himalaya, though
in the older sequences they are difficult to
identify and have been noted particularly in
Kumaon. The prominent Panjal Traps of Kashmir,
and the less-known Abor Volcanics of Arunachal
Himalaya are in a different category as these
Phanerozoic rocks are clearly associated with
stratigraphically identifiable horizons. So
too are the basic and ultrabasic rocks of the
'exotic zone' of the Tethyan Himalaya.

Although isotopic age data are intrinsically
useful, no attempt to date the sequences, deformations and metamorphism is worth making unless some
hypothesis for the formation of the Himalaya
exists, to which the combined isotopic age and
other data can be applied in the hope of verifying
it or falsifying it. Many workers agree that the
Himalaya as a whole are not derived from a former
geosyncline. Only the Tethyan zone is regarded
as originally geosynclinal by this group; the

Lower Himalaya rocks, and much of the material in the basal part of the Higher Himalaya, being modified parts of the Peninsular Platform; with, it is generally agreed, considerable crustal shortening because of the numerous thrusts. It follows that as that Platform includes rocks of various ages and degrees of metamorphism, any Precambrian metamorphism in the Himalaya has to be looked at as possibly entirely Peninsular, perhaps occurring long before the Himalaya existed.

Some workers think in terms of a long-persistent geosyncline to the north of the Peninsula. On this hypothesis the Precambrian rocks in the Himalaya, although once in stratigraphic continuity in the broadest sense, were of different facies and not comparable with those even in the northern part of the Peninsula. These workers reject the hypothesis of continental drift and collision. They think in terms of a 'proto-Himalaya' which was in existence even in Precambrian times, with a geanticline now forming an 'axis' and represented by the crystalline rocks in the central part of the broad mountain belt.

It is because of such differences of approach to the fundamental problem of origin that any interpretation of the isotopic data in respect of ages of metamorphism is apt to produce controversy.

The valuably provocative and at present extremely fashionable hypothesis of plate tectonics has many advocates who cite the Himalaya as the most impressive example of a collision-type mountain belt. Many of these advocates are unfamiliar with the Himalaya. This concept may be correct, but it is heavily dependent upon what Carey (1976) has described as an obsession on the part of the English-speaking geologists with compressional orogenesis. It is also desirable to remember that to interpret the Lower Himalayan and other crystalline rocks as non-geosynclinal does not necessarily imply acceptance of a collision origin. It is fairly plain that there remains much uncertainty about the origin of the Himalaya, at least among those who work there.

Punjab Himalaya

Few data yet exist for the area west of the Indus, such as the Safed Koh, in particular, which are geologically very poorly known. For the extreme west, on the boundary of the Himalaya, four samples from the great disturbed zone in the gorges of the Kabul River, collected by the late A.F.de Lapparent, were analysed by Rb-Sr as total rocks. These banded biotite gneisses gave model ages of 640-680 Ma (for three samples) and 470+20 Ma. All are well enriched in radiogenic strontium. It has not been possible to identify exactly where the samples lay in relation to structures, and further samples have yet to be analysed, but it seems very probable that the age is Precambrian, perhaps with one rejuvenated (Crawford, unpub. data).

East of the Indus, mapping is much further advanced (see Calkins et al. 1975 for detailed references). The Hazara Group is now shown to be at least in part Precambrian, Rb-Sr total-rock ages of 740+20 and 930+20 having been found (Crawford and Davies 1975). The extensive non-foliated Mansehra Granite has recently been dated by the workers at the Rennes Laboratory and has an age of about 510 Ma (Le Fort, pers. comm.).

It is regrettable that no data exist for the unique Nanga Parbat massif, the region of the classic investigations of Misch (1949). He recognised only one metamorphism, that of the Salkhala Group, which outcrops extensively and is generally thought of as the oldest part of the sequence. Gansser (1964) pointed out that as grade increases along the strike of these steeply-dipping rocks, the metamorphism is unrelated to burial, but to the early Tertiary granitization. The area is one of abnormal height, suggesting very young uplift, and it appears to be tectonically distinct from that west of the Indus, which here cuts southwards across the Himalaya.

The rest of the Himalaya in Kashmir are untypical, with a basic of Tethyan sediments separated from the Sub-Himalaya only by the narrow Pir Panjal. The Salkhala and succeeding groups of rocks were sampled for Rb-Sr work, but in general are poorly enriched in radiogenic strontium; a suite of 14 Dogra Slates samples from the Banihal Pass road between the pass and the Chenab, analysed as total-rocks showed bad scatter and poor enrichment in radiogenic strontium, making age assessment difficult. The rocks are probably no older than about 800 Ma.

From the crystalline areas between the Tethyan basins of Kashmir and the larger one, the southern part of which is the famous Spiti region, age determinations have been made only in small areas, though these have been selected as crucially important for structural studies by a number of workers. An accumulation of isotopic age data of varied value exists. This is by K-Ar and Rb-Sr and is from several different laboratories.

Bhanot et al. (1974) obtained an apparent age of 633+100 Ma for a single sample of gneiss from near the Rohtang Pass, at 32°17', 77°10', with a present-day $^{87}Sr/^{86}Sr$ ratio of 0.80+0.02. This ratio is too low to permit placing much reliance on the apparent age. Though the real age is unlikely to be Cenozoic, Bhanot et al. suggest that the rock could be of similar age as the Mandi Granite (see Definition of Himalaya; Rb-Sr total-rock isochron age of 500+100 Ma (Jäger et al., 1971)).

Pande and Kumar (1974) quote K-Ar biotite ages (Soviet IGEM laboratory analyses) for the same region and the area 25 km to the north. These are mainly Cenozoic, including the age for a migmatite from the Rohtang Pass (38+3 Ma). The Jaspa Granite 25 km north gave 43+3 and 54+3 Ma, and a migmatite beyond its northern margin, 22+2 Ma. To the south of the Rohtang Pass ages

of 61±3 and 38±3 were obtained for biotites
from a schist and a migmatite, and 76±5 for a
migmatised biotite schist. However a migmatised
garnetiferous mica schist east of Kothi gave the
very different age of 756±20 Ma.

Frank et al. (1973) and Frank (1977) gave a
total-rock Rb-Sr isochron of 512±16 Ma for
granites in the gneisses and schists immediately
south of the Chandra Valley. These are distinguished by high initial $^{87}Sr/^{86}Sr$ ratios, which
suggest an old crustal origin. From the Central
Gneiss of Kulu these workers obtained 16.5±2 Ma
for 10 biotites.

Mehta and Rex (1977) obtained a suite of K-Ar
muscovite and biotite ages for the Central
Gneiss of Manali-Rohtang Pass. These number
eight and range from 20.1±0.9 to 35.9±1.8 Ma.
For Kulu, biotite ages of migmatitic gneiss,
schist and amphibolite range from 50-75 Ma,
with a 30 Ma muscovite. The total rock analyses
gave 277±8 and 315±9 Ma.

Mehta (1977) made numerous Rb-Sr analyses of
samples from the Central Gneiss of the same
area and from migmatitic gneisses of Kulu. The
Rohtang Pass-Manali rocks gave a five-sample
total-rock isochron of 600±9 Ma, with muscovite
ages of 25±1 to 29±1 Ma, and biotite ages of
17-16.5 Ma. The Kulu gneisses gave a 5-sample
isochron of 517±8 Ma, while the muscovite (one
sample) gave 21.2±1 and three biotites range
from 10.7-14 Ma.

Following stratigraphic and structural work by
Powell and Conaghan (1973), Powell et al. (1979)
gave data for seven samples from the Himalayan
Central Gneiss of Lahaul. These were analysed
by Rb-Sr and include total-rocks, muscovite and
biotite. The total-rock suite does not give a
satisfactory isochron, but the data indicate
strongly a late Precambrian age for at least
some of the group. Of the mineral ages, the
biotites combine with their total rocks to give
Cenozoic ages of 16-19 Ma and 26 and 12 Ma.
Total-rock-muscovite pairs from two samples give
late Palaeozoic ages of 345 and 390 Ma. The
biotite-total rock ages of 16-19 agree with those
of Frank et al. (1977) from Kulu.

Nagpal et al. (1973) made a fission-track
analysis of an apatite from the Dharamsala
granite, which gave 4.7±1.1 Ma. Nagpal and
Nagpaul (1975) reported fission-track ages of
muscovites from Himachal Pradesh at 20±5 Ma, 38±
24 Ma and 29±11 Ma.

The accumulation of these data has not wholly
clarified our understanding of the Punjab
Himalaya. This is plainly to be seen from the
lengthy discussion by Powell and Conaghan (1978)
and Mehta (1977, 1978). Mehta argued that the
data reflect at least three major events in the
evolutionary history of the Himalaya. The
large-scale recumbent folding, high-grade metamorphism and synkinematic granite intrusion
took place, according to him between 600 and
500 Ma ago. This event produced a 'protoform'
of the Central Crystalline axis, later rejuven-

ated. There was then a "Hercynian Magmatic-
Epeirogenic Cycle", indicated by ages of 360-
290 Ma. The 75-10 Ma mineral ages are those of
the 'Himalayan Orogeny', which is associated
with open folding, low-grade metamorphism, uplift,
thrusts, nappes and 'regional retrogression'.

Powell and Conaghan argue that the pre-Cenozoic
(or pre-late Mesozoic) ages do not relate to the
major folding and Himalayan regional metamorphism
but to granites incorporated in thrust sheets
where regional metamorphism was weak and affected
only the biotites. Palaeozoic muscovite ages
they regard as reflecting age of crystallisation.
They insist that the earliest folding in the
Chandra Valley is post-Callovian, as it deforms
fossiliferous rocks near Tandi (Pickett et al.,
1975), and that this folding has been traced
by Frank et al. (1973) to the limit of the metamorphic terrain. They do not deny the existence
of old granites and metamorphism, but they do
not accept that the metamorphism is regional
and related to the Himalayan orogeny.

These matters are discussed further in the
final section.

Kumaon Himalaya

The Kumaon Himalaya extend for 320 km from the
Sutlej River to the Nepal border on the Kali
River. The four main divisions are all present
here: the sub-Himalayan Siwalik belt with a normal section of Cenozoic rocks, sharply distinguished from a tectonically very complex Lower
Himalaya still with many problems of stratigraphy
unsolved, and a comparably complex High Himalaya
beyond the Main Central Thrust. The Tethyan
Himalaya of Kumaon are famous for having both a
normal sequence of Phanerozoic rocks from
Ordovician to Cretaceous and an 'exotic' region,
with material which is probably derived from the
north, and thought to be relics of a vanished sea
floor (Gansser 1964).

The region is one where the first detailed
mapping of the Himalaya was started in the 1930s.
That it is still the subject of controversy is
an indication of the difficulties of interpretation. There is a voluminous literature (see
Valdiya, this volume; or Fuchs & Sinha, 1978).

Age dating of the Lower Kumaon Himalaya is at
an early stage. Very limited work has been done
so far. Jäger et al. (1971) analysed the Chor
granite and the Mandi granite (the latter
strictly lies in the Punjab Himalaya as it lies
just north of the Sutlej) by Rb-Sr. Only one
sample was from the Chor granite, a biotite
which gave 50±10 Ma. Four total-rock samples of
Mandi granite gave 500±100 while three biotites
gave 24-31 Ma. The Mandi granite was regarded
by Fuchs (1967) as the front of the crystalline
sheet.

Ashgirei et al. (1975) quote five K-Ar ages
of Simla area rocks. A biotite-chlorite schist
gave a total-rock age of 23 Ma; a chlorite-
graphitic schist, 42 Ma. Three total rock

analyses of a biotitic, a muscovitic and a biotite-tourmaline 'Boileaugunge quartzite' (i.e., Jutogh Series rocks) gave 86, 120 and 123 Ma respectively. The same authors quote three K-Ar ages ranging from 28+2 to 89+6 Ma.

Ashgirei et al. also mention ages of the Chor granite (total rocks, 295+13 Ma) and a granitized Chail schist (biotite, 338+0 Ma(!)) which appear to be by K-Ar, as a doubt is expressed that these ages may be too high.

Sinha (1975) quotes K-Ar ages of Soviet origin in a paper on magmatic rocks in the Simla Hills region. It is difficult to understand the data, which include some from other work, but he appears to have one age of 42+2 Ma for an amphibolitized gabbro-diabase from the Giri River, in the Shali Formation; and another of 410+10 for amygdaloidal diabase from Darla Ghat (Sutlej River); and a third for a similar rock from the same locality, which is 1190+35 Ma. From these and other data he argues for three magmatic episodes: one Precambrian ('Riphean'); a second from 140+10 to 228+10, i.e. late Palaeozoic; and a third in the early Tertiary.

Bhanot et al. (1976) dated granitic porphyroids (Koidal gneiss) at 100-1200 Ma which Valdiya (this volume) regards as intrusive into a Ramgarh metasedimentary group in a nappe beneath the Almora group.

Sinha and Bagdasarian (1977) made 10 K-Ar analyses of a variety of samples from Kumaon. Those of the Lower Kumaon Himalayan rocks include a diabase from the Krol Series at Naini Tal which gave 51 Ma. It was regarded by them as representative of the last phase of major Himalayan orogenic movement. Reports of similar ages are said to be in press. Another basic igneous rock from the Yamuna Valley at Kuwa gave an age of 710 Ma and is compared with an age of 1190+35 Ma for a similar rock from the Simla Hills. A mylonitized granitoid intruding phyllites at Chandpur gave 325 Ma and an altered quartz diorite from Kuwa, 414 Ma. Sinha and Bagdasarian compare these with ages of 410 and 338+12 for similar rocks from Simla. A mylonitized granite from Barkot in the Yamuna Valley gave 325 Ma which the authors compare with a 338 Ma age of a metamorphosed rock from the Chail Series near Pandoh. All are regarded as evidence of a Hercynian cycle of magmatism and metamorphism. They mention also a 191 Ma age for a gneissose quartz porphyry between Pauri and Lansdowne, occurring within phyllites, which they stress shows no contact. They regard the dated rock as probably a metamorphosed acid volcanic.

Powell et al. (1979) give an isochron from analyses of four Rb-rich total-rock samples of intensely foliated and lineated gneiss of the Almora-Askot area. This gives an age of 1620+90 Ma, with an initial $^{87}Sr/^{86}Sr$ ratio of 0.749+0.007. Micas from one of the samples give somewhat uncertain results but suggest Cenozoic resetting.

A suite of 14 samples, including some of Simla rocks, all collected with the help of geologists of the Geological Survey of India, were analysed as total-rocks by Rb-Sr (Crawford, unpub. data). These gave maximum apparent ages of 930-980 Ma. Assessment of real age is difficult, as the present-day $^{87}Sr/^{86}Sr$ ratios are all less then 0.85, with the exception of one sample collected from the Chail Series outcrop, which has 1.0729. This suggests that its age is about 650 Ma, but there is no certainty that the Simla Slate rocks are in fact older.

Fission track ages also exist. Nagpal et al. (1973) gave one for the apatite of the Mandi granite of 36.0+1.2 Ma, contrasting with one for apatite from the Chor granite of 15+0.2 Ma. Nagpal and Nagpaul (1975) examined a muscovite from the Chor granite, which gave 48+24 Ma. Other muscovites from pegmatites in the Chor area near Naura gave a mean age of 38+24 Ma. These indicate a limited value for the technique.

The reliability of these data varies greatly, and the data are of very uneven value. Many are poorly documented. There is a great range of apparent ages, often of uncertain meaning.

We have almost no isotopic age data for the Higher Kumaon Himalaya. Bhanot et al. (1977) dated two total-rock samples of granitic augen gneiss from within the main body of metasediments of amphibolitic facies in the Munsaru area of Pithoragarh (Kalamuni Pass, 2 km north of, at 30°3'N, 80°12'E). This is a 'Central Crystalline Gneiss'. They gave a two point Rb-Sr isochron age of 1830+200 Ma, with an initial $^{87}Sr/^{86}Sr$ ratio of 0.725. This is scarcely an isochron but at least indicates a probable Precambrian age.

For the Tethyan zone in Kumaon very few age determinations have been made. Seitz et al. (1971) gave a K-Ar age for the tourmaline granite of the Arwa Valley, Garhwal (part of the Badrinath suite) as 18 Ma. Sinha and Bagdasarian (1977) have however made K-Ar analyses of spilites associated with the 'exotic blocks' of the Balcha Darwa Pass area. These are the first such analyses. Of five rocks collected, only three proved suitable for analysis. They gave 107, 75 and 73 Ma. The ages are appropriate in so far as the igneous rocks are closely associated with fossiliferous sediments with cephalopod and ammonoid faunas. Farther to the southwest an aplitic granite, rich in tourmaline as is common, and from Malari gave 73 Ma. The authors regard the contemporaneity of this and some of the exotic spilites as significant, as to them the granite is to be regarded as associated with a deep fracture zone dividing the central crystalline zone from the Tethyan Himalaya.

Nepal Himalaya

The long 800 km stretch of Nepal Himalaya was almost entirely unknown until 1950. Since then much work has taken place but the usual difficulties in the Lower Himalaya have led to a variety of interpretations and it is still too early to be able to outline satisfactorily the geology, which presents problems exactly comparable with those of Kumaon and the Darjeeling area. Main references are Bordet (1961, 1977), Bordet et al. (1967), Gansser (1964), Frank and Fuchs (1970), Bodenhausen and Egeler (1971), Ohta and Akiba (1973) and Le Fort (1973, 1975).

Krummenacher (1961) made the earliest of Himalayan age determinations, on rocks essentially from the Higher Himalaya. One sample from the Nawakot nappe was a quartzite from which muscovite thought to be detrital was analysed by K-Ar and gave 728+12 Ma. From this result Krummenacher inferred a Precambrian metamorphism. All his other analyses, numbering 14, are of biotites and gave ages of less than 20 Ma; these were of gneisses, schists and granites, and one diorite and one arkose collected from the Kathmandu and Khumbu nappes and the 'Tibetan Series' (i.e. 'Slab'). Krummenacher argued that the end of the Himalayan metamorphism in Nepal was in the late Miocene or early Pliocene at about 13 Ma. Gansser (1964) expressed interest in the apparent absence of any metamorphism between this and the Precambrian one postulated by Krummenacher. Krummenacher made 25 more analyses (Krummenacher 1966) of rocks from central Nepal. These are of material from samples from the Siwalik Sub-Himalaya along a section across the whole belt to the Mustang granite in the far north on the Tibetan border. The one Siwalik analysis is of a detrital biotite in a micaceous sandstone and gave 18.5 Ma. From the middle part of Nepal eight samples came from sedimentary, metamorphic and igneous rocks. Of these, the oldest (1280 Ma) was a detrital sericite from a carbonaceous schist near Tensing; a detrital muscovite from the base of the Kunchla Series gave 872 Ma; a uralite from an eruptive rock gave 819+80 Ma. These confirmed Krummenacher in his belief in a Precambrian metamorphism, though it was no more clearly defined, he thought, than between 1250 and 800 Ma. In the same group of samples he obtained an age of 354 Ma for a microcline in a granite pebble in schists 500 m from the 1280 Ma sample site, and 381 Ma for phlogopite from a marble 1 km north of a contact with the Phalung granite. This granite, between Kathmandu and Etora, gave a biotite age of 48 Ma and the same age was given by phlogopite only 500 m from the contact.

From the 'écailles de Tatopani-Dana', Krummenacher obtained ages of 152 Ma (phlogopite from crystalline limestone), 126 Ma (biotite and fine sericite, carbonaceous schists), 53 Ma (biotite from base of gneiss sequence between Tatopani and Pokhara) and ages of 22, 19 and 15.5 Ma from minerals in the same area.

From the 'Tibetan Series' detrital sericite gave 291, 214, 119 and 114 Ma. A detrital muscovite gave 147 Ma, this being from a Lower Cretaceous sandstone, and a glauconite from an Aptian-Albian sequence gave 139 Ma. The rest of this group gave much younger ages. A muscovite from the base of the 'Infracambrian' crystalline limestone 3 km south of Tukche gave 24 Ma; another from a pebble of the granite itself gave 15 Ma.

Krummenacher regards the 354 Ma age as due to argon loss, and the rock as Precambrian. The 381 Ma biotite age of the Phalung granite and the 48 Ma phlogopite ages from the intruded rocks he regards as 'mixed ages', influenced by two successive metamorphisms, as he does the 152, 126 and 53 Ma ages from the 'écailles de Tatopani-Dana'.

In respect of the Tibetan Series he points out that the minimum age of a detrital mineral must be that of the sediment in which it is found. Thus the 119 and 114 Ma ages for example, which are from Triassic rocks, cannot be those of the sediments and must have been those of rejuvenation. This applies also to the 219, 214 and 147 Ma ages, which he implies are not as old as they should be.

The difference between the Mustang granite ages he attributes to the fact that the pebbles both came from terraces and the older pegmatite pebble from a high terrace and the younger granite pebble from a 'very recent' terrace.

For these reasons Krummenacher suggests only two metamorphisms are certainly detected: the ill-defined Precambrian one, and one between 25 and 15 Ma. He points out that this Himalayan metamorphism is slightly older than that in the Everest region to the east which is between 18 and 10 Ma.

Bodenhausen and Egeler (1971) refer to age determinations made for them by Priem, not fully reported. For samples from the Dumphu gneiss, thought by the authors to be Precambrian and part of the Annapurna Complex, an 'average for muscovite' was 12.9+1 (Rb-Sr) and 12.4+1 (K-Ar), while biotite gave 6.7+1 (Rb-Sr) and 16.0+3 (K-Ar). The total-rock gave an Rb-Sr apparent age of 483+24 Ma, and the authors regard this age as one affected by loss of radiogenic strontium during Himalayan metamorphism.

The leucogranite of Manaslu in central Nepal (Le Fort 1973) was dated very accurately by Hamet and Allègre (1976, see also Bordet et al. this volume), who obtained an Rb-Sr isochron of 27.4+0.5 Ma, i.e. late Oligocene. Two aspects of this work are important: that the K-Ar mineral ages for the same granite (Krummenacher 1971) are younger (14-17 Ma) and that the initial $^{87}Sr/^{86}Sr$ ratio is 0.7408, 'the highest recorded for this massif'. Hamet and Allègre therefore suggest quick cooling and a remelted origin. Fourcade et al. (1977) reiterate this argument, following geochemical work.

It should be noted that Kumar et al. (1978) give evidence of an angular unconformity within

the Precambrian in central Nepal and suggest a
Precambrian deformation.

Darjeeling-Sikkim-Bhutan and Arunachal Himalaya

Though the western part of this section was
explored very early (Mallet 1875) and is famous
for von Loczy's delayed publication (1907) of
the Kanchenjunga-Darjeeling section as a great
recumbent fold, no modern map of the whole part
exists. For further east Gansser, who is almost
the only worker, gives the best possible
account (Gansser, 1964).

For Darjeeling, Auden (1935) is a useful
reference. Mukopadhyay and Gangopadhyay (1971)
detected multiple deformation in the Kalimpong
district, which if widespread, as seems likely,
still further complicates the problem of reversed
metamorphism. Roy (1974) has also discussed
the problem.

There are no age determinations fully reported.
Acharyya (1973) refers to 10 analyses by K-Ar,
done by Hamrabaeb and colleagues on his behalf,
of samples from the 'frontal zone of the
Darjeeling-Duars area'. These gave 35-40 Ma.

In 1967, with the aid of S. Saha of the
Geological Survey of India, I collected 18
samples from the Darjeeling area. These range
from the least-metamorphosed Daling Schists
collected at the lowest possible level to the
high-grade gneisses at the highest level near
Ghum; some were collected also along the
Darjeeling-Tista River bridge road. Analyses
were made by Rb-Sr on both total-rocks and
minerals, though no satisfactory minerals could
be extracted from the least-metamorphosed
rocks. The results are extremely interesting
but the data need replication. Five total-rock
muscovite pairs from the highest topographic
levels (and thereby highest grade of metamorphism) indicate an age of about 15 Ma.
This was perhaps to be expected. More curiously, five total-rocks (two from low-grade
schists, as near as possible to the base of the
exposed Dalings, and three gneisses) fit a
straight line suggesting an isochron at about
180-190 Ma, i.e. late Triassic. There is a
further less satisfactory three-point fit of
total-rocks from the lowest levels near the
Tista River; these give an age of about
840 Ma, but with a fairly high present-day
$^{87}Sr/^{86}Sr$ ratio of about 0.87. Each of these
three samples could be regarded as having a
considerably higher age as their present-day
ratios are 1.01, 1.02 and 1.47. The last of
the three gives a model age of 1130 Ma (at
0.700) and 1100 Ma (at 0.720). Unfortunately
the highest grade rocks when analysed are
individually uninformative because of low
present-day ratios, and at the same time it is
questionable to what extent the samples may
validly be grouped to make a statistical population. One muscovite analysis, from a muscovite
schist in the low-grade rocks, gave an age of

2460 Ma at 0.700, which reduces to 2090 at
0.750. Whatever its real age this appears
likely to be an inherited one, as does that
of a muscovite from a sample collected nearby
which gave 1080 (1010) Ma (Crawford, unpub.
data).

Lack of a published detailed geological map,
the need to repeat the analyses and preferably
to complement the suite of samples, makes it
undesirable to draw firm conclusions about the
principal problem, the age of the metamorphism.
Moreover, it would be wise to combine further
work with structural studies in this crucially
important area.

Northwards the Darjeeling gneiss pass into
the Higher Himalaya, forming in the northwest
the synclinal massif of Kanchenjunga, consisting
mostly of granitic augen gneisses. Beyond are
carbonate rocks of the Tethyan sequence. In the
northeast Auden (1935) noted a predominance of
gneisses, intruded by tourmaline granites which
are apparently post-orogenic. In the northernmost area the sediments included some regarded
by Auden as of Gondwana facies.

No age data are available in fully reported
form. According to Acharyya (1973) two samples
of augen gneiss from the Paro Formation (locally,
Chungthan Formation) from East and North Sikkim,
analysed by K-Ar by Hamrabaeb and his group,
as total-rocks gave 178-213 Ma, and 'disclose
a broadly Triassic phase of metamorphism'. For
Bhutan Gansser (1964) quotes three K-Ar mineral
ages, as preliminary determinations made on
his behalf by Siegner. A muscovite quartzite
from north of Paro yielded a muscovite which
gave 9+2 Ma; another muscovite from a
muscovite pegmatite west of Tongsa gave 12+ Ma,
and a biotite from a pegmatite at the same
locality gave 11+2 Ma. These appear to be the
only data. It is noteworthy that the Soviet
K-Ar total-rock age for two samples of augen
gneiss from the 'Paro Formation' is older.
There appear to be no age data for the still
poorly-known Arunachal Himalaya.

Summary and Discussion of Data for Himalayan Rocks

The data are still very few, are of very variable quality, and the lack of attribution of many
items to some established rock sequence makes for
difficulties both in summarising and in interpretation.

First, reliable Precambrian ages have been
found. Some of these are of low-grade metasediments, but can be accepted as those of the time
of formation of the rock. It seems probable that
along the entire length of the Lower Himalaya
there are Upper Precambrian rocks. The oldest
reliable age is for the Kumaon Almora-Askot
gneisses, 1620+90 Ma (Powell et al., 1979). These
rubidium-strontium ages accord with the 1000-
2000 Ma age for the granitic porphyroid intrusions
from the same area. It is also relevant that for
the 'Central Crystalline Gneiss' of the Pitho-

ragarh region of the Kumaon Higher Himalaya Bhanot et al. (1977) obtained an Rb-Sr age of 1830±200, though this is from only two analyses and hardly an isochron age.

The time span of the Precambrian rocks of about 1000 million years to those who regard the Lower Himalayan material as largely Peninsular suggests that we might hope to find still older representatives in the Himalaya. They are known to be as old as 2600 Ma in the northern part of the Peninsula exposures (Crawford 1969). It would be reasonable generally to expect the oldest to be more likely to be found in Kumaon and western Nepal rather than elsewhere. That the oldest age so far found in the Punjab Himalaya is less than 1000 Ma accords well with the absence of rocks older than that west of the Aravalli belt of the Peninsula. In that belt and Bundelkhand there are very old rocks, perhaps even older than 2600 Ma in Bundelkhand, though not yet dated isotopically. The Aravalli belt itself has a suite of rocks ranging from 2600 to about 700 Ma. In the Darjeeling Himalaya and areas east, it might be reasonable to expect ages of those rocks in the Eastern Ghats Belt (which could be as old as 2600 Ma, or more) but the proximity of the Shillong Plateau block suggests that the ages to be expected in the nearby Himalaya would be rather younger. Though we have no detailed knowledge of the block itself, the granites there (and those in the easternmost part of the main Peninsula outcrop in Bengal) are about 750 Ma old, and the metamorphic basement is probably less than 1200 Ma old.

If these Precambrian ages are regarded as those of metamorphism then their interpretation in terms of orogeny becomes a question of whether they are associated with Himalayan deformation. This is a matter to be solved by field work.

The second conclusion is that plenty of evidence exists about the Cenozoic age of the High Himalayan (and Tethyan Himalayan) granites, the ubiquitous tourmaline-rich granites which are a conspicuous feature. We do lack a close study of even one intrusive body in relation to its contacts: such a study would be valuable and should be made using both Rb-Sr and K-Ar techniques on the same rock samples.

There is sufficient evidence to suggest that some granitic bodies in the Lower Himalaya have early Palaeozoic ages, and these ages also come from 'Central Gneisses' (Jäger et al., 1971; Bhanot et al., 1974; Frank et al., 1973). Such ages are not usual in the Peninsula as those of actual intrusion, except of some pegmatites in the Aravalli belt. However, among the Malani suite of acid volcanics and peralkaline granites, which as a whole has a well-established age of 745±10 Ma, a single, well-enriched sample of Jalor granite was much younger, its maximum age being 430 Ma (Crawford 1970). The Malani granites are non-orogenic. Are the

Himalayan ones relics of late-intruded parts of the suite? The Mandi granite with an age of 500±100 (Jäger et al. 1971) is exactly on prolongation northeastwards, parallel to the Aravalli trend. If so, then they are surely incorporated in the Himalaya rather than related to its formation.

Ages between about 400 Ma and 20 Ma seem to include most of those either of doubtful quality or inadequately reported, and those affected by isotopic loss or gain. There is little point in discussing most of them. Two aspects however do need comment. One is the limited evidence for ages of between 360 and 290 Ma (Ashgirei et al. 1975; Sinha 1975; Mehta and Rex, 1977; Sinha and Bagdasarian 1977; Powell et al. 1979 in press). As this is regarded by some workers, e.g. Mehta, as evidence for a 'Hercynian' event, it needs further work.

The other is the unfortunately indecisive evidence, as yet, for ages of about 180-190 Ma in several sections of the Himalaya; Kumaon, Lansdowne area (Sinha and Bagdasarian 1977); Darjeeling (Crawford, unpub. data) and north and east Sikkim (Acharyya 1973). Three of these can be thought of as possible metamorphic events affecting older rocks. If correctly dated as late Triassic this is very important, for such a 'Cimmerian' event is common in the Northern Domain and in the Central Domain (Stöcklin 1977) and in the latter has been found isotopically (Crawford 1977 and below,"Western Sector"). Stöcklin regards the Southern Domain as lacking it. Demonstration of its existence in the Himalaya would provide a strong argument for their development by processes other than Cenozoic collision.

Axial Ophiolite Belt

This belt is regarded by Stoneley (1974) and Stöcklin (1977) as dual. The eastern part in Chinese territory is not well-known. The remainder consists of a western section only the southern element of which passes into Oman from Iran, across the Straits of Hormuz; the northern element continues through Makran and Baluchistan. There is then a prominent belt extending northwards from Karachi to the western end of the Himalaya, passing along their northern boundary, through Kashmir and into Indian Tibet (Gansser 1977).

Stöcklin further subdivides the belt. The severe tectonization makes it questionably suitable for easy elucidation by isotopic techniques and its predominantly basic igneous rocks are of no use for Rb-Sr work. Their association with fossiliferous rocks, and with some granitoid intrusions permits dating, but very little isotopic work has been done so far, or at least published. The belt is of great economic interest in so far as it is commonly rich in chromite.

In Oman (Glennie 1977) Precambrian basement in coastal Dhofar and the adjacent Kuria Muria Islands is best regarded as part of the Arabian Platform, but similar rocks are known in Jebel Ja'alan in the easternmost part of Oman, and these gave, according to Glennie, K-Ar and Rb-Sr ages of about 860 Ma. Pre-Permian rocks (Amdeh Formation) in Jebel Akhdar (Amdeh Formation) were metamorphosed and an event at about 330 Ma has been detected. This activity in central Oman is interesting in so far as igneous rocks brought up in salt plugs in the Zagros near Bandar Abbas were analysed by Rb-Sr and gave a crude total-rock isochron of about the same age, though the rocks were, on general geological grounds, regarded as probably much older (Crawford 1977).

The Hawasina Nappe of Oman includes igneous rocks, and a sample of tuff was dated by K-Ar at 92±6 Ma, while dykes in the massive peridotite -serpentine complex of the Semail Nappe gave an age of about 86 Ma (Alleman and Peters 1972).

Not much is yet known in detail of the northern part of the Axial Belt in Makran and western Baluchistan. For the stretch north of Karachi, several valuable new papers appeared in *Geodynamics of Pakistan* (Farah and De Jong, eds, 1979), though isotopic ages are so far lacking.

The part of the belt north of the Himalaya in Pakistan and India is wider than elsewhere. In Pakistan to the north of Hazara there is an extensive outcrop of greenschists, amphibolites and gabbros, the 'Upper Swat Hornblendic Group', which Stöcklin regards as extending westwards into the Jalalabad and Kabul areas of Aghanistan. A hornblende from a pegmatite was analysed by Snelling and gave a K-Ar age of 67 Ma (Jan and Kempe 1973). Kempe (1973) suggests a consanguinous relationship with alkaline granites of Warsak, 30 km westnorthwest of Peshawar. A riebeckite from them was also dated by Snelling, at 41 Ma, and a biotite from an associated nepheline syenite gave 50 Ma. Desio (in Spencer 1974) regards the Group as extending along the middle Indus to the west flank of Nanga Parbat (itself a massif of metamorphosed Precambrian rocks) and Stöcklin continues this zone from the northeastern side of that massif through Dras and eastwards along the upper Indus. Sinha and Bagdasarian (1977) mention K-Ar analyses of spilites from the 'exotic' zone of the Kumaon Tethyan Himalaya which are relevant; these ages are 107, 75 and 73 Ma. Between this and the northernmost subdivision of the Axial Belt, the Chitral-Skardu greenschists, is a wide outcrop of post-ophiolitic diorites and granites, including the Ladakh granite. There is little doubt that these rocks are correctly thought to be late Cretaceous-early Tertiary (Frank et al. 1977) though we lack much isotopic dating. The Ladakh granite gave a Rb-Sr age of 45 Ma (Desio and Zanettin 1970) and a similar Eocene age of 48 Ma was found for the Satpura

granodiorite southeast of Gilgit (Desio et al., 1964).

Central Domain

General

This is a vast area. Stöcklin (1977) points out its unequal division into the western sector in central (and northern) Iran and central Afghanistan, separated just north of Kabul from the eastern sector of Nuristan, the region south of the Hindu Kush proper, Pamir and Karakorum. This eastern sector appears to be displaced northwards along the Chaman Fault-Wanch Fault system in relation to the western one. Though there is undoubted major Cenozoic sinistral displacement along the Chaman Fault system, the mechanism of movement further north is still obscure, and the eastern sector is considerably narrower and very much higher.

Western Sector

The western sector is now widely accepted to be, following the arguments of Stöcklin (1968), a much modified extension of the Arabian Platform. A prominent feature of it is the existence of many inliers of Precambrian rocks, conformably below Cambrian, which appear most commonly in horsts, and between which are grabens with abnormally thick Palaeozoic sequences elsewhere common but epicontinental. The discovery of these sequences came from careful reconnaissance and systematic mapping of selected regions. It led to a need for isotopic age dating. In particular it became desirable to date the widespread leucogranite type which intrudes the older, but not the younger, Precambrian. The results of systematic sampling for Rb-Sr work have been disappointing; nearly all Precambrian rocks have been rejuvenated and lost radiogenic strontium. A summary is given in Crawford (1977), since when there has been little if any addition to published data. One age which is prominent in northwestern Iran is 175±10 Ma, the age of biotite from the undoubtedly Precambrian granite, here the Doran granite of the Soltanieh Mountains (Stöcklin and Eftekhar-nezhad 1969) near Zanjan. It appears also as the age of intrusion of tourmaline granite in the Talesh Hills (northwestern Alborz) in the Northern Domain, where a metamorphic event is prominent also at about 375 Ma.

In the Kerman region there is some evidence of disturbance at about 190 Ma, affecting probable Precambrian rhyolites. This is, so far, the only isotopic evidence adding to the very strong geological evidence for a widespread late Triassic (Cimmerian) event in Central Iran. Stöcklin (1977) relates this to 'detachment of the Central Domain from the main continental mass to the south - its fragmentation into a great number of blocks, and, thus, the destruction of its platform character'. It is plain that future work should aim to extend recognition of this event. One of the problems which has appeared as a result of the Iranian reconnaissance dating is that rejuvenation is expressed at different times in different places.

In eastern Iran information is lacking apart from biotite ages of (Rb-Sr) from gneisses on the margins of the Lut Block. One gave 160+10 Ma (from a garnetiferous mica schist) and two from biotite gneisses gave ages between 48 and 44 Ma (Crawford 1977).

For the large central and western area of Afghanistan we lack data. In the more mountainous section approaching the Kabul node Blaise et al. (1970) report K-Ar and Rb-Sr ages. These include isochron ages for areas strictly in the Hindu Kush and ages for the area south are 'model' ages which are in the region of 210 Ma and 65 Ma. Sampling by the writer under the guidance of de Lapparent and his colleagues was for Rb-Sr work, but this has been held up. Some K-Ar analyses of the same material made by Adams indicate reliable late Precambrian ages for gneisses and schists at Kabul and 25 km to the southwest. These ages are rejuvenated to less than 20 Ma for samples of similar rocks along the sharply-defined Arghendeh (Chaman) Fault.

Eastern Sector

Data do not exist for the geologically still unknown area of Nuristan (mainly gneisses). For the formidably difficult terrain of the Karakorum, an area of very young uplift, we depend heavily upon the work of Desio and his collaborators (see Desio, 1979, for general account, excellent map and full references). Desio divides the area into five zones, of which the central one is a granite batholith. The southern belt is that of Ladakh-Deosai, and the northern is less continuous. A particular feature of the Karakorum is that the granites occur as extremely elongated bodies, parallel to each other and to tectonic axes. From the limited isotopic work done so far (Desio et al. 1964; Desio and Zanettin 1970) they appear not to be of the same age. The Ladakh belt (mentioned above) is early Tertiary. The axial belt appears to range from about 25 Ma to 8 Ma, while the northern belt has been sampled only once and gave an early Tertiary age.

The limited isotopic dating has given Miocene-Pliocene ages for the 'axial granite batholith' but early Cenozoic and late Triassic granites exist to north and west.

The Pamir proper is exclusively worked on by Soviet geologists, and the main part of Tibet, in Chinese territory, is equally inaccessible to non-nationals.

Northern Domain

The only part of the Northern Domain of Stöcklin within the range of activities of the ICG Working Group 6 is that in Iran and Afghanistan. The two areas lie on opposite sides of the meridional Harirud structure which broadly coincides with the political division.

The Iranian sector is dominated by the enigmatic South Caspian depression, unique in the belt. As Stöcklin states, many Soviet workers regard this zone of very thick sediments lying apparently on material with the geophysical characteristics of basalt, as an incipient geosyncline. He prefers to regard the thin basement as a relic of oceanic crust, perhaps of a 'Palaeotethys'. This area adjoins the Alborz, which displays a thinning of the Infracambrian-Cambrian as older Precambrian rises in horsts. In contrast, the Ordovician-Carboniferous rocks are here thicker than to the south, and contain basic volcanics (Davies et al. 1972). It was not possible to date these by Rb-Sr but there is good evidence of an event at about 375 Ma (total-rock isochron from five samples), and here also tourmaline granites at 175+10 Ma occur.

Attempts to date old rocks in horsts further east (Gorgan area) failed because of poor strontium enrichment in the schists. Little isotopic work has yet been done on the equally interesting Afghan sector, which has recently been mapped by Soviet, Afghan and French geologists (see Boulin and Bouyx 1977; Stöcklin 1977). The Hindu Kush has been a major structural element since Precambrian times (de Lapparent, pers. comm.). Evidence exists of Hercynian movements; the Lower Palaeozoic rocks are strongly folded and unconformably overlapped by Upper Carboniferous-Permian rocks, though in the western region (so-called Paroparinsus) the unconformity is at the base of the Permian. Samples from the Hindu Kush north and south of the Herat Fault collected by the writer with the French group still await analysis, but many are unsuitable for Rb-Sr work. This region is markedly affected by late Triassic Cimmerian deformation, with granitic intrusion.

General Discussion and Conclusions

Since its vigorous development as a technique in the 1950s and 1960s, isotopic age dating has been especially applied for two particular purposes: clarifying Precambrian stratigraphy, previously in chaos, and in obtaining actual ages for Phanerozoic sequences commonly well-known palaeontologically. It has also been used in more recent years to give information on timing of uplift. This period since 1950 saw the first real mapping of much of the western part of the belt. Isotopic techniques proved highly successful in their application to Precambrian shields and stable platforms. In particular, a much clearer pattern emerged of the Indian Peninsular Precambrian. Knowledge of Arabia advanced more slowly. In so far as mapping of the Iranian section of the belt itself revealed widespread Precambrian, it was desirable to date it. Equally, that much of the Himalaya had long been thought to be Precambrian, a need arose there too, quite apart from a need to date the time of uplift.

The limited work so far has shown that application of reconnaissance techniques to orogenic belts is less immediately successful than to stable areas. This is particularly evident in the greatly fractured terrain of Iran, the problem being one of isotopic disturbance. By contrast, in the Himalaya, the problem has been rather more one of lack of agreement on the geological succession and structure (especially in the Lower Himalaya), and profound differences of opinion on the interpretation of data. Though there is much evidence of isotopic disturbance, scarcely surprisingly, it has (in contrast to Iran) been possible to obtain several reliable Precambrian ages, and also to get dated confirmation of Cenozoic events. But we still have far to go in understanding Himalayan metamorphism, particularly 'reversed metamorphism'.

Further work is suggested as follows:

i combined isotopic and structural studies in one area of reversed metamorphism, using two dating methods. Refinement of techniques since 1970 should give excellent data. Darjeeling is the obvious first choice, but a 1:50,000 scale geological map (preferably 1:25,000) is essential.

ii Rb-Sr and K-Ar techniques need to be used more widely in combination, both on total-rocks and minerals on the same samples, to elucidate further the Cenozoic evolution.

iii Continued studies need to be made on the basal 'slab' of the High Himalaya, preferably by combined techniques.

iv 'Reconnaissance' age dating should if practicable be continued in Afghanistan and areas north with emphasis on probable Precambrian terrain.

References

Acharyya, S.K., Late Palaeozoic Glaciation versus Volcanic Activity along the Himalayan Chain with Special Reference to the Eastern Himalaya. Himalayan Geol. 3, 209-230, 1973.

Alleman, F. and T. Peters, The Ophiolite-radiolarite belt of the North Oman Mountains. Eclog.geol.Helv. 65, 657-697, 1972.

Ashgirei, G.D., A.K. Sinha, I.C. Pande and B.C. Mallik. A Contribution to the Geology, Geochronology and History of Regional Metamorphism of Himachal Himalaya. Chayanica Geologica, 1, 143-151, 1975.

Auden, J.B. Traverses in the Himalaya. Rec. Geol.Surv.India, 69, 123-167, 1935.

Bhanot, V.B., J.S. Gill, R.P. Arora and J.K. Bhalla. Radio-metric dating of the Dalhousie Granite. Current Science 43, 208, 1974.

Bhanot, V.B., A.K. Bhandari, V.P. Singh and A.K. Goel. The petrographic studies and age determination of Koidal Gneiss, Kumaon, Himalaya. Current Science, 45, 18, 1976.

Bhanot, V.B., V.P. Singh, A.K. Kausal and V.C. Thakur. Early Proterozoic Rb-Sr whole-rock age for Central Crystalline Gneiss of higher Himalaya, Kumaon. J.Geol.Soc.India, 18, 90-91, 1977.

Blaise, H., P. Bordet, J. Lang, A.F. de Lapparent, F. Leutwein and J. Sonet. Mesures géochronologiques de quelques roches cristallines d'Afghanistan central. C.R. Acad.Sci.Paris, ser. D, 270, 2772-2775, 1970.

Bodenhausen, J.W.A. and C.G. Egeler. On the geology of the upper Kali Gandaki Valley, Nepalese Himalayas. K.Nederlandse akad. wetensch., proc. acad. sci., B, 74, 526-546, 1971.

Bordet, P. Recherches géologiques dans l'Himalaya du Népal, region de Makalu. Paris. C.R.N.S., 275 p. 1961.

Bordet, P. Géologie de la dalle du Tibet (Himalaya Central). Soc. géol. France, mem. hors-série no. 8, 235-250, 1977.

Bordet, P., M. Colchen, P. Le Fort, R. Mouterde and M. Remy. Données nouvelles sur la géologie de la Thakkola (Himalaya du Népal). Bull.Soc.géol. France ser. 7, v.9, 883-896, 1967.

Boulin, J. and E. Bouyx. Introduction à la géologie de l'Hindou Kouch occidental. Soc.géol.France, mém. hors-serie No. 8, 87-105, 1977.

Calkins, J.A., T.W. Offield, S.K.M. Abdullah and S. Tayyab Ali. Geology of the Southern Himalaya in Hazara, Pakistan and Adjacent Areas. U.S.Geol.Surv.Prof.Paper 716-C, 29 p. 1975.

Carey, S.W. The Expanding Earth. Amsterdam, Elsevier, 488 p., 1976.

Crawford, A.R. India, Ceylon and Pakistan: New Age Data and Comparisons with Australia. Nature 223, 380-384, 1969.

Crawford, A.R. The Precambrian geochronology of Rajasthan and Bundelkhand, northern India. Canad.J.Earth Sci., 7, 91-110, 1970.

Crawford, A.R. A Summary of Isotopic Age Data for Iran, Pakistan and India. Soc.géol. France, mém. hors-serie no. 8, 252-260, 1977.

Crawford, A.R. and R.G. Davies. Ages of Pre-Mesozoic formations of the Lesser Himalaya, Hazara District, Northern Pakistan. Geol.Mag. 112, 509-514, 1975.

Davies, R.G., C.R. Jones, B. Hamzepour and G.C. Clarke. Geology of the Masuleh Sheet, 1:100,000, Northwest Iran. Geol.Surv.Iran Rep. 24, 110 p., 1972.

Desio, A. Geologic Evolution of the Karakorum. In, Farah and De Jong, eds, Geodynamics of Pakistan, Quetta, Geol.Surv. of Pakistan, 111-124, 1979.

Desio, A., E. Tongiorgi and G. Ferrara. On the Geological Age of Some Granites of the Karakorum, Hindu Kush and Badakhshan (Central Asia). Internat.Geol.Cong., 22nd, New Delhi 1964, Pt XI, 479-496, 1964.

Desio, A. and B. Zanettin. Geology of the Baltoro Basin: Desio's Italian Expeditions to the Karakorum and Hindu Kush: Scientific Reports, 3, 309 p., Leiden, Brill, 1970.

Farah, Abul and De Jong, K.A. Geodynamics of Pakistan. Quetta, GSP, 1979.

Fourcade, S., J. Hamet and C.-J. Allègre. Données de la méthode rubidium-strontium et détermination des Terres Rares dans le leucogranite du Manaslu: implications pour l'orogenèse himalayenne. C.R.Acad.Sci.Paris, 284, série D, 717-720, 1977.

Frank, W. Geochemistry and isotopic geochemistry in the Himalaya - Discussion, in, Colloques internationaux du CRNS No 268, Himalaya, Sciences de la Terre, Paris, 1977.

Frank, W. and G.R. Fuchs. Geological investigations in West Nepal and their Significance for the Geology of the Himalayas. Geol.Rdsch. 59, 552-580, 1970.

Frank, W., G. Hoinkes, C. Miller, F. Purscheller, W. Richter and M. Thotti. Relations between Metamorphism and Orogeny in a Typical Section of the Indian Himalayas. Tschermaks Min. Petrol.Mitt. 20, 303-332, 1973.

Frank, W., A. Gansser and V. Trommsdorff. Geological Observations in the Ladach Area (Himalayas). A Preliminary Report. Schweiz. mineral.petrogr.Mitt. 57, 89-113, 1977.

Fuchs, G. and A.K. Sinha. The Tectonics of the Garhwal-Kumaon Lesser Himalaya. Jahr.Geol. B.-A. 121, 219-241, 1978.

Gansser, A. Geology of the Himalayas. London, Interscience, 289 p., 1964.

Gansser, A. The Great Suture Zone Between Himalaya and Tibet. A Preliminary Account. Colloques internationaux du CRNS No 268, Himalaya, Sciences de la Terre, Paris, 181-191 (with map), 1977.

Glennie, K.W. Outline of the Geology of Oman. Soc.géol.France Mem. hors-série no. 8, 25-31, 1977.

Hamet, J. and C.-J. Allègre. Rb-Sr Systematics in Granite from Central Népal (Manaslu): Significance of the Oligocene Age and High $^{87}Sr/^{86}Sr$ ratio in the Himalayan Orogeny. Geol. 4, 470-472, 1976.

Jäger, E., A.K. Bhandari and V.B. Bhanot. Rb-Sr Age Determinations on Biotites and Whole-rock Samples from the Mandi and Chor Granites, Himachal Pradesh, India. Eclog.geol. Helv. 64, 521-527, 1971.

Jan, M. Quasim and D.R.C. Kempe. The Petrology of the Basic and Intermediate Rocks of Upper Swat, Pakistan. Geol.Mag. 110, 285-300, 1973.

Kempe, D.R.C. The Petrology of the Warsak Alkaline Granites, Pakistan, and their Relationship to Other Alkaline Rocks of the Region. Geol.Mag. 110, 385-404, 1973.

Krummenacher, D. Déterminations d'âge isotopique faites sur quelques roches de l'Himalaya du Népal par la méthode potassium-argon. Bull.suisse Min.Petr. 41, 273-283, 1961.

Krummenacher, D. Népal Central: géochronométrie des séries de l'Himalaya. Bull.suisse Min. Petr. 46, 43-54, 1966.

Kumar, R., A.N. Shah and D.K. Bingham. Positive Evidence of a Precambrian Tectonic Phase in Central Nepal, Himalaya. J.Geol.Soc.India 19, 519-522, 1978.

Lapparent, A.F. de. Esquisse géologique de l'Afghanistan. Rev.Géogr.phys.Géol.dyn. 14, 327-344, 1972.

Le Fort, P. Les leucogranites à tourmaline de l'Himalaya sur l'example du granite de Manaslu (Népal central). Bull.Soc.géol.France, 7e Serie, tome 15, 555-561, 1973.

Le Fort, P. Himalayas: The collided range. Present knowledge of the continental arc. Amer.J.Sci. 275A (Rodgers Volume), 1-44, 1975.

Loczy, L. von. Beobachtungen im östlichen Himalaya (vom 8, bis 28 Febr. 1878). Földr. Közlem. 35, 1-14, 1907.

Mallet, F.R. On the geology and mineral resources of the Darjeeling District and the Western Duars. Mem.Geol.Surv.India 11, 50 p., 1875.

Mehta, P.K. Rb-Sr geochronology of the Kulu-Mandi Belt: Its implications for the Himalayan tectogenesis. Geol.Rdsch. 66, 156-175, 1977.

Mehta, P.K. Rb-Sr geochronology of the Kulu-Mandi Belt: Its implications for the Himalayan tectogenesis - A Reply. Geol. Rdsch. 68, 383-392, 1978.

Mehta, P.K. and D.C. Rex. K-Ar geochronology of the Kulu-Mandi Belt, NW Himalaya, India. N.Jb.Miner.Mh, Jg. 1977, H.8, 343-355, 1977.

Misch, P. Metasomatic granitization of batholithic dimensions. Amer.J.Sci. 247, 209-245, 1949.

Mukopadhyay, M.K. and P.K. Gangopadhyay. Structural Characteristics of Rocks Round Kalimpong, West Bengal. Himalayan Geol. 1, 213-230, 1971.

Nagpal, K.K., M.L. Gupta and P.K. Mehta. Fission-track ages of some Himalayan granites. Himalayan Geol. 3, 249-261, 1973.

Nagpal, M.K. and K.K. Nagpaul. Fission-track ages of some Himalayan muscovites. Himalayan Geol. 4 (for 1974), 447-452, 1975.

Ohta, Y. and C. Akiba, eds. Geology of the Nepal Himalayas. Himalayan Committee of Hokkaido University, Sapporo, 1973, 292 p. Plus volume of maps.

Pande, I.C. and S. Kumar. Absolute age determinations of crystalline rocks of Manali-Jaspa Region, Northwestern Himalaya. Geol. Rdsch. 63, 539-558, 1974.

Pickett, J., J. Jell, P. Conaghan and C. McA. Powell. Jurassic Invertebrates from the Himalayan Central Gneiss. Alcheringa 1, 71-85, 1975.

Powell, C. McA. and P.J. Conaghan. Polyphase deformation in Phanerozoic rocks of the Central Himalayan Gneiss, northwest India. J.Geol. 81, 127-143, 1973.

Powell, C. McA. and P.J. Conaghan. Rb-Sr geochronology of the Kulu-Mandi Belt: Its Implications for the Himalayan tectogenesis - a Discussion. Geol.Rdsch. 68, 380-383, 1978.

Powell, C. McA., A.R. Crawford, R.L. Armstrong, R. Prakash and H.R. Wynne-Edwards. Reconnaissance Rb-Sr Dates for the Himalayan Central Gneiss, northwest India. Indian J.Earth Sci. 6, 139-151, 1979.

Roy, S.S. Polymetamorphism in Daling rocks from a part of Eastern Himalaya and some Problems of Himalayan metamorphism. Himalayan Geol. 4, 74-101, 1974.

Rupke, J. Stratigraphic and Structural Evolution of the Kumaon Lesser Himalaya. Sedim.Geol. 11, 81-265, 1974.

Seitz, J.F., A.P. Tewari and H. Obradovich. A note on the Absolute Age of the Tourmaline-granite, Arwa Valley, Garhwal Himalaya. Abs.Sem.Recent Geol.Stud. in Himalaya, Geol. Surv.India, Calcutta, p. 43, 1971.

Sinha, A.K. Geochronology, Petrography-Petrochemistry and Tectonic Significance of Basic Rock Suites of North-Western Himalaya with Special Reference to Himachal Himalaya, India.

Chayanica Geologica 1, 478-493, 1975.

Sinha, A.K. and G.P. Bagdasarian. Potassium-argon Dating of some Magmatic and Metamorphic Rocks from Tethyan and Lesser Zones of Kumaon and Garhwal Indian Himalaya and its Implication in the Himalayan Tectogenesis. Colloques internationaux du CRNS, no. 268, Himalaya, Sciences de la Terre, Paris, C.R.N.S., 387-393, 1977.

Spencer, A.M. (ed.) Mesozoic-Cenozoic Orogenic Belts/Data for Orogenic Studies. London, Geological Society, 809 p., 1974.

Stöcklin, J. Structural History and Tectonics of Iran - a Review. Am.Assoc.Petr.Geol.Bull. 52, 1229-1258, 1968.

Stöcklin, J. Structural Correlation of the Alpine Ranges Between Iran and Central Asia. Soc.géol.France. mem. hors-série no. 8, 333-353, 1977.

Stöcklin, J. and J. Eftekhar-nezhad (compilers). Explanatory Text of the Zanjan Quadrangle Map, 1:250,000. Geol.Surv.Iran, Geol.Quad No D4, 1969.

Stoneley, R. Evolution of the Continental Margins Bounding a Former Southern Tethys. In, C.A. Burk and C.L. Drake (eds). The Geology of Continental Margins. New York, Springer Verlag, 889-903, 1974.

BASIC IGNEOUS EPISODES IN THE HIMALAYA AND THEIR TECTONIC SIGNIFICANCE

V. Divakara Rao

National Geophysical Research Institute, Hyderabad, India

Abstract. Caledonian, Hercynian and Tertiary orogenic movements are imprinted on the Himalayan sector of the Indian subcontinent and are manifested in particular widespread basic magmatic intrusive/extrusive activity. These basic rocks extend all the way from Pirpanjal, Mandi, Darla, Zanskar in the west, through the basics at Bhowali, Askot and Bageswar in Kumaon to the Abor volcanics in the Siang District, Arunachal (NEFA). Analyses of samples from these areas clearly demonstrate that the basic rocks vary from tholeiitic to high alumina and calc-alkali types. On the basis of their chemistry and lithostratigraphic association, it is suggested that these rocks belong to more than two episodes of magmatic and tectonic activity and have been brought into juxtaposition by later tectonic movements. The mantle under the Himalayan belt was more heterogenous and less fractionated in the Carboniferous when tholeiites were formed compared to the Cretaceous when alkali basalts appeared.

Introduction

The Himalaya are one of the youngest mountain chains and yet are one of highly complicated history. An attempt is made in this paper to decipher the nature of basic igneous activity in different parts of the Himalaya and to deduce the conditions existing during the tectonic history.

Basic Magmatism in the Himalaya

Distinct pre-Himalayan orogenies in the Himalayan belt are yet to be established although such orogenies have been identified by Valdiya (1964) and Saxena (1971). Lefort (1975) envisaged only epirogenic movements and limited volcanism in the Himalaya during Hercynian and Caledonian movements. The nature of Precambrian magmatism in the Himalaya is not clear. Chakrabarty and Mithani (1972) suggested that the talc-chlorite schists in the Salkhala series and the Dogra slates of Kashmir and Himachal-Himalaya are the metamorphosed equivalents of coeval amygdaloidal lavas. Similarly, the chlorite-hornblende schists and the epidiorites associated with the Jutoghs in the North Western

Himalaya and Darjeeling and Daling series and their equivalents on the Eastern Himalaya may also be taken as representatives of Precambrian basic igneous activity. The chloritic tuffs, tuffaceous slates and pyroclastics interbedded with phyllites and quartzites in the Chandpurs of the Kumaon Himalaya are the manifestations of the early Paleozoic volcanics according to Nair and Mithal (1977). The spilitic keratophyres of the Bafliaz volcanics of Kashmir are considered to be Lower Paleozoic (Wakhaloo and Shah, 1968). Basalts, epidiorites and amphibolites (amygdaloidal and vesicular) associated with Nagthat quartzites and their equivalents in Bhimtal, Bhowali, Askot and Bageswar of Kumaon are considered to be products of penecontemporaneous volcanism. The thick volcanic suites represented by chlorite-amphibolite schists, associated with quartzites, between Abdari and Karanprayag and Cahmoli in the Alaknanda valley also appear to represent penecontemporaneous volcanics of early Caledonian age.

Hercynian volcanism was intense in the Kashmir region. Thick lava flows occur at Pirpanjal, the Liddar and Narbat valleys, the Bamhal area, Mandi and Darla. These volcanics consist of basalts, andesites and andesite-basalts (Pareekh 1973). Generally, it is believed that this volcanic activity commenced in the Middle Carboniferous and continued during Upper Carboniferous. Mandi, Darla, Dalhousie and Chamba volcanics of Himachal Himalaya and the Abor volcanics of the Eastern NEFA Himalaya are considered to be of Precambrian age.

The Abor volcanics, consisting of altered basalts, tuffs and pyroclastics interbedded with the Siang group of metasediments in the Siang District of Arunachal Himalaya, are of late Paleozoic age (Murthy, 1970). Crawford (1974) attributed a Gondwana age to these volcanics. The earlier upheaval of the Himalaya is marked by the late Cretaceous-early Eocene ophiolite suite in NW Himalaya. This ophiolite sequence continues through Dras and Kargil into the upper Indus valley. The Dras volcanics occurring from south of Astor to the Dras region, consist of a suite of laminated ash beds, red cherts and agglomeritic slates with flows of augite andesite and augite basalts.

Table 1. Average chemical composition of basaltic rocks from different parts of the Himalaya

	I		II						III						IV
	PT(9)	AR(6)	K1(23)	K2(8)	K3(3)	K4(4)	K5(2)	K6(2)	T(10)	Z(8)	TZ(10)	MV(7)	THV(8)	YV(6)	SK(8)
SiO_2	56.36	54.89	52.27	49.96	48.18	51.99	50.32	50.80	48.36	51.73	47.00	50.87	46.91	48.65	53.98
TiO_2	1.30	1.44	1.16	0.56	0.56	1.03	2.13	1.12	1.38	0.51	0.83	1.27	0.94	1.28	0.47
Al_2O_3	14.69	13.18	13.94	15.13	15.89	14.80	14.08	14.93	14.64	16.46	14.95	14.35	15.92	14.29	17.67
Fe_2O_3	2.92	10.62	2.06	1.96	1.77	2.68	1.78	2.33	5.52	4.82	4.29	4.89	4.06	3.56	3.23
FeO	6.13		8.99	8.00	9.00	8.51	10.82	9.50	5.82	5.61	5.64	7.28	6.63	6.71	4.72
MgO	3.31	8.41	8.52	9.06	9.10	8.15	7.10	9.01	7.80	6.81	11.86	4.88	7.10	5.35	4.26
CaO	7.60	5.25	5.26	6.77	7.80	4.94	6.69	5.01	6.18	6.26	9.51	9.68	9.85	11.74	9.00
Na_2O	2.94	3.68	3.83	4.72	2.60	4.64	3.90	4.43	4.70	4.44	2.89	2.07	3.59	2.60	3.82
K_2O	1.35	1.15	1.68	1.69	1.38	1.01	1.40	1.00	0.48	0.62	0.30	0.80	0.74	0.69	0.55
P_2O_5	0.08	0.21	0.26	0.25	0.30	0.46	0.49	0.33	0.26	0.62	0.06	0.21	0.17	0.07	0.04
MnO	0.30	0.17	0.21	0.25	0.34	0.16	0.25	0.19	0.22	0.26	0.20	0.15	0.22	0.52	0.16
H_2O	2.21	1.94	1.94	1.75	2.27				3.44	1.49	3.24	3.45	3.50	2.71	1.93
Co	56	37	25	50	42	22	32	24	62	46	39	43	79	104	24
Ni	80	178	22	75	29	118	35	54	101	49	80	28	83	58	91
V			205	269	159	474	1550	260	210	54	106	57	220	203	139
Cr	78	31	169	113	287	50	10	100	144	344	340	298	109	57	357
Ga			12	12	44	18	27	18	26	22	14	30	14	43	42
Pb	87	22	28	BDL	BDL	ND	ND	ND	215	BDL	ND	644	250	227	161
Cr	308	340	33	240	139	525	1400	135	101	168	346	253	409	422	62
CaO/TiO_2	5.85	4.34	4.53	12.09	13.93	4.80	3.14	4.47	4.48	12.27	11.46	7.62	10.48	9.17	19.15
Al_2O_3/TiO_2	11.30	9.15	12.02	27.02	28.38	14.37	6.61	13.33	10.61	32.27	18.01	11.30	16.94	11.16	37.60
S.I.	19.88	35.25	33.97	35.63	38.16	32.61	28.40	34.30	32.07	30.54	47.48	24.60	32.10	28.29	25.69
FeO(T)/MgO	2.645	1.272	1.272	1.077	1.164	1.34	1.749	1.287	1.383	1.46	0.801	2.378	1.448	1.853	1.79
FeO(T)/FeO+MgO	0.727	0.56	0.56	0.519	0.538	0.573	0.636	0.563	0.580	0.594	0.445	0.704	0.592	0.65	0.642

AR:Abor (Eastern Himalaya);K1 to K6: Kumaon; PT:Panjal Traps; T: Techonala; Z: Zanskar; TZ:Tangze;MV: Marling (W. Himalayan) THV: Thidsi; YV: Yugar; SK: Sanko-Kargil

Table II

	(1) Carboniferous	(2) Permo-Carboniferous		(3) Permian				(4) Cretaceous
	PT(9)	AR(6)	K(42)	T(10)	Z(8)	TZ(10)	MV YV THV(21)	SK(8)
SiO_2	56.36	54.89	50.60	48.36	51.73	47.00	48.81	53.98
TiO_2	1.30	1.44	1.09	1.38	0.51	0.83	1.16	0.47
Al_2O_3	14.69	13.18	14.80	14.64	16.46	14.95	14.85	17.67
Fe_2O_3	2.92	4.41	2.10	5.52	4.82	4.29	4.14	3.23
FeO	6.13	5.65	9.14	5.82	5.61	5.64	6.87	4.72
MgO	3.31	8.41	8.49	7.80	6.81	11.86	5.78	4.26
CaO	7.60	6.25	6.08	6.18	6.26	9.51	10.42	9.00
Na_2O	2.94	3.68	4.02	4.70	4.44	2.89	2.75	3.82
K_2O	1.35	1.15	1.36	0.48	0.62	0.30	0.74	0.55
P_2O_5	0.08	0.21	0.35	0.26	0.62	0.06	0.15	0.04
MnO	0.30	0.17	0.33	0.22	0.26	0.20	0.30	0.16
H_2O	2.21	–	–	3.44	1.49	3.24	3.22	1.93
Co	56	37	33	62	46	39	75	24
Ni	80	178	56	101	49	80	56	91
V	ND	ND	486	210	54	106	160	139
Cu	78	31	121	144	344	340	155	357
Ga	ND	ND	22	26	22	14	29	42
Pb	87	22	–	215	BDL	ND	373	161
Cr	308	340	412	101	168	346	361	62
CaO/TiO_2	5.85	4.34	5.58	4.48	12.27	11.46	8.98	19.15
Al_2O_3/TiO_2	11.30	9.15	13.58	10.61	32.27	18.01	12.80	37.60
$\dfrac{MgO \times 100}{FeO+Fe_2O_3+MgO+Alk.}$	19.88	35.25	33.81	32.07	30.54	47.48	28.50	25.69
$\dfrac{FeO(T)}{MgO}$	2.65	1.14	1.32	1.38	1.46	0.80	1.89	1.79
$\dfrac{FeO(T)}{FeO(T)+MgO}$	0.77	0.53	0.57	0.58	0.59	0.45	0.65	0.64

The Panjal volcanic series (Middlemiss, 1910; Bhat and Ziauddin, 1978) form an important part of the stratigraphic column of Kashmir. Both the base and the top of this series are well dated by the occurrence of fossiliferous beds of upper Middle Carboniferous to Upper Permian age. Volcanic activity seems to have variously started from upper Middle Carboniferous to Permian in different areas and ceased at different horizons from lower Upper Permian to Upper Triassic.

Srikantia and Sarma (1969) have differentiated the Mandi-Darla volcanics from the Panjal ones and suggested that the former belong to the Shali basin and the Panjal traps to the Tethys basin. These Mandi-Darla volcanics are readily divisible into two well marked series, the Lower Permian agglomeritic slate and the Upper Permian lava flows flows of Panjal.

In the Bhowali-Bhimtal areas of Kumaon Himalaya, the volcanic rocks lie beneath the quartzites and occupy the core of the anticline (Shah and Merh, 1976). The calcite and quartz porpyries continue up to Amitpur-Rannibag (Raina and Dungrakote, 1975). These porphyries merge with the spilitic lavas of Bhimtal (Divakara Rao et al., 1974). Most previous workers correlate the quartzites and slates of Bhowali with pre-Tertiary Nagthat, considering the basic volcanics to be contemporaneous lava flows and tuffs. However, this is modified by Shah and Merh (1976) who have divided the Nagthat into two; the Upper Paleozoic sequence which is divided into two lithostratigraphic units, the lower part containing the volcanics. Volcanogenic rocks and associated sediments, including the keratophyres and the limestones of Ramgarh, are designated as the Bhimtal-Ramgarh formation and equated with the Blaini. The upper quartzites and variegated slates are the Nainital formation. The general succession is suggested by Shah and Merh (1978).

The volcanics occur within the core of the main anticline and also as two distinct horizons. They are classified into three main types i) spilite diabases, ii) spilite basalt and iii) tuffs. Divakara Rao et al. (1974) and Vardarajan (1974) have described the spilitic nature of these rocks. In the core, all three types contain coarse diabase

in a central position. The basaltic type is next in abundance.

The volcanic rocks of the Arunachal Himalaya were first reported from the Dafla hills by Goodwin (1875). Coggin Brown (1912) classified them as 'Abor volcanics'. Subsequent work by Mullick and Basuchandary (in Chatterjee, 1972) established the continuity of these volcanics along the Dihang valley. These authors also observed undoubted lateral facies variation of the Buxa with which the Abor volcanics are interstratified and the Gondwana rocks east of the Dihang river near Pasighat; the entire sequence was therefore considered late Paleozoic (Murthy, 1970).

Recent work has revealed four main litho-tectonic units delineated by three large-scale thrusts in the Siang District (Jain and Tandon, 1974). The Precambrian to Middle Paleozoic Siang metamorphic group and the sedimentary Miri group, including the Abor volcanics, are thrust over the metamorphic Gondwana rocks along the Tsangpo valley. The Abor volcanics are mainly exposed along the Dihang valley in NNW-SSE trending, linear, bodies, associated with the Miri and Siang groups. To the east, the volcanics are known to occur in the Yemene river and in the west they are associated with the volcano-clastic sediments and amygdaloidal basaltic flows in the youngest slate unit of the Siang group. Lava flows are intercalated in the sediments of the Miri group at the Yeanbag stream and west of

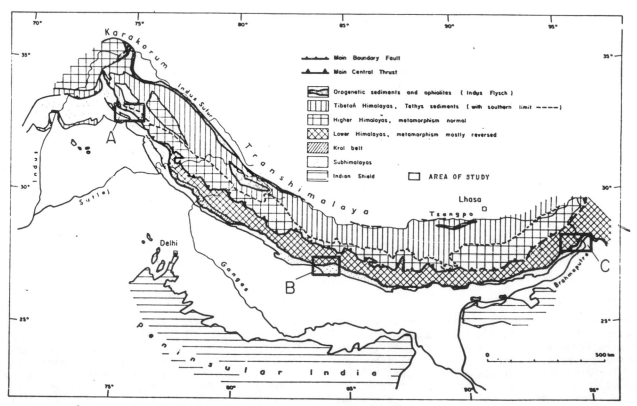

Fig. 1. Generalized geological map of the Himalaya showing areas of work.

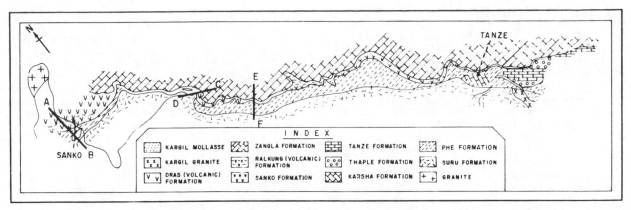

Fig.IA. GEOLOGICAL MAP OF THE ZASKAR VALLEY AND ADJOINING AREAS
AFTER NANDA AND SINGH (1976)

the Siang river, at Sikemen Nadi. Narrow linear outcrops of these volcanics also occur in the Bame-Along, Bame-Daparizo, Along-Kaying and Basor-Dali regions. Nearly 1500 m thick, monotonously westward dipping, the basaltic rocks between Dosing and Mibang Nadi have been identified as the type section of the Abor volcanics.

Petrography and Chemical Characteristics

The Panjal traps of the western sector are hard, compact, massive and generally fine-grained lavas. They normally contain elongated vesicles of which

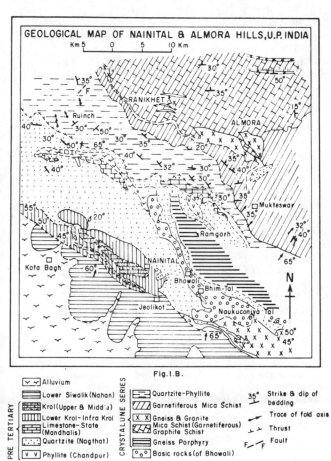

Fig.I.B.

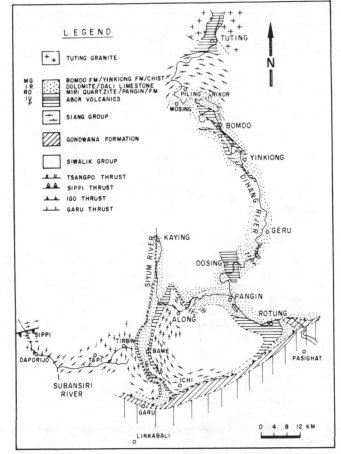

Fig.IC. GEOLOGICAL MAP OF PARTS OF SIANG DISTRICT, ARUNACHAL HIMALAYA SHOWING DISTRIBUTION OF THE LITHOTECTONIC UNITS. AFTER JAIN AND THAKUR (1978)

the filling is more or less selective: elongated vesicules contain chlorite and the more rounded ones, quartz. Under the microscope, the Panjal trap is a hemicrystalline aggregate of plagioclase and pyroxene, with mesostasis of microlites of these minerals and their alteration products and devitrified glass in the groundmass. Opaque minerals are quite common. Ilmenite, commonly altered to leucoxene is abundant. Epidotization and chloritization are predominant. Biotite and albite occur along cleavage and fracture lines.

In Kumaon, spilitic diabase contains mostly laths of fresh and partly altered plagioclase in a chloritic groundmass. Uralitization and chloritization of pyroxene (augite) is partial and the relicts show ophitic to sub-ophitic texture. The glassy variety contains an aphanitic groundmass with rare phenocrysts of sodic plagioclase. The main mass is mostly isotropic and consists of sheaf-like uralite needles and radiating minilites. Amygdales contain chalcedonic silica and chlorite. The intersertal type is coarse-grained and contains laths of plagioclase in a groundmass of either uralite-chlorite or dusty brown opaque mineral. Tuffs and tuffites contain fine-grained, dark green to greyish green layers of finely crystalline, cherty silica. Chlorite is predominant in some samples. Occasionally, the volcanic layers contain palagonite. In keratophyres of Bhimtal, phenocrysts of sodic plagioclase are imbedded in a micro-groundmass of quartz and felspar with sericite, chlorite and biotite.

The Abor volcanics of the Siang District are mostly basalt and tuffs. Under the microscope, they show porphyroblasts of sodic plagioclase in a groundmass of fine-grained quartz and plagioclase. Sericitization and chloritization are quite common and the rocks exhibit green schist facies metamorphism. Opaque minerals occur in a few sections. Secondary calcite, that is replaced by quartz, is observed in a few samples. Tuffs are mostly fine-grained with small laths of plagioclase in the groundmass.

Chemistry

Major, minor and trace elements were analysed from some 150 samples of various basic rocks from Bhowali, Bhimtal, Raiaugar, Askot and Bageswar from Kumaon Himalaya (Divakara Rao et al., 1974), from Mandi, Dras, Kargil, Sancho, Tangze, Marling, Thidsi and Yugar Nalas from Kashmir Himalaya (Divakara Rao et al., 1979; Radhakrishna and Divakara Rao, 1979) and from Along-Yinkiang, Daparizo-Along and Zero sections from the Siang District of Arunachal, representing the Pirpanjal, Phe volcanic sequences of western Himalaya, the basic volcanics of Kumaon and the Abor volcanics of NEFA (see Table 1).

The Table shows that in different basic rocks of the western sector at Tangze, Marling, Thidsi, Yugar, Sancho-Kargil and Pirpanjal, SiO_2 varies from 47% at Tangze to 56.36% in Pirpanjal. Other

elements also exhibit a limited variation while NaO_2 content indicates that Thidsi, Tangze, Sancho and Zanskar samples are spilitic. The H_2O content shows a positive relationship with silica. Al_2O_3 varies little, from 15.20 to 16.46%. MgO is substantially higher in Tangze and lower in the Sancho and Pirpanjal traps. There is a two-fold variation of TiO_2 from 0.5 to 1.3% in the western sector. Among the trace elements, Pb shows particular enrichment in Marling (644 ppm) and is lowest in Zanskar. V and Cr show substantial variations.

The average value of SiO_2 in the Kumaon basic rocks (Table 1) appears to be low compared to the western sector. However, of the eight groups of basic rocks sampled in Kumaon (Table 2), silica varies from 48.48 to 50.27%, being slightly lower than in the western sector. The TiO_2 content in the central Kumaon basic rocks is higher than in those of the Bhimtal-Bhowani area. Al_2O_3 and FeO are almost similar. Na_2O is higher, indicating the undoubted spilitic nature of most of the Kumaon basic rocks. K_2O appears to be slightly higher compared to the western sector and H_2O lower. Among the trace elements, Co is lower, Ni almost similar and V substantially higher, whilst Cu and Pb contents are low.

The Abor volcanics show two types of basic rocks, tuffs and basalts. The basalts have 50.07 - 57.59% SiO_2 while the tuffs have as much as 72.78% (Table 2). The alkali content varies widely, in particular sodium varies from 1.75 to 5.78%. Al_2O_3 also exhibits about 4% variation while TiO_2 is almost constant. CaO, MgO are almost similar except for one or two samples which exhibit higher values. Among the trace elements, Ni varies greatly, from 81 to 434 ppm, and Cr from 145 to 605 ppm, while Cu, Co, Pb are low. There is not much variation in major and trace elements concentrations in tuffs.

Discussion

A comparison of the average chemical composition of basaltic rocks from the western Pirpanjal (PT), Tangze (TZ), Techonala (T), Sancho-Kargil (Sk) units, Kumaon (K) and NEFA Himalaya (AR) clearly show (Table 1) that they vary in nature from typical tholeiitic to calc-alkaline and high alumina types. The tholeiitic and alkali basalt series have been long recognized (Kennedy, 1933). The third and other significant group that is found to occur abundantly is the high alumina basalt. Tilley (1950) recognized such basalts as a distinct type and Kuno (1968) proposed that they represent a primary magma series of similar status to the alkali basalt and tholeiitic magma. He suggested that high alumina basalt magma is generated by partial melting of the mantle at depths intermediate between the depths of origin of tholeiite and alkali basalt. Experimental results by Green et al. (1967) have shown that quartz

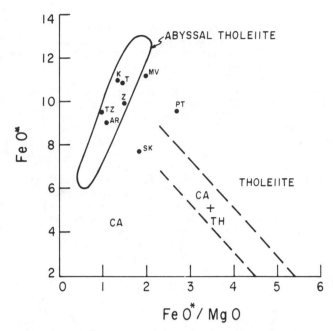

Fig. 2. FeO/MgO versus FeO indicating the abyssal tholeiitic nature of TZ, AR,K and T basalts.

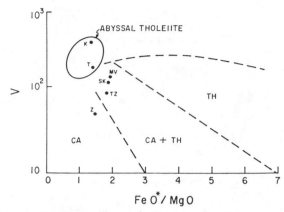

Fig. 4. FeO/MgO versus V diagram exhibiting CA+TH and abyssal tholeiitic nature of basic rocks from various areas.

normative tholeiites from olivine normative parent magma form at depths less than 15 km while at depths of 15 - 35 km, high alumina basalts slightly enriched in silica can form. Deeper, at 35 - 60 km, the derivative liquids tend to be alkali basalts.

However, Green et al. (1967) have also suggested that the relationship between the tholeiite, high alumina and alkali basalt in a given volcanic province depends also on depth of magma segregation or fractional crystallization; thus alkali basalts are derived at 15 - 35 km depths and tholeiites at depths less than 15 km.

A study of different trace elements and major oxides in various basaltic rocks from the Himalayan sector indicate that Kumaon (K), Techonala (TZ), Zanskar (Z), Abor (AR) and Marling (MV) fall in abyssal tholeiitic, oceanic tholeiite

fields with respect to their FeO (T), FeO/MgO (T) (Fig 2 and 3) while with respect to their V content, Sancho-Kargil, Marling and Tangze tend towards calc-alkaline/tholeiite trend and Zanskar towards the calc-alkaline field (Fig 4). In the case of Cr, Kumaon (K), Zanskar (Z) and Marling (MV) fall in abyssal tholeiite, while TZ, T, SK fall in the field of active continental margin volcanics (Fig 5). Similarly with respect to Ni versus FeO/MgO, all the samples fall within the field of volcanics of active continental margins (Fig 6).

This suggests that most of the basaltic rocks from different parts of the Himalaya, though exhibiting inhomogeneity and a wide range in different elemental concentrations, are of tholeiitic type except for the high alumina basalts and calc-alkali types that are found locally. When the available chronostratigraphic data is taken into consideration, the basic volcanics spread from the Carboniferous to the Cretaceous. A comparison of the composition, according to their approximate ages (Fig 7) indicates that the TiO_2 varies from high to low from Carboniferous to

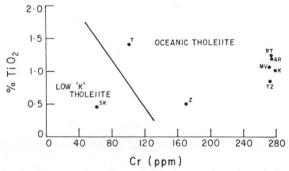

Fig. 3. Cr versus TiO_2 diagram showing low 'K' tholeiitic nature of SK and the oceanic nature of other basic rocks.

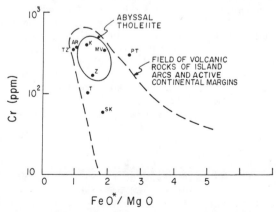

Fig. 5. FeO/MgO versus Cr diagram indicating continental tholeiitic nature of different basalts.

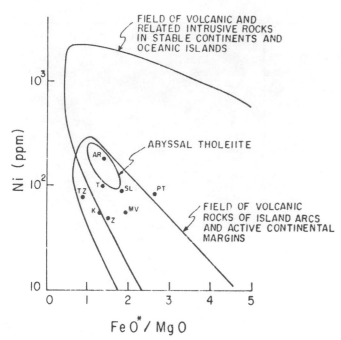

Fig. 6. FeO/MgO versus Ni.

tionization of the magma in the earlier period. Many of these samples also exhibit characteristics of active continental margin volcanics (Fig 5 and 6) suggesting their possible formation along the margin of a thin continental crust. The occurrence of the ophiolite sequence of Sancho-Kargil-Dras in the Western Himalaya (Srikantia et al., 1976; Radhakrishna and Divakara Rao, 1979) and at Arunachal (GSI, unpublished reports) clearly support the possible existence of a plate boundary in this region. Also Divakara Rao et al. (1976) have suggested, from the compositional characteristics of basic rocks from the north and south of the crystalline axis, the existence of two distinct tectono-lithological basins that were probably separated by a subduction zone.

The changing of the nature of the magma from tholeiitic in the Permo-Carboniferous to calc-alkaline in the Cretaceous clearly suggests that either the crustal thickness increased substantially during this period or that the depth of magma generation increased. Association of shallow

Cretaceous, Na_2O varies and increases in general with a maximum in the Permo-Carboniferous and remains almost constant in later periods. K_2O decreases, suggesting high K basalts in the Carboniferous to low K tholeiites in the Cretaceous.

P_2O_5 and MgO vary erratically irrespective of time, while Cr shows a steady decrease from older to younger rocks.

The variations of Na_2O can be explained by the spilitic nature of the Kumaon basalts which were suggested by Divakara Rao et al. (1974) to be tectonic spilites. Many of these basaltic rocks, which show spilitic character from the Permo-Carboniferous onwards, appear to have been spilitized because of the pre-tectonic disturbances of the Himalayan orogeny. The magmas seem to have been compositionally affected by tectonic events before their extrusion/intrusion.

If the experimental results of Green et al. (1967) and Kuno (1968) on the origin of basaltic rocks of different composition are accepted, most of the Himalayan basalts are tholeiitic in nature and must have been formed at shallow depths, i.e. less than 15 km. Fractionization of the magma at greater depths must have given rise to high alumina and alkali basalts.

Significantly, the calc-alkaline and high alumina basalts are from the Cretaceous, Sancho-Kargil (SK) group, suggesting that the conditions of the mantle or the magma source changed from Carboniferous to Permian and Cretaceous.

Compositional differences, especially in TiO_2, Al_2O_3, and SiO_2 content and of some trace elements in the tholeiites from different places, ranging in age from Carboniferous (PT) to Permian (TZ, Z, T, MV) indicate the heterogeneity and lower frac-

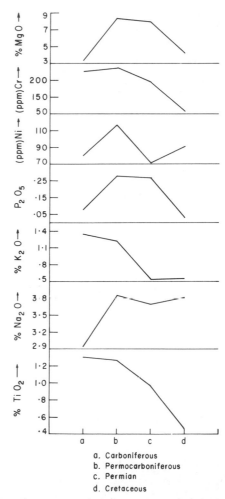

a. Carboniferous
b. Permocarboniferous
c. Permian
d. Cretaceous

Fig. 7. Diagram exhibiting compositional variations in basic rocks from Carboniferous to Cretaceous.

water marine sediments with the abyssal tholeiites at Tangze, Zanskar and elsewhere, indicate that these volcanics must have formed at a thin plate margin where mantle is at shallower depth.

Al_2O_3/TiO_2 and CaO/TiO_2 ratios increase with the degree of partial melting, as TiO_2 decreases (Sun and Nesbitt, 1978). Low Ti (0.6%) basalts from ophiolite complexes, island arcs and inter-arc basins are characterized by higher Al_2O_3/TiO_2 and CaO/TiO_2 ratios. A comparison of these ratios in different basalts from the Himalayan sector shows that the Cretaceous Sancho-Kargil basic rocks have higher values than the very low value obtained from Permo-Carboniferous and Permian volcanics. Such low Ti basalts could be formed by remelting of a severely depleted source that had experienced a previous episode of magma extraction. This remelting can be induced by the introduction of water from the subducted oceanic crust. This clearly shows that the heterogeneous mantle below the Himalayan sector had been depleted in some of those elements by the last Cretaceous phase of volcanic activity and subduction was coeval with the first early Cretaceous phase of the Himalayan orogeny. Al_2O_3/TiO_2 and CaO/TiO_2 ratios also indicate that the degree of partial melting increased from the Carboniferous to the Cretaceous (11.3% and 5.82% in PT to 37.6% and 19.5% in SK; Sun and Nesbitt, 1977). Increase in degree of partial melting indirectly reflects the higher amount of water content in the source magma which is related, in turn, again, to the rate of subduction or the amount of material subducted.

Studies of the composition of the Quaternary circum-Pacific volcanic rocks reveal a systematic compositional variation along continental margins and island arcs (Kuno, 1968; Sugimura, 1968), Tatsumoto, 1969); for instance in areas like Kamchatka and NE Japan, rocks of basaltic series change continuously across the arc from tholeiitic on the oceanic side through high alumina basalt to more alkali types on the continental side.

Kuno (1959, 1960) pointed out that there is a definite relation between the depth of earthquake foci and the type of basalt magma produced. The predominant tholeiitic nature of the basic volcanics in many parts of the Himalaya and the frequent occurrence of shallow focus earthquakes along the belt also support a thin plate margin theory.

The increase in Fe_2O_3 from the Carboniferous to the Cretaceous indicates the increasing effect of partial pressure of oxygen, which indirectly points to the increasing role of water in the magma. Similarly, traces of Cu, Ni, V indicate progressive increase of differentiation of the mantle magma source with time. A two-fold increase in the concentration of Cu during the evolution of the tholeiitic melt is established by Oleymikov (1974).

The nature of metamorphosed sediments and the composition of the different basalts asociated with them indicate deep-water, thin crustal conditions in early Carboniferous to Permian, whilst thick, shallow water sediments in the Cretaceous support the view that basin conditions and nature of deposits were gradually changing.

Paleo-current analysis and fault plane solutions for earthquake foci (Valdiya, 1975; Chandra, 1979) indicate the sub-surface continuation of the Aravalli trend of Peninsular India and of its structural grain upto/into the middle Himalaya.

An analysis of the chemical composition of different basic volcanics, their lithostratigraphic setting, together with the earthquake foci and paleo-current data thus indicate that the basaltic rocks of the middle and higher Himalaya ranges, associated with the ophiolite sequences, have probably originated at the Indian and Asian plate boundary mainly in the subduction zone, prior to the first phase of the Himalayan orogeny, from the Carboniferous to the Cretaceous. The mantle composition and fractionation trend also have changed with time.

Conclusions

Study of the compositional characteristics of the basic magmas in the Himalayas and their lithostratigraphic associations indicate that:
1. The basic volcanics vary in composition from tholeiite in the Carboniferous to high alumina basalts and alkali basalts during the Permian - Cretaceous.
2. The magma chamber was more heterogenous in the Carboniferous and fractionation started in the Permian and reached equilibrium in the Cretaceous.
3. The basalts are probably the products of subduction processes along the Indian/Asian plate oundary formed earlier and which continued upto the first phase of the Himalayan orogeny, i.e. Permo-Cretaceous.

References

Bhat, M.I., and S.M. Ziauddin, Geochemistry of the Panjal traps of Mount Kayol, Lidderwatt, Pahalgam, Kashmir, Jour. Geol. Soc. India, 19, 403-410, 1978.

Chakrabarthy, S., and J.G.P. Mithani, Basic igneous igneous activities in NW and Central Himalaya. Their implications on the study of the ages and structural trends of the orogenies, Geol. Surv. India, Misc. Publ. 15: Himalayan Geology, 321-334, 1972.

Chatterjee, G.C., General Report of the Geological Survey of India for the year 1967-'68. Rec. Geol. Surv. India, 102, 135-136, 1972.

Coggin Brown, J., A geological reconnaissance through Dihang valley, being the geological report of the Abor expedition, Rec. Geol. Surv. India, 42, 231- 264, 1912.

Crawford, A.R., The Indian Suture line, the Himalaya Tibet and Gondwanaland, Geol. Mag., 111, 369-380, 1974.

Divakara Rao, V., Compositional characteristics of basic magmatism in Kashmir and Kumaon Himalaya and their tectonic significance. In: Structural geology of the Himalaya, ed. P.S. Saklani, New

Delhi, Today and Tomorrow Publishers, 201-220.

Divakara Rao, V., K. Satyanaraya, S.M. Naqvi, S.M. Hussain and V.D. Kumar, Geochemistry and petrogenesis of basic rocks from Bhowali, U.P., Geophysics Res. Bull., 12, 63-74, 1974.

Divakara Rao, V., S.M. Naqvi, S.V. Srikantia and B.D. Dungrakote, Geochemistry and tectonic significance of spilitic rocks from parts of Himalaya. Proc. Intern. Sem. Him. Geol. 1968, Delhi, in press.

Divakara Rao, V., M.V. Subba Rao and B. Ashalata, Geochemistry and origin of the Marling, Thidsi and Yugar Nala basics of the Phe volcanic sequence, Zanskar. Geophys. Res. Bull., 18, 2, pp. 57-66.

Goodwin Austen, Notes on the geology of part of the Dafla Hills, Assam, lately visited by the force under Brigadier General Stattin, C.B., Jour. Asiatic Soc. Bengal, 44, 35-41, 1875.

Green, T.H., D.H. Green and A.E. Ringwood, The origin of high alumina basalts and their relationship to quartz tholeiite and alkali basalts, Ear. Pl. Sci. Lett., 2, 41-51, 1967.

Jain, A.K. and S.K. Tandon, Stratigraphy and structure of the Siang District, Arunachal (NEFA) Himalaya. Him. Geol., 4, 28-60, 1974.

Jain, A.K. and V.C. Thakur, Abor Volcanics of the Arunachal Himalaya. Jour. Geol. Soc. India., 19, 335-349, 1978.

Kennedy, W.Q., Trends of differentiation in basaltic magmas, Am. Jour. Sci., 25, 239, 1933.

Kuno, H., Origin of Cenozoic Petrographic Provinces of Japan and surrounding areas. Bull. Volc. Ser. 2, 20-37, 1959.

Kuno, H., High alumina basalts. Jour. Petrol., I, 121, 1960.

Kuno, H., Differentiation in basaltic magmas, In: 'Basalts', The Poldervaart treatise on Rocks of basaltic composition, ed. H.H. Hess, 2, Interscience, 633-688, 1968.

Lefort, P., Himalayas. The collision range. Present knowledge of the continental arc. Am. Jour. Sci., 275,A, 1-44, 1975.

Middlemiss, C.S., A revision of the Silurian-Trias sequence in Kashmir. Paleont.ind., 29,1-79,1910.

Murthy, M.V.N., Tectonics and mafic igneous activities in North East India in relation to upper mantle. Proc. Sec. Symp. Upper Mantle Proj., NGRI, Hyderabad, 289-304, 1970.

Nair, N.G.K. and R.S. Mithal, The Himalaya and the Caucasus. A comparison of their magmatism and metallogeny. Him. Geol., 7, 175-197, 1977.

Oleymikov, M.M., Geochemical trends of behaviour of Cu, Pb and Au during evolution of tholeiitic melts in intermediate chambers. Doklady, 218, 225-226, 1974.

Pareekh, M.S., Geological setting, petrography and petrochemistry of the Darla trap and its comparative study with Mandi and Panjal traps,

Jour. Geol. Soc. India, 14(4), 355-368, 1973.

Radhakrishna, T and V. Divakara Rao, Permian volcanism in Tethys zone of Kashmir Himalaya and its tectonic significance, in press.

Raina, B.N. and B.D. Dungrakote, Geology of the area between Nainital and Champwat, Kumaon Himalaya. Him. Geol., 5, 1-28, 1975.

Saxena, M.N., Geological classification and tectonic history of Himalaya. Proc. Ind. Nat. Sci. Acad., 37A (1), 28-54, 1971.

Shah, O.K. and S.S. Merh, Spilites of the Bhimtal and Bhowali area, District Nainital, U.P., Him. Geol., 6, 423-448, 1976.

Shen Susun and W. Nesbitt, Chemical heterogeneity of the Earth and mantle evolution, Ear. Planet. Lett., 35, 429-448, 1977.

Shen Susun and W. Nesbitt, Geochemical regularities and genetic significance of ophiolitic basalts. Geology, 6, 11, 689-693, 1978.

Srikantia, S.V., The tectonic and stratigraphic positions of 'Panjal Volcanics' in the Kashmir Himalaya, a reappraisal. Him. Geol., 3, 59-71, 1973.

Srikantia, S.V. and R.P. Sarma, Shali formation a note on the stratigraphic sequences. Bull. Geol. Soc. India, 6 (3), 93-97, 1969.

Srikantia, S.V. et al., Geology of the part of Zanskar mountains, Ladakh Himalaya with special. reference to late Caledonian 'Kerghiasc orogeny', Symp. Min. Res. and Power Development of the Himalaya with particular reference to Kashmir, G.S.I., 1976.

Sugimura, A., Spatial relations of basaltic magmas in Island arcs, In: 'Basalts', The Poldervaart treatise on Rocks of basaltic composition, ed. H.H. Hess, 2, 537-572, 1968.

Tatsumoto, M., Lead isotopes in volcanic rocks and possible ocean floor thrusting beneath Island arcs, Ear. Pl. Sci. Lett., 6, 369-376, 1969.

Tilley, C.E., Some aspects of magmatic evolution, Quart. Jour. Geol. Soc. Lond., 106, 37, 1950.

Umesh Chandra, Focal mechanisms and their tectonic implications for Alpine Himalayan region, East, Report Geodynamics Project, in press.

Valdiya, K.S., A note on the tectonic history and evolution of the Himalaya, C.R. 22 Intern. Geol. Congr., II, New Delhi, 269-282, 1964.

Valdiya, K.S., Simla slates, the Precambrian flysch of the lesser Himalaya, its turbidites, sedimentary structures and paleo-currents, Geol. Soc. Am. Bull., 81, 451-468, 1975.

Varadaraja, S., Prehnite-Pumpellyite Metagreywacke facies of metamorphism of the meta-basites of Bhimtal-Bhowali area, Nainital District, Kumaon, India, Him. Geol., 4, 581-599, 1974.

Wakhaloo, S.N., and S.K. Shah, A note on the Bafliaz volcanics of Western Pir-Panjal, Publ. Cent. Adv. Stdy., Geol. Punjab Univ., 5, 53-64, 1968.

SEISMICITY AND CONTINENTAL SUBDUCTION IN THE HIMALAYAN ARC

Leonardo Seeber, John G. Armbruster and Richard C. Quittmeyer[*]

Lamont-Doherty Geological Observatory of Columbia University
Palisades, New York 10964

*Also with the Department of Geological Sciences, Columbia University

Abstract. A subducting slab, the Indian shield; an overriding slab, the Tethyan slab, and a sedimentary wedge contained between and decoupled from the two converging slabs are the main elements of the active tectonics of the Himalaya in a scheme deduced primarily from earthquake data. The fundamental active thrust fault coincides with the upper surface of the subducting slab. The Detachment is the portion of this fault between the subducting slab and the sedimentary wedge. The Basement Thrust is the portion of this fault between the interacting slabs. The Basement Thrust Front (BTF) marks the line separating the very shallow-dipping Detachment from the steeper-dipping Basement Thrust.

The BTF is associated with a narrow belt of thrust earthquakes and with the very steep gradient that forms a topographic front between the Lesser and the High Himalaya. The thrust earthquake belt and the topographic front are continuously correlated over the entire 2,500 km long Himalayan arc. The central 1,700 km of this arc is found to fit closely a small circle. The radius of the small circle (1,700 km) predicts a 30° dip of the thrust at the BTF in order to minimize strain energy during slab deformation at the downward bend.

The great Himalayan earthquakes occur south or up-dip from the BTF, and are interpreted as Detachment events. In the interseismic periods between the great events, moderate-magnitude thrust earthquakes are concentrated in the narrow belt down-dip from the BTF and the Detachment appears to be aseismic. This observation is crucial for correctly estimating the seismic hazard along the Himalayan arc. Down-dip from the narrow belt of thrust earthquakes, the Basement Thrust appears to be aseismic.

At the northwestern terminus of the Himalayan arc, in Hazara, locally monitored seismicity is highest on the Basement Thrust near the BTF. However, south of the BTF little seismicity can be associated with the Detachment. In the basement below the Detachment the seismicity is mostly on tear faults and on transcurrent faults, repectively transverse and parallel to the BTF. Within the sedimentary wedge the seismicity is concentrated in a relatively small and sharply defined volume.

While the seismicity within the basement and the sedimentary wedge in Hazara can be representative of the rest of the Himalayan arc, the mode of slip on the Detachment in Hazara is probably anomalous since in this area the Detachment is extensively associated with a thick evaporite layer which is not found in the central Himalaya. The surface features that characterize the Hazara arc region, including the Hazara-Kashmir syntaxis, may be the result of unusually effective decoupling caused by the salt layer associated with the Detachment in this region.

Introduction

Earthquake data have provided so far only weak constraints to models of the active tectonics of the Himalayan arc (Figure 1). The data confirm that the Himalaya is a zone of current thrusting limited to crustal depths [Gutenberg and Richter, 1954; Fitch, 1970], that the crust is unusually thick [Gupta and Narain, 1967; Menke, 1977; Chun and Yoshii, 1977], and that secondary structures involving normal and strike-slip faulting are also present [Valdiya, 1976; Banghar, 1974; Chandra, 1978; Molnar et al., 1973]. A detailed understanding of the active tectonic processes, however, has not yet been forthcoming. Since the precision and accuracy of the hypocentral data were not well defined, conclusions based on them were made with caution. Telemetered networks of high-gain seismic stations have been operating in the Hazara arc region of northern Pakistan since 1973 (Figure 2). The numerous and well-located hypocenters obtained from these networks provide a detailed view of the ongoing crustal deformation and lead to a model of the active tectonics in this area. It is evident from the network data that the active basement structures associated with the Himalayan front in Kashmir continue toward the northwest beyond the syntaxis and into

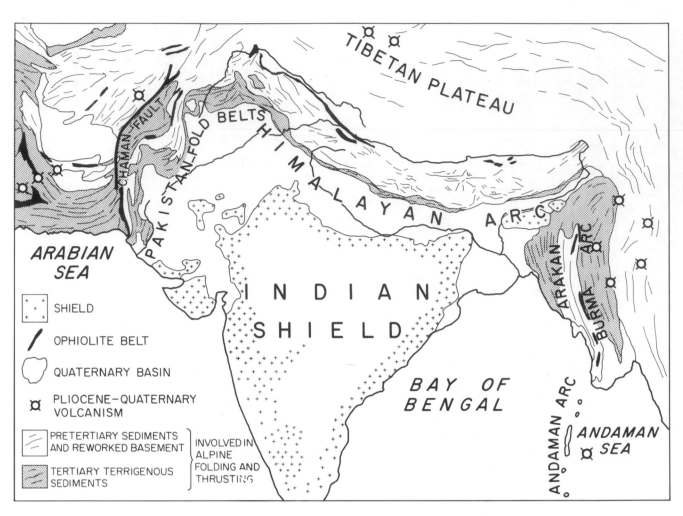

Fig. 1. Regional tectonic setting of the Himalayan arc at the northern margin of the Indian subcontinent [based on Gansser, 1964]. The seismic networks are located where the Himalayan arc meets the Pakistan fold belts. Recent volcanic activity [Katsui, 1971] and some of the tectonic features discussed in the text are indicated.

the Hazara arc region [Armbruster et al., 1978]. This suggests that the detailed knowledge of the tectonics in the Hazara arc may be applicable to the rest of the Himalayan front.

Pennington [1979] showed that at least for the northwestern terminus of the Himalaya (the region covered by the Tarbela network), there is no systematic bias of locations determined from teleseismic data with respect to those determined from local network data. (The errors in hypocenters computed using data from a local network are relatively minor). The major differences in the location pairs are in the depth of focus. Thus, the large lateral changes in velocity that are suspected to characterize the Himalaya [e.g. Menke and Jacob, 1976], do not appear to introduce a detectible bias into tele-seismic epicenters. As the major basement struc-tures beneath the Tarbela region are also found

along the central portion of the Himalayan arc, it is likely that teleseismic epicenters from the central Himalaya are similarly free of any systematic bias.

A re-examination of some of the deatils of the Himalayan seismicity should, therefore, provide new constraints to models of the active Himalayan structure. The task is undertaken here using primarily teleseismic and intensity data, in combination with results from the local network in Hazara. In the tectonic scheme which is developed we attempt to incorporate the important results from other geologic studies of the Himalaya.

The Hazara Arc- A Thin-Skin Structure Revealed by Microearthquake Data

The Hazara arc borders the extreme northwestern portion of the Indian shield and is the transition

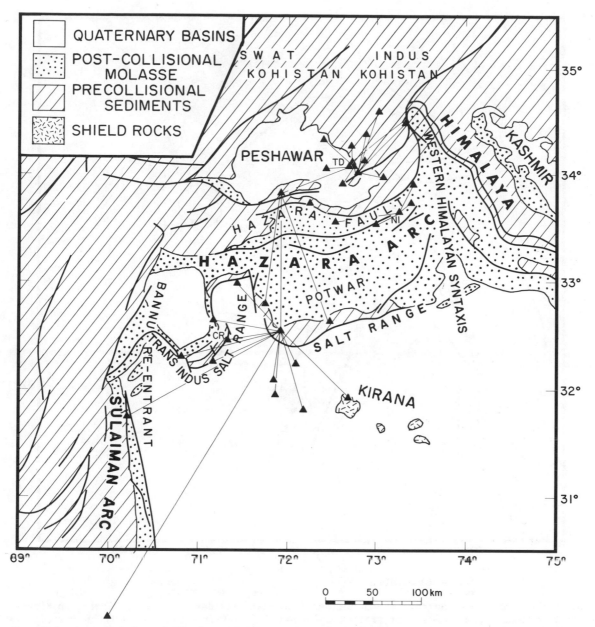

Fig. 2. Seismic networks of Northern Pakistan (stations indicated by triangles) are shown with structural features and the basic geologic units. Compare to Figure 3. This area is boxed in Figure 10. TD = Tarbela Dam.

between the Himalayan front and the Pakistan fold belt (Figure 1). It is bounded on the east by the Jhelum reentrant, at the western Himalayan syntaxis (Figure 2), and on the west by the less developed Bannu reentrant at the northern end of the Sulaiman arc [Sarwar and DeJong, 1979]. Some of the structures and stratigraphic units characteristic of the Hazara arc are not found outside of this region. Others show Himalayan affinity and still others show an affinity with the Pakistan fold belt.

The data from the local seismic network in the Hazara arc (Figure 2) indicate two northwest trending active fault zones: the Indus-Kohistan seismic zone (IKSZ), primarily a thrust fault dipping northeast, and the Hazara lower seismic zone (HLSZ), primarily a steeply dipping right-lateral strike-slip fault (Figure 3). These northwest trending structures are buried and contrast sharply with the southwest trending surface structures of the eastern Hazara arc. While neither of these fault zones can be

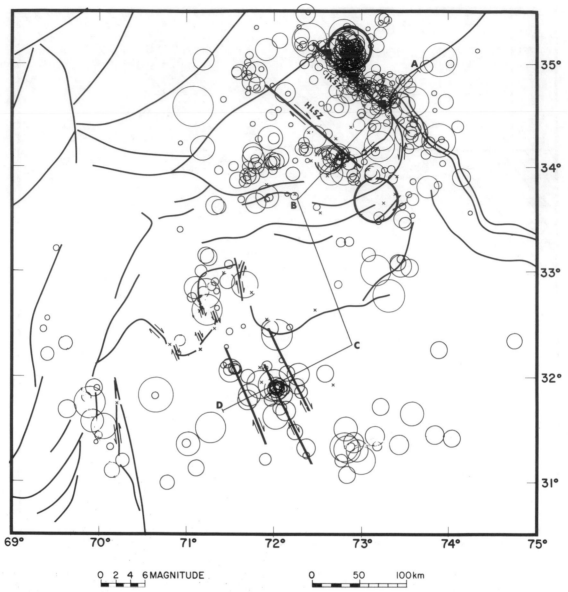

0 2 4 6 MAGNITUDE   0   50   100km

Fig. 3. Six months of epicentral data in northern Pakistan, all depths (area indicated in Figure 10). Only earthquakes of magnitude 0.5 or greater are plotted to obtain a picture of the seismicity within the area of the network unbiased by station sensitivity. This short sample is comparable to the longer term seismicity [see Seeber and Jacob, 1977, Figure 2]. The main structural features known from geologic work are shown in thinner lines; major fault zones inferred from the seismicity in the southern part of the Hazara arc are indicated in thicker lines. [From Seeber and Armbruster, 1979]. The Indus Kohistan Seismic Zone (IKSZ) and Hazara Lower Seismic Zone (HLSZ) are active basement structures which extend the main Himalayan trends past the western Himalayan syntaxis (see Figure 4).

associated with surface faulting, the IKSZ is clearly associated with a prominent topographic step in which the northeast side is elevated with respect to the southwest side. The IKSZ and this topographic step are contiguous and aligned with the thrust earthquake belt and the topographic step of the Himalaya (Figure 10; see discussion later) and are interpreted as forming a continuous

feature [Armbruster et al., 1978; Seeber and Armbruster, 1979]. Thus, the IKSZ and the HLSZ extend the basement structures of the Himalaya northwest of the Hazara-Kashmir syntaxis for at least 100 km to the Swat valley where a major structural boundary has been recognized [Desio, 1979; Tahirkheli et al., 1979].

The lack of structural correlation between the

shallow and the deep crustal active zones (Figure 3) strongly suggests a decoupling layer between these zones. The seismicity provides further indirect and direct evidence of a decoupling layer. Concentrated seismic zones are characterized by sharp quasi-horizontal boundaries (Figures 4 and 5). Moreover, where the pattern of active faulting can be resolved by fault-plane solutions over a large range of crustal depth, the fault geometry and slip vectors in the deep and in the shallow crust are found to be discontinuous. In some cases they are drastically different. Finally, direct evidence of horizontal slip on the inferred decoupling layer is also available and will be discussed later.

In Figure 4 the most accurate hypocenters obtained by the seismic networks in the Hazara arc region are shown in cross-section together with a tectonic model deduced primarily from this seismicity [Armbruster et al., 1978; Seeber and Jacob, 1978; Seeber and Armbruster, 1979]. In this model, the shallow angle thrust recognized long ago at the base of the Salt Range [Lehner, 1945; Wadia, 1961, p. 141-142] is extended northward below the Potwar Plateau [Exploration data, S.M. Hussain, personal communication] and is connected to the decoupling layer deduced from the seismicity north of the Hazara thrust (the boundary fault of the Hazara arc) forming a major detachment. Thus, the sedimentary layers of the fold-thrust belt extend northward to include the

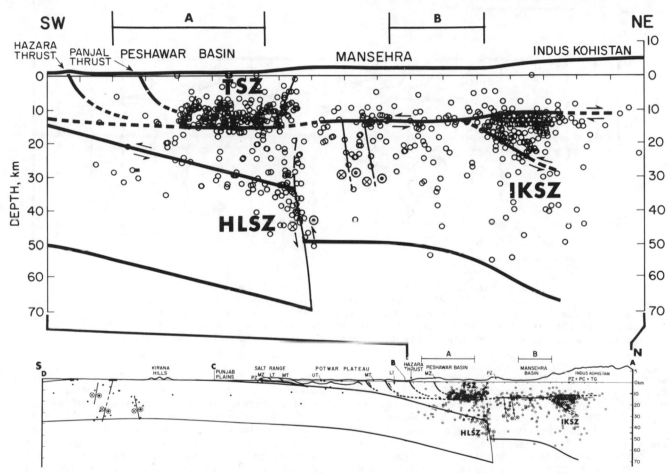

Fig. 4. The active tectonic structure in the Hazara arc region as tentatively deduced from the seismicity detected by the Tarbela network. The hypocentral data from a strip along this section is also plotted; the section is located in Figure 3. The sense of movement on the faults in the portion A-B of the section is known from composite fault-plane solutions; ⊗ = motion away from viewer; ⊙ = motion towards viewer; UT, MT, LT = Upper, Middle, Lower Tertiary; MZ = Mesozoic; PZ = Paleozoic; PC = Precambrian; TG = Tertiary granite; HLSZ = Hazara Lower seismic zone; IKSZ = Indus-Kohistan seismic zone; TSZ = Tarbela seismic zone. The Detachment, the quasi-horizontal fault that extends from the base of the Salt range to the IKSZ and beyond, is active even where seismicity is presently absent. On these portions of the Detachment slip may occur either aseismically or by rare large earthquakes (compare with Figure 11). The Moho is arbitrarily drawn at 35 km below the upper surface of the basement. No vertical exaggeration.

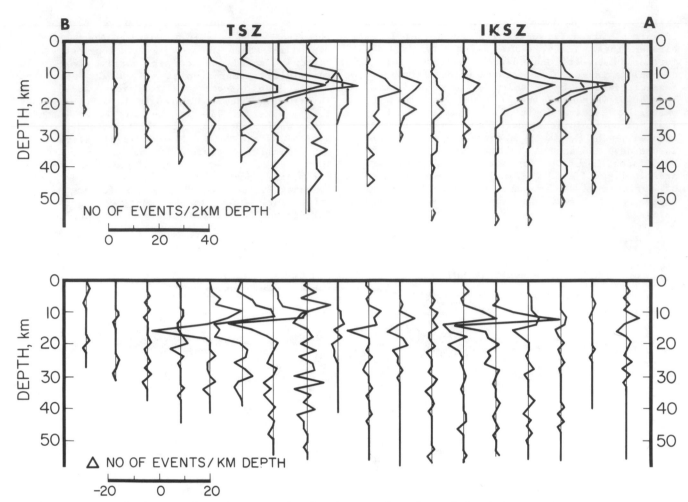

Fig. 5. The enlarged portion of the section in Figure 4 (section AB) is divided at intervals of 10 km and the rate of seismic activity with depth as well as the change (Δ) in rate of activity with depth is plotted for each 10 km. The sharpest changes in the seismicity as a function of depth are an increase at the top of the IKSZ and a decrease at the bottom of the TSZ. The detachment plane is identified with these sharp quasi-horizontal boundaries to the seismic zones. Compare to Figure 4.

metasedimentary and granitic rocks of Hazara. Moreover, this layer extends north to and beyond the IKSZ, the Himalayan basement thrust (discussed later). The metasedimentary layer above the detachment is not involved in the basement thrust and the resulting configuration resembles "flake tectonics" as observed in the eastern Alps [Oxburgh, 1972]. Accurate hypocentral locations from local networks are not yet available for the area north of the IKSZ and the configuration and extent of the detachment in this area are still unknown.

The deformational front along the Hazara arc is currently advancing into the foredeep, progressivly involving younger sediments. The transition from depositional to erosional conditions follows the beginning of deformation, so the last phases of deposition are syndeformational [Johnson et al., 1979]. The folding and

thrusting and the shortening of the sedimentary wedge along its front are the most prominent surface expressions of the ongoing southward slip of the decollement toward the foreland. Faults transverse to this deformational front have been documented in the Mianwali reentrant between the Salt Range and the Trans-Indus Salt Range [Figure 2; Hemphill and Kidwai, 1973]. These transverse faults are seismically active and occur in the basement extending as deep as the lower crust [Seeber and Armbruster, 1979].

Evidence for the southward slip of the decollement and for deep-seated or transverse faults is also available from seismicity for the area just south of the IKSZ, about 200 km north of the Salt Range [Figure 4, Area "B"). In Figure 6 the relatively low seismicity between the IKSZ and the Tarbela seismic zone (TSZ) is projected on a vertical plane striking northwest (parallel to the

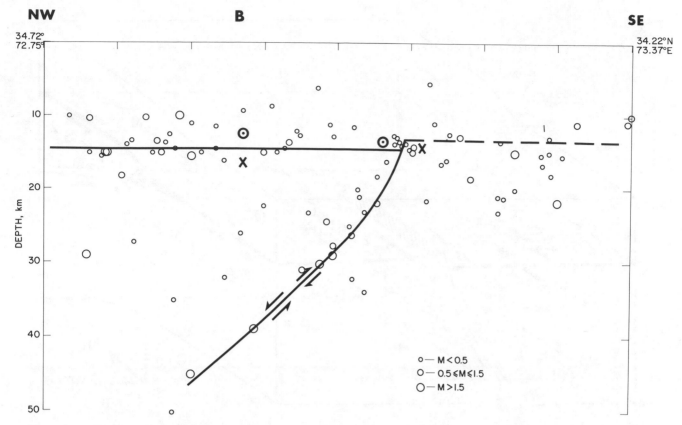

Fig. 6. A cross-section through the underthrusting crust with the detachment and overriding sediments above it. This view is toward the IKSZ from the HLSZ (looking in the direction of motion of the underthrusting block) and sees only the area where the level of activity is generally low (part B of section A-B in Figure 4). Motions as determined by composite fault plane solutions (solid lines) reveal the horizontal plane of detachment (☉ = motion toward viewer, x = motion away from viewer) which is offset by a tear fault. The dip-slip motion on the tear fault is in a different sense in different areas of the fault surface. At the offset in the detachment there is concentrated left-lateral strike-slip activity which may occur where blocks are in contact without lubrication by salt. The sense of the offset on the detachment is deduced from the polarity of the strike-slip motion. No vertical exaggeration.

IKSZ). The tectonic interpretation of the hypocentral and first motion data is schematically shown in Figure 7. The horizontal seismic zone in the depth range between 10 and 15 km is associated with the detachment. The first-motion data indicate that the decollement above the detachment moves to the southwest relative to the lower layer, presumably the basement. The deeply reaching fault perpendicular to the IKSZ is interpreted as a tear fault in the subducting slab. First-motion data indicate that different portions of this fault are offset in the depth direction with opposite sense of slip, the expected "scissoring" of a tear fault. Along the intersection between the detachment and the tear fault a narrow alignment of hypocenters (perpendicular to the section in Figure 6) is associated with a composite fault solution indicating left-lateral strike-slip. A step on the detachment (southeast side up) produced by the tear fault

would be associated with the observed pattern of slip.

Near the two very active zones of seismicity, the TSZ and IKSZ (Figure 5), the detachment may also be active; but the relatively meager seismicity from the detachment cannot be discriminated in these areas. The detachment is here recognized as the upper and the lower quasi-horizontal boundary to the IKSZ and the TSZ, respectively.

Figure 4 indicates that within the decollement there is little seismicity except for one very concentrated region, the TSZ. The seismicity within this zone occurs mostly on steeply dipping faults that experience reverse or strike-slip motion. The overall effect of these active faults is approximately a north-south shortening of the decollement [Armbruster et al., 1978]. The sharp northeast and southwest boundaries of the seismicity (Figure 4) do not correspond to any of the

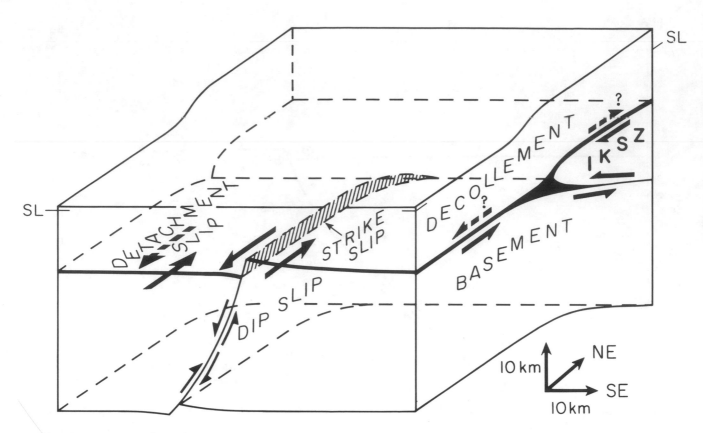

Fig. 7. A three dimensional representation of the Siran River tear fault and surrounding structures. Compare Figure 6 to the front (southwest) face and part B of Figure 4 to the right (southeast) face. Motions which have not been confirmed by composite fault plane solutions are indicated with a question mark, in particular the upper surface of the IKSZ.

faults deduced from the seismicity, or to tectonic boundaries observed at the surface. It is difficult to accept the hypothesis that internal deformation within the decollement is limited to the TSZ; this peculiar zone of activity is tentatively interpreted as a zone of seismic (brittle) deformation within a field where the deformation is primarily aseismic (plastic). Such differences in rheologic behavior may correspond to differences in rock type. In one speculative sequence of tectonic event that can account for the structure presently observed in the Hazara arc (Figure 14), the TSZ is associated with a block of basement rock that originated from the upthrown side of the basement thrust (the IKSZ) and is now trapped in the decollement.

The detachment beneath the outer portion of the Hazara arc, i.e. the Potwar Plateau, coincides with the thick Infracambrian salt formation that outcrops at the Salt Range [S.M. Hussain, Amoco Pakistan Exploration Company, personal communication]. Preliminary results from a study of apparent velocity vs. hypocentral depth (not presented here) indicate that in the inner portion of the Hazara arc, north of the Hazara thrust, the detachment coincides with a low velocity layer

[also A. Marussi, personal communication]. Recently, a significant quantity of gypsum (or anhydrite) was found in the Salkhala (Precambrian?) formation outcropping in the uplifted block along the axis of the Indus reentrant at the Tarbela dam site [Calkins et al., 1975; J. Simpson, personal communication]. These results suggest that the salt layer beneath the Potwar Plateau extends north and coincides with the detachment beneath the area of the Hazara arc covered by the Tarbela network (Figures 2 and 4).

The low-strength properties of salt are well known and the pattern of deformation in the Hazara arc may be largely effected by the presence of salt at the detachment. Unequivocal reports of major earthquakes, similar to those associated with the Himalayan detachment (see later discussion), are not documented in the historical record for the Hazara arc region [Quittmeyer and Jacob, 1979]. It is possible that most of the slip on the detachment in this region occurs aseismically at the very low stress levels characteristic of salt tectonics [Seeber et al., 1979].

The extension of the northwest trending Himalayan basement structures beyond the syntaxis indicates that the eastern portion of the Hazara arc should be considered part of the Himalayan front. Then, this extreme northwestern portion of the Himalayan front is characterized by a very wide thrust and fold belt, involving terrigenous sediments, and by the Salt Range thrust at the forward end of the belt where the sedimentary sequence above and including the Infracambrian salt is exposed. This peculiar structure, is nowhere duplicated along the forward edge of the Himalayan deformation.

Even a thin and weak sediment layer can transfer sufficient force to propagate the deformation forward in a thrust-and-fold belt if the sediment layer is very weakly coupled to the basement [Chapple, 1978]. Thus a wide belt of deformation can be expected where the detachment occurs at a weak salt layer. This suggests that the shaping of the western terminus of the Himalayan deformational front into the Hazara arc and the location of the Hazara-Kashmir syntaxis reflects the distribution of very weak coupling at the detachment associated with the Infracambrian salt layer. In this hypothesis the western Himalayan syntaxis and the related Jhelum reentrant mark the boundary between strong basement-sediment coupling with presumably no salt on the east side, and weak coupling with salt on the west side (Figure 8b).

The available data on the distribution of the basal salt support this hypothesis. On the west side of the syntaxis the presence of salt on the detachment beneath the Potwar Plateau is well documented by petroleum exploration. In the Sub-Himalaya east of the syntaxis no salt is reported in association with the Jammu stromatolitic limestones (Upper Precambrian?) on the upthrown (northern) side of the Riasi thrust (Figure 2; Gansser, personal communication). Further east, in the Kangra Sub-Himalaya about 300 km from the syntaxis, exploration wells show no salt at the base of the sediments [Mathur and Kohli, 1964]; however, salt is found along the trace of the MBT above the salt-free detachment [Srikantia and Sharma, 1972].

It is possible that Infracambrian salt was present in the sedimentary pile along large portions of the passive margin of India that subsequently became involved in the Chaman transform boundary and the Himalayan convergent boundary [Gansser, 1964, p. 26-28]. The sediment pile of the former inner margin of India is now contained in the Himalayan nappes. The extensive underthrusting has brought at the base of the sedimentary wedge, portions of the shield originally distant from the margin and formerly covered by little or no pre-collisional sediments. Thus, in the central portions of the Himalayan front we may find the Infracambrian salt at the base of the nappes, at the MBT, whereas at the base of the sedimentary wedge, at the detachment, we would expect the terrigenous sediments to rest directly on the Precambrian basement.

On the other hand, in Hazara, near the transition from the Himalayan convergent plate boundary to the Chaman transform boundary, convergence will bring to the Himalayan front portions of the India margin facing the Chaman transform. Thus in this area the youngest terigenous sediments continue to be deposited on the pre-collisional marginal sediment pile and the detachment still occurs at the Infracambrian salt layer at the base of this pile.

A Seismo-Tectonic Prospective of the Himalaya

The Himalayan Arc

Figures 9 and 10 show the salient features of topography, structure and seismicity along the Himalayan front and surrounding region. The epicenters shown in Figure 10 represent a reasonable compromise between reliability and completeness (U.S. Geological Survey; 1963-1977; 20 or more P-arrival times per determination); the fault plane solutions comprise a representative set; only the data from the better known great earthquakes are shown. The topography is represented by the deposition-erosion boundary, the southern limit of the Siwalik outcrop (near sea level), and by the series of points at the 4 km contour nearest the foreland (4 km thresholds in Figure 10). These points marks the steep topographic gradient (topographic front) between the Lesser Himalaya (average elevation near this steep gradient is 3 km) and the High Himalaya [average elevation is 5 km; Bird, 1978].

Structure, topography and seismicity are quite uniform along the Himalayan front. The most prominent seismic feature is a narrow earthquake belt where all available fault-plane solutions indicate thrusting. This belt can be easily traced along the entire Himalaya. The dashed curve in both Figures 9 and 10 is a portion of a small circle centered at 42.25°N and 91.10°E with a radius of 1695 km. This geometric arc provides the best visual fit to the narrow belt of thrust earthquakes. The fit is very good for about 1,700 km between 76°E and 92°E along the central Himalaya (see also Figure 11), but the earthquake belt diverges from the circle along the more complex eastern and western terminations of this structure. The topographic front, as defined above, follows very closely the seismicity, not only in the central portion of the Himalaya, where they both fit the small circle, but also in the extreme eastern and northwestern portion of the Himalaya where they both deviate from the small circle. The MCT, where mapped [Gansser, 1977], also follows closely this seismo-topographic feature in all regions except in Hazara (see previous section; compare Figures 9 and 10).

In conclusion, the most striking seismic, topographic and surface structural features of the Himalaya are closely associated in space over the entire arcuate megastructure. Moreover, along the central portion of the Himalaya these features fit very closely to a small circle and define the

(A)

(B)

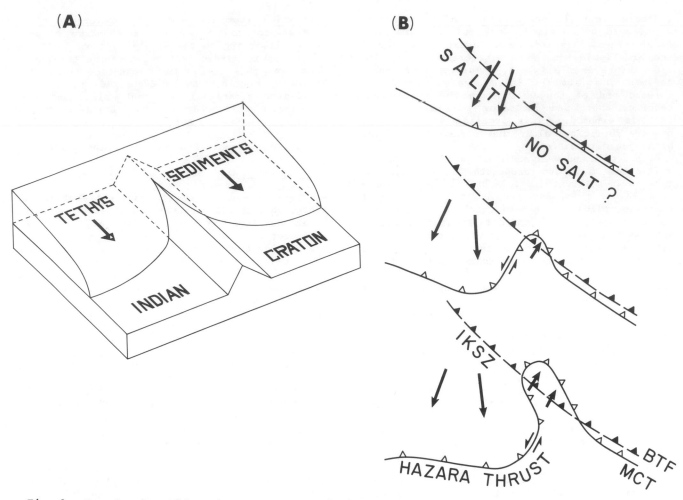

Fig. 8. Two sketches of how the northwestern Himalayan syntaxis could have been formed. A: Wadia [1931] suggested that a pre-existing projection or horst on the Indian craton has divided the Himalayan front into two components, molding the structures around the obstruction. B: Our model for the formation of the syntaxis begins with two contrasting portions of the continental shelf of India, one with a thick section of sediments underlain by salt, the second with less sediment and no major effect of salt in the tectonics. Basement thrusting (solid symbol) and thrusting at the surface (open symbol) remain together where there is no effect of salt, e.g. the Basement Thrust Front (BTF) and the Main Central Thrust (MCT) in the central Himalaya. Where salt provides a plane of weakness, gravitational force and/or a force generated at a convergent zone further to the northeast push the surface features (e.g. the Hazara thrust, Figures 2 and 4) southward from the basement structure (e.g. the Indus-Kohistan seismic zone, IKSZ). Figure 14 is a cross sectional view of the structure of the Himalayan front with and without salt.

Himalayan arc. This suggests that these features are related to the same fundamental element of Himalayan tectonics (discussed in a later section).

Great Himalayan Earthquakes. Great earthquakes ($M_s \geq 7.8$) which occurred along the Himalayan arc during the past 90 years provide crucial information on the tectonic processes currently affecting the Himalaya. The available data (discussed below) indicate these great earthquakes are not the result of slip on the main boundary thrust or any other of the imbricate thrusts of the Sub-Himalaya, but rather they rupture a quasi-horizontal surface that extends south of the MBT and is analagous to the detachment documented in the Hazara region (see earlier section). Previous workers have implied that the earthquakes comprising the band of activity located north of the MBT are tectonically similar to the great Himalayan shocks. In the model presented below, we suggest that the great earthquakes form a seismic belt distinct from the belt of moderate-magnitude thrust-earthquakes and that the difference in the style of seismic energy

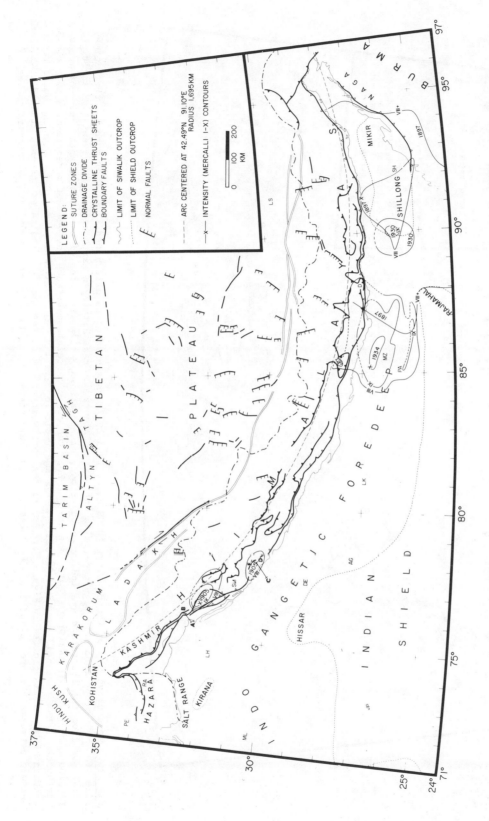

Fig. 9. Structural map of the Himalayan arc and Tibet [sources: Gansser, 1977; Molnar and Tapponnier, 1978] with intensity contours of great earthquakes along the Himalayan front. A-B and C-D-E are leveling lines shown in Figure 12. Note the relation between a small circle fitted to the Himalayan arc and the crystalline thrust sheets. Cities: PE = Peshawar, RA = Rawalpindi, LH = Lahore, ML = Multan, JP = Jodhpur, SM = Simla, DE = Delhi, AG = Agra, LK = Lucknow, PA = Patna, MZ = Muzaffarpur, D = Darjeeling, LS = Lhasa, SH = Shillong.

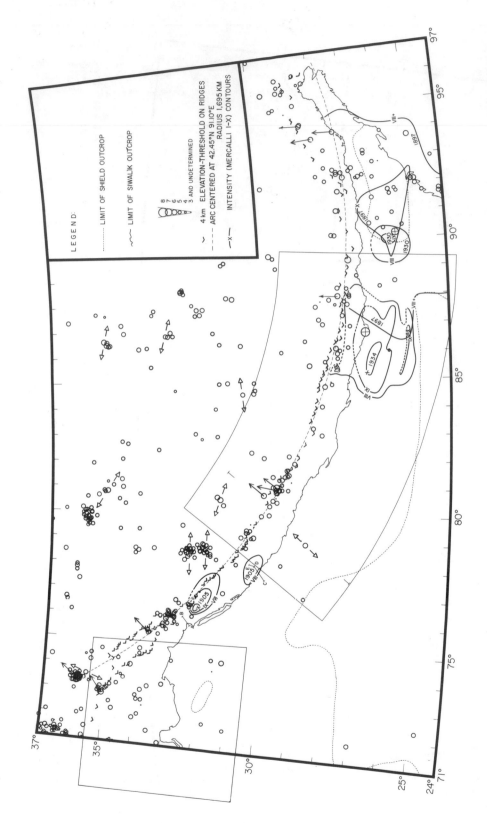

Fig. 10. Epicentral map of the Himalaya (USGS data, 1963-1977; only epicenters determined with 20 or more P arrival-times), size of circle proportional to magnitude, with intensity contours of Great Himalayan earthquakes. Representative fault plane solutions from the literature are indicated by arrows, single arrows = slip direction of thrusting events, pairs of arrows = tension axes of normal faulting events (see text). Note the close correlation between the belt of moderate magnitude thrust-earthquakes, the threshold of 4 km elevation and a small circle fitted to the Himalayan arc. Data from the boxed central section of the arc is plotted in Figure 11. Box at the northwestern end of the arc is studied with local seismic networks (Figures 2 and 3). ⊕ = instrumental epicenters of the 1934 Bihar and the 1930 Dubri earthquakes.

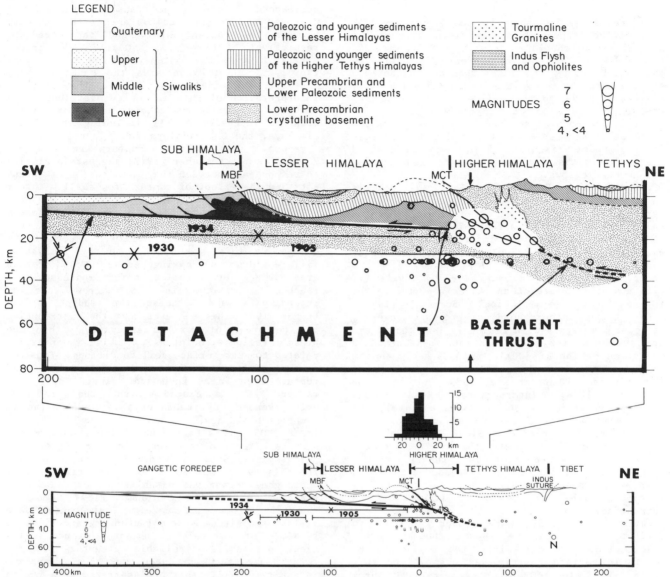

Fig. 11. The most reliable seismic data for the central Himalaya are shown with geology. Hypocentral data from 1963 to 1977 as in Figure 10 (boxed area) are projected along the Himalayan arc (open circles). The preferred nodal planes for representative thrust solutions (as in Figure 10) are shown as line segments drawn across the hypocenter symbol. N marks a normal faulting event with tension axis into and out of the figure. The concentration of earthquakes at 33 km is not real: this depth is arbitrarily assigned to events for which a reliable depth determination is not available. The epicenter and extent of intensity VIII associated with the 1905 Kangra (M = 8; no reliable epicenter), 1930 Dhubri (M = 7.1), and 1934 Bihar (M = 8.3) earthquakes are indicated by crosses and horizontal lines, respectively. Surface geology is from Gansser [1964, plate II, section A]. Below sea-level the structures are somewhat modified to fit the seismicity. The thicker line is the master thrust. The largest earthquakes seem to be associated with the shallower portion of this thrust, the Detachment, while moderate-sized earthquakes occur down-dip on the Basement Thrust. The projected location of the topographic front as indicated by the 4 km thresholds (Figure 10) is given in the histogram.

release between these two belts has a physical basis in the tectonic structures associated with each belt.

Figures 9 and 10 show intensity contours for four of the largest Himalayan earthquakes of the past 90 years; the 1897 Assam (M_s = 8.7), the 1905 Kangra (M_s = 8.0), the 1934 Bihar (M_s = 8.3) and the 1930 Dubri (M_s = 7.1) earthquakes. (The 1950 Assam earthquake is ommitted because its intensity distribution is poorly defined and because the eastern extremity of the Himalayan arc may be tectonically complex and atypical). Data from two of these events (1934 and 1905) play a major role in the model discussed below and a detailed discussion of the pertinent facts related to these two shocks follows.

The 1934 Bihar Earthquake. This earthquake is in many ways the best known of the great Himalayan earthquakes because it affected highly populated and easily accessible parts of India, and because it occurred at a time when instrumental seismology was relatively well developed. The epicenter shown in Figure 10 (26.77°N, 86.69°E) is a relocation we obtained using 109 P arrivals (77 with a weight greater than .4) as reported in the International Seismological Summary [1934]. Although the southern quadrant is poorly covered, at least 3 arrivals with residuals less than 3 sec are available for each quadrant. The standard deviation of the residuals is 3.29 sec when the depth is held at 33 km. Other determinations of the epicenter [Dunn et al., 1939; Gutenberg and Richter, 1954; International Seismological Summary, 1934] fall close to (within 50 km), but to the southwest of the relocated position.

The epicenter for the Bihar earthquake is situated about 100 km toward the foreland from the belt of moderate-magnitude thrust events (Figure 10). Even allowing for reasonable errors, it does not seem possible that the 1934 rupture initiated within this belt as defined by the current seismicity.

Dunn et al. [1939] were probably correct in their conclusion that the Bihar earthquake did not result from movement along the MBT since no coseismic surface rupture was observed on this fault. The relocated epicenter of the 1934 shock falls slightly to the south of the MBT (Figures 9 and 10). The large moment of 1934 event [1.6 x 10^{28} dyne-cm; Chen and Molnar, 1977] implies rupture dimensions of the order of hundreds of km and displacements of several meters. It seems very unlikely that a rupture with these characteristics which nucleates near the trace of the MBT does not reach this trace.

The meizoseismal zone lies primarily south of the MBT. The region of intensity greater or equal to VIII extends about 300 km along the strike of the Himalaya and 250 km perpendicular to the strike (the intensity scale used by the Geological Survey of India in this study is a modified form of the Rossi-Forel scale, the Mercalli scale, with ten intensity categories). In the transverse direction it covers entirely the Lesser Himalaya, the Sub-Himalaya, and the foredeep. The northern and southern boundaries of the intensity VIII area correspond closely to the transition between the High and the Lesser Himalaya and to the boundary between the Indian shield and the foredeep, respectively (Figure 10). Outside the intensity VIII area the intensities fall off very rapidly. Inside this area the zones of high intensity are widely distributed. The simplest interpretation of this intensity distribution is that the seismic source extends under most of the area of intensity $\geq$ VIII including the Lesser Himalaya, the Sub-Himalaya and a portion of the foredeep. This implies a rupture area of about 75 x 10^3 km^2. Kelleher [1972] suggests a similar correspondence between the distribution of intensity and the extent of rupture for shallow thrust earthquakes along the Andes.

The largest region of intensity X is closely associated with a "slump belt" and a zone of soil liquefaction [Dunn et al., 1939]. Much of the damage in this area is caused by foundation failure and not to shaking per se. South of this area damage by actual shaking increases and reaches a maximum in the narrow zone of intensity IX and X near the Indian shield (Figure 10). This suggests that the distribution of the areas of maximum intensity within the broader region with intensity $\geq$ VIII is primarily related to near-surface soil conditions. Thus the central area of intensity X does not delimit the rupture area [cf. Chen and Molnar, 1977], but rather is associated with the relative effectiveness of liquefaction versus shaking in damaging buildings.

The 1905 Kangra Earthquake. The report on the surface effects of the Kangra earthquake of 1905 [Middlemiss, 1910; the Rossi-Forel scale is used for this event] was compiled in the tradition of thoroughness that was established for the Geological Survey of India by Oldham's [1899] Memoir on the 1897 Assam earthquake. Instrumental data yields a magnitude M_s = 8 [Gutenberg and Richter, 1954]. Unfortunately these data are not sufficient to establish a reliable epicenter.

The Kangra shock produced two zones of maximum intensity ($\geq$ VIII) that are separated by a region of lower intensity about 100 km long. The available data suggest this intensity field was generated by a single great earthquake [Richter, 1958, p. 63; Quittmeyer and Jacob, 1979]. This implies that the intensity VIII isoseismal may be a conservative indication of the area of rupture of the detachment earthquakes of the Himalaya. Thus, both the 1905 and 1934 events appear to have ruptured sections of the Himalayan arc approaching 300 km in length, however, the extent in a transverse direction of the intensity VIII area, and presumably of the rupture, is much less for the 1905 earthquake than for the 1934 shock. This difference may be related to the relatively shallow depth-to-basement in the foredeep associated with the 1905 event [Mathur and Kohli, 1964].

The axis of maximum intensity in the 1905 earthquake follows approximately the trace of the MBT, and the MBT is usually considered the fault associated with this event [e.g. Krishnaswamy et al., 1970]. However, an extensive search [Middlemiss, 1910] found no evidence for coseismic displacement along the trace of the MBT or any of the other imbricate thrusts of the Sub-Himalaya. As with the 1934 event, this is a strong argument against a major subsurface rupture on the MBT during the 1905 event.

Leveling data provide additional evidence that the main rupture associated with the 1905 earthquake is not on the MBT. A level line through Dehra Dun that crosses the MBT (line CDE in Figure 9) was surveyed in 1904 and resurveyed in 1906 following the Kangra earthquake [Middlemiss, 1910]. The observed changes in elevation associated with the 1905 event, presumably mostly coseismic, indicate less uplift on the northeast side than on the southwest side of the MBT (Figure 12). This would suggest normal movement (south up, north down) during the 1905 shock if slip occurred predominantly on the MBT. Such movement is in an opposite sense to the Quaternary displacements observed on the MBT in the same area. Furthermore, Molnar et al. [1977] interpret recent changes in elevation observed along a releveled line about 50 km northwest of the 1905

meizoseismal area [Chugh, 1974; Figure 12; Line AB, Figure 9] as evidence for interseismic reverse slip that is occurring aseismically on the MBT. Our preferred explanation is that the 1905 rupture is not on the MBT but rather on a buried thrust fault that dips shallowly under the entire 1905 earthquake region. Such a rupture can account for both the interseismic and coseismic level changes (assuming that these profiles, AB and CDE in Figure 10, at the opposite ends of the 1905 intensity VIII area are representative).

This point is best illustrated by comparing elevation changes during the 1905 Kangra event and the 1964 Alaska earthquake (Figures 15 and 16). Elevation changes during the Alaskan shocks were successfully explained by a simple model involving slip on a shallow-dipping thrust [Savage and Hastie, 1966]. Coseismic uplift occurred above the shallow portion of this rupture nearest to the foreland; coseismic subsidence occurred above the deeper portion of this thrust toward the hinterland [Plafker, 1965]. An opposite pattern of uplift and subsidence is observed in the interseismic period [Plafker and Rubin, 1978]. The leveling data for the area affected by the Kangra earthquake fit qualitatively into a similar pattern; the boundary between uplift and subsidence corresponds approximately with the trace of the MBT.

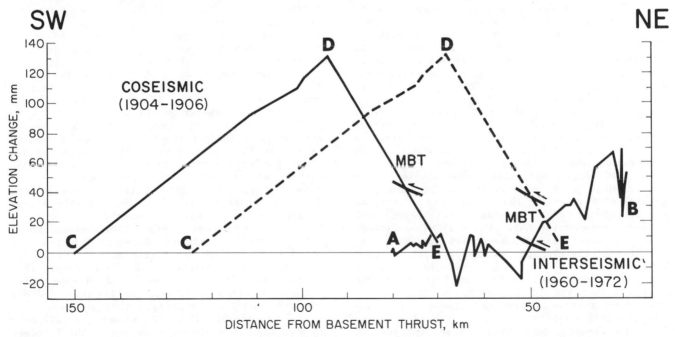

Fig. 12. Releveling data in the area of the 1905 Kangra earthquake - the profiles are located in Figures 9 andn 10. Profile A-B (reference point at A) [Chugh, 1974] and C-D-E (reference point at C) [Middlemiss, 1910] are near the northwestern and southeastern ends of the inferred rupture, respectively. The solid plots are aligned according to the distance from the Basement Thrust (BTF) taken as the small circle in Figures 9 and 10. The dashed C-D-E is aligned to A-B according to the distance to the Main Boundary Thrust (MBT). The coseismic elevation changes are not consistent with thrust slip on the MBT. Both coseismic and interseismic level changes can be interpreted as slip and relative strain accumulation, respectively, on a buried shallow dipping detachment (see Figure 16).

The coseismic change in elevation in the Dehra Dun profile (CDE in Figure 9) is relative to point C. Since point C itself may have been uplifted coseismically, the actual uplift may be considerably greater than shown in Figure 12. Thus the actual difference between the maximum uplifts in 1964 and 1905 may be less than it appears in Figure 16 (the ratio is about 10 to 1 if the local effects from surface faulting in 1964 are neglected).

Other Large Earthquakes. The region with "severe destruction" (intensity $\geq$ VIII) in the 1897 Assam earthquake has an area of 1.8×10^5 km^2 [length = 600 km x width = 300 km; Oldham, 1899]. If this region is approximately equivalent to the rupture zone, this earthquake has a rupture size similar to the greatest earthquakes on record (e.g. Alaska, 1964; Kamchatka, 1952). The distribution of intensities for the 1897 earthquake Figure 9 is consistent with a rupture that extends far into the foredeep region.

The events in 1803 and 1833 [Oldham, 1882] are the only known earthquakes prior to 1897 in the central portion of the Himalayan front that can possibly be categorized as great (M $\geq$ 7.8). The maximum intensity areas for these events are poorly defined, but they are consistent with ruptures filling the gap between the 1905 and the 1934 ruptures.

The Dubri earthquake of 1930 [Gee, 1934] (M_s = 7.1) has a fairly reliable epicenter in the foredeep within the meizoseismal area of the 1897 earthquake. It clearly is not associated with the MBT (Figures 9 and 11). However, in view of the relatively small magnitude, it is not possible to preclude a steeply-dipping rupture for this event. A mechanism of faulting other than a shallow thrust is possible for the Dubri event.

Summary. All the well-surveyed meizoseismal areas of great Himalayan earthquakes are characterized by the absence of significant coseismic fault offset at the surface of the earth. This constraint and various additional constraints from leveling, intensity distribution, and instrumental locations indicate that these earthquakes (1897, 1905, 1934) ruptured a shallow dipping fault that extends south beyond the surface trace of the MBT. The 1897 event extends south the most, the 1905 event extends the least. Northward, these ruptures appear to continue beneath both the Lesser and Sub-Himalayas. Thus, the great Himalayan earthquakes form a seismic belt distinct and to the south of the belt of moderate-magnitude thrust-earthquakes and they are probably associated with a buried shallow-dipping thrust that lies below the Lesser Himalaya, the Sub-Himalaya and part of the foredeep.

Tectonic Scheme

The main features of the tectonic scheme developed here and some of the data that fits this scheme are shown in Figure 11. These features are

(1) the Detachment, a thrust with very shallow dip that extends beneath the foredeep and the Lesser and Sub-Himalayas, and that separates the sedimentary wedge from the basement; and (2) the Basement Thrust, a more steeply dipping thrust that juxtaposes basement of the Indian shield with the Tethyan slab, the pre-collisional leading edge of the Indian shield [Powell, 1979]. The Detachment merges downdip into the Basement Thrust and the transition between the two is labelled the Basement Thrust Front (BTF). This front is a fundamental feature of Himalayan tectonics, and it is the underlying cause of the three prominent features observed at the surface along the Himalayan arc: the high topographic gradient between the Lesser and High Himalaya, the belt of moderate-magnitude thrust earthquakes, and the trace of the Main Central Thrust (Figures 9 and 10).

The structure above sea level in Figure 11 is taken from the section of the Kumaon by Gansser [1964, plate IIA; see Figure 10 for location]; below sea level, Gansser's section is modified and expanded to explain the seismic data. The seismic data in Figure 11 include all USGS hypocenters with 20 or more P-wave arrival times from 1963 to 1977, in the 1,200 km long portion of the Himalaya boxed in Figure 10. Also shown in Figure 11 are the intensity VIII distribution and epicenters (not available for the 1905 event) for the great earthquakes of 1905 and 1934 as well as for the 1930 Dubri earthquake.

The great earthquakes and the moderate-magnitude thrust earthquakes stand out as distinct belts of seismicity (Figure 11). The scatter of the moderate-magnitude events about the best-fitting small circle (0 km on the horizontal coordinate in Figure 11) is quite small considering the length of the section (1,200 km). On the other hand, the great earthquakes clearly are caused by fault ruptures that extend far south of the small circle. Thus, the strain release along the Detachment occurs during great earthquakes, while more frequent but smaller earthquakes characterize the strain release on the Basement Thrust at the BTF. The possible significance of aseismic movement on both the Detachment and Basement Thrust is discussed later.

The northward dip of the Basement Thrust at the BTF is constrained by fault-plane solutions for four moderate-magnitude earthquakes (Figure 11). The average dip of the northward dipping nodal plane is 30° [Molnar et al., 1977]. Thus, the dip of the BTF is significantly steeper than 1.5°-3.0°, the dip of the Detachment as indicated by the basement-sediment interface in the foredeep [Mathur and Kohli, 1964].

It is useful to devise a new subdivision of the Himalaya based upon the tectonic model in Figure 13. The grounds for this subdivision are the most general characteristics of the active tectonic structure; purposely omitted are subdivisions related to older structures that now do not have a major role. In this scheme the BTF

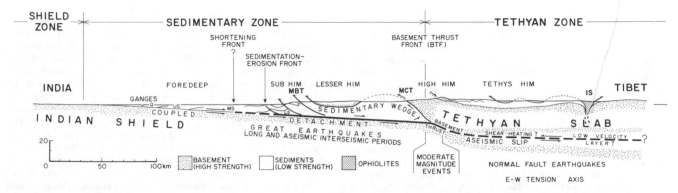

Fig. 13. Model (to scale) of the Himalayan continental subduction structure in the central portion of the arc (compare with Figure 11). The progressive stages of the upper surface of the subducting slab are: (1) subaerial in the shield zone, (2) buried below the sedimentary wedge, first as a non-tectonic contact, then as the Detachment with great earthquakes, (3) buried below the Tethyan slab, first as the seismically active Basement Thrust down-dip from the BTF, then as the aseismic but still active contact between basement slabs. The arrows indicate qualitatively the relative velocity in a frame fixed to the Tethyan slab. Q = Quaternary; US, MS, LS = Upper, Middle, Lower Siwalik; MBT = Main Boundary Thrust; MCT = Main Central Thrust; IS = Indus Suture.

is the fundamental boundary between the Tethyan zone to the north and the Sedimentary zone to the south. Since the Indian shield to the south of the Sedimentary zone is also involved in the present deformation a Shield zone is also designated.

The Tethyan zone includes the High and the Tethys Himalaya as well as an indefinite portion of the Tibetan plateau. The most striking feature of this zone is the uniformly high topography with an average elevation of about 5 km. In the Tethyan zone seismicity is widespread (Figure 10). Fault plane solutions indicate that normal dip-slip motion predominates, although some strike-slip motion is also seen. These fault movements are generally consistent with east-west extension in this zone.

LANDSAT imagery reveals a widespread system of normal faults and grabens striking north-south within the Tibetan zone [Figure 9; Molnar and Tapponnier, 1978; Ni and York, 1978]. Among these structures those that are better known from surface investigations are south of the Indus suture in the Tethys and High Himalaya. The Thakkola graben in western Nepal is a classical example [29°N, 84°E in Figure 9; Bordet et al., 1971]. These prominent graben structures do not extend south of the BTF (the southern boundary of the Tethyan zone). They appear to be the latest phase of deformation in the Tethyan zone.

The Indus-Tsangpo suture also lies within the Tethyan zone (Figure 13). This feature separates terrane of Gondwana affinity, with a pre-collisional sedimentary character, on the Indian side, from terrane of still mostly unknown affinity, but possibly also Gondwana, on the Tibetan side. The suture presumably marks the site of the first collisional contact between India and some continental or island arc terrane [Stocklin, 1977; Powell, 1979; Klootwijk, 1979].

Presently the Indus-Tsangpo suture appears neither to be tectonically active, nor to serve as a tectonic boundary. There is no recognizable earthquake zone (Figures 10 and 11) nor a topographic break [Bird, 1978] nor any other geomorphic indication of recent differential movement. Furthermore, it is "interrupted by north-south aligned structural anomalies in the Tibetan as well as the Himalayan arc" [Gansser, 1977].

Thus, seismicity, surface faulting, and topography, all very sensitive indicators of current tectonism, are remarkably uniform over the Tethyan zone. This suggests a uniform tectonic process in this zone, and poses difficulties to the "waterbed" hypothesis in which the exterior horizontal forces on the system are balanced by the lithostatic pressure associated with the Tibetan plateau north of the suture and by the shear stress on the Himalayan thrust south of the suture [Tapponnier and Molnar, 1976]. In their hypothesis the equilibrium is static north of the suture and dynamic south of the suture. This fundamental difference is incompatible with the lack of recent and contemporary tectonism at the suture [Gansser, 1977].

The tectonic process active in the Tethyan zone, as indicated in Figure 13, is very similar to the one suggested by Powell and Conaghan [1973]. The subducted portion of the Indian shield lies under a crust of normal thickness in the Tibetan zone and thus produces an effective thickness of the crust that is twice the normal value. Powell and Conaghan suggest that the entire Tibetan plateau is underlain by the subducted Indian slab. This interpretation of the uniform character of the morpho-tectonic features of the Tibetan plateau is both appealing and simple. However, the above discussion does not necessarily support their model and a detailed discussion of the Tethyan zone far from the Basement Thrust Front (BTF) is beyond the scope of this paper.

The Basement Thrust down-dip from the BTF, including the presumed quasi-horizontal extension of the Basement Thrust located between the upper and lower crustal layers in the Tethyan zone, should experience a uniform rate of long-term slip. However, none of the fault-plane solutions obtained for the seismicity scattered over the portion of the Tethyan zone north of the narrow belt of thrust earthquakes near the BTF are consistent with seismic slip on the Basement Thrust extension. Since there is no record of large, infrequent earthquakes in the Tethyan zone, the model we have developed requires that slip on the Basement Thrust be primarily aseismic, except near the BTF.

This conclusion is reinforced by the micro-earthquake data from the Hazara region (Figures 3 and 4). The Indus-Kohistan seismic zone (IKSZ) is the northwestern termination of the Basement Thrust. The seismicity in the IKSZ associated with the thrust does not extend more than 20 km down-dip from the BTF, from a depth of 13 km to about 25 km. Further down-dip the Basement Thrust is not recognized in the seismicity and presumably slip occurs here in an aseismic manner. The Basement Thrust in oceanic subduction zones (see later section) is also characterized by a transition from seismic to aseismic slip at a relatively short distance down-dip from a Basement Thrust Front.

In the scheme of Figure 13 the BTF is not permanently fixed with respect to the Tethyan slab (the hanging wall of the thrust). The rate of erosion in the Tethyan zone near the BTF can be expected to increase rapidly toward the south and reach a maximum at the high Himalaya. In order to maintain the equilibrium of this dynamic system, the BTF moves north relative to the hanging wall and the Tethyan slab slides up along the MCT. Thus, the High Himalaya are uplifted to counteract the effect of erosion. In this process portions of the Tethyan slab originally down-dip from the BTF are gradually brought to the surface. The ductile shear zone associated with the MCT [e.g. LeFort, 1975], which does not show signs of recent near-surface activity, can be interpreted as an uplifted portion of the Basement Thrust. The MCT as mapped by Valdiya [1977], south of the ductile shear zone, can be interpreted as the active fault between the rising Higher Himalaya and the Lesser Himalaya. The peculiar metamorphic imprint across the ductile MCT and the recent anatectic leuco-granites of the High Himalaya (Figure 11) can then be related to shear heating and/or the presence of water-rich sediments on the Basement Thrust [LeFort, 1975]. The pressure required by the metamorphism on the MCT is consistent with the depth of the Basement Thrust as indicated by the seismicity. The MCT as a shear zone is consistent with aseismic slip on the Basement Thrust, down-dip from the belt of moderate-magnitude earthquakes.

The Indian Shield Zone is the most stable unit

in the Himalayan orogen. Nevertheless, a gentle gravity high indicates that the shield bulges upward forming a broad rise south of the foredeep. This feature is similar to the outer rise of the oceanic basement that is located oceanward of a subduction trench [Molnar et al., 1973]. North of this rise, the Indian shield dips gently beneath the foredeep and forms the basement of the sedimentary trough. The pronounced gravity low associated with the Himalaya south of the Basement Thrust Front (BTF), including the foredeep, can be modeled most simply by a rigid lithospheric slab that is held down below the level at which it would be in hydrostatic equilibrium [Warsi and Molnar, 1977]. The limited gravity data available from the area north of the BTF [Marussi, 1976; Kono, 1974; Qureshy, 1974] indicate a gravity high in that region. This would be expected if the area below the equilibrium level south of the line of contact between the overriding and underriding slabs is balanced by an area above hydrostatic equilibrium north of this line. In our model (Figure 13), this line corresponds with the BTF.

The Sedimentary Zone is located between the Tethyan zone and the Indian shield; it contains the Himalayan sedimentary wedge. Conforming to a typical wedge configuration [e.g. Chapple, 1978], the apex of the wedge is toward the foreland and the thick end is toward the hinterland. The Sedimentary zone includes the foredeep and the Sub-Himalayas, with terrigenous sediments of authochthonous and allochthonous character, as well as the Lesser Himalaya nappe-zone involving pre-collisional sediments of the inner shelf and crystalline sheets from the outer shelf, the forward portion of the Tethyan slab. Authochthonous shield rocks are not found anywhere in the Sedimentary zone.

The topography averaged along strike rises northard from near sea level, at the erosion-deposition boundary between the Sub-Himalaya and the foredeep, to about 3 km just south of the Basement Thrust Front (BTF) [Bird, 1978; Figure 13]. Most of the high and currently rising area of erosion was never covered by terrigenous sediments, whereas the subsiding area (foredeep) has long been the site of deposition. This pattern appears to have persisted through Siwalik time. Vergence is consistently down-slope, toward the foreland.

The structure and stratigraphy of the foredeep, the outer portion of the sedimentary wedge, are relatively well known from exploration data, including a number of deep wells [Mathur and Kohli, 1964]. In the foredeep the basement at the lower boundary of the wedge is usually the Indian shield; isolated pockets of pre-collisional sediments are occasionally found. This boundary dips northward at an angle of 1.5° to 3.0°. Subsurface data other than seismicity are not available in the Lesser Himalaya. Here the sediment-basement boundary identified with the Himalayan detachment is extended somewhat speculatively to the BTF at the same angle of dip (approximately 2.5° in Figure 13). In this model the depth to basement is about 10 km at the Sub-Himalaya, in agreement with aeromagnetic data [Mathur and Kohli, 1964], and about 15 km at the BTF. At the BTF the depth of basement is weakly constrained by the seismicity on the Basement Thrust. Most of the hypocenters for which the depth is not arbitrarily fixed at 33 km, cluster from 10 to 20 km. Preliminary results from a detailed modeling of the P and S waveforms for the larger earthquakes in this zone confirms that hypocenters in the central portion of the belt of thrust earthquakes are in this depth range. In Hazara the depth to basement at the BTF is well constrained by microearthquake data at about 12 km (Figure 4). In our model, the pressure required for the metamorphism on the MCT [LeFort, 1975] is also a constraint for the depth of the BTF (see above) and is in agreement with the depth obtained from the seismicity.

The thinnest portion of the wedge near the shield zone consists entirely of Quaternary alluvium that rests in a non-tectonic contact directly on the basement. The Upper, Middle and Lower Siwaliks are first found in the sedimentary column further down-dip along the lower boundary of the wedge. The spacing in this progression is such that the Lower Siwaliks are usually found in the imbricate thrust zone of the Sub-Himalaya but not south of this zone [Mathur and Kohli, 1964]. This implies that the volume of Lower Siwalik deposits located south of the trace of the MBT is relatively small (Figure 13). Most of these sediments must be either carried away by erosion and/or underthrust at the MBT [Gansser, 1964, p. 246]. A comparison with the Appalachians (see later section) suggests that the Siwaliks have been extensively thrust under the Lesser Himalaya and that on most of the Detachment the basement is in contact with terrigenous sediments. Thus, the older sediments and the crystalline rocks forming the nappes of the Lesser Himalaya may be riding on a substratum of Siwalik sediments, and the rheology of these molasse-like sediments may be the most important factor in determining the bulk strength of the sedimentary wedge.

The asymmetrical pattern of molasse deposition in the foredeep, whereby the older Siwaliks are only found on the northern and deeper part of the basin, suggests both that the basement has been steadily subducting northward below the sedimentary wedge during the deposition of the Siwaliks and that it has dragged these sediments down at least part of the way. The shear stress applied on the lower face of the wedge, dragging the wedge toward the hinterland, is counteracted by the push toward the south applied on the back side of the wedge by the Tethyan slab. Thus sediments are dragged toward the hinterland at the bottom of the wedge and flow toward the foreland near the upper surface of the wedge. Both the southward dipping slope and the vergence at the upper surface support this model.

To clarify the above discussion consider the following simple system which has a similar dynamic equilibrium: a conveyor belt (the subducting basement) passes below a scraper (the Tethyan slab). A load of sand (the sedimentary wedge) is trapped between the belt and the scraper. As the sand is dragged down toward the scraper by the belt, it piles up against the scraper and then flows down-slope, away from the scraper, near the free surface [Wiltschko, 1979]. In the real situation sediments are transported in and out of the system by erosion and deposition. The perturbing effects of this external mass transport on the simple closed system may be quite significant. The rather uniform sedimentation rate during the Siwalik period [Johnson et al., 1979; Mathur and Kohli, 1964; Curray and Moore, 1971] suggests that this external mass transport is in fact incorporated in the dynamic equilibrium of the system.

In general, the deformation in the sedimentary zone is entirely confined to the wedge, and does not involve the basement. Thus folding in the Sub-Himalaya implies that the Detachment extends at least as far south as the sedimentation-erosion boundary between the Sub-Himalaya and the foredeep. In the Siwaliks the onset of folding usually occurs before the cessation of deposition [Johnson et al., 1979] and shortening of the wedge must also occur to the south of the sedimentation-erosion boundary. Recently, intergranular deformation and solution-cleavage were found to have contributed significantly to shortening in the fossil wedge of the Appalachians [Engelder and Engelder, 1977]. By these methods, appreciable shortening may occur in the outer portion of the wedge where large structures contribute little to the shortening. This implies that detachments can in general extend further toward the foreland than the most forwardly located folds and thrusts.

The above arguments lead to the conclusion that the Himalayan Detachment is characterized by a variable displacement, maximum at the BTF, the deeper end of the detachment, and decreasing gradually up-dip to a non-tectonic contact between sediments and basement. Thus, independent evidence from seismicity and structure indicates that the transition between non-tectonic and tectonic basement-sediment contact is somewhere under the foredeep, i.e. south of the sedimentation-erosion boundary. The displacement field in the great detachment earthquakes of the Himalaya would probably reflect the long term displacement pattern, i.e. larger displacements at depth to the north and smaller displacements near the surface to the south. Realistic modeling of these events must account for this pattern of deformation as well as for the large difference in rigidity between the basement and the sediments.

In the central portion of the Himalayan arc, the one reliable fault-plane solution in the sedimentary zone is a normal fault with a north-south trending T-axis (Figures 10 and 11). Molnar et al. [1973] interpret this as a basement fault associated with the downward bending of the lithosphere as it subducts toward the north.

The IKSZ, the extension into Hazara of the thrust-earthquake belt associated with the Himalayan BTF (and discussed in a previous section; see Figures 3 and 4), is surrounded by a broad zone of scattered seismicity in the underthrusting crustal layer. Armbruster et al. [1978] suggest that this scattered seismicity is related to the high stresses induced by the bending to which the crust is subjected near the BTF. A similar zone of deep crustal seismicity may surround the thrust earthquake belt at the BTF in the central portion of the Himalayan arc. Some of the seismicity near the BTF in Figure 11 may be related to such a seismic zone, and not to the Basement Thrust itself.

Valdiya [1976] finds a correlation between transverse, steep-angle faults in the Sub and Lesser Himalaya and earthquake zones that extend into the foredeep. These features may be associated with tear faults in the basement. A well documented example of such a fault is the Siran river tear fault in the Hazara area (Figures 6 and 7, discussed in a previous section).

Earthquakes can also occur within the sedimentary wedge. The TSZ in Hazara (Figure 4) is an example. In this region the distribution of hypocenters suggests that deformation within the wedge occurs mostly in an aseismic manner, except in the TSZ. Blocks of relatively brittle rock, perhaps basement, contained in the softer matrix of the sedimentary wedge may be a typical source of seismicity within the wedge (Figure 14).

In conclusion, the contrast between the very active belt that presumably bounds to the north the ruptures of the great earthquakes, and the low interseismic seismicity on the Detachment is even more striking than it appears in Figures 10 and 11. In fact, most of the moderate and smaller magnitude earthquakes in the sedimentary zone are probably not located on the Detachment, and occur either above it, in the sedimentary wedge or below it in the basement.

Why the Arcuate Shape of the Himalaya?

A thin spherical shell of radius R can bend, conserving its original surface area, along a small circle of radius r. The angle of dip θ at this bend is uniquely determined by

$$\theta = 2 \arcsin r/R$$

Frank [1968] suggests that the arcuate shape of oceanic subduction zones and the dip-angle of the subducting lithospheric slabs may be determined by this simple geometric relationship. However, the observed dips match very poorly with the dips predicted from the radii of the arcs. Thus, while the original slope of the arc probably requires conservation of the lithospheric surface area, the dip of the subducting slab appears to be primarily determined by other causes.

In the Himalayan arc r = 1700 km and R = 6400 km, yielding a value for θ of 30°. The average dip of the nodal plane dipping towards the north in the four thrust solutions shown in Figure 11 is also about 30°. However, considering the number of solutions and the scatter (from 15 to 41°) this remarkable agreement is more fortuitous than significant.

On the other hand, it is difficult to explain the nearly perfect geometric shape of the Himalayan arc as fortuitous. It is also unlikely that this characteristic is inherited from the shape of the island arc which presumably collided with India along the Indus suture about 40 m.y. ago [Molnar and Tapponnier, 1975]. The perfect shape of the Himalayan arc suggests that the active process of continental subduction itself requires, or at least maintains this shape.

While in oceanic subduction zones the lithosphere sinks deeply into the upper mantle and is forced to assume a shape according to the velocity field in the mantle [Uyeda and Kanamori, 1979], in continental subduction the crust, and perhaps the entire subducting lithosphere, remain above the asthenosphere and can be relatively unaffected by mantle convection. Thus, lithospheric surface area is likely to be conserved, and the related geometric requirement (eq. 1) should be satisfied to a greater degree in continental than in oceanic subduction zones.

The Termination of the Himalayan Arc

The thrust earthquake belt and the topographic break are the features of the Himalayan arc that fit most precisely a small circle in the central portion of the arc (Figure 10). According to the tectonic model of Figure 13, these two features are indicators of the BTF. Near the two ends of the Himalayan arc, both these features, and presumably the BTF, deviate (in a similar manner) from the small circle.

At about 92°E, both BTF indicators deviate towards the east-northeast from the east-west strike of the Himalayan small circle. Also, the MBT and perhaps the MCT follow approximately this east-northeast strike. The narrow alluvial plain of eastern Assam is the foredeep for both the eastern portion of the Himalayan arc, striking ENE, and the northern portion of the Burma arc with approximately the same strike.

The extension into the Burma arc of the structures associated with the northeasterly directed subduction of oceanic lithosphere along the Andaman arc (Figure 1) is indicated by the continuity of the volcanic belt [Katsui, 1971; Curray et al., 1979] and of the subcrustal seismicity [Fitch, 1970; Molnar et al., 1973; Chandra, 1975; Verma et al., 1976]. These belts extend uniformly from the Sunda arc in Indonesia to Burma and continue north to near the eastern terminus of the Himalayan arc. Unlike the oceanic portions of the Sunda arc which border an oceanic basin, the Burma arc is bordered on the outer side

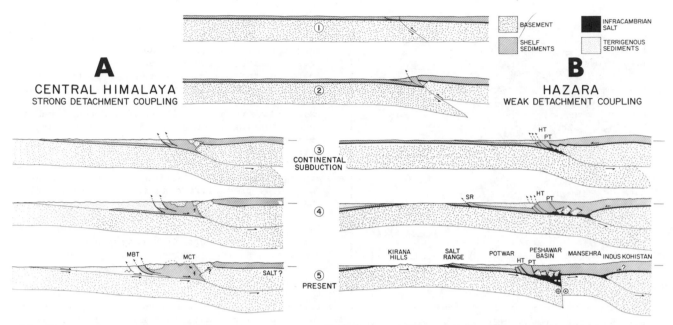

A

CENTRAL HIMALAYA
STRONG DETACHMENT COUPLING

B

HAZARA
WEAK DETACHMENT COUPLING

Fig. 14. Sequence of tectonic events in the development of the Himalayan continental subduction structure. In this scheme the difference between the Hazara and the central Himalaya sections are primarily a consequence of the coupling at the Detachment. In Hazara the Detachment can develop along the thick Infracambrian evaporite layer and the sediment-basement coupling is weak. In the central Himalaya the Detachment is mostly between the basement and the terrigenous sediments and the coupling is relatively strong. The distribution of the salt layer in stages 1 and 2 is pertinent to Hazara. In the central Himalaya the Infracambrian salt, if present at all, must have been less extended toward the foreland. Compare stage 5A with Figure 13; stage 5B with Figure 4. HT = Hazara Thrust; PT = Panjal Thrust; MBT = Main Boundary Thrust; MCT = Main Central Thrust.

by a subsiding trough filled with terrigenous sediments. These sediments are progressively involved in the deformation associated with a typical fold and thrust belt. This non-marine molasse-like sedimentation appears to be continuous during the Upper Tertiary [Mathur and Evans, 1964]. The seismicity indicates a Benioff zone that is continuous from crustal to intermediate depths (200 km); thus significant portions of oceanic lithosphere must have subducted since the mid-Tertiary. It follows that at least some of the non-marine sediment must have been deposited on oceanic crust converging toward the Burma subduction zone. This raises the possibility that parts of eastern Assam and eastern Bengal are still underlain by oceanic crust [Desikachar, 1974].

The meizoseismal area of the 1897 earthquake extends over the Shillong Plateau (Figure 9) a large area where Precambrian shield rocks are at the surface. It was argued above that the lack of surface rupture during the 1897 event indicates a rupture on a shallow-dipping fault buried below the mezoseismal area. In this case, the shield rocks in Shillong and Mikir (Figure 9) are part of an allochthonous sheet detached along this fault, and the portion of Assam and Bengal underlain by oceanic crust could extend below this allochthonous sheet.

Thus, the deviation of the Himalayan arc from the small circle and other complexities near the eastern syntaxis may result from interaction of the Himalayan front with oceanic or transitional (continent-ocean) crust in eastern Assam. This interaction could be a late development of the Himalaya evolution, depending on the original shape of pre-collisional greater India [Powell, 1979].

West of approximately 76°E, in the Kashmir Himalaya, both the seismicity and the topography cannot be interpreted in terms of a single morpho-tectonic lineament with an arcuate shape. The topographic break is taken at 4 km ("thresholds" in Figures 9 and 10). The highest peaks or the Pir Panjal range are somewhat higher than 4 km and the thresholds mark approximately the topographic axis of this range on the southwest side of the Kashmir valley (compare with Figures 9 and 10). On the northeast side of the valley the topography is more typical of the High Himalaya and a topographic front is found at about the 4 km contour. However, the MBT (Murree thrust) and the MCT (Panjal thrust) extend subparallel and closely spaced on the southwest flank of the Pir Panjal, far toward the foreland from this topographic front. Teleseismic epicenters in Kashmir are rather sparce and scattered [Quittmeyer et al., 1979] and the relationship between topography,

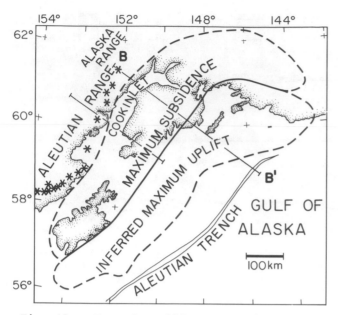

Fig. 15. Tectonic uplift and subsidence in south-central Alaska associated with the 1964 earthquake [Plafker, 1965]. Active or dormant volcanoes are shown by stars. The two section lines are related to Figure 16.

surface faulting, and seismicity is not well defined.

Further northwest, in Hazara, the IKSZ forms a very prominent seismic zone aligned with the MBT and MCT of Kashmir (compare Figures 3 and 10). The IKSZ is associated with a clear topographic front and is interpreted as the northwestern extension of the Himalayan Basement Thrust Front (BTF) that presumably is associated with the Pir Panjal structure (Figures 3 and 4, cf. Armbruster et al., [1978]). This branch of the BTF is somewhat displaced toward the foreland from the Himalayan small circle (Figure 10).

A prominent earthquake sequence forms the cluster located about 100 km north-northeast of the IKSZ (36.0°N; 73.5°E, Figure 10) and is approximately aligned with the topographic break northeast of the Kashmir valley and with the Himalayan small circle. The fault plane solution of the main shock indicates thrusting with a poorly determined strike parallel to the small circle [Chandra, 1978]. This suggests that a second branch of the BTF extends along the east side of the Kashmir valley to the 1972 earthquake cluster along a path quite close to the Himalayan small circle.

Tectonic Development: Salt vs. No Salt

One possible scheme for the development of the main features of the continental subduction zone in the central and the northwestern portion of the Himalayan arc is schematically illustrated in Figure 14. The development of the Basement Thrust

(stage 1 and 2) is probably quite similar in both sections of the Himalaya. Sometime after subduction begins (stage 3A), but well before the present, the pre-collisional sedimentary pile south of the BTF in the central portion of the Himalaya becomes allochthonous. From this point on the Detachment develops at or near the interface between the terrigenous sediments and the underthrusting basement in this section of the arc (stage 5A). On the other hand, during the corresponding stages of development in the northwestern Himalayan terminus, the basement approaching the subduction structure is still covered by pre-collisional sediments that include the thick basal salt layer (stage 3B; see section on the Hazara arc). In this case the Detachment is at the salt layer. Thus, in a relative sense, the sedimentary wedge is strongly coupled to the basement in the central Himalaya and weakly coupled in Hazara.

In both areas, the horizontal force applied to the sedimentary wedge through the Detachment is initially balanced by the force on the wedge from the overriding crustal block (the Tethyan slab). In the central Himalaya this force is high and the portion of the Tethyan slab overhanging above the deeper portions of the sedimentary wedge is supported or even pushed back by this force (3A). In Hazara where the force is low, the upper part of the Tethyan slab collapses for lack of support and fragments of this slab are incorporated in the sedimentary wedge and are displaced toward the foreland (4B). Fragments of the Tethyan slab are carried south of the sedimentary wedge also in the central Himalaya. However, in this case they are in the form of crystalline sheets forming the structurally highest nappes of the Lesser Himalaya (5A).

In Hazara (stages 3B to 5B) the sedimentary pile on the upthrown side of the Basement Thrust is detached from the basement, possibly along the same evaporite layer that decouples the sedimentary wedge south of the BTF. Thus the sedimentary wedge is expanded to include the sediments north of the BTF and there is no MCT serving as a surface expression of the present position of the BTF (see also Figure 4; the BTF is on the southwest side of the IKSZ). This sedimentary wedge has moved south 100-150 km with respect to the overriding basement slab north of the BTF. In order to account for this displacement, either the sedimentary layer must be extended or the basement must be shortened somewhere north of the BTF by a corresponding amount. It was suggested above that a second Basement Thrust is located on the northeast side of the Kashmir Valley and may extend northwest to the Gilgit area. This thrust would be subparallel and 100 km northeast of the IKSZ and could provide the required shortening of the basement with respect to the sediments.

The fine structure of the IKSZ (Figure 4) suggests that the sedimentary wedge above the IKSZ is now moving backward (north) relative to the overriding basement slab [Seeber and Armbruster, 1979; Figure 14, 5B]. This may be the result of increased sediment-basement coupling after the formation of the HLSZ and/or the uplift of the Salt Range (Figure 4).

More obvious consequences of the high and low basement-sediment coupling in the Central Himalaya and in Hazara, respectively, are the extent of the Detachment and the style of deformation within the sedimentary wedge. In the central Himalaya the transition between the detached and the autochthonous part of the sedimentary wedge is located somewhere in the foredeep between the sedimentation-deposition front and the thin leading edge of the sedimentary wedge (Figure 13). This transition is gradual since the shortening in the wedge progressively increases toward the north. Instead, in Hazara almost the entire sedimentary wedge is a decollement since the Detachment surfaces at the base of the Salt range, about 50 km from the outcrops of the basement in the Kirana hills (Figures 2 and 10). In this case the shortening in the wedge occurs primarily by faulting at the southern limit of this deformation. At the eastern end of the Salt range there is a gradual transition between an outcropping and a buried southern termination of the Detachment.

Continental Subduction in the Himalaya Compared with the Fossil Continental Collision in the Appalachians and the Active Oceanic Subduction in Alaska

The fossil continental collision zone of the southern Appalachians and the active oceanic subduction zone of Alaska are compared to the continental collision and subduction zone of the Himalaya in Figure 16. The main features of the three structures are remarkably similar. In all three cases there is a sedimentary wedge (usually referred to as an accretionary prism in the case of oceanic subduction) detached from the underlying basement. This detachment merges down-dip into a basement thrust. The basement block forming the hanging wall of this thrust provides the support on the hinterland side of the wedge. In the southern Appalachian section there is a suggestion that the Basement Thrust Front (BTF) is at the base of the Kings Mt. crystalline belt [R.D. Hatcher, personal communication]. An alternative interpretation of the subsurface data [Cook et al., 1979] puts the BTF at an undetermined distance further southeast.

In the Himalaya the MBT is the boundary between terrigenous and pre-collisional shelf sediments. Thus most of this wedge is composed of these terrigenous sediments. Instead, in the southern Appalachians the terrigenous sediments are a relatively minor component of the wedge. (Note that in Figure 16 the "younger sediments" in the Himalayan section are all terrigenous while in the Appalachian section only a portion of the "younger sediments" are terrigenous). As continental

subduction continues, the ratio between terrigenous and pre-collisional sediments in the central Himalayan wedge increases because erosion effects primarily the pre-collisional sediments of the Lesser Himalaya and the eroded material is replaced exclusively by terrigenous sediments (assuming that the total volume of the wedge remains constant). This suggests that in the Appalachians, continental subduction [Bally, 1975], if it followed continental collision at all, may have been shorter-lived than in the Himalaya.

Probably most of the Himalayan Detachment occurs at the boundary between the terrigenous sediments and the crystalline basement. In the southern Appalachians the Detachment is believed to coincide mostly with a layer of low shear-strength, the Rome shale [Hatcher, 1978]. The large width of the Appalachian fold and thrust belt may be symptomatic of the effectiveness of the decoupling layer. In fact, the dimensions and structure of the Valley and Ridge province are similar to those of the Potwar in the outer wedge of Hazara (Figure 4). Here the wedge is well decoupled from the basement by the evaporite layer and contains a large proportion of pre-collisional sediments.

In the Gulf of Alaska oceanic lithosphere of the Pacific plate subducts below the lithosphere of the North American plate. Although the accretionary wedge is unusually wide, probably because of the large input of sediments into the Gulf of Alaska, the Alaska subduction zone in Figure 16 is a rather typical example of this type of structure. The mode of slip along the main fault is remarkably similar in the Alaska and Himalayan subduction structures. In each, three portions of the fault can be distinguished: (1) a detachment; (2) a seismically active basement thrust; and (3) an aseismic portion of this basement thrust. In both cases the detachment characteristically slips by great earthquakes. During these events the extent of rupture and the displacement are typically measured in hundreds of kilometers and tens of meters, respectively. The interseismic periods on the detachment are typically about 35 years to a few hundred years at oceanic subduction zones [Sykes and Quittmeyer, 1979]. In the Himalaya they are undetermined, but probably at least 100-200 years. During these periods the detachment is characterized by relatively low or, possibly, no seismicity (Figure 16 for Alaska, Figure 11 for the Himalaya).

The down-dip end of the detachment in the Gulf of Alaska subduction zone is marked by a BTF. As in the Himalaya structure, this BTF is recognized as the up-dip (forward) limit of an intense seismic zone of thrust earthquakes. In contrast with the seismicity on the detachment, the seismicity down-dip from the BTF is relatively continuous and characterized by moderate magnitude and smaller events [Jacob et al., 1977; Davies and House, 1979]. It is this seismicity that forms the narrow earthquake belt of the

SOUTHERN APPALACHIANS

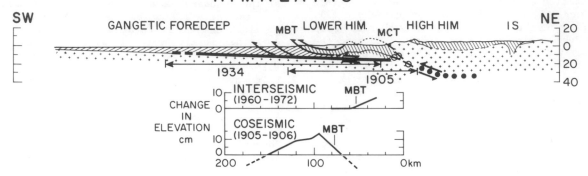

HIMALAYAS

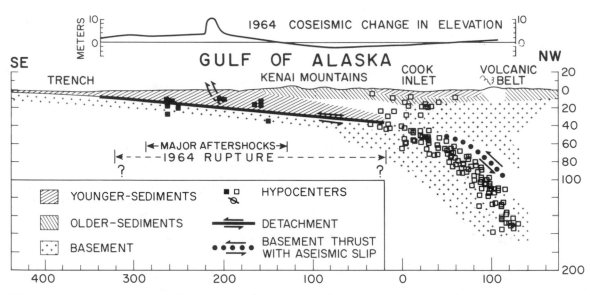

Fig. 16. The southern Appalachians, a continental collision structure now inactive, and the active oceanic subduction structure in the Gulf of Alaska are compared to the Himalayan continental subduction structure. In the Alaskan section aftershocks (filled hypocenters) and inferred rupture (slightly modified) of the great 1964 earthquake (M_s = 8.4) are from Plafker [1965] (section B-B' in Figure 15); hypocenters during the interseismic period (April-June, 1972) are from Lahr et al. [1974] (section located in Figure 15 by the unmarked line). The Southern Appalachian section is from recent deep seismic reflection results [COCORP; from Cook et al., 1979]. Cook et al. [1979] prefer to extend the detachment to the SE beyond the Kings Mt. belt, but they give as a plausible alternative the interpretation adopted here (see text). In the Himalayan section the extent of intensity $\geq$ VIII (arrows) of two great earthquakes (1905, M_s = 8.0 and 1934 M_s = 8.3) indicate approximately the respective ruptures (see text). Note that the great earthquakes are associated with the detachment in both oceanic and continental subduction structures. The configuration of the deeper portion of

the basement thrust slipping aseismically is conjectural [cf. Engdahl and Scholz, 1977]. The four thrust earthquakes in the Himalayan section between the aseismic thrust and the detachment (the bar through the hypocenters indicate the dip of the inferred rupture plane; see Figure 11) indicate the portion of the thrust slipping by intermediate-magnitude earthquakes. Thrust earthquakes in oceanic subduction structures are similarly located [cf. Isacks and Barazangi, 1977]. In the Himalayan and Alaskan sections the "younger" sediments are terrigenous sediments (Tertiary and Quaternary) in the Appalachians "younger" sediments include terrigenous and metamorphosed continental platform and shelf sediments. In all these sections the "older" sediments include metasediments. The 1905-1906 releveling data are from Middlemiss [1910]; the 1960-1972 data arefrom Chugh [1974] (see Figure 12); the 1964 data are from Plafker [1965]. There is evidence that in Alaska the motions reverse in the interseismic period [Plafker and Rubin, 1978]. A similar pattern is observed in the area of the 1905 event.

Himalaya (Figures 10 and 11). Also in the oceanic subduction zone this seismicity forms a narrow belt at the shallow end of the Benioff zone, which, otherwise, consists mostly of hypocenters within the sinking slab [Isacks and Barazangi, 1977]. In the Himalayan structure the BTF as defined by the seismicity is interpreted as the up-dip terminus of the interplate basement contact. The BTF in the Alaskan subduction structure can be similarly interpreted.

The sedimentary wedge is decoupled from both plates interacting at the subduction zone. The detachment is the active boundary between the wedge and the subducting plate. The boundary with the overriding plate at the back of the wedge is not as clearly defined. However, this boundary should also be active since it divides a wedge of weak sediments that deform plastically from a slab of strong crystalline rocks that behaves more rigidly (cf. Figures 13 and 14). In the Himalaya this boundary is at or near the MCT (Figure 13). Shallow seismicity above the BTF, is found in many other closely monitored subduction zones [Davies and House, 1979; Isacks and Barazangi, 1977; Engdahl, 1977; Uyeda, 1977]. In our interpretation this seismicity marks the boundary between the weak sedimentary wedge and the strong overriding plate and the aseismic front adjacent to this seismicity [Yoshii, 1975] is the consequence of the relative strength of the overriding plate. Thus, the great detachment earthquakes are not true interplate earthquakes since the sedimentary wedge above the detachment is not rigidly connected with either of the converging plates. Rather it is the belt of moderate-magnitude thrust-earthquake that marks the contact between the two plates and is the most fundamental seismic expression of the plate boundary.

Down-dip from the seismically active portion of the Basement Thrust, this fault usually becomes steeper in oceanic subduction zones, but presumably becomes less steep, bending back to a quasi-horizontal orientation, in the Himalayan subduction zone. However, the slip on these boundaries becomes similarly aseismic (see earlier discussion for the Himalayan case). Isacks and Barazangi [1977], Engdahl and Scholz [1977], and Jacob et al. [1977], among others, discuss the oceanic subduction case. In both continental and oceanic subduction the transition between seismic and aseismic slip is ascribed to the increase in temperature and/or to the availability of water down-dip along the thrust. The increase in temperature can result from shear heating or from heat transferred from the overriding plate or from both; the water is obtained from chemical dewatering of the subducted crustal rocks.

Implication for Seismic Hazard. The model which we have presented describes the active tectonic processes of the Himalayan and provides a framework within which seismic hazard can be better estimated. Although locally destructive earthquakes can occur on the Basement Thrust, the greatest hazard is from the much larger detachment earthquakes. In the Hazara area the Detachment is associated with a thick layer of salt and the lack of seismicity on the Detachment may be associated with aseismic slip on the salt. In the remainder of the Himalayan arc, lack of seismicity on the Detachment is associated with strain accumulation between great earthquakes. These events can be expected to affect large areas of the Indo-Gangetic plain and Lower Himalaya, as they have in the past.

Acknowledgements

Klaus Jacob gave us helpful suggestions during many stimulating discussions. Terry Engelder, David Simpson, Klaus Jacob, and Omar Perez reviewed the manuscript. We are grateful to the Water and Power Development Authority (WAPDA, Pakistan) and to Tippetts-Abbett-McCarthy-Stratton (TAMS, New York) for their continuing support for our field operations in Pakistan. This work was sponsored by Contracts USGS 16749 and NSF EAR 77-15187. Lamont-Doherty Geological Observatory Contribution Number 3046.

References

Armbruster J., L. Seeber, and K.H. Jacob, The northwestern termination of the Himalayan mountain front: Active tectonics from micro-earthquakes, J. Geophys. Res., 83, no. B1, 269-282, 1978.
Bally, A.W., A geodynamic scenario for hydrocarbon occurrences, in Global Tectonics and Petroleum

Occurrences, also in Proceedings 9th World Petroleum Congress, Tokyo, 2, Geology, 1975.

Banghar, A.R., Focal mechanisms of earthquakes in China, Mongolia, Russia, Nepal, Pakistan, and Afghanistan, Earthquake Notes, 45, 1-11, 1974.

Bird, P., Initiation of intracontinental subduction in the Himalaya, J. Geophys. Res., 83, 4975-4987, 1978.

Bordet, P., M. Colchen, D. Krummenacher, P. LeFort, R. Mouterde, and J.M. Remy, Recherches geologiques dans l'Himalaya du Nepal, region de la Thakkhola, Ed. C.N.R.S., Paris, 270 p., 1971.

Calkins, J.A., T.W. Offield, S.K.M. Abdullah, and S.T. Ali, Geology of the southern Himalaya in Hazara, Pakistan and adjacent areas, U.S. Geol. Surv. Prof. Paper 716-C, Washington, D.C., 1975.

Chandra, U., Seismicity, earthquake mechanisms and tectonics of Burma 20°N-28°N, Geophys. J. Roy. astro. Soc., 40, 1-15, 1975.

Chandra, U., Seismicity, earthquake mechanisms and tectonics along the Himalayan mountain range and vicinity, Phys. Earth Planet. Inter., 16, 109-131, 1978.

Chapple, W.M., Mechanics of thin-skinned fold-and-thrust belts, Geol. Soc. Amer. Bull., 89, 1189-1198, 1978.

Chen, W.P., and P. Molnar, Seismic moments of major earthquakes and the average rate of slip in central Asia, J. Geophys. Res., 82, 2945-2969, 1977.

Chugh, R.S., Study of recent crustal movements in India and future programs, paper presented at the International Symposium on Recent Crustal Movements, Zurich, 1974.

Chun, K.Y., and T. Yoshii, Crustal structure of the Tibetan Plateau: A surface wave study by a moving window analysis, Bull. Seismol. Soc. Amer., 67, 735-750, 1977.

Cook, F.A., D.S. Albaugh, L.D. Brown, S. Kaufman, J.E. Oliver, and R.D. Hatcher, Thin-skinned tectonics in the crystalline southern Appalachians; COCORP seismic-reflection profiling of the Blue Ridge and Piedmont, Geology, 7, 563-567, 1979.

Curray, J.R., and D.G. Moore, Growth of the Bengal deep-sea fan and denudation of the Himalayas, Geol. Soc. Amer. Bull., 82, 563-572, 1971.

Curray, J.R., D.G. Moore, L.A. Lawver, F.J. Emmel, R.W. Raitt, M. Henry, and R. Kieckhefer, Tectonics of the Andaman Sea and Burma, in Geological and Geophysical Investigations of Continental Margins, Memoir 29, edited by J.S. Watkins, L. Montadert, and P.W. Dickerson, p. 189, Amer. Assoc. Petrol. Geol., Tusla, Oklahoma, 1979.

Davies, J.N., and L. House, Aleutian subduction zone seismicity, volcano-trench separation and their relation to great thrust-type earthquakes, J. Geophys. Res., 84, 4583-4591, 1979.

Desikachar, S.V., A review of the tectonic and geological history of eastern India in terms of "plate tectonics" theory, Jour. Geol. Soc. India, 15, 211-249, 1974.

Desio, A., Geologic evolution of the Karakorum, in Geodynamics of Pakistan, edited by A. Farah and K. DeJong, p. 111, Geol. Surv. Pakistan, Quetta, 1979.

Dunn, J.A., J.B. Auden, A.M.N. Ghosh, S.C. Roy, and D.N. Wadia, The Bihar-Nepal earthquake of 1934, Mem. Geol. Survey India, 73, 1939.

Engdhal, E.R., Seismicity and plate subduction in the central Aleutians, in Island Arcs, Deep Sea Trenches and Back-Arc Basins, Maurice Ewing Series 1, edited by M. Talwani and W.C. Pitman III, 259-272, AGU, Washington, D.C., 1977.

Engdahl, E.R., and C.H. Scholz, A double Benioff zone beneath the central Aleutians: An unbending of the lithosphere, Geophys. Res. Lett., 4, 473-476, 1977.

Engelder, T., and R. Engelder, Fossil distortion and decollement tectonics of the Appalachian Plateau, Geology, 5, 457-460, 1977.

Fitch, T.J., Earthquake mechanisms in the Himalayan Burmese and Andaman regions and continental tectonics in Central Asia, J. Geophys. Res., 75, 2699-2709, 1970.

Frank, F.C., Curvature of Island Arcs, Nature, 220, 363, 1968.

Gansser, A., Geology of the Himalayas, Inter-Science Publishers, John Wiley and Sons, London, 1964.

Gansser, A., The great suture zone between Himalaya and Tibet a preliminary account, Editions du C.N.R.S., 268, 181-191, 1977.

Gee, E.R., The Dhubri earthquake of 3rd July 1930, Mem. Geol. Survey India, 65, 1-106, 1934.

Gupta, H.K., and M. Narain, Crustal structure in the Himalayan and Tibet Plateau region from surface wave dispersion, Seismol. Soc. Amer. Bull., 57, 235-248, 1967.

Gutenberg, B., and C.F. Richter, Seismicity of the Earth and Associated Phenomena, 273 p., Princeton University Press, Princeton, New Jersey, 1954.

Hatcher, R.D., Jr., Tectonics of the western Piedmont and Blue Ridge, southern Appalachians: Review and Speculation, Amer. Jour. Sci., 278, 276-304, 1978.

Hemphill, W.R., and A.H. Kidwai, Stratigraphy of the Bannu and Dera Ismail Khan areas, Pakistan, U.S. Geol. Surv. Prof. Paper 716-B, 1973.

International Seismological Summary, Kew Observatory, Richmond, Surrey.

Isacks, B.L., and M. Barazangi, Geometry of Benioff zones: Lateral segmentation and downwards bending of the subducted lithosphere, in Island Arcs, Deep Sea Trenches and Back-Arc Basins, Maurice Ewing Ser. 1, edited by M. Talwani and W.C. Pitman III, AGU, Washington, D.C., 1977.

Jacob, K.H., K. Nakamura, and J.N. Davies, Trench-volcano gap along the Alaska-Aleutian arc: Facts and speculations on the role of terrigenous sediments for subduction, in Island Arcs, Deep Sea Trenches and Back-Arc Basins, Maurice Ewing Ser. 1, edited by M. Talwani and W.C. Pitman III, AGU, Washington, D.C., 1977.

Johnson, G.D., N.M. Johnson, N.D. Opdyke, and R.A.K. Tahirkheli, Magnetic reversal stratigraphy and sedimentary tectonic history of the Upper Siwalik Group, Eastern Salt Range and Southwestern Kashmir, in Geodynamics of Pakistan, edited by A. Farah and K. DeJong, p. 249, Geol. Surv. Pakistan, Quetta, 1979.

Katsui, Y. (ed.), List of the World Active Volcanoes, Volcanological Society of Japan, 1971.

Kelleher, J.A., Rupture zones of large South American earthquakes and some predictions, J. Geophys. Res., 77, 2087-2103, 1972.

Klootwijk, C.T., A review of palaeomagnetic data from the Indo-Pakistan fragment of Gondwanaland, in Geodynamics of Pakistan, edited by A. Farah and K. DeJong, p. 41, Geol. Surv. Pakistan, Quetta, 1979.

Kono, M., Gravity anomalies in east Nepal and their implications to the crustal structure of the Himalayas, Geophys. Jour. Roy astr. Soc., 39, 283-299, 1974.

Krishnaswamy, V.S., S.P. Jalote, and S.K. Shome, Recent crustal movements in north-west Himalaya and the gangetic foredeep and related patterns of seismicity, Proceedings 4th Symposium on Earthquake Engineering, University of Roorkee, 1970.

Lahr, J.C., R.A. Page, and J.A. Thomas, Catalog of earthquakes in South Central Alaska, April-June 1972, U.S. Geol. Surv. Open-File Report, 1974.

LeFort, P., Himalayas: The collided range. Present knowledge of the continental arc, Amer. J. Sci., 275A, 1-44, 1975.

Lehner, E., The age of the Punjab saline series: Possible reconciliation of the opposing views, Proc. Nat. Acad. Sci. India, 14, no. 6, 261-266, 1945.

Mathur, L.P., and P. Evans, Oil in India, Sp. Brochure, Inter. Geol. Congr. 22nd Session, New Delhi, 64-79, 1964.

Mathur, L.P., and G. Kohli, Exploration and development for oil in India, World Petroleum Congress, 6th, Frankfurt an Main, Pr. Sec. 1, 633-658, 1964.

Marussi, A., Gravity in the Karakorum, Atti dei Convegni Lincei, 21, 131-139, 1976.

Menke, W., and K.H. Jacob, Seismicity patterns in Pakistan and northwestern India associated with continental collision, Seismol. Soc. Amer. Bull., 66, 1695-1711, 1976.

Menke, W.H., Lateral inhomogeneities in P velocity under the Tarbela array of the Lesser Himalaya, Bull. Seismol. Soc. Amer., 67, 725-734, 1977.

Middlemiss, C.S., The Kangra earthquake of 4th April 1905, Mem. Geol. Survey India, 38, 1910.

Molnar, P., T.J. Fitch, and F-T Wu, Fault-plane solutions of shallow earthquakes and contemporary tectonics in Asia, Earth Planet. Sci. Lett., 19, 101-112, 1973.

Molnar, P., and P. Tapponnier, Cenozoic tectonics of Asia: Effects of a continental collision, Science, 189, 419-426, 1975.

Molnar, P., W.P. Chen, T.J. Fitch, P. Tapponnier, W.E.K. Warsi, and F-T Wu, Structure and tectonics of the Himalaya: A brief summary of geophysical observations, Editions de C.N.R.S., 268, 269-294, 1977.

Molnar, P., and P. Tapponnier, Active tectonics of Tibet, J. Geophys. Res., 83, 5361-5375, 1978.

Ni, J., and J.E. York, Late Cenozoic extensional tectonics of the Tibetan Plateau, J. Geophys. Res., 83, 5377-5384, 1978.

Oldham, T., A catalogue of Indian earthquakes from the earliest time to the end of A.D. 1869, Geol. Surv. India Mem., 19, part 3, 163-215, 1882.

Oldham, R.D., Report on the great earthquake of 12 June 1897, Mem. Geol. Surv. India, 29, 1899.

Oxburgh, E.R., Flake tectonics and continental collision, Nature, 239, 202-204, 1972.

Pennington, W.D., A summary of field and seismic observations of the Pattan earthquake, 28 December 1974, in Geodynamics of Pakistan, p. 143, edited by A. Farah and K. DeJong, Geol. Survey Pakistan, Quetta, 1979.

Plafker, G., Tectonic deformation associated with the 1964 Alaska earthquake, Science, 148, 1675, 1965.

Plafker, G., and M. Rubin, Uplift history and earthquake recurrence as deduced from marine terraces on Middleton Island, Alaska, U.S. Geol. Surv. Open-File Report 78-943, Menlo Park, California, 1978.

Powell, C.McA., and P.J. Conaghan, Plate tectonics and the Himalayas, Earth Planet. Sci. Lett., 20, 1-12, 1973.

Powell, C.McA., A speculative tectonic history of Pakistan and surroundings: some constraints from the Indian Ocean, in Geodynamics of Pakistan, p. 5, edited by A. Farah and K. DeJong, Geol. Surv. Pakistan, Quetta, 1979.

Quittmeyer, R.C., and K.H. Jacob, Historical and modern seismicity of Pakistan, Afghanistan, Northwestern India and Southeastern Iran, Bull. Seismol. Soc. Amer., 69, 773-823, 1979.

Quittmeyer, R.C., A. Farah, and K.H. Jacob, The seismicity of Pakistan and its relation to surface faults, in Geodynamics of Pakistan, p. 271, edited by A. Farah and K. DeJong, Geol. Surv. Pakistan, Quetta, 1979.

Qureshy, M.N., S.V. Venkatachalam, and C. Subrahmanyam, Vertical tectonics in Middle Himalayas: An appraisal from recent gravity data, Bull. Geol. Soc. Amer., 85, 921-926, 1974.

Richter, C.F., Elementary Seismology, W.H. Freeman and Company, San Francisco and London, 768 p., 1958.

Sarwar, G., and K.A. DeJong, Arcs, oroclines, syntaxes: The curvatures of mountain belts in Pakistan, in Geodynamics of Pakistan, p. 341, edited by A. Farah and K. DeJong, Geol. Surv. Pakistan, Quetta, 1979.

Savage, J.C., and L.M. Hastie, Surface deformation associated with dip-slip faulting, J. Geophys. Res., 71, 4897-4904, 1966.

Seeber, L., and K.H. Jacob, Microearthquake survey of northern Pakistan: Preliminary results and tectonic implications, in Proc. C.N.R.S. Coloquium on the Geology and Ecology of the Himalayas, Paris, 347-360, 1978.

Seeber, L., and J. Armbruster, Seismicity of the Hazara arc in northern Pakistan: Decollement vs. basement faulting, in Geodynamics of Pakistan, 131-142, edited by A. Farah and K. DeJong, Geol. Surv. Pakistan, Quetta, 1979.

Seeber, L., R. Quittmeyer, and J. Armbruster, Seismotectonics of Pakistan: A review of results from network data and implications for the Central Himalaya, Structural Geology of the Himalaya, P.S. Saklani (ed.), Univ. of Delhi, 1979.

Srikantia, S.V., and R.P. Sharma, The Precambrian salt deposit of the Himachal Pradesh Himalaya: Its occurrence, tectonics, and correlation, Himalayan Geology, 2, 222-238, 1972.

Stocklin, J., Structural correlation of the Alpine ranges between Iran and Central Asia, Mem. h. ser. Soc. Geol. Fr., no. 8, 333-353, 1977.

Sykes, L.R., and R.C. Quittmeyer, Recurrence time of great earthquakes along convergent plate boundaries, EOS, 60, 884, November, 1979.

Tahirkehli, R.A.K., M. Mattauer, F. Proust, and P. Tapponnier, The India-Eurasia suture zone in Northern Pakistan: Synthesis and interpretation of recent data at plate scale, in Geodynamics of Pakistan, p. 125, edited by A. Farah and K. DeJong, Geol. Surv. Pakistan, Quetta, 1979.

Tapponnier, P., and P. Molnar, Slip-line field theory and large-scale continental tectonics, Nature, 264, 319-324, 1976.

U.S.G.S. National Earthquake Information Center, Denver Federal Circle, Denver, Colorado 80225.

Uyeda, S., Some basic problems in the trench-arc back-arc system, in Island Arcs, Deep Sea Trenches and Back-Arc Basins, Maurice Ewing Series 1, edited by M. Talwani and W.C. Pitman III, AGU, Washington, D.C., 1977.

Uyeda, S., and H. Kanamori, Back-arc opening and the mode of subduction, J. Geophys. Res., 84, 1049-1061, 1979.

Valdiya, K.S., Himalayan transverse faults and folds and their parallelism with subsurface structures of north Indian Plains, Tectonophysics, 32, 353-386, 1976.

Valdiya, K.S., Structural set-up of the Kumaun Lesser Himalaya, Editions du C.N.R.S., 268, 449-462, 1977.

Verma, R.A., M. Mukhopadhyay, and M.S. Ahluwalia, Seismicity gravity and tectonics of northeast India and northern Burma, Bull. Seismol. Soc. Amer., 66, 1683-1694, 1976.

Wadia, D.N., The syntaxis of the northwest Himalaya - its rocks, tectonics, and orogeny, India Geol. Surv. Record, 65, Part 2, 189, 1931.

Wadia, D.N., The Geology of India, Third Edition, MacMillan, London, 1961.

Warsi, W.E.K., and P. Molnar, Gravity anomalies and plate tectonics in the Himalaya, Editions du C.N.R.S., 268, 463-478, 1977.

Wiltschko, D.V., A mechanical model for thrust sheet deformation at a ramp, J. Geophys. Res., 84, 1091-1104, 1979.

Yoshii, T., Proposal of the "Aseismic Front", Zishin, 20, 365-367, 1975.

FOCAL MECHANISM SOLUTIONS AND THEIR TECTONIC IMPLICATIONS FOR THE EASTERN ALPINE-HIMALAYAN REGION

Umesh Chandra

Ebasco Services Inc., 2211 West Meadowview, Greensboro, North Carolina 27407, U.S.A.

Abstract. Focal mechanism solutions of earthquakes in conjunction with regional seismicity maps were used to investigate the tectonics of the eastern sector of the Alpine-Himalayan Region. A catalog of available fault plane solutions of the region, giving trend and plunge of the P, T and B axes, and the dip direction and dip of the nodal planes, was prepared. Selected earthquake mechanisms from this catalog were used in a region by region discussion of tectonics.

Focal mechanism solutions show thrust faulting along the general trend of the Elburz ranges. Thrusting is also noted along northwest faults in the region of convergence of Elburz and Kopet Dagh ranges. The northwest direction of faults predominates throughout the entire length of the Zagros Folded Belt. The seismic activity, west of 49°E, occurs mainly along the northwest striking, right lateral Main Recent Fault rather than along the nearby Main Zagros Thrust. The arching up of geological formations along a NNE axis, NNE alignment of epicenters and focal mechanism solutions near Bandar Abbas and north of it, suggest that the colliding edge of the Arabian plate in the region north of Oman Peninsula is in the form of a projected promontory. The focal mechanism solutions confirm a trench-arch subduction zone along the Makran continental margin.

The focal mechanism solutions for earthquakes near the Quetta transverse zone, 30°N parallel, show right lateral strike slip along west to northwest striking faults. The geometry of the Sulaiman Range Festoon is such that left lateral movement along Ornach-Nal and Chaman faults causes north-south compression in the southern part of the Sulaiman Range. The continued convergence of the Indian plate with Eurasia is accommodated partly by crustal shortening, as indicated by thrust focal mechanism solutions throughout the entire Himalayan and Burmese ranges along faults striking parallel to the local structural trend, and partly by lateral movement at the northwest and northeast Indian plate margins. The movement is left lateral along the Chaman Fault, right lateral along the Karakoram Fault, left lateral northeast of the Himalayan flank of the Assam syntaxis and right lateral along the northern part of the Naga Hill flank of the syntaxis. The eastern Afghanistan block appears to be in a state of northwest compression. Spatial distribution of earthquake foci and focal mechanism solutions, together with the geological data, suggest the existence of a remnant sinking slab of oceanic lithosphere in the mantle under the Hindu Kush and Pamir ranges. Two normal fault plane solutions, with northeast striking nodal planes, near Gartok, Tibet, and two normal solutions, with north striking nodal planes, near Kinnaur are interpreted to indicate a subsurface continuation of the Aravalli Range of peninsular India. Focal mechanism solutions indicate that the activity along the Dauki Fault was rejuvenated by the continental collision and as a result the Shillong Plateau was block uplifted as a horst.

Introduction

Although plate tectonics has been quite successful in explaining the occurrence of earthquakes and other tectonic phenomena within the oceanic and island arc regions, it has not met with the same success in explaining the seismicity and tectonic deformation within the continental regions. Thus, while in the oceanic areas, with a few exceptions such as the Hawaiian islands, the earthquakes occur along long narrow regions, the earthquakes within the continents define a broad zone of tectonic deformation. The eastern part of the Alpine-Himalayan Region is a striking example of this pattern of wide zone of deformation. This study considers the region between the fold belts of Iran and Burma.

The general tectonic setting, major structural trends and fault zones in the region are shown in Figure 1. Epicenter maps of Iran and the region surrounding the Himalaya are presented in Figures 2 and 3, respectively. Most of the earthquakes, throughout the region under study, follow the general trend of the mountains. Three major mountain belts (Caucasus-Kopet Dagh, Elburz and Zagros ranges) and associated seismicity trends occur in Iran. The relationship of these mountain belts to those of eastern Turkey is not

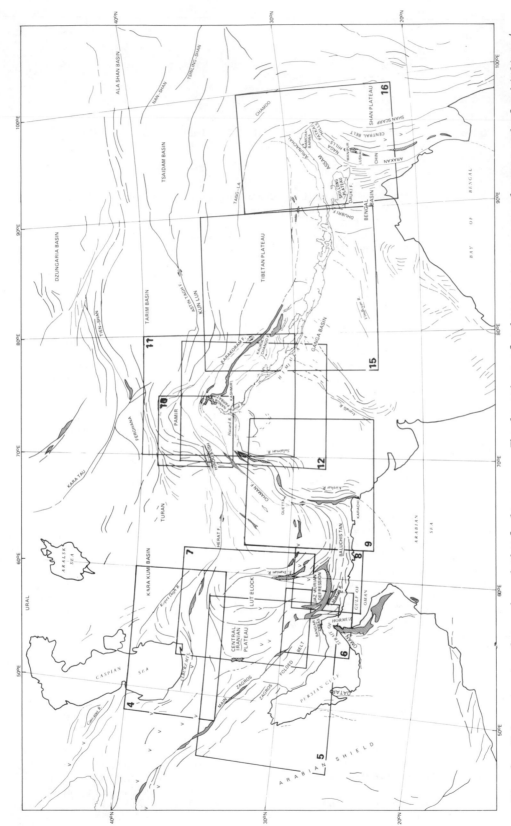

Fig. 1. Major structural trends and the fault zones. The locations of subrecent to recent volcanoes and of ophiolites/colored melange are shown by v and stippled regions, respectively. Thick lines demarcate regions for which maps showing focal mechanism solutions are presented in this paper. The corresponding figure numbers are also shown. (adapted from Gansser, 1964).

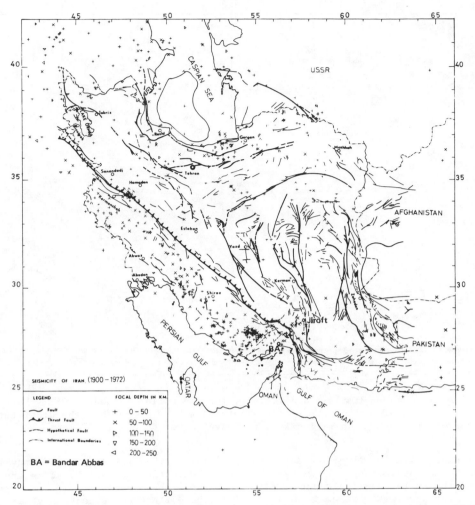

Fig. 2. Epicenter map of Iran for the period 1900-1972 (after Nowroozi, 1976).

quite clear. Toward the east, the Elburz belt merges with the Kopet Dagh ranges. The structural trends of these ranges further curve around toward northeast in such a way that the Hindu Kush and Pamir ranges appear to be their continuation. In the southeast, the Zagros ranges of southern Iran are separated from the Makran ranges by the Minab Fault (or the Oman Line in the description of some authors, e.g., Farhoudi and Karig, 1977; White and Ross, 1979). Although continental collision between the Arabian and Eurasian plates has occurred along the Zagros Suture Zone (identified by a suite of ophiolites), collision is pending in the Gulf of Oman. Continued convergence has brought the opposing continental margins of Oman (Arabian plate) and Makran (Eurasian plate) into contact. This provides a site of initial stages of a continent-continent type collision (White and Ross, 1979).

The seismic activity is by no means confined to the above three mountain belts. A northwest trend across the Central Iranian Plateau, with no known structural relationship is observed (Figure 2). There is some suggestion of possible trends transverse to the Zagros trend. The most notable of these is a NNE trend (about 50 km wide and 250 km long) whose southern end lies in the strait of Hormuz. Significant earthquake activity occurs in the northern part of Lut Block and along its eastern boundary.

Several authors have suggested the Makran region of Iran and Pakistan to be a trench-arc system with active plate subduction, although important differences from a typical Pacific trench-arc system have been noted (Farhoudi and Karig, 1977; White and Ross, 1979; Jacob and Quittmeyer, 1979). The Makran ranges are terminated in the east by the Ornach-Nal and the Chaman Faults. South of the Karachi coast, the boundary between the Indian and Eurasian plates is defined by the Murray Ridge and the Owen Fracture Zone (see, for example, Jacob and Quittmeyer, 1979).

The Himalayan mountain range is generally considered to extend from Nanga Parbat (8125 m) in the west to Namcha Barwa (7755 m) in the east

FOCAL MECHANISM SOLUTIONS   245

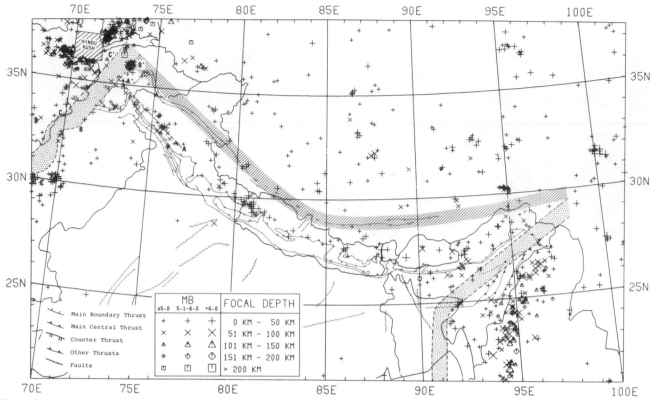

Fig. 3. Epicenter map of the Himalaya and vicinity for the period, January 1963 - March 1974, on a Lambert conformal conic projection with standard parallels 24°N and 36°N. The distribution of earthquake epicenters in the Hindu Kush region is very dense (after Chandra, 1978).

(Figure 1). Throughout its length of about 2,400 km, the trend of the mountain range changes very gradually from northwest in Kashmir to northeast in Arunachal. At the northwest extremity of the Himalaya, however, the geological structures bend sharply toward the south, then southwest and curve toward the west to form the Kashmir syntaxis. The southwest flank of this syntaxis curves further toward the south and continues as the southward trending Sulaiman Range. At a latitude of about 30°N, the formations of the Sulaiman Range turn toward the west with an east-west segment of about 200 km in length. Near Quetta the mountain ranges turn again toward the south, where they are described as the Kirthar ranges. This north-south trend continues toward the Arabian Sea, intersecting the coast line near Karachi.

The Karakoram and Pamir ranges are located north of the western extremity of the Himalaya.

At the northeastern extremity of the Himalaya, the convergence of the mountain ranges of the Himalayan Arc and the Burmese Arc forms the socalled Assam syntaxis. The Burmese Arc consists of the Patkai, Naga, Manipur, Lushai, Chin, Arakan and other hill ranges. It continues toward the south through the Andaman and Nicobar Islands and curves eastward from Sumatra toward Java.

The northeast trend of the Aravalli mountains of Peninsular India intersects the Himalaya nearly at right angles.

The earthquake epicenters generally follow the trends of Kirthar, Sulaiman, Hindu Kush, Pamir, Himalayan and Burmese mountains. However, the greatest concentration of earthquake activity occurs along the Hindu Kush and Pamir ranges; Quetta, Kashmir and Assam syntaxes. Throughout Tibet and southwest China, the earthquakes occur in an irregular manner with no clear trends.

McKenzie (1972) and Nowroozi (1972) proposed models involving a number of small plates to explain the observed seismicity and tectonics of the Middle East.

Following Argand's (1924) idea that the continental regions behave plastically, Molnar and Tapponnier (1975, 1977, 1978) and Tapponnier and Molnar (1976, 1977) suggested that the continental lithosphere of Asia behaves like a rigid plastic medium indented by India. They explained the orientation and sense of motion of large strike slip faults in central Asia, the convergence at the Burma Arc, the existence of the Assam syntaxis, the conjugate strike-slip faults in Mongolia, and the extension at the Baikal rift and Shansi graben by analogy with the slip-line field in the indented plastic material, for different indentation geometries.

However, they considered the Indian continent as a whole to be a rigid indenter.

Contrary to what one would normally expect about the level of earthquake activity along plate boundaries, it is interesting to note that along the suture zone between the Indian and Eurasian continents (Quetta Line, Indus Tsangpo suture zone and the suture in northeast India, see, e.g., Stoneley, 1974) and along the suture zone between the Arabian and Iranian continents (Zagros suture zone), the seismicity is very small, almost nil (Figures 2 and 3). In Figure 3, the aseismic lineament regions are shown by shaded and stippled regions. Chandra (1979a) suggested that these aseismic lineaments of widths of about 100 km define a Rigid Boundary Zone. He presented a model for the deformation of the Indian continent within this Rigid Boundary Zone, since the continental collision. The model explains the origin of the Himalayan syntaxes, their re-entrant character, geological similarity of the two flanks of the Kashmir syntaxis, lack of similarity between the Himalayan and Naga Hill flanks of the Assam syntaxis, and the observation that most of the Himalaya are made up of rocks belonging geologically to the Indian peninsular shield (Gansser, 1966). Chandra (1977) suggested that the high stresses generated by the continental collision may be very extensive spatially and that the entire Indian Peninsula may be in a state of left lateral shear along NNE vertical planes.

It is clear that continental tectonics are considerably more complicated than the tectonics of oceanic regions. Focal mechanism solutions of earthquakes provide important information about the orientation of regional stress, nature of faulting and sense of motion on faults. In this paper, the tectonic implications of fault plane solutions for the region between the fold belts of Iran and Burma are examined.

Catalog of Focal Mechanism Solutions

A number of authors have published focal mechanism solutions of earthquakes in the region under study. From these solutions a catalog, presented in Table 1, was compiled. In some cases, several authors determined solutions for the same earthquake. The multiple solutions are indicated by a, b, c, etc. next to the "Hr" column. The solutions are presented in terms of the orientation (trend, ϕ, measured clockwise from the north and plunge, θ, measured from the horizontal) of the mechanism axes P, T and B, and the dip and dip direction of the nodal planes. P is the pressure axis or the axis of maximum compression, T is the tension axis or the axis of least compression, and B is the intermediate or null axis. Unfortunately, the solutions derived by different authors were not published in a consistent form. Therefore, for a large number of earthquakes, the orientations of P, T and B

axes, and of the nodal planes were computed from the parameters of published solutions.

Although, no attempt has been made to distinguish between the reliable and unreliable solutions in Table 1, only those solutions should be considered reliable for which first motion data were read from seismograms, preferably those obtained from the World Wide Standardized Stations Network (WWSSN), by the investigator(s) reporting the solutions. First motion data reported in routine seismological bulletins, such as the Earthquake Data Report (EDR) of the USGS and the International Seismological Center Bulletin (ISC), are of doubtful quality and, therefore, these data should not be used in determining the solutions (see also, Isacks and Molnar, 1971; Das and Filson, 1975; Chandra, 1978).

Fault Plane Solutions and
Their Tectonic Implications

In this section, a region by region discussion of the fault plane solutions and their tectonic implications is presented. The solutions used in the following discussion and in the various figures are identified by an asterisk preceding the 'Date' column in Table 1. Many of these earthquakes have multiple solutions derived by different investigators from WWSSN data. These alternate solutions, whether or not they agree with the solutions used in the following discussion, are also just as reliable.

Elburz Range

The Elburz mountains extend from north-central Iran to northeast Iran where they appear to merge with the Caucasus-Kopet Dagh ranges. In this region, the Elburz ranges bend toward southeast, paralleling the trend of the Caucasus-Kopet Dagh ranges. Near the northern border of Iran and Afghanistan, the trend of these mountain ranges further curves around toward northeast linking them with the Hindu Kush and Pamir mountain chains.

The earthquake epicenters generally follow the trend of the Elburz ranges. The epicenters do not show any clear trend in northwest Iran. However, McKenzie (1972) assumed that the Elburz seismicity trend is a continuation from the eastern end of the North Anatolian Fault. There is a weak indication for a possible continuation of the Elburz seismicity trend along the southern Caspian Sea coast joining the Caucasus range (Figure 2).

The focal mechanism solutions of earthquakes in the vicinity of Elburz ranges are shown in Figure 4. The earthquake of 3-2-76 occurred near the border of northwest Iran and southern U.S.S.R. The focal mechanism solution of this earthquake shows normal faulting along northwest striking nodal planes. The focal depth was reported to be 58 km (Preliminary Determination of Epicenters by U.S.G.S.). The earthquake may have been

						P		T		B		Nodal Planes				
												1		2		
Date	Hr.	Lat. (°N)	Long. (°E)	Depth (km)	Mag.	φ	θ	φ	θ	φ	θ	Dip dir.	Dip	Dip dir.	Dip	Reference
15 JAN 1934	8	26.5	86.5	25	8.4	46	23	276	57	146	23	190	30	64	72	SINGH AND GUPTA(1980)
30 MAY 1935	21	29.5	66.8	20	7.6	278	16	165	53	19	32	303	68	60	40	SINGH AND GUPTA(1980)
* 5 OCT 1948	20a	37.8	58.4	5	7.3	20	40	200	50	290	0	200	5	20	85	MCKENZIE(1972)
5 OCT 1948	20b	37.6	58.4		7.0	20	40	200	50	290	0	20	85	200	5	SHIROKOVA(1967)
15 AUG 1950	a	28.5	98.7			180	60	0	30	90	0	0	75	180	15	TANDON(1955)
15 AUG 1950	14b	28.7	96.6			199	14	293	18	72	67	65	67	157	88	BEN-MENAHEM ET AL.(1974)
15 AUG 1950	c	28.3	96.8			171	33	351	57	81	0	171	78	351	12	CHEN AND MOLNAR(1977)
12 MAR 1951	14	27.0	95.0		5.0	335	10	225	50	75	30	120	40	0	70	SHIROKOVA(1967)
8 JUN 1952	18	29.5	54.8			285	72	151	13	58	13	347	34	140	59	SCHAFFNER(1959)
12 FEB 1953	8a	35.2	55.0		6.3	140	20	30	50	245	25	170	70	285	40	SHIROKOVA(1967)
12 FEB 1953	8b	35.1	54.7			121	23	213	4	312	67	349	71	255	77	WICKEN AND HODGSON(1967)
12 FEB 1953	8c	35.1	54.7			269	10	156	65	3	23	64	40	288	59	CANITEZ AND UCER(1967)
22 FEB 1953		35.8	55.0			32	23	122	3	220	65	253	72	165	76	INST. OF GEOPHYSICS(1966)
23 JAN 1954	16	37.4	72.5		5.0	10	25	140	50	270	25	350	75	235	30	SHIROKOVA(1967)
23 FEB 1954	6	37.0	92.0		5.0	210	15	90	70	305	15	10	35	225	60	SHIROKOVA(1967)
18 AUG 1954	23	39.1	70.6		5.0	345	5	250	75	75	15	355	50	150	40	SHIROKOVA(1967)
28 JAN 1955	17	33.0	83.0		6.0	190	0	105	55	280	35	345	55	225	50	SHIROKOVA(1967)
15 APR 1955	3	39.9	74.6		7.0	170	5	70	60	260	30	190	55	325	50	SHIROKOVA(1967)
27 JUN 1955	10	32.5	78.5		6.0	20	30	115	10	240	55	155	80	250	60	SHIROKOVA(1967)
19 AUG 1955	8	39.7	68.0		5.5	130	15	245	50	30	35	350	40	105	70	SHIROKOVA(1967)
3 FEB 1956	13	32.5	46.0		5.0	210	15	105	50	315	40	240	70	355	45	SHIROKOVA(1967)
29 FEB 1956	20	23.0	54.5		6.0	240	30	40	60	145	10	230	75	90	20	SHIROKOVA(1967)
5 MAR 1956	7	37.8	77.1		5.0	330	5	235	70	60	20	345	50	130	40	SHIROKOVA(1967)
11 APR 1956	1	39.0	70.3		5.0	335	30	160	60	75	5	340	75	150	15	SHIROKOVA(1967)
12 APR 1956	22a	37.0	50.0		5.0	0	15	250	50	105	35	30	70	140	45	SHIROKOVA(1967)
12 APR 1956	22b	37.3	50.2			122	10	216	18	6	69	348	70	80	85	CANITEZ(1969)
12 APR 1956	c	37.1	50.1	25		355	35	85	50	263	58	213	38	317	80	INST. OF GEOPHYSICS(1966)
13 MAY 1956	7	30.0	70.0		6.0	310	0	220	45	35	45	95	60	345	60	SHIROKOVA(1967)
8 JUN 1956	4	35.2	67.5		6.0	145	60	0	20	265	15	350	70	200	25	SHIROKOVA(1967)
9 JUN 1956	23	35.0	67.5		7.3	330	0	240	40	60	50	110	65	10	60	SHIROKOVA(1967)
16 JUL 1956	15	22.2	95.7	40		319	52	205	17	104	33	180	70	64	40	ICHIKAWA ET AL.(1972)
16 SEP 1956	8	34.0	69.5		6.0	120	10	345	75	205	15	280	35	125	55	SHIROKOVA(1967)
10 OCT 1956	15	28.3	77.8	20	6.5	331	5	63	21	229	68	290	79	194	72	TANDON(1975)
31 OCT 1956	14a	27.0	54.0		6.0	20	65	230	25	135	10	220	70	75	25	SHIROKOVA(1967)
31 OCT 1956	14b	27.3	54.4			24	3	121	26	294	64	249	69	345	75	CANITEZ AND UCER(1967)
31 OCT 1956	14c	27.3	54.4			239	16	138	32	353	42	13	55	276	80	CANITEZ AND UCER(1967)
31 OCT 1956	14d	27.3	54.4			184	4	91	46	278	43	328	56	218	62	CANITEZ(1969)
1951 - 1957		36.5	70.5			170	10	30	85	260	10	175	55	340	35	SHIROKOVA(1967)
16 MAR 1957	0	33.5	52.0		5.0	200	15	330	65	105	20	185	60	45	35	SHIROKOVA(1967)
14 APR 1957	7	30.0	84.0		6.0	330	10	110	80	235	10	325	55	80	35	SHIROKOVA(1967)
1 JUL 1957	19	24.4	93.7	41		33	3	301	33	127	56	72	69	172	65	ICHIKAWA ET AL.(1972)
* 2 JUL 1957	0 a	36.1	52.7	10	7.3	22	0	289	82	112	8	30	46	194	45	MCKENZIE(1972)
2 JUL 1957	0 b	36.0	52.5		7.3	20	10	260	80	115	5	190	40	30	50	SHIROKOVA(1967)
2 JUL 1957	0 c	36.0	52.5			30	0	298	79	120	11	199	46	41	44	WICKEN AND HODGSON(1967)
2 JUL 1957	0 d	36.0	52.5			209	26	300	2	34	63	78	70	342	73	SOBOUTI(1963)
2 JUL 1957	0 e	36.0	52.5			202	22	58	65	297	13	359	26	213	68	CANITEZ AND UCER(1967)
30 AUG 1957	16	39.3	72.9		5.0	335	25	190	60	70	15	345	65	130	25	SHIROKOVA(1967)
1 SEP 1957	12	38.9	74.0		5.0	160	10	55	50	255	30	185	60	300	50	SHIROKOVA(1967)
*13 DEC 1957	1 a	34.4	47.7	40	7.2	71	1	339	62	162	28	226	50	96	53	MCKENZIE(1972)
13 DEC 1957	1 b	34.0	47.0		7.0	50	5	310	70	150	20	215	45	70	50	SHIROKOVA(1967)
13 DEC 1957	1c	34.1	47.1			94	23	334	50	199	30	231	35	118	75	WICKEN AND HODGSON(1967)
13 DEC 1957	1d	34.1	47.1			81	4	340	69	173	20	240	45	99	52	CANITEZ AND UCER(1967)
22 MAR 1958	11	35.0	67.0		5.0	325	0	235	70	55	20	345	50	125	50	SHIROKOVA(1967)
13 AUG 1958	7	37.2	66.5		5.0	140	5	45	55	230	30	165	55	290	50	SHIROKOVA(1967)
14 AUG 1958	11a	34.0	47.5		5.0	190	15	95	15	320	70	230	90	320	70	SHIROKOVA(1967)
14 AUG 1958	11b	34.0	47.5			145	27	239	7	345	62	15	66	279	76	CANITEZ(1969)
14 AUG 1958	15	34.0	47.5			157	23	61	14	301	63	287	63	20	84	CANITEZ(1969)
16 AUG 1958	19a	34.5	47.5		6.0	190	10	280	20	75	70	55	70	145	80	SHIROKOVA(1967)
16 AUG 1958	19b	34.5	47.5			84	33	350	4	253	56	212	64	312	71	CANITEZ AND UCER(1967)
28 DEC 1958	5	30.0	79.8	25	6.5	192	9	12	81	102	0	12	36	192	54	TANDON(1975)
1 MAY 1959	8a	36.5	51.5		5.0	40	20	160	55	285	30	15	70	250	40	SHIROKOVA(1967)
1 MAY 1959	8b	36.5	51.5			240	4	333	34	145	56	101	64	202	70	CANITEZ(1969)
1 MAY 1959	8c	36.5	51.5			173	11	264	4	8	79	39	79	308	85	CANITEZ AND UCER(1967)
9 JAN 1960	7	36.4	70.1	234		150	15	260	50	49	36	120	70	10	50	ROECKER ET AL.(1980)
8 FEB 1960	18	36.2	70.6	175		143	10	46	33	247	55	280	60	160	75	ROECKER ET AL.(1980)
19 FEB 1960	10	36.5	71.1	211		95	5	195	80	4	10	80	50	285	45	ROECKER ET AL.(1980)
23 FEB 1960	2	36.5	71.1	194		310	5	135	80	40	1	310	50	140	40	ROECKER ET AL.(1980)
24 APR 1960	12 a	28.0	54.4			187	6	300	74	94	15	23	41	174	53	CANITEZ AND UCER(1967)
24 APR 1960	12 b	28.0	54.4			193	12	86	52	291	35	337	45	220	66	CANITEZ(1969)
2 JUN 1960	12	33.5	49.0			100	39	10	0	280	51	228	64	332	64	CANITEZ(1969)
4 FEB 1961	8 a	24.9	95.3	141		216	6	36	84	306	0	36	39	216	51	ICHIKAWA ET AL.(1972)
4 FEB 1961	8 b	24.9	93.3	141	7.6	59	54	157	4	250	35	185	59	303	50	VERMA ET AL.(1976)
20 MAR 1961	3	36.8	71.3	75		335	0	65	60	245	30	310	55	180	50	ROECKER ET AL.(1980)
6 APR 1961	18	27.8	56.7		6.0	180	25	350	70	270	5	345	20	185	70	SHIROKOVA(1967)
13 APR 1961	16	39.7	77.6	38		345	4	165	86	75	0	165	41	345	49	ICHIKAWA ET AL.(1972)

TABLE 1 (continued)

						P		T		B		Nodal Planes				
												1		2		
Date	Hr.	Lat. (°N)	Long. (°E)	Depth (km)	Mag.	φ	θ	φ	θ	φ	θ	Dip dir.	Dip	Dip dir.	Dip	Reference
28 APR 1961	5	36.6	71.3	218		40	0	310	40	130	50	180	65	80	60	ROECKER ET AL.(1980)
4 JUN 1961	7	34.2	81.9	11		198	76	18	14	108	0	18	59	198	31	ICHIKAWA ET AL.(1972)
11 JUN 1961	5 a	27.9	54.6		7.0	145	20	345	65	240	10	310	25	150	65	SHIROKOVA(1967)
11 JUN 1961	5 b	27.9	54.6		7.0	145	20	250	45	40	40	110	70	5	45	SHIROKOVA(1967)
11 JUN 1961	5 c	27.9	54.6			81	9	313	75	173	11	247	37	91	55	WICKEN AND HODGSON(1967)
14 JUN 1961	0	24.6	94.7	91		148	6	245	54	54	35	120	80	1	50	ICHIKAWA ET AL.(1972)
19 JUN 1961	17	36.5	70.9	197		30	20	170	70	296	12	20	65	230	30	ROECKER ET AL.(1980)
20 JUL 1961	0	38.4	72.4	120		15	61	126	12	223	24	278	40	146	59	ROECKER ET AL.(1980)
17 AUG 1961	13	37.5	71.7	113		105	60	315	26	220	15	164	22	304	72	ROECKER ET AL.(1980)
18 AUG 1961	7	38.7	72.7	110		142	2	234	79	52	10	334	44	131	46	ROECKER ET AL.(1980)
21 AUG 1961	7	36.5	71.7	108		130	0	220	45	40	45	95	60	345	60	ROECKER ET AL.(1980)
6 SEP 1961	13	36.5	70.6	204		225	30	45	60	135	0	225	75	45	15	ROECKER ET AL.(1980)
28 OCT 1961	10	33.6	48.5			43	8	143	53	308	36	257	49	15	62	CANITEZ(1969)
5 JAN 1962	4	36.5	71.4	104		166	9	67	40	266	48	305	55	200	70	ROECKER ET AL.(1980)
8 JAN 1962	22	36.4	70.8	212		182	6	84	54	276	35	330	50	210	60	ROECKER ET AL.(1980)
27 FEB 1962	5	36.5	71.5	101		355	0	90	75	265	15	340	50	190	50	ROECKER ET AL.(1980)
28 MAR 1962	0	36.6	71.5	102		307	22	78	59	208	21	290	70	160	30	ROECKER ET AL.(1980)
17 JUN 1962	4	33.7	75.8	88		23	21	285	20	156	60	154	60	245	89	ICHIKAWA ET AL.(1972)
* 6 JUL 1962	a	36.5	70.3	204		189	35	344	52	90	12	178	81	54	16	STEVENS(1966)
6 JUL 1962	23 b	36.5	70.4	208	6.8	180	33	0	57	90	0	180	78	0	12	CHANDER((1965)
6 JUL 1962	23 c	36.5	70.4	208		172	14	59	58	270	29	320	40	195	65	ROECKER ET AL.(1980)
3 AUG 1962	18	36.5	71.1	203		340	5	245	60	75	30	135	50	5	55	ROECKER ET AL.(1980)
* 1 SEP 1962	19 a	35.6	49.9	27	7.2	216	4	318	71	125	18	55	44	200	52	MCKENZIE(1972)
1 SEP 1962	19 b	35.6	49.9			240	5	345	72	148	17	78	43	224	52	WICKEN AND HODGSON(1967)
1 SEP 1962	19 c	35.6	49.9			236	11	46	79	146	2	58	34	234	56	CANITEZ AND UCER(1967)
22 SEP 1962	6	26.5	96.8	0		273	29	93	61	184	5	113	16	273	74	RASTOGI ET AL.(1973)
9 OCT 1962	15	36.4	71.2	238		40	25	235	65	133	6	205	25	45	70	ROECKER ET AL.(1980)
12 JAN 1963	6	36.1	69.1	122		155	0	345	85	245	1	150	45	340	40	ROECKER ET AL.(1980)
17 FEB 1963	5	36.5	70.6	201		255	0	160	55	345	25	280	50	50	50	ROECKER ET AL.(1980)
18 FEB 1963	14	36.5	70.8	218		165	5	70	55	258	35	320	50	195	55	ROECKER ET AL.(1980)
7 MAR 1963	21	36.5	71.4	96		135	0	235	80	45	10	120	50	325	45	ROECKER ET AL.(1980)
24 MAR 1963	12 a	34.3	47.8	10	5.3	131	35	311	55	221	0	311	10	131	80	CHANDRA(1979E)
24 MAR 1963	12 b	34.3	47.8	10	5.3	97	19	352	36	209	48	230	50	131	79	CHANDRA(1979E)
*24 MAR 1963	12 c	34.3	47.8	10	5.3	168	16	268	32	55	54	131	80	34	55	CHANDRA(1979B)
24 MAR 1963	12 d	34.4	47.9			337	23	225	41	88	40	109	42	7	79	CANITEZ AND UCER(1967)
24 MAR 1963	12 e	34.4	47.9			356	37	247	22	134	44	116	46	215	81	CANITEZ AND UCER(1967)
23 APR 1963	9	25.7	99.6	93		86	43	266	47	176	0	266	2	86	88	RASTOGI ET AL.(1973)
29 MAY 1963	a	27.0	59.4	52		33	0	220	55	112	12	263	13	49	79	AKASCHEH(1972)
*29 MAY 1963	8 b	27.0	59.4	52	5.2	158	63	316	27	48	0	316	72	158	18	CHANDRA(1979B)
29 MAY 1963	8 c	27.0	59.4	52	5.2	59	11	150	1	243	79	285	82	134	83	CHANDRA(1979B)
1 JUN 1963	10	36.1	71.2	96		100	5	355	63	193	26	120	55	250	45	ROECKER ET AL.(1980)
11 JUN 1963	3	37.1	70.1	24		295	5	185	70	27	19	100	40	310	55	ROECKER ET AL.(1980)
*19 JUN 1963	10 a	25.0	92.1	51	5.9	171	20	351	70	81	0	351	25	171	65	CHANDRA(1978)
19 JUN 1963	10 b	25.0	92.1	44		183	55	3	55	273	0	3	0	183	80	RASTOGI ET AL.(1973)
*21 JUN 1963	15 a	24.9	92.1	53	5.7	167	33	5	55	262	8	316	14	174	79	CHANDRA(1978)
21 JUN 1963	15 b	25.1	92.1	47		91	5	271	85	1	0	91	50	271	40	ICHIKAWA ET AL.(1972)
26 JUN 1963	14	36.4	76.6	89		25	43	205	47	115	0	205	2	25	86	RASTOGI(1974)
10 JUL 1963	2	36.4	71.6	87		275	5	180	55	9	35	305	60	60	50	ROECKER ET AL.(1980)
29 JUL 1963	16 a	27.8	55.6			231	3	131	38	323	54	21	52	209	37	CANITEZ AND UCER(1967)
29 JUL 1963	16 b	27.8	55.6			187	3	92	60	279	30	339	50	213	55	CANITEZ(1969)
13 AUG 1963	7	36.6	71.0	245		255	10	80	85	344	1	255	50	75	30	ROECKER ET AL.(1980)
29 AUG 1963		39.6	74.2			0	15	180	75	90	0	180	30	0	60	MOLNAR ET AL.(1973)
* 2 SEP 1963	1 a	33.9	74.7	44	5.1	235	25	55	65	325	0	235	70	55	20	CHANDRA(1978)
2 SEP 1963	1 b	34.0	74.7	23		98	1	6	46	187	44	131	60	241	59	ICHIKAWA ET AL.(1972)
29 SEP 1963	10	36.5	70.3	205		145	0	55	54	235	36	295	55	175	55	ROECKER ET AL.(1980)
14 OCT 1963	21	37.5	71.9	113		107	0	197	43	17	47	325	60	70	60	ROECKER ET AL.(1980)
16 OCT 1963		38.6	73.4			144	7	234	7	9	80	9	80	279	90	MOLNAR ET AL.(1973)
28 DEC 1963	1	36.6	70.1	209		170	8	68	53	266	35	316	48	199	65	ROECKER ET AL.(1980)
19 JAN 1964	9	26.9	54.0			176	30	67	29	302	46	301	46	32	89	CANITEZ AND UCER(1967)
*22 JAN 1964	15 a	22.4	93.6	88	6.1	266	63	120	23	24	14	326	25	109	70	CHANDRA(1975B)
22 JAN 1964	15 b	22.3	93.6	60		21	0	291	29	111	61	160	70	62	70	ICHIKAWA ET AL.(1972)
23 JAN 1964	15	36.6	71.2	76		110	15	270	75	15	1	100	60	300	30	ROECKER ET AL.(1980)
*28 JAN 1964	14 a	36.5	70.9	207	6.1	124	25	280	63	29	9	116	71	327	21	HEDAYATI AND HIRASAWA(1966)
28 JAN 1964	14 b	36.5	71.0	197		150	20	335	70	241	2	330	25	150	65	ROECKER ET AL.(1980)
28 JAN 1964	c	36.5	71.0	197		140	26	320	64	50	0	140	68	320	18	CHATELAIN ET AL.(1980)
16 FEB 1964	0 a	30.1	51.2			214	36	314	10	57	52	91	57	349	73	CANITEZ AND UCER(1967)
16 FEB 1964	0 b	30.1	51.2			214	3	311	66	122	24	57	47	193	53	CANITEZ AND UCER(1967)
*18 FEB 1964	3	27.5	91.1	30	5.6	212	5	32	85	302	0	212	50	32	40	CHANDRA(1978)
18 FEB 1964	17	36.5	70.7	202		162	3	69	40	264	44	310	50	195	65	ROECKER ET AL.(1980)
* 27 FEB 1964	15 a	21.7	94.4	102	6.4	265	26	95	63	357	4	268	71	75	19	CHANDRA(1975B)
27 FEB 1964	15 b	21.7	94.4	102	6.4	208	30	90	40	322	34	236	86	333	36	FITCH(1970)
27 FEB 1964	15 c	21.7	94.4	91		59	25	284	42	150	38	170	40	68	80	ICHIKAWA ET AL.(1972)
23 MAR 1964	8	38.3	73.6	125		355	41	345	18	138	44	110	47	216	75	ROECKER ET AL.(1980)
* 27 MAR 1964	23	27.2	89.3	32	6.3	346	26	166	64	256	0	346	71	166	19	CHANDRA(1978)
16 MAY 1964	8	36.4	71.4	110		9	6	271	56	103	33	160	50	35	60	ROECKER ET AL.(1980)

TABLE 1 (continued)

Date	Hr.	Lat. (°N)	Long. (°E)	Depth (km)	Mag.	P φ	P θ	T φ	T θ	B φ	B θ	Plane 1 Dip dir.	Plane 1 Dip	Plane 2 Dip dir.	Plane 2 Dip	Reference
17 MAY 1964	11	36.5	70.5	226		22	26	149	51	278	27	0	75	250	30	ROECKER ET AL.(1980)
3 JUN 1964	2a	25.9	95.8	100	5.5	293	39	113	51	203	0	113	6	293	84	RASTOGI ET AL.(1973)
3 JUN 1964	2b	25.9	95.7	121	5.4	270	73	90	17	180	0	90	62	270	28	VERMA ET AL.(1976)
13 JUN 1964	17	23.0	94.0	60		133	14	227	14	0	70	90	90	0	70	ICHIKAWA ET AL.(1972)
*12 JUL 1964	20	24.9	95.3	155	6.7	211	24	340	55	110	24	192	73	70	30	CHANDRA(1975b)
*13 JUL 1964	10	23.7	94.7	117	6.5	272	14	71	75	181	5	268	60	100	31	CHANDRA(1975b)
19 AUG 1964	a	28.2	52.6	50		140	20	2	68	236	15	299	29	152	65	AKASCHEH(1972)
19 AUG 1964	b	28.2	52.7	52		168	18	305	68	72	16	9	30	156	64	AKASCHEH(1972)
* 1 SEP 1964	13	27.2	92.3	33	5.7	134	25	314	65	44	0	314	20	134	70	CHANDRA(1978)
10 SEP 1964	13	27.1	92.3	33		237	7	331	22	130	66	102	69	196	89	ICHIKAWA ET AL.(1972)
*26 SEP 1964	0 a	30.1	80.7	50	6.2	207	28	19	62	115	3	204	73	35	17	CHANDRA(1978)
26 SEP 1964	0 b	27.0	80.5	50		29	21	291	20	162	60	160	60	251	89	ICHIKAWA ET AL.(1972)
26 SEP 1964	c	30.1	80.7			220	15	40	75	130	0	220	60	40	30	MOLNAR ET AL.(1973)
26 SEP 1964	d	30.1	80.7	50	6.2	250	5	74	85	342	2	70	40	254	50	VERMA ET AL.(1977)
28 SEP 1964	6	36.4	71.5	77		105	10	0	65	200	24	125	55	260	45	ROECKER ET AL.(1980)
*13 OCT 1964	23	35.8	71.1	120	5.8	123	0	214	54	33	36	334	55	93	55	CHANDRA(1978)
*21 OCT 1964	23 a	28.1	93.8	37	5.9	174	42	354	48	84	0	354	3	174	87	CHANDRA(1978)
21 OCT 1964	b	28.1	93.8			177	40	35	50	87	0	177	85	357	5	MOLNAR ET AL.(1973)
8 NOV 1964	10 a	29.7	51.0			31	1	120	57	300	33	239	53	2	54	CANITEZ AND UCER(1967)
8 NOV 1964	10 b	29.7	51.0			334	7	222	72	66	17	136	41	348	54	CANITEZ AND UCER(1967)
8 NOV 1964	10 c	29.7	51.0			10	9	156	79	279	6	197	36	5	54	CANITEZ AND UCER(1967)
9 NOV 1964	8 a	39.8	48.4			131	27	27	25	261	52	258	52	349	89	CANITEZ AND UCER(1967)
9 NOV 1964	8 b	39.8	48.4			165	60	20	24	257	14	230	24	8	71	CANITEZ AND UCER(1967)
27 NOV 1964	11	36.4	70.7	211		181	5	283	65	92	24	160	55	25	45	ROECKER ET AL.(1980)
2 DEC 1964	8	29.6	81.1	3		343	14	77	14	210	70	120	90	210	70	ICHIKAWA ET AL.(1972)
*22 DEC 1964	4 a	28.2	57.0	42	5.5									182	76	MCKENZIE(1972)
22 DEC 1964	4 b	28.2	57.0			181	46	86	5	351	44	302	55	53	63	CANITEZ AND UCER(1967)
24 DEC 1964	1	36.4	70.9	127		350	15	180	75	81	2	170	30	350	60	ROECKER ET AL.(1980)
*12 JAN 1965	13 a	27.6	88.0	23	6.1	181	33	1	57	271	0	181	78	1	12	CHANDRA(1978)
12 JAN 1965	b	27.6	88.0			180	30	0	60	90	0	180	75	0	15	MOLNAR ET AL.(1973)
12 JAN 1965	13 c	27.4	87.8	23		3	19	96	8	206	69	138	82	231	71	ICHIKAWA ET AL.(1972)
12 JAN 1965	13 d	27.4	88.0	23	6.1	9	47	99	0	9	2	189	88	RASTOGI(1974)
12 JAN 1965	13 e	27.4	87.8	23	5.8	121	4	14	25	216	64	258	70	156	76	CHOUHAN AND SRIVASTAVA(1975)
*29 JAN 1965	20	35.6	73.6	33	5.7	253	56	19	22	120	25	162	32	39	71	CHANDRA(1978)
* 2 FEB 1965	15	37.5	73.4	33	5.8	161	2	70	11	261	79	205	84	296	81	CHANDRA(1978)
18 FEB 1965	4 a	25.0	94.2	45	5.4	22	67	288	1	199	22	269	50	128	50	VERMA ET AL.(1976)
18 FEB 1965	4 b	25.0	94.2	45	5.4	42	8	141	46	303	44	259	53	9	67	CHOUHAN AND SRIVASTAVA(1975)
18 FEB 1965	4 c	25.0	94.2	45		221	29	311	0	41	61	90	70	352	70	ICHIKAWA ET AL.(1972)
25 FEB 1965	10 a	23.6	94.6	94	5.2	243	18	139	24	359	55	278	84	12	56	CHOUHAN AND SRIVASTAVA(1975)
25 FEB 1965	10 b	23.6	94.7	94		54	26	317	13	205	59	278	80	182	61	ICHIKAWA ET AL.(1972)
14 MAR 1965	15 a	36.3	70.7	219	6.6	196	17	62	66	291	16	209	64	352	31	CHANDRA(1970)
14 MAR 1965	15 b	36.3	70.7	219	6.6	215	15	35	75	125	0	215	60	35	30	RITSEMA(1966)
*14 MAR 1965	15 c	36.3	70.7	219	6.6	201	25	20	65	111	0	201	70	21	20	NOWROOZI(1972)
14 MAR 1965	15 d	36.7	70.7	205		40	14	220	76	130	0	40	59	220	31	ICHIKAWA ET AL.(1972)
14 MAR 1965	15 e	36.4	70.7	205		220	15	40	75	130	0	40	30	220	60	ROECKER ET AL.(1980)
14 MAR 1965	f	36.4	70.7	205		219	15	2	72	127	8	212	60	61	32	CHATELAIN ET AL.(1980)
10 APR 1965	21	37.3	71.9	129		290	16	195	18	58	65	64	67	332	86	ROECKER ET AL.(1980)
11 APR 1965	22	26.8	92.3	70		60	7	329	8	189	79	104	89	194	79	ICHIKAWA ET AL.(1972)
30 MAY 1965	11	36.4	70.1	234		268	58	116	28	20	14	329	22	106	75	ROECKER ET AL.(1980)
1 JUN 1965	4 a	20.3	95.0	33	5.5	190	69	10	21	100	0	10	66	190	24	RASTOGI ET AL.(1973)
1 JUN 1965	4 b	20.2	94.8	81		339	61	204	21	106	19	55	29	189	69	ICHIKAWA ET AL.(1972)
1 JUN 1965	7 a	28.5	83.2	20	5.3	197	9	84	70	290	18	357	40	213	56	RASTOGI(1974)
1 JUN 1965	7 b	28.6	83.1	20		0	15	94	14	224	69	137	89	227	69	ICHIKAWA ET AL.(1972)
*15 JUN 1965	7	29.6	95.6	30	5.6	88	14	354	14	221	70	221	70	131	90	CHANDRA(1978)
*18 JUN 1965	8 a	25.0	93.7	66	5.8	180	32	53	44	290	30	204	83	306	31	CHANDRA(1978)
18 JUN 1965	8 b	24.9	93.7	48	5.2	190	20	95	5	328	67	49	86	321	67	CHOUHAN AND SRIVASTAVA(1975)
21 JUN 1965	a	28.1	56.0	28		30	15	210	75	120	5	210	30	30	60	AKASCHEH(1972)
*21 JUN 1965	0 b	28.1	55.9	40	6.0	30	13	259	70	124	14	192	34	43	60	MCKENZIE(1972)
21 JUN 1965	0 c	28.1	55.9	40	6.0	154	6	252	54	60	36	126	60	6	50	NOWROOZI(1972)
*22 JUN 1965	5 a	36.2	77.6	107	5.7	219	48	39	42	129	0	219	3	39	87	CHANDRA(1978)
22 JUN 1965	5 b	36.2	77.6	107	5.7	48	11	228	79	138	0	228	34	48	56	RASTOGI(1974)
20 JUL 1965	7	36.7	71.3	191		308	30	62	35	188	40	181	42	274	86	ROECKER ET AL.(1980)
22 SEP 1965	4	20.8	99.4	5		34	0	304	14	124	76	170	80	78	80	ICHIKAWA ET AL.(1972)
16 NOV 1965	1	36.4	71.1	242		341	19	89	43	233	41	310	75	205	45	ROECKER ET AL.(1980)
5 DEC 1965	22 a	23.3	94.5	97	5.0	213	5	300	12	101	79	76	80	170	86	CHOUHAN AND SRIVASTAVA(1975)
5 DEC 1965	22 b	23.3	94.5	97		31	8	121	6	250	80	166	89	256	80	ICHIKAWA ET AL.(1972)
9 DEC 1965	20 a	27.4	92.5	29	5.3	11	19	191	71	101	0	11	64	191	26	RASTOGI ET AL.(1973)
9 DEC 1965	20 b	27.4	92.5	29	5.2	204	15	296	12	87	75	69	76	161	86	CHOUHAN AND SRIVASTAVA(1975)
15 DEC 1965	4 a	22.0	94.5	109	5.2	347	42	231	25	120	37	100	39	203	80	ICHIKAWA ET AL.(1972)
15 DEC 1965	4 b	22.0	94.5	109	5.2	208	50	28	40	299	1	29	84	218	6	VERMA ET AL.(1976)
*24 JAN 1966	7 a	29.9	69.7	26	5.6	155	24	19	58	254	20	300	27	171	72	CHANDRA(1978)
24 JAN 1966	7 b	29.9	69.7			25	2	118	62	296	23	5	53	225	70	MOLNAR ET AL.(1973)
24 JAN 1966	7 c	29.9	69.7	26	5.6	148	15	57	2	319	75	282	78	14	81	NOWROOZI(1972)
* 7 FEB 1966	4 a	29.9	69.7	10	6.0	147	0	–	90	57	0	327	45	147	45	CHANDRA(1978)
7 FEB 1966	b	29.9	69.7			27	2	123	59	296	22	9	55	225	50	MOLNAR ET AL.(1973)

TABLE 1 (continued)

| | | | | | | P | | T | | B | | Nodal Planes 1 | | 2 | | |
Date	Hr.	Lat. (°N)	Long. (°E)	Depth (km)	Mag.	φ	θ	φ	θ	φ	θ	Dip dir.	Dip	Dip dir.	Dip	Reference
7 FEB 1966	4 c	29.9	69.7	10	6.0	149	0	239	0	-	90	194	90	284	90	NOWROOZI(1972)
7 FEB 1966	4 d	29.9	69.7	38	6.0	338	8	242	48	75	43	122	54	12	64	BANGHAR(1974A)
*7 FEB 1966	23 a	30.3	69.9	11	5.8	163	6	58	68	256	21	182	55	321	43	CHANDRA(1978)
7 FEB 1966	b	30.3	69.9			178	3	83	59	270	31	330	50	204	55	MOLNAR ET AL.(1973)
*6 MAR 1966	2	31.6	80.6	12	5.7	135	80	315	10	45	0	315	55	135	35	CHANDRA(1978)
*6 MAR 1966	2 a	31.5	80.5	50	6.0	-	90	135	0	45	0	315	45	135	45	CHANDRA(1978)
6 MAR 1966	b	31.6	80.6			298	85	118	5	28	0	298	40	118	50	MOLNAR ET AL.(1973)
*6 JUN 1966	a	36.4	71.1	221		182	11	348	79	90	4	178	56	6	34	BILLINGTON ET AL.(1977)
6 JUN 1966	7 b	36.3	71.1	225	6.3	175	7	322	82	84	4	0	38	171	52	CHANDRA(1971)
6 JUN 1966	7 c	36.4	71.1	235	6.3	166	4	270	75	74	13	152	50	0	44	BANGHAR(1974A)
6 JUN 1966	7 d	36.4	71.1	214		178	18	322	68	86	13	170	65	20	30	ROECKER ET AL.(1980)
6 JUN 1966	e	36.4	71.1	221		180	5	0	85	90	0	180	50	0	40	CHATELAIN ET AL.(1980)
27 JUN 1966	a	29.6	80.9			207	4	94	86	297	10	216	49	16	41	MOLNAR ET AL.(1973)
*27 JUN 1966	10 b	29.7	80.9	37	6.1	212	18	9	71	120	7	206	63	43	28	CHANDRA(1971)
27 JUN 1966	10 c	29.7	81.0	13	6.0	207	43	27	47	117	0	27	2	207	88	RASTOGI(1974)
27 JUN 1966	10 d	29.7	80.9	36		329	25	214	42	80	38	100	40	338	80	ICHIKAWA ET AL.(1972)
7 JUL 1966	19	36.6	71.1	79		286	8	30	59	192	24	260	59	138	47	ROECKER ET AL.(1980)
*27 JUL 1966	14	32.6	48.8	33	5.3	183	13	311	69	89	16	170	59	24	34	MCKENZIE(1972)
* 1 AUG 1966	21 a	30.1	68.6	33	6.0	322	1	53	11	229	79	187	82	276	83	CHANDRA(1978)
1 AUG 1966	b	29.9	68.7			210	8	313	59	114	34	181	60	66	45	MOLNAR ET AL.(1973)
1 AUG 1966		30.0	68.5			180	5	0	85	90	0	180	50	0	40	MOLNAR ET AL.(1973)
*15 AUG 1966	2 a	28.7	78.9	53	5.6	191	75	11	15	101	0	191	30	11	60	CHANDRA(1978)
15 AUG 1966	2 b	28.6	78.9	22	5.8	305	41	47	13	151	46	180	50	83	70	TANDON(1975)
18 SEP 1966	20 a	27.9	54.3	18	5.9	175	11	336	78	84	4	172	56	0	34	MCKENZIE(1972)
*18 SEP 1966	20 b	27.9	54.3	18	5.9	222	5	41	85	132	0	222	52	42	42	NOWROOZI(1972)
*24 SEP 1966	10 a	27.4	54.6	38	5.3	206	35	26	55	296	0	206	80	26	10	CHANDRA(1979B)
24 SEP 1966	10 b	27.4	54.6	38	5.3	15	2	284	24	110	66	57	75	152	72	CHANDRA(1979B)
24 SEP 1966	c	27.4	54.5	33		212	12	325	60	120	27	62	40	191	62	AKASCHEH(1972)
*26 SEP 1966	5 a	27.5	92.6	19	5.5	163	65	343	65	253	0	343	20	163	70	CHANDRA(1978)
26 SEP 1966	5 b	27.5	92.6	19	5.5	204	27	24	63	114	0	24	18	204	72	RASTOGI ET AL.(1973)
26 SEP 1966	5 c	27.5	92.6	20	5.4	73	4	336	52	164	38	101	60	218	53	CHOUHAN AND SRIVASTAVA(1975)
2 OCT 1966	4	24.4	94.8	75	4.9	56	9	154	41	315	47	277	54	21	69	TANDON AND SRIVASTAVA(1975)
22 OCT 1966	3	23.0	54.3	72	5.1	138	3	34	20	228	71	262	74	169	80	CHOUHAN AND SRIVASTAVA(1975)
2 DEC 1966		28.2	53.2	40		182	18	295	54	78	32	39	39	157	69	AKASCHEH(1972)
15 DEC 1966	2 a	21.5	94.4	98	5.6	255	25	64	65	162	0	72	20	252	70	RASTOGI ET AL.(1973)
15 DEC 1966	2 b	21.5	94.4	84	5.4	258	18	116	66	353	14	269	66	57	30	CHOUHAN AND SRIVASTAVA(1975)
*16 DEC 1966	20 a	29.7	80.9	15	5.8	190	6	10	84	280	0	190	51	10	39	CHANDRA(1978)
15 DEC 1966	b	29.7	80.9			166	15	53	56	265	31	310	40	190	66	MOLNAR ET AL.(1973)
16 DEC 1966	c	29.6	80.8	15	5.7	218	51	38	39	128	0	38	84	218	6	VERMA ET AL.(1977)
16 DEC 1966	20 d	29.6	80.8	68	5.9	190	23	11	67	101	0	10	22	191	68	BANGHAR(1974A)
* 2 JAN 1967	13 a	30.6	50.4	47	5.2	220	27	40	63	310	0	220	72	40	18	CHANDRA(1979B)
2 JAN 1967	13 b	30.6	50.4	47	5.2	223	34	320	11	65	54	97	58	357	71	CHANDRA(1979B)
*11 JAN 1967	11	34.1	45.7	34	5.6	236	5	2	81	145	7	230	50	54	40	MCKENZIE(1972)
*25 JAN 1967	1 a	36.7	71.6	283	5.7	55	24	256	65	148	8	62	69	218	22	NOWROOZI(1972)
25 JAN 1967	1 b	36.7	71.6	281		48	11	265	76	138	7	216	35	54	57	ROECKER ET AL.(1980)
25 JAN 1967	c	36.7	71.6	281		60	24	276	60	155	17	212	26	72	68	CHATELAIN ET AL.(1980)
30 JAN 1967	21	26.1	96.1	39	5.4	143	85	316	13	45	0	315	60	134	30	VERMA ET AL.(1976)
11 FEB 1967	8	36.7	71.1	88		2	9	249	62	96	25	22	60	154	42	ROECKER ET AL.(1980)
15 FEB 1967	8	20.8	94.1	19	4.8	134	20	30	11	288	67	359	78	268	60	RASTOGI ET AL.(1973)
*20 FEB 1967	15 a	33.7	75.3	18	5.6	52	0	145	83	322	7	239	45	45	45	CHANDRA(1978)
20 FEB 1967	b	33.7	75.3			10	14	123	57	272	28	348	65	222	40	MOLNAR ET AL.(1973)
20 FEB 1967	15 c	33.6	75.4	25	5.7	242	12	49	78	152	3	66	33	240	57	TANDON(1975)
21 FEB 1967	12	33.7	75.4	41	5.1	190	15	10	75	100	0	10	30	190	60	RASTOGI(1974)
11 MAR 1967	16	28.4	94.4	12	5.3	183	15	3	75	93	0	3	30	183	60	RASTOGI ET AL.(1973)
*14 MAR 1967	6 a	28.5	94.3	12	5.8	186	39	6	51	96	0	6	6	186	84	CHANDRA(1978)
14 MAR 1967	b	28.5	94.3			183	40	3	50	93	0	183	80	3	10	MOLNAR ET AL.(1973)
11 MAY 1967		39.4	73.7			177	0	87	0	-	90	222	90	312	90	MOLNAR ET AL.(1973)
*15 SEP 1967	10	27.4	91.8	57	5.8	173	15	353	75	83	0	353	30	173	60	CHANDRA(1978)
10 DEC 1967	18	22.5	94.9	153		27	1	297	0	207	89	162	89	252	89	ICHIKAWA ET AL.(1972)
28 DEC 1967	20	37.3	71.9	147		350	35	123	54	230	12	318	79	193	14	ROECKER ET AL.(1980)
11 FEB 1968	20	34.2	78.7	24	5.1	34	37	238	56	48	10	136	80	0	15	TANDON AND SRIVASTAVA(1975A)
30 MAY 1968	a	27.8	54.0	27		240	26	354	41	128	38	211	81	110	40	NOWROOZI(1972)
*30 MAY 1968	b	27.8	54.0	27		210	35	21	55	117	4	207	80	50	11	NOWROOZI(1972)
12 JUN 1968	4	24.9	91.9	44	5.3	116	8	15	55	329	37	36	62	278	50	TANDON AND SRIVASTAVA(1975A)
*23 JUN 1968		23.8	51.1	33		197	26	357	63	103	8	191	71	37	20	NOWROOZI(1972)
* 2 AUG 1968	13 a	27.5	60.9	65	5.7	193	65	347	26	89	13	147	22	355	70	JACOB AND QUITTMEYER(1979)
2 AUG 1968	13 b	27.5	60.9	62	5.7	135	26	5	53	238	24	156	74	270	28	NOWROOZI(1972)
*31 AUG 1968	10 a	34.0	59.0	13	6.0	206	11	116	3	12	78	341	80	72	60	MCKENZIE(1972)
31 AUG 1968	10 b	34.1	59.0	14		50	0	140	14	320	76	5	80	275	80	NIAZI(1969)
31 AUG 1968	10 c	34.0	59.0	13	6.0	224	0	314	0	-	90	179	90	269	90	NOWROOZI(1972)
* 1 SEP 1968	7 a	34.0	58.2	15	5.9	256	16	17	61	159	24	107	36	238	65	MCKENZIE(1972)
1 SEP 1968	7 b	34.0	58.2	15	5.9	237	18	351	38	121	59	207	85	108	40	NOWROOZI(1972)
1 SEP 1968	7 c	34.0	58.2	15	5.9	202	35	21	55	112	0	202	80	22	10	NOWROOZI(1972)
1 SEP 1968	7 d	34.1	58.2	14		245	25	340	10	90	63	115	65	20	80	NIAZI(1969)
* 4 SEP 1968	23 a	34.0	58.2	15	5.4	237	20	56	70	326	0	237	65	57	25	MCKENZIE(1972)

TABLE 1 (continued)

Date	Hr.	Lat. (°N)	Long. (°E)	Depth (km)	Mag.	P φ	P θ	T φ	T θ	B φ	B θ	Nodal Planes 1 Dip dir.	Dip	Nodal Planes 2 Dip dir.	Dip	Reference
4 SEP 1968	b	33.9	58.2	15		210	41	30	49	120	0	210	86	30	4	NOWROOZI(1972)
*14 SEP 1968	13 a	28.4	53.1	33	5.8	198	15	18	75	287	0	18	30	198	60	MCKENZIE(1972)
14 SEP 1968	13 b	28.4	53.1	33	5.8	197	40	359	49	99	9	189	85	72	10	NOWROOZI(1972)
5 NOV 1968	2	32.3	76.5	33	4.5	248	16	68	74	158	0	68	35	248	67	CHAUDHURY ET AL.(1974)
* 3 JAN 1969	3 a	37.1	57.9	11	5.6	225	15	30	74	134	4	222	60	51	30	MCKENZIE(1972)
3 JAN 1969	3 b	37.1	57.9	11	5.6	242	26	356	41	130	39	213	81	112	40	NOWROOZI(1972)
3 JAN 1969	3 c	37.1	57.9	11	5.6	214	25	33	65	124	0	214	70	35	20	NOWROOZI(1972)
3 MAR 1969	6	30.2	79.9	18	5.3	41	5	268	83	131	5	50	32	208	22	SRIVASTAVA(1973)
* 5 MAR 1969	19 a	36.4	70.7	208	5.9	209	33	28	57	119	0	209	78	25	12	NOWROOZI(1972)
5 MAR 1969	b	36.4	70.7	208		205	20	25	70	115	0	205	65	25	25	CHATELAIN ET AL.(1980)
5 MAR 1969	19 c	36.4	70.7	211	5.9	212	25	32	65	121	0	32	20	212	70	BANGHAR(1974B)
29 APR 1969	4 a	29.6	51.5	36	5.6	178	26	273	11	24	62	313	60	48	64	CHANDRA(1979B)
*29 APR 1969	4 b	29.6	51.5	36	5.6	177	25	357	65	267	0	177	70	357	20	CHANDRA(1979B)
*15 MAY 1969	20	34.6	70.9	22	5.6	160	30	353	60	264	5	324	16	165	75	CHANDRA(1978)
21 JUN 1969	16 a	27.4	57.5	65	5.3	310	9	55	58	215	30	161	45	286	60	CHANDRA(1979B)
*21 JUN 1969	16 b	27.4	57.5	65	5.3	264	40	24	31	138	34	150	35	52	84	CHANDRA(1979B)
22 JUN 1969	1	30.6	79.4	19	5.4	53	26	218	63	320	6	48	72	246	20	CHAUDHURY ET AL.(1974)
30 JUN 1969	8	26.9	92.7	44	5.0	319	1	66	70	311	21	242	48	22	50	TANDON AND SRIVASTAVA(1975A)
* 8 AUG 1969	6 a	36.4	70.9	198	5.8	175	13	345	76	88	4	176	58	4	32	NOWROOZI(1972)
8 AUG 1969	b	36.4	70.9	196		16	7	196	83	106	0	196	38	16	52	CHATELAIN ET AL.(1980)
29 AUG 1969	10	26.3	96.1	72	5.2	256	3	56	66	194	24	267	56	130	46	TANDON AND SRIVASTAVA(1975A)
14 SEP 1969	a	39.7	74.9			157	6	261	54	65	19	145	60	352	50	MOLNAR ET AL.(1973)
14 SEP 1969	16 b	39.7	74.9		5.5	180	15	0	75	90	0	180	60	0	30	DAS AND FILSON(1975)
*17 OCT 1969	1 a	23.1	94.7	134	6.0	354	31	197	57	91	10	145	17	3	76	CHANDRA(1975H)
17 OCT 1969	1 b	23.1	94.7	134	6.0	0	29	180	61	90	0	0	74	180	16	DAS AND FILSON(1975)
17 OCT 1969	1 c	23.1	94.7	136	6.0	354	28	206	52	93	16	124	20	6	80	BANGHAR(1974B)
* 7 NOV 1969	18 a	27.8	60.0	74	6.5	216	48	348	34	95	25	115	26	8	82	JACOB AND QUITTMEYER(1979)
7 NOV 1969	18 b	27.9	60.1	35	6.1	212	33	327	33	85	40	90	40	180	50	NOWROOZI(1972)
*24 NOV 1969	17	37.2	71.7	123	5.6	321	50	141	40	231	0	141	85	321	5	NOWROOZI(1972)
*19 FEB 1970	7 a	27.4	94.0	18	5.5	145	5	239	40	49	49	4	59	109	67	CHANDRA(1978)
19 FEB 1970	7 b	27.4	94.0	18	5.5	77	10	337	46	176	42	219	51	109	67	CHANDRA(1978)
19 FEB 1970	c	27.4	93.9			167	40	347	50	77	0	167	85	347	5	MOLNAR ET AL.(1973)
19 FEB 1970	7 d	27.4	94.0	52	5.3	220	6	318	56	126	34	70	50	193	60	CHAUDHURY ET AL.(1974)
*23 FEB 1970	11	27.8	54.5	20	5.5	195	7	15	83	285	0	195	52	15	38	CHANDRA(1979E)
26 FEB 1970	19	27.6	85.7	96	5.0	158	40	54	15	254	48	264	52	160	76	TANDON AND SRIVASTAVA(1975A)
*28 FEB 1970	19	27.8	56.3	35	5.5	188	10	8	80	278	0	188	55	8	35	CHANDRA(1979E)
25 JUL 1970		39.9	77.8										70	0	30	MOLNAR ET AL.(1973)
*29 JUL 1970	10 a	26.0	95.4	59	6.5	265	39	135	38	20	28	19	28	110	89	CHANDRA(1975F)
29 JUL 1970	10 b	26.0	95.4	59	6.5	232	22	133	20	5	60	95	90	359	60	DAS AND FILSON(1975)
29 JUL 1970	10 c	26.0	95.4	68	6.4	344	56	160	34	253	2	163	90	330	10	VERMA ET AL.(1976)
29 JUL 1970	10 d	26.0	95.4	67	6.5	301	39	121	51	32	0	121	6	301	84	BANGHAR(1974B)
29 JUL 1970	10 e	26.0	95.4	59		242	13	339	29	131	58	107	60	203	79	ICHIKAWA ET AL.(1972)
*30 JUL 1970	0	37.8	55.9	19	5.7	165	50	61	12	322	38	278	47	32	68	CHANDRA(1979E)
25 OCT 1970	11 a	36.8	45.1	19	5.5	147	61	260	12	356	26	50	40	281	62	CHANDRA(1979B)
*25 OCT 1970	11 b	36.8	45.1	19	5.5	-	90	58	0	328	0	238	45	58	45	CHANDRA(1979B)
* 9 NOV 1970	17	29.5	56.9	106	5.5	157	55	17	35	107	0	17	80	197	10	CHANDRA(1979H)
*14 FEB 1971	16	36.6	55.6	39	5.2	340	1	72	63	250	27	316	52	185	50	CHANDRA(1979B)
6 APR 1971	6 a	29.8	51.9	10	5.2	197	11	288	2	28	79	63	81	332	84	CHANDRA(1979B)
* 6 APR 1971	6 b	29.8	51.9	10	5.2	205	19	319	50	102	34	179	72	66	40	CHANDRA(1979B)
*12 APR 1971	19	26.3	55.6	44	6.0	191	10	11	80	101	0	11	35	191	55	CHANDRA(1979H)
3 MAY 1971		30.8	84.3			101	77	281	13	11	0	281	58	101	32	MOLNAR AND TAPPONNIER(1978)
*26 MAY 1971	2	35.5	58.2	26	5.4	69	24	249	66	339	0	69	69	249	21	CHANDRA(1979B)
*30 MAY 1971	15 a	25.2	96.4	15	5.8	261	11	352	3	97	78	36	84	127	80	CHANDRA(1975E)
30 MAY 1971	15 b	25.2	96.4	15	5.8	83	0	353	0	-	90	308	90	38	90	DAS AND FILSON(1975)
31 MAY 1971	5	25.2	96.5	33		63	14	157	14	290	70	20	90	290	76	ICHIKAWA ET AL.(1972)
17 JUL 1971	15	26.4	93.2	52	5.4	80	23	331	37	194	44	110	84	217	44	CHAUDHURY ET AL.(1974)
4 AUG 1971	0	36.4	70.7	207	5.6	194	30	25	63	253	5	342	76	181	16	TANDON AND SRIVASTAVA(1975A)
6 AUG 1971		36.4	70.7	207		198	35	18	55	108	0	198	80	18	10	CHATELAIN ET AL.(1980)
10 OCT 1971	18	23.0	95.9	48	4.9	98	50	202	13	237	38	310	70	196	46	TANDON AND SRIVASTAVA(1975A)
14 OCT 1971	12	23.1	95.9	47	5.1	313	27	212	10	82	63	106	65	8	82	TANDON AND SRIVASTAVA(1975A)
* 8 NOV 1971	3	27.1	54.5	36	5.6	29	0	-	90	119	0	209	45	29	45	CHANDRA(1979B)
9 DEC 1971	1 a	27.2	56.4	15	5.3	203	9	96	60	298	28	226	60	354	43	CHANDRA(1979B)
* 9 DEC 1971	1 b	27.2	56.4	15	5.3	182	3	273	3	48	85	48	65	318	90	CHANDRA(1979B)
*29 DEC 1971	22 a	25.1	94.7	33	5.5	214	26	337	48	107	30	190	78	81	33	CHANDRA(1975B)
29 DEC 1971	22 b	25.2	94.7	46	5.0	90	88	180	55	181	36	242	56	120	56	TANDON AND SRIVASTAVA(1975A)
*12 JAN 1972	18	37.7	75.1	84	5.6	200	11	292	11	66	75	336	90	66	75	CHANDRA(1978)
*20 JAN 1972	a	36.4	70.7	214		188	19	8	71	98	0	188	64	8	26	BILLINGTON ET AL.(1977)
20 JAN 1972	b	36.4	70.7	214		228	10	339	64	134	22	208	60	76	40	CHATELAIN ET AL.(1980)
22 FEB 1972	1	36.5	70.5	213	5.2	200	13	75	70	247	17	327	60	182	35	TANDON AND SRIVASTAVA(1975A)
10 MAR 1972	14	33.9	72.7	33	4.9	140	70	3	14	272	12	342	32	188	62	TANDON AND SRIVASTAVA(1975A)
10 APR 1972	2 a	28.4	52.8		6.9	195	25	330	60	95	20	42	49	195	45	AKASCHEH(1973)
*10 APR 1972	2 b	28.4	52.8	33	6.1	227	6	47	84	317	0	227	51	47	35	CHANDRA(1979E)
*12 APR 1972	23	28.4	53.0	33	5.0	194	0	-	90	284	0	194	45	14	45	CHANDRA(1979E)
*14 JUN 1972	4	33.0	46.1	33	5.3	33	15	227	75	124	3	36	60	208	30	CHANDRA(1979B)
24 JUN 1972		36.3	69.7	47		276	17	152	62	14	20	292	68	68	34	CHATELAIN ET AL.(1980)

TABLE 1 (continued)

Date	Hr.	Lat. (°N)	Long. (°E)	Depth (km)	Mag.	P φ	P θ	T φ	T θ	B φ	B θ	1 Dip dir.	1 Dip	2 Dip dir.	2 Dip	Reference
*6 AUG 1972	1	25.1	61.2	33	5.5	184	24	342	64	90	9	177	70	22	22	CHANDRA(1979B)
*8 AUG 1972	19	25.0	61.1	41	5.5	190	40	10	50	280	0	190	85	10	5	CHANDRA(1979B)
*3 SEP 1972	16 a	36.0	73.4	36	6.3	40	16	228	74	131	2	217	29	42	61	CHANDRA(1978)
3 SEP 1972	16 b	36.0	73.4	36	6.3	0	5	180	50	90	0	180	40	0	50	MOLNAR ET AL.(1977)
3 SEP 1972	16 c	36.0	73.4	36	6.3	36	15	216	75	126	0	216	30	36	60	DAS AND FILSON(1975)
*3 SEP 1972	23 a	35.9	73.3	33	5.6	38	15	218	75	308	0	218	30	38	60	CHANDRA(1978)
3 SEP 1972	23 b	35.9	73.3	33	5.6	153	6	271	77	62	11	346	40	143	52	CHANDRA(1978)
*4 SEP 1972	13 a	35.9	73.4	33	5.8	37	10	217	80	307	0	217	35	37	55	CHANDRA(1978)
4 SEP 1972	13 b	35.9	73.4	33	5.8	323	0	-	90	233	0	323	45	143	45	CHANDRA(1978)
16 NOV 1972		35.7	69.9	120		166	18	346	72	76	0	166	63	346	27	CHATELAIN ET AL.(1980)
*17 NOV 1972	9 a	27.3	59.1	65	5.4	192	60	12	30	102	0	192	15	12	75	CHANDRA(1979B)
17 NOV 1972	9 b	27.3	59.1	65	5.4	115	5	22	32	213	58	154	72	253	64	CHANDRA(1979B)
*28 DEC 1972	16	34.7	70.4	63	5.6	148	42	328	48	58	0	328	3	148	87	CHANDRA(1978)
*16 JAN 1973	21 a	37.2	75.7	42	5.1	25	3	290	59	117	31	177	50	51	55	CHANDRA(1978)
16 JAN 1973	b	33.3	75.8	39	5.1	270	5	90	85	180	0	90	40	270	50	VERMA ET AL.(1977)
16 JAN 1973	21 c	33.2	75.7	42	5.5	99	11	350	58	196	29	120	66	252	45	CHAUDHURY ET AL.(1974)
*20 JAN 1973		29.3	68.7	19	5.6	165	11	74	1	339	77	300	80	30	82	QUITTMEYER ET AL.(1979)
*24 FEB 1973	0	28.6	52.6	27	5.2	222	15	42	75	312	0	222	60	42	30	CHANDRA(1979B)
*12 MAR 1973	13	32.1	49.3	62	4.9	159	25	339	65	69	0	339	20	159	70	CHANDRA(1979B)
*22 APR 1973	21	30.7	49.8	57	5.0	19	11	157	75	287	10	11	57	212	35	CHANDRA(1979B)
12 OCT 1973		37.7	71.9	35		180	10	0	80	90	0	180	55	0	35	CHATELAIN ET AL.(1980)
17 OCT 1973		36.4	71.1	211		168	15	322	75	76	10	162	60	356	30	CHATELAIN ET AL.(1980)
24 OCT 1973	5	33.1	75.9		5.4	82	5	304	78	4	7	89	56	284	34	TANDON AND SRIVASTAVA(1975A)
11 NOV 1973	7 a	30.6	52.9	11	5.5	81	21	195	46	334	36	305	40	53	75	CHANDRA(1979B)
*11 NOV 1973	7 b	30.6	52.9	11	5.5	135	15	243	49	33	37	107	70	335	44	CHANDRA(1979B)
24 MAR 1974		27.7	86.1			187	43	7	47	97	0	187	88	7	2	MOLNAR AND TAPPONNIER(1978)
13 MAY 1974		36.5	71.0	197		124	35	304	55	34	0	124	80	304	10	CHATELAIN ET AL.(1980)
*30 JUL 1974	a	36.4	70.8	211		196	25	16	65	106	0	196	70	16	20	BILLINGTON ET AL.(1977)
30 JUL 1974	b	36.4	70.8	209		190	24	35	63	285	10	199	70	348	23	CHATELAIN ET AL.(1980)
11 AUG 1974		39.4	73.8	7		183	0	273	36	92	53	144	66	43	66	NI(1978)
11 AUG 1974	1	39.4	73.7	41		183	1	272	46	92	45	148	61	39	59	NI(1978)
11 AUG 1974	20	39.5	73.6	26		34	66	227	22	135	5	222	68	56	22	NI(1978)
27 AUG 1974	21	39.5	73.8	14		355	10	245	4	147	78	201	85	110	80	NI(1978)
* 4 OCT 1974		26.3	66.5	33	5.9	2	6	98	46	263	42	328	65	218	52	QUITTMEYER ET AL.(1979)
* 2 DEC 1974	9	28.0	55.8	36	5.4	105	13	10	22	223	64	239	65	146	84	CHANDRA(1979B)
10 DEC 1974		36.5	70.5	213		137	10	317	80	47	0	137	55	317	35	CHATELAIN ET AL.(1980)
*28 DEC 1974	12 a	35.1	72.9	22	6.0	151	23	345	67	243	5	320	23	155	68	CHANDRA(1975A)
28 DEC 1974	12 b	35.1	72.9	22	6.0	204	2	105	78	295	12	14	49	214	53	PENNINGTON(1979)
*19 JAN 1975	a	32.5	78.4			90	85	270	5	0	0	270	50	90	40	MOLNAR AND TAPPONNIER(1978)
19 JAN 1975	8 b	32.5	78.4		6.2	180	72	45	14	312	11	209	32	36	60	BANGHAR(1976)
19 JAN 1975	8 c	32.5	78.3	37	6.8	224	36	318	6	56	54	97	61	356	70	KHATTRI ET AL.(1975)
19 JAN 1975	8 d	32.5	78.3	37	6.8	100	76	272	14	2	1	90	31	272	59	KHATTRI ET AL.(1978)
3 MAR 1975		36.5	70.9	187		337	15	157	75	67	0	337	60	157	30	CHATELAIN ET AL.(1980)
* 7 MAR 1975	7	27.5	56.3	27	5.8	160	20	340	70	70	0	160	65	340	25	CHANDRA(1979B)
28 APR 1975		35.8	79.9			26	39	116	0	206	51	259	64	155	64	MOLNAR AND TAPPONNIER(1978)
*28 APR 1975	2	33.3	54.8	42	5.3	133	37	313	53	223	0	133	82	313	8	CHANDRA(1979B)
14 MAY 1975		36.1	70.9	97		161	42	18	42	270	20	180	90	270	20	CHATELAIN ET AL.(1980)
*17 MAY 1975	16 a	27.5	57.7	33	4.9	153	0	244	54	63	36	4	55	123	55	CHANDRA(1979B)
17 MAY 1975	16 b	27.5	57.7	33	4.9	600	0	140	85	230	0	320	80	140	40	CHANDRA(1979B)
19 MAY 1975		35.2	80.8			205	40	299	5	36	50	335	68	80	60	MOLNAR AND TAPPONNIER(1978)
4 JUN 1975		35.9	79.9			140	42	233	3	327	48	270	64	18	60	MOLNAR AND TAPPONNIER(1978)
*19 JUL 1975		31.9	78.6			90	85	270	5	0	0	270	50	90	40	MOLNAR AND TAPPONNIER(1978)
*29 JUL 1975		32.6	78.3			90	83	270	7	0	0	270	52	90	38	MOLNAR AND TAPPONNIER(1978)
* 3 OCT 1975		30.3	66.3	11	6.7	347	2	77	2	212	87	302	90	212	87	QUITTMEYER ET AL.(1979)
*24 DEC 1975	11	27.0	55.5	33	5.5	158	23	338	67	68	0	338	22	158	68	CHANDRA(1979B)
* 3 FEB 1976	16	39.9	48.4	58	5.2	25	83	205	7	295	0	205	52	25	38	CHANDRA(1979B)
*16 MAR 1976	7	27.3	55.1	33	5.4	132	2	33	73	222	17	295	45	147	50	CHANDRA(1979B)
*22 APR 1976	17	28.7	52.1	24	6.0	206	5	26	85	116	0	26	40	206	50	CHANDRA(1979B)
*26 APR 1976	4	28.7	52.1	29	5.2	12	0	-	90	102	0	192	45	12	45	CHANDRA(1979B)
*14 FEB 1977	0	33.6	73.3	33	5.2	24	30	275	30	150	45	150	45	240	90	SEEBER AND ARMBRUSTER(1979)
23 MAY 1977	2	35.0	69.1	29		163	0	73	0	-	90	298	90	208	90	PREVOT ET AL.(1980)
3 JUN 1977	12	34.6	70.7	5		335	5	117	83	245	6	163	40	332	50	PREVOT ET AL.(1980)
10 JUN 1977	1	34.6	70.6	25		306	10	212	16	70	70	80	70	348	86	PREVOT ET AL.(1980)
17 JUN 1977	15	36.5	70.4	197		155	14	263	48	52	38	16	44	127	70	CHATELAIN ET AL.(1980)
18 JUN 1977	21	36.5	70.3	212		154	14	268	46	62	40	24	48	133	70	CHATELAIN ET AL.(1980)
18 JUN 1977	23	36.0	70.6	112		22	3	286	64	114	26	46	53	178	49	CHATELAIN ET AL.(1980)
19 JUN 1977	5	36.5	70.6	209		152	7	252	54	56	35	6	48	124	62	CHATELAIN ET AL.(1980)
19 JUN 1977	13	36.3	70.1	117		160	26	340	65	70	0	160	70	344	18	CHATELAIN ET AL.(1980)
19 JUN 1977	22	36.3	70.7	135		177	14	282	42	70	42	40	48	146	76	CHATELAIN ET AL.(1980)
20 JUN 1977	1	36.1	70.4	99		358	5	260	56	90	34	24	58	145	50	CHATELAIN ET AL.(1980)
20 JUN 1977	15	36.5	70.4	220		174	20	286	44	66	38	146	76	39	42	CHATELAIN ET AL.(1980)
20 JUN 1977	18	36.2	69.4	119		285	62	189	2	98	26	37	48	167	54	CHATELAIN ET AL.(1980)
20 JUN 1977	20	36.7	71.1	233		152	14	264	56	52	30	6	40	128	66	CHATELAIN ET AL.(1980)
22 JUN 1977	2	36.5	70.3	209		186	21	296	40	74	42	52	44	154	78	CHATELAIN ET AL.(1980)
23 JUN 1977	15	36.8	71.2	260		164	18	274	44	60	39	126	75	28	44	CHATELAIN ET AL.(1980)

TABLE 1 (continued)

| | | | | | | | P | | T | | B | | Nodal Planes | | | | |
| | | | | | | | | | | | | | 1 | | 2 | | |
Date	Hr.	Lat. (°N)	Long. (°E)	Depth (km)	Mag.	φ	θ	φ	θ	φ	θ	Dip dir.	Dip	Dip dir.	Dip	Reference
23 JUN 1977	15	36.6	70.3	212		159	32	273	32	36	42	36	40	136	90	CHATELAIN ET AL.(1980)
23 JUN 1977	22	36.0	70.7	97		340	82	63	1	164	8	246	46	82	46	CHATELAIN ET AL.(1980)
24 JUN 1977	2	36.7	71.1	217		144	22	270	54	40	26	1	32	122	70	CHATELAIN ET AL.(1980)
24 JUN 1977	20	36.1	70.7	107		136	52	275	30	18	20	292	78	54	24	CHATELAIN ET AL.(1980)
26 JUN 1977	9	35.9	69.3	97		246	8	146	52	330	36	274	61	33	51	CHATELAIN ET AL.(1980)
26 JUN 1977	15	36.6	71.0	212		299	16	188	52	40	32	326	70	82	40	CHATELAIN ET AL.(1980)
27 JUN 1977	16	36.5	70.3	209		166	14	272	46	64	40	28	46	136	70	CHATELAIN ET AL.(1980)
1 JUL 1977	5	36.5	71.0	221		145	8	248	56	51	34	0	46	120	62	CHATELAIN ET AL.(1980)
2 JUL 1977	21	36.0	70.7	94		86	65	266	25	176	0	86	20	266	70	CHATELAIN ET AL.(1980)
3 JUL 1977	18	36.4	71.5	96		308	8	188	73	40	15	321	55	111	39	CHATELAIN ET AL.(1980)
4 JUL 1977	20	36.2	69.4	128		48	14	146	28	296	58	9	80	274	60	CHATELAIN ET AL.(1980)
7 JUL 1977	16	36.1	69.1	112		188	9	282	34	82	56	50	60	149	74	CHATELAIN ET AL.(1980)
9 JUL 1977	17	36.6	71.0	197		288	37	177	26	60	41	49	42	144	84	CHATELAIN ET AL.(1980)
11 JUL 1977	11	36.4	71.3	104		178	8	297	72	84	16	165	56	16	40	CHATELAIN ET AL.(1980)
*13 JUL 1977	8	29.9	67.5	10		158	2	108	19	292	72	334	76	242	78	ARMBRUSTER ET AL.(1979)

Notes:

1. Abbreviations: Hr. - Hour, Lat. - Latitude, Long. - Longitude, Mag. - Magnitude, dir. - direction
2. φ, trend measured clockwise from the north; θ, plunge measured from the horizontal

caused by tension within a sinking lithospheric slab in the region.

The Buyin-Zara earthquake of 1-9-62 was accompanied by surface faulting. From field mapping, Ambraseys (1963) showed that the faulting had a large thrust component with also some left lateral strike slip, and occurred on a fault striking N80°W. The focal mechanism solution of this earthquake, assuming the south dipping nodal plane to be the fault plane, is in good agreement with Ambraseys' (1963) field observations, and also with the left lateral sense of motion postulated by Nowroozi (1972) and McKenzie (1972) for the boundary of Caspian and Persian plates. The solution for event 2-7-57 is similar to that of event 1-9-62, but the strike slip component of motion in this case is very small.

The trend of the Elburz range in the vicinity of event 14-2-71 becomes northeast. The focal mechanism solution of this earthquake has a component of thrust as well as strike slip faulting. The northeast nodal plane paralleling the general strike of the Elburz range in the epicentral region shows a left lateral component of motion.

In the northeast part of Iran, the Elburz mountains converge with the northwest trending Kopet Dagh ranges. The earthquake of 30-7-70 occurred within the wedge, between the two ranges, near the region of their convergence. The focal mechanism solution shows normal faulting with a component of left lateral strike slip motion along a north trending nodal plane. The axis of tension trends northeast. It appears that a right lateral motion along the Main Kopet Dagh Fault (Berberian, 1976) and a left lateral component of motion along the general strike of the Elburz range causes a local zone of tension in the region of their convergence.

The earthquakes of 3-1-69 and 5-10-48 occurred in the vicinity of Kopet Dagh range. The focal mechanism solutions show thrust faulting along northwest nodal planes, parallel to the general structural trend of the Kopet Dagh range.

Zagros Folded Belt

The Zagros Folded Belt lies southwest of the Main Zagros Thrust. In the southeast it continues up to the straits of Hormuz and in the northwest it merges with the mountains of eastern Turkey. This region has been a zone of quiet and conformable sedimentation from Cambrian to Pliocene times with only gentle epeirogenic movements and salt diapirism (Takin, 1972). However, since the Late Tertiary Zagros Orogeny, the sediments were folded into a series of parallel northwest trending anticlines and synclines. The amplitudes of the folds decrease toward the southwest with increasing distance from the Main Zagros Thrust.

Seismically, the Zagros Folded Belt is the most active region in Iran. The earthquakes throughout this region occur in a diffuse pattern in a band about 200 km wide.

Focal mechanism solutions for earthquakes in the Zagros Folded Belt are presented in Figure 5. Almost all of the earthquakes in this belt (8-11-71, 24-9-66, 23-2-70, 18-9-66, 30-5-68, 14-9-68, 12-4-72, 10-4-72, 24-2-73, 22-4-76, 26-4-76, 29-4-69, 23-6-68, 2-1-67, 22-4-73, 27-7-66, 14-6-72 and 11-1-67) have thrust fault plane solutions. The strikes of the northeast dipping nodal planes in these solutions vary from NW to WNW. The solution for 6-4-71 earthquake shows strike slip faulting with a component of thrust motion. The NNW trending nodal plane shows a right lateral strike slip motion. The mechanism solution for 12-3-73 earthquake shows thrust faulting along ENE trending nodal planes. This departure in the strike direction of the

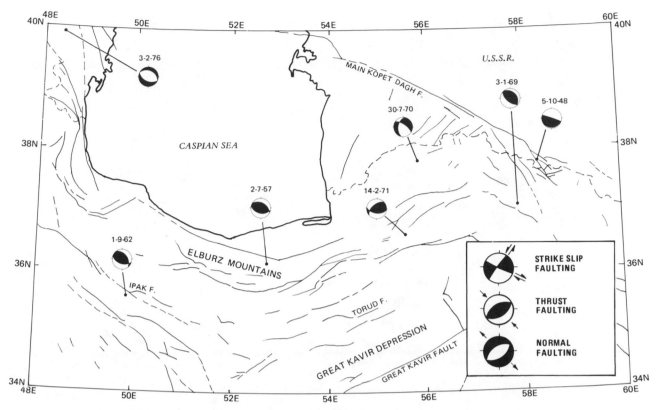

Fig. 4. Focal mechanism solutions for earthquakes in the vicinity of Elburz and Kopet Dagh ranges. P-wave compression quadrants are shown in dark.

nodal planes from the strikes of the nodal planes for other mechanism solutions in the Zagros Folded Belt is related to a local structure. A thrust fault in the epicentral region curves around toward ENE at its eastern end (Berberian, 1976).

From the focal mechanism solutions of earthquakes in the Zagros Folded Belt, it may be concluded that the continued convergence of the Arabian and Iranian continents is being absorbed by folding and crustal shortening, and to a lesser extent by right lateral motion along faults in the Zagros Folded Belt.

Zagros Thrust Zone

The Main Zagros Thrust is a major structural discontinuity. Immediately southwest of it lies an imbricated belt containing strongly tectonized series with radiolarites and ophiolites. Further southwest lies a simply folded belt of sedimentary rocks. In the Kermanshah and Neyriz regions, Braud and Ricou (1976) mapped another major structure which is very close and generally parallel to, but quite distinct from, the Main Zagros Thrust. This structure is younger than the Main Zagros Thrust and transects it in several places. It appears as a succession of right lateral vertical fault segments.

Tchalenko and Braud (1974) termed it the Main Recent Fault.

Tchalenko and Braud (1974) suggested that the intense seismic activity west of 49°E, along the trend of the Main Zagros Thrust, is related to the Main Recent Fault (with branches, Sahneh, Nahavand, Garun and Dorud faults). In this region major earthquakes of Silakhor, 23 January 1909; Farsinaj, 13 December 1959; and Nahavand, 16 August 1958 occurred. Each of these earthquakes were accompanied by surface faulting or ground deformation. Along the Main Zagros Thrust itself, southeast of 49°E longitude, the earthquake activity is remarkably low. However, the Qeshlaq earthquake of 11 November 1973 occurred in this region. Near the southeast part of the Main Zagros Thrust, north of Bandar Abbas and somewhat west of it, seismic activity is greater.

The focal mechanism solutions of earthquakes along the Main Zagros Thrust region are presented in Figure 5.

The earthquakes of 13-12-57 and 24-3-63 occurred near Garun Fault. The focal mechanism solution for 13-12-57 earthquake shows a predominantly thrust faulting with a small component of right lateral motion along the north-south nodal plane. The solution for 24-3-63 earthquake is predominantly strike slip with right lateral motion along the northwest striking nodal plane.

FOCAL MECHANISM SOLUTIONS   255

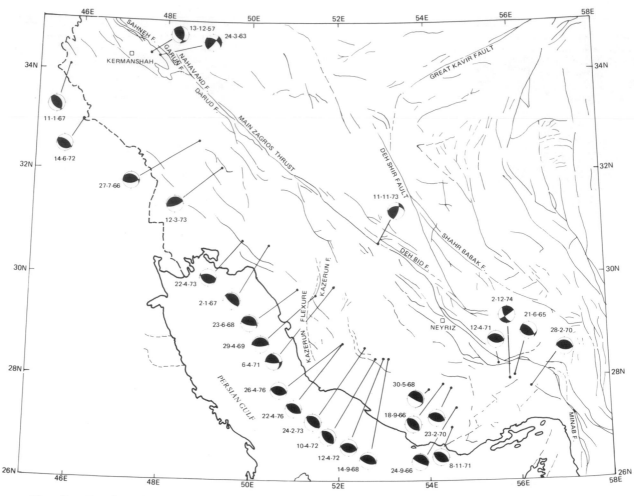

Fig. 5. Focal mechanism solutions for earthquakes in the Zagros Folded Belt and vicinity.

The earthquake of 11-11-73 occurred near Qeshlaq village about 40 km west of Dehbid. The fault plane solution is strike slip with a component of thrusting. The strike slip component of motion of the east-west trending nodal plane is right lateral. This strike direction is in agreement with the strikes of two small faults near Dehbid, shown by Berberian (1976).

In the southeast part of the Zagros Thrust the earthquakes of 28-2-70, 21-6-65 and 12-4-71 have thrust solutions with west to northwest trending nodal planes. The strikes of the nodal planes are parallel to the strikes of local thrust faults in the area. The earthquake of 2-12-74 shows a predominantly strike slip motion. The northwest trending nodal plane shows a left lateral motion and the northeast nodal plane has a right lateral motion. The left lateral motion along the northwest nodal plane may have been caused by a displacement of material of the Eurasian plate toward northwest along its boundary with the Arabian plate, as a consequence of continental collision in the southeast.

Bandar Abbas Region

The Bandar Abbas region located north of the Strait of Hormuz is characterized by a high level of seismicity. During recent years, several damaging earthquakes, some of which caused fatalities, have occurred in the region. An epicenter map of Iran (Figure 2) delineates a NNE trending seismic zone, about 50 km wide and 250 km long, beginning southwest of Bandar Abbas and continuing up to Jiroft in the northeast. Alternatively, this zone could be looked upon as the western limb of an arc-like pattern of epicenters. The alignment of epicenters along Bandar Abbas-Jiroft zone and arching up of the geological formations north of the Zagros Thrust along a NNE axis suggest the colliding edge of the Arabian plate in this region, north of Oman peninsula, to be in the form of a projected promontory.

White and Ross (1979) detected a large basement ridge on multichannel seismic reflection and gravity profiles to the west of the Oman Line.

They suggested that this ridge could be a sub-surface continuation of the Musandam peninsula beneath the Strait of Hormuz.

Figure 6 shows the focal mechanism solutions of earthquakes in the Bandar Abbas region. The fault plane solution for the earthquake of 17-5-75 shows thrusting along northeast striking nodal planes. No faults or tectonic structures with northeast strike have been mapped in this region. The cause of northwest-southeast compression giving rise to this earthquake remains unexplained. The solution for the 21-6-69 earthquake shows a left lateral strike slip motion along a southeast striking nodal plane, parallel to the strike of the local faults in the epicentral region. Therefore, this nodal plane may be chosen as the fault plane.

The first motion data for the 22-12-64 earthquake are consistent with thrusting along east-west nodal planes.

The earthquake of 9-12-71 shows a pure strike slip faulting. On the basis of local structural trends, the northeast striking nodal plane with left lateral motion, may be selected as the fault plane.

The focal mechanism solutions for 7-3-75, 24-12-75 and 16-3-76 earthquakes show thrust faulting along ENE nodal planes.

Thus, it may be interpreted that toward the northeast, the collision of a projected promontory of the Arabian plate, north of Oman, with the Eurasian continent results in a mass movement of the Iranian block giving rise to a right lateral movement along the south striking Minab

Fault and left lateral movement along east or southeast striking faults terminating in the east at the southern margin of the Jaz Murian depression. Northwest of Oman, the convergence of this promontory against the Iranian block is taken up, at least in part, by deformation of the rocks of Zagros Folded Foothills, which are part of the Arabian plate.

Central and East Iran

This region includes the Central Iranian Plateau, Lut Block and the Jaz Murian Depression. The region is bounded in the east by the East Iranian ranges, in the northwest by the Great Kavir Fault, in the west by the Deh Shir Fault, in the southwest by Shahr Babak Fault and in the south by elongated east-west trending granite and diorite bodies (Figure 7). Colored melange is present at many places throughout the entire boundary zone of this region (Stocklin, 1974). The Central Iranian Plateau is separated from the Lut Block by the north-south trending Nayband Fault.

From the presence of colored melange, Takin (1972) inferred Central and East Iran to be a microcontinent, previously surrounded by narrow Red Sea type ocean basins. These basins began to form in late Jurassic time and reached their full development in the Late Cretaceous (Stocklin, 1974). The initial compression during the convergence of Afro-Arabia toward Eurasia, in the early Tertiary, was taken up by the closing of these young small ocean basins.

The focal mechanism solution, Figure 7, for the earthquake of 26-5-71 shows thrust faulting along NNW nodal planes. Thus, the strike directions of both nodal planes are at right angles to the general strike of the Great Kavir Fault in the immediate vicinity of the epicenter.

The seismicity map of Iran, Figure 2, shows a northwest trend of epicenters through the central part of the region. This trend appears to extend toward the Elburz range. The earthquake of 28-4-75 occurred on this trend near the Biabanak Fault in the Central Iranian Plateau. The focal mechanism solution shows thrust faulting. The northeast strike of the nodal planes is close to the general strike of the Biabanak Fault. The region is under northwest-southeast compression. This stress may be related to the compression during the closing phase of the oceanic trough, identified by colored melange, around the crescent shaped Central Iranian Plateau.

The earthquake of 9-11-70 occurred in the southern part of the Central Iranian Plateau. This is an intermediate depth, 106 km, earthquake and has a normal fault plane solution with WNW nodal planes. The focal mechanism and intermediate depth of this earthquake suggest the existence of a remnant slab of the oceanic lithosphere of the Arabian plate which was subducted along the Zagros suture zone prior to the continental collision. The earthquake was caused by tension within this sinking slab.

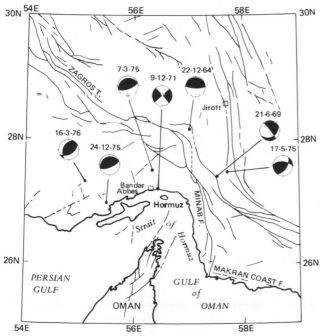

Fig. 6. Focal mechanism solutions for earthquakes in the vicinity of Bandar Abbas.

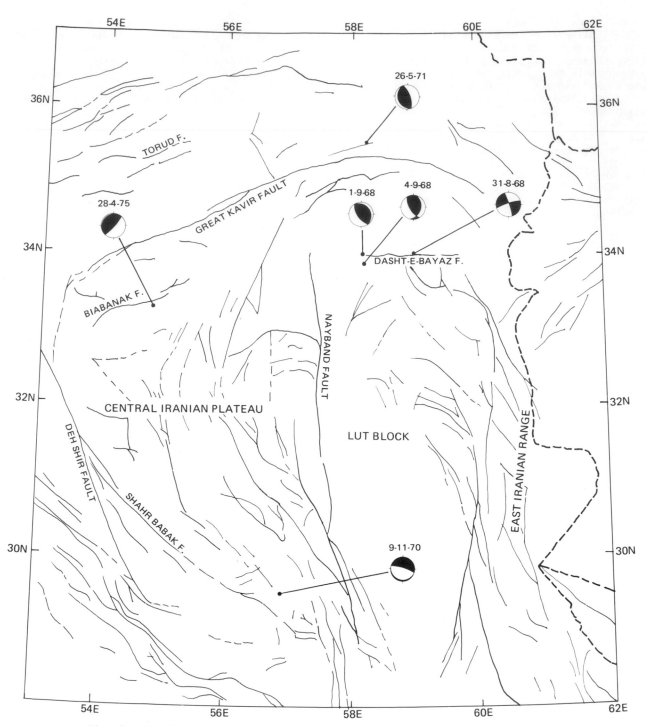

Fig. 7. Focal mechanism solutions for earthquakes in Central and East Iran.

The Dasht-e Bayaz earthquake of 31-8-68 occurred in the northern part of the Lut block and was accompanied by an east-west fresh fault rupture with a 4.6 m left lateral displacement (Niazi, 1968). The focal mechanism solution of this earthquake and two of its larger aftershocks (1-9-68 and 4-9-68) are presented in Figure 7. The main shock has a strike slip mechanism in which the ENE trending nodal plane shows left lateral motion in agreement with the field observations. The two aftershocks at the western end of the Dasht-e Bayaz Fault show thrust faulting along northwest to north striking nodal planes, in contrast to the strike slip mechanism for the main shock. McKenzie (1972) suggested that the thrust solutions for the aftershocks could explain the observed branched nature of the surface break, since the east-west left lateral motion on the main break could be taken up by thrusting on a fault with a more northerly strike. He also suggested that the nature of this faulting during the aftershocks, could permit a connection with the main active zone further north through a thrust fault striking northwest. The similarity of the focal mechanism solution for the earthquake of 26-5-71, occurring north of the two aftershocks at the northern border of Central and East Iran, with the solutions for the two aftershocks suggests that the above interpretation may be correct.

Makran Range

The Makran continental margin of Iran and Pakistan is a region of active subduction where continental collision is still pending. It is terminated in the west by the Minab Fault or Oman Line, and in the east by the Ornach-Nal and Chaman Faults. The Minab Fault is a major facies divider between the Zagros Folded Belt in the west and the Makran Range in the east (Berberian, 1976). The Makran ranges of western Baluchistan are not a structural continuation of the Kirthar ranges and are separated from these ranges by a

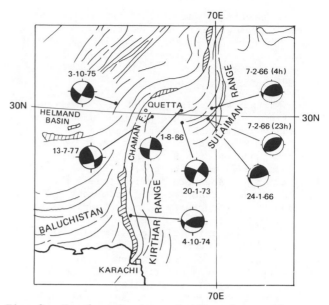

Fig. 9. Focal mechanism solutions for earthquakes in the vicinity of the Kirthar Range and Quetta syntaxis. The locations of ophiolites are shown by stippled regions.

major fault zone (Ornach-Nal and Chaman Faults) which is also a major facies boundary.

The Landsat images of the region indicate geomorphic features typical of arc-trench systems (Farhoudi and Karig, 1977). Compared to most other subduction zones, the dipping slab is not very sharply defined by the earthquake foci in the Makran region. Jacob and Quittmeyer (1979), however, discussed the presence of a dipping seismic zone which extends to a depth of at least 80 km. They also suggested that the location of the volcanic arc, in the northern Jaz Murian and Baluchistan region, coincides with the 100 km depth contour of the down-dip projected extension of the dipping seismic zone beneath and beyond where seismicity data are presently available.

The focal mechanism solutions of earthquakes in the Makran region are shown in Figure 8. Two shallow focus earthquakes, which occurred on 6-8-72 and 8-8-72 near the Makran coast, show thrust faulting mechanism along east-west nodal planes dipping gently toward the north. The mechanism solutions for four earthquakes (2-8-68, 7-11-69, 29-5-63 and 17-11-72) in the Jaz Murian region show normal faulting. One of the nodal planes in each case dips steeply toward the north, except for the earthquake of 29-5-63 in which case the corresponding nodal plane strikes toward northeast and dips steeply toward northwest. This change in strike may be related to some local geological inhomogeneity. In general, the T axes for the earthquakes in the Jaz Murian region are oriented parallel to the dipping zone of seismic activity. These solutions, therefore, support the existence of postulated zone of

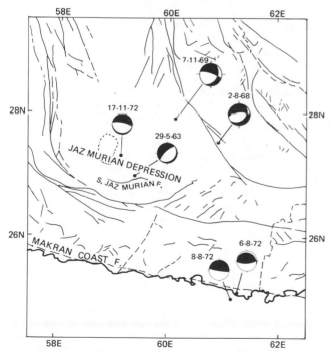

Fig. 8. Focal mechanism solutions for earthquakes in southeast Iran.

active subduction, wherein, deeper earthquakes (focal depths greater than about 50 km) are caused by down-dip extension within the sinking slab.

Kirthar Range

The Kirthar mountains extend in a north-south direction, for a distance of about 550 km, from Quetta toward Karachi on the Arabian Sea coast. The Murray Ridge at the northern part of the Owen Fracture zone and the Rann of Kutch seismic zone (Chandra, 1977) meet the coastal region near Karachi at the southern termination of the Kirthar ranges. The tectonic relationship between these structures is not clearly understood.

The focal mechanism of the earthquake of 4-1-74, in the southern part of the Kirthar Range (Figure 9), was determined by Quittmeyer et al. (1979). This solution has a component of thrust faulting, and shows left lateral motion along a northeast nodal plane and right lateral motion along a northwest striking nodal plane. On the basis of mapped surface faults in the area with NNW strike, Quittmeyer et al. (1979) considered the choice of northwest trending nodal plane as fault plane appropriate.

Quetta Syntaxis

At a latitude of about 30°N, the Sulaiman ranges curve westward in a festoon-like structure. The curve is convex toward the south. The west trend continues for a distance of about 200 km where, in the vicinity of Quetta, the formations bend sharply south.

The seismic activity occurs along and south of the southern margin of the Sulaiman Range festoon.

Focal mechanism solutions (Figure 9) for three earthquakes, 24-1-66, 7-2-66 (4h) and 7-2-66 (23h), at the eastern margin of the Sulaiman Range festoon show thrusting along east-west or northeast striking nodal planes. The trend of the axis of pressure in each of these solutions is slightly west of north. The geometry of the structural trend in this region is such that the left lateral movement along the Chaman Fault would cause north-south compression in the epicentral region of these three earthquakes.

The mechanism solutions for events 20-1-73, 1-8-66 and 13-7-77 in the central part of this east-west transverse structure, and for 3-10-75 earthquake, at the western extremity of the east-west trending ranges, are predominantly strike slip. These solutions indicate right lateral movement along west to northwest striking nodal planes, or alternately, left lateral motion along north to northeast striking nodal planes. The west to northwest strike is generally in agreement with the seismicity trend in the epicentral region of these earthquakes (see also, Quittmeyer et al., 1979).

East Afghanistan Range

The suture zone at the northwest margin of the Indian shield is identified with the Quetta Line described by Gansser (1966). The Chaman Fault, a large left lateral strike slip fault, is related to the suture. A gap in seismicity, about 100 km wide, is observed on the epicenter map of the region (shaded area in Figure 3). Northwest of the suture, in Afghanistan, widespread orogeny occurred in the Oligocene (Schreiber et al., 1972).

The fault plane solutions for the earthquakes of 13-10-64, 15-5-69 and 28-12-72, in east Afghanistan, are shown in Figure 10. These solutions show thrusting along northeast striking nodal planes. The earthquakes may be related to the stresses, caused by the colliding continents, transmitted across the aseismic lineament region (Figure 3). Chandra (1978) considered this aseismic lineament region to be rigid. A stress system similar to the contemporary stress, northwest-southeast compression indicated by the focal mechanism solutions, may have been responsible for the Oligocene orogeny in Afghanistan.

Recently Prevot et al. (1980) reported on microearthquake studies in four separate parts of eastern Afghanistan and discussed in detail the pattern of seismicity and active tectonics in the region.

Hindu Kush - Pamir

The Hindu Kush and Pamir ranges are located northwest and north of Nanga Parbat at the western extremity of the Himalaya. The general structural trend of the Pamir ranges makes a sharp curve convex toward the north. At its southwest margin it merges with the northeast trending Hindu Kush ranges.

The spatial distribution of earthquake foci in the region has been considered in detail by Nowroozi (1971) and Billington et al. (1977). Nowroozi (1971) identified a sharply defined E-W alignment, about 120 km long and about 25-30 km wide, of deeper earthquakes. Most earthquakes along this alignment occur in a vertical slab within 175-250 km depth. However, Billington et al. (1977) interpreted the spatial distribution of earthquakes as an evidence for the existence of a contorted Benioff zone. They presented a map of isodepth contours of the seismic zone, indicating that the direction of dip of the seismic zone reverses from about 70° toward the north in the western part of the region (west of 71.5°E) to about 50° toward southeast in the eastern part of the region.

Citing the north convexity of the Pamir Arc, occurrence of Upper Cretaceous and Lower Cenozoic marine sedimentary rocks in the northernmost part of the Pamirs, late Cenozoic volcanism in the central Pamirs, and occurrence of mantle earthquakes as evidences, several authors have

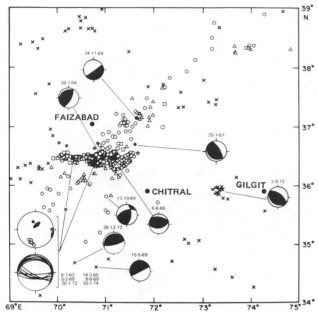

Fig. 10. Focal mechanism solutions for earthquakes in the vicinity of the East Afghanistan,
Hindu Kush and Pamir ranges. Hypocenter depth
symbols: cross = 0 to 65 km; circle = 66 to
125 km; triangle = 126 to 175 km; square = 176
to 225 km; hexagon = 226 to 270 km; wide +
symbol = 271 to 282 km; star = 277 km. (adapted
from Billington et al., 1977)

suggested the consumption of an oceanic basin in
this region by a southward subduction of the
Eurasian plate under the Indian plate (Molnar
et al., 1973; Molnar and Tapponnier, 1975;
Chandra, 1978). The apparent lack of mantle
earthquakes along the main Himalayan Arc led
Billington et al. (1977) to suggest that the
Hindu Kush-Pamir region is the site of the final
stage of subduction of suboceanic lithosphere
along the collision boundary between the Indian
and Eurasian plates.

Focal mechanism solutions of all earthquakes in
the Hindu Kush-Pamir region (Figure 10) show
thrusting. The mechanism solutions for the
earthquake of 6-6-66 and six other earthquakes
(shown as a combined plot in the lower left part
of Figure 10) indicate thrusting along E-W nodal
planes. The strikes of nodal planes in the
focal mechanism solutions for the earthquakes of
28-1-64 and 24-11-69 are due northeast parallel
to the general trend of the Hindu Kush mountains.
On the other hand, the nodal planes in the
solution for 25-1-67 earthquake strike toward
southeast paralleling the general trend of the
Karakoram ranges. From the very steep dip of the
Benioff zone indicated by the spatial distribution of earthquake foci (Nowroozi, 1971;
Billington et al., 1977) and the near vertical
orientation of the T-axes in these solutions, it
may be interpreted that the earthquakes were

caused by down dip extension within a sinking
slab.

Roecker et al. (1980) and Chatlein et al.
(1980) have studied in detail the configuration
of Hindu Kush-Pamir seismic zone, microearthquake
seismicity and fault plane solutions, and their
tectonic implications.

Karakoram Range

The Karakoram ranges lie northeast of the Indus
suture zone, southeast of the Pamirs. Focal
mechanism solutions of two earthquakes, 2-2-65
and 12-1-72, in this region show right lateral
strike slip motion along southeast striking
nodal planes (Figure 11). The P-axes in these
solutions trend toward north, thus suggesting
that the contemporary stress in the region is
related to the northward collision of the Indian
continent with Eurasia. The right lateral strike
slip motion along the Karakoram Fault, indicated
by the fault plane solutions, suggests that at
the northwest margin of colliding continents, the
movement of the Indian plate, besides being
absorbed by crustal shortening, is taken up, at
least partly, by right lateral movement along
the Karakoram Fault (see also, Peive et al.,
1964; Desio, 1973).

The earthquake of 22-6-65, which occurred
within the region between the Karakoram and Astin
Tagh faults close to their junction, has a normal

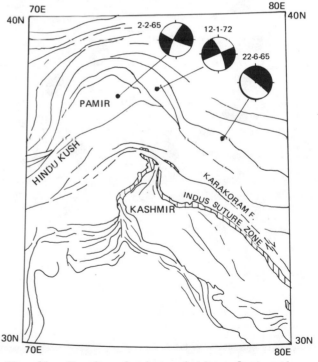

Fig. 11. Focal mechanism solutions for earthquakes in the vicinity of the Pamir and
Karakoram ranges. The locations of ophiolites
are shown by stippled regions.

fault plane solution. The axis of tension in this solution trends toward northeast. It appears that right lateral motion along the Karakoram Fault and left lateral motion along the Astin Tagh Fault cause a local zone of tension in the region of their convergence.

Kashmir Syntaxis

At the northwest extremity of the Himalaya, the geological formations bend sharply toward the south, then southwest and curve toward the west to form the Kashmir syntaxis. The southwest flank of this syntaxis curves further toward the south and continues as the southward trending Sulaiman Range. Geologically the two flanks of the Kashmir syntaxis are quite similar. The formations on the northeast flank of the syntaxis are repeated with their tectonic relationships on the southwest flank (Wadia; 1931, 1936).

A seismicity map of the region, Figure 3, shows a northwest linear trend, AA', in the alignment of epicenters between the points $32.3^{\circ}N$, $76.6^{\circ}E$ and $35.3^{\circ}N$, $72.8^{\circ}E$. Although, most of the epicenters on this trend occur close to the surface trace of the Main Central Thrust (MCT), the earthquakes are largely related to the northeast dipping imbricate structure of the Main Boundary Thrust

(MBT). A NNE trend, BB', in the alignment of epicenters may be identified between the points $33.0^{\circ}N$, $71.0^{\circ}E$ and $36.0^{\circ}N$, $73.0^{\circ}E$. Further northwest, a parallel NNE alignment of epicenters, CC', the northern part of which demarcates the eastern margin of the Hindu Kush region, occurs. There is about 100 km wide gap in seismicity between the alignments BB' and CC'. Almost all intermediate focus earthquakes in this region occur northwest of this aseismic lineament.

Figure 12 shows the focal mechanism solutions of earthquakes in the vicinity of the Kashmir syntaxis. P-wave first motion and S-wave polarization data together with the P nodal planes corresponding to several of these solutions, previously unpublished, are presented on an equal area projection of the lower focal hemisphere in Figure 13.

Seeber and Armbruster (1979) determined the focal mechanism solution for the 14-2-77 earthquake which occurred near Rawalpindi close to the Hazara Fault. On the basis of the strike direction of the Hazara Fault in the vicinity of the epicenter and distribution of aftershocks, the northeast trending nodal plane showing left lateral motion was identified as the fault plane. No satisfactory explanation for the left lateral sense of motion was proposed. However, Chandra (1978) presented a qualitative model for the origin of the Himalayan syntaxes which suggested that, following the continental collision, the entire region may have been in a state of left lateral shear along NNE vertical planes.

The earthquake of 28-12-74 occurred at the intersection, A', of the northwest alignment of epicenters AA' and NNE alignment of epicenters BB' in Figure 3. The focal mechanism has a thrust solution with northeast striking nodal planes and axis of pressure, P, trending in the northwest-southeast direction. This identifies the earthquake with the NNE alignment, BB', of epicenters and relates to the Hazara Thrust system of southwest flank of the Kashmir syntaxis.

The earthquakes of 3-9-72 (at 16h and 23h) and 4-9-72 occurred at the northeastern extremity, B', of the NNE alignment, BB', of epicenters. This region corresponds to the inner portion of the intersection of the Indus Suture zone and the NNE aseismic lineament shown in Figure 3. Two alternate solutions were determined for each of the events 3-9-72 (23h) and 4-9-72. One of these alternate solutions shows thrusting along northwest striking nodal planes with the axis of pressure trending in the northeast direction The other solution shows thrusting along northeast striking nodal planes with the P axis trending northwest. However, the larger event, 3-9-72 (16h), for which the mechanism was determined by using both P and S-wave data has a unique solution with the nodal planes striking in the northwest direction. The high level of seismicity in the vicinity of point B' is, therefore, related to the fault system of the Indus suture zone.

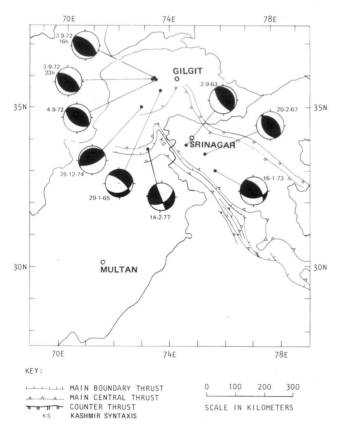

Fig. 12. Focal mechanism solutions for earthquakes near Kashmir syntaxis.

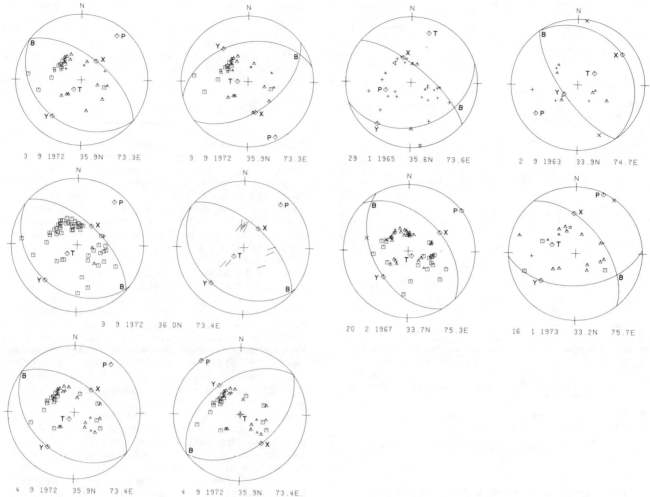

Fig. 13. P-wave first motion and S-wave polarization data, for earthquakes near Kashmir syntaxis, together with the nodal planes on an equal area projection of the lower focal hemisphere. The diamond symbols indicate P, T, X and Y axes of the mechanism solution. B axis is determined by the intersection of the two nodal planes shown by solid lines. P is the pressure axis or axis of maximum compression, T is the tension axis or axis of least compression, B is the intermediate or null axis, X and Y are the poles of the two nodal planes. Square and triangular symbols represent compressions read from long period and short period records, respectively. Cross and plus symbols represent dilatations read from long period and short period records, respectively. Less reliable data are shown by smaller symbols. S polarization directions are shown by small line segments. If P-wave first motion directions were read on both long period as well as short period records at any particular station, the short period observation was ignored.

The earthquake of 29-1-65, which occurred about 50 km southeast of the three September 1972 events, has a normal fault plane solution. The axis of tension is nearly horizontal and dips toward NNE. The earthquake is interpreted as resulting from the tension caused by the bending of the lithospheric plate as it dips into the zone of underthrusting (Stauder, 1968).

The earthquakes of 2-9-63, 20-2-67 and 16-1-73 occurred further southeast along the northeast flank of the Kashmir syntaxis and are spatially correlated with the MCT-MBT system. Fault plane solutions are of thrust type and the strike of one of the nodal planes in each of these solutions is northwest conforming with the local strike of the MCT-MBT.

Northwest Termination of the Himalaya (Composite Fault Plane Solutions from Microearthquakes

Armbruster et al. (1978) studied the microearthquake data obtained from a telemetered network (comprising initially of 6 stations and later expanded to 10 stations) in northern

TABLE 2. Composite fault plane solutions of microearthquakes recorded by the Tarbela Network. Orientation of the mechanism axes for each earthquake were computed from the strike and dip of the nodal planes published by Armbruster et al. (1978).

CFPS No.	P φ	P θ	T φ	T θ	B φ	B θ	Nodal Planes 1 Dip dir.	1 Dip	2 Dip dir.	2 Dip
1	148	41	328	49	238	0	328	4	148	86
2	296	6	30	37	197	52	259	70	156	60
3	194	32	301	25	62	48	336	86	70	48
4	230	15	50	75	140	0	230	60	50	30
5	187	14	93	14	320	70	320	70	50	90
6	288	29	108	61	198	0	288	74	108	16
7	24	75	204	15	114	0	24	30	204	60
8	174	25	272	16	32	59	312	84	45	60
9	355	7	85	7	220	80	310	90	220	80
10	195	24	287	5	27	66	328	77	64	70
11	230	28	133	13	20	58	358	60	94	80
12	76	29	227	58	339	14	290	20	65	75
13	346	16	87	32	233	53	308	79	214	55
14	230	15	50	75	140	0	230	60	50	30
15	345	0	75	0	165	90	300	90	30	90
16	58	20	238	70	148	0	238	25	58	65
17	226	25	46	65	136	0	226	70	46	20

Notes:

1. Abbreviations: CFPS – Composite fault plane solution, dir. – direction

2. φ, trend measured clockwise from the north; θ, plunge measured from the horizontal

3. CFPS numbers correspond to those shown in Figure 14.

Pakistan centered at Tarbela dam on the Indus River, during an 11-month period prior to impounding of the Tarbela reservoir. Because of a limited distribution of stations in a network, with respect to azimuth and distance from the focus of an earthquake, in microearthquake surveys it is customary to derive a composite fault plane solution (CFPS) by superimposing data from a number of events projected onto a common focal sphere. From an analysis of the data for 304 earthquakes, Armbruster et al. (1978) derived CFPS for 17 groups of events for which the spatial distribution suggested a common mechanism of faulting. The orientation of the mechanism axes, and dip direction and dip of the two nodal planes in these solutions are presented in Table 2. Upper hemisphere equal area projections of these solutions are presented in Figure 14.

The CFPS, shown in Figure 14, along with the associated seismic patterns, identify a set of perpendicular (striking northwest and northeast) steeply dipping faults in the region west of the Kashmir syntaxis. They identify the existence of several unmapped faults and the subsurface continuation of some large mapped faults, for example, the Main Boundary Thrust and the Riasi

Thrust. Such a fine resolution is not possible from the teleseismic data.

The axes of maximum compressive stress in these solutions trend generally due north-south, thus indicating that the convergence of the Indian and Eurasian continents is the principal cause of crustal deformation and shortening in this region.

Central Himalaya

Between the peaks of Nanga Parbat and Namcha Barwa, the trend of the mountains changes gradually from northwest to northeast. It is interesting to note that the trend of the Aravalli ranges, which disappear in the northeast under the sedimentary cover of the Ganga Basin, intersects the Himalaya nearly at right angles. A northeast trending ridge structure, named as the Delhi-Hardwar Ridge, has been discovered under the sediments of Ganga Basin by the geophysical investigations (Valdiya, 1973). The ridge structure occurs along the northeast extension of the Aravalli mountains. Figure 3 shows a northeast alignment of epicenters along this extension.

The seismic activity along the Indus-Tsangpo suture zone is very small. Most of the activity occurs at distances more than about 100 km south or southwest of the suture zone. The epicenters generally follow the trend of the mountains.

The focal mechanism solutions of earthquakes in the central Himalaya are shown in Figure 15. Four of the earthquakes 19-1-75, 19-7-75, 6-3-66 (2h 10m) and 6-3-66 (2h 15m) have normal fault plane solutions with the trends of the nodal planes varying between north and northeast. Because these earthquakes occurred directly along the northeast extension of the Aravalli ranges and the Delhi-Hardwar Ridge, it is interpreted that the Aravalli structure continues toward northeast under the sedimentary cover of the Ganga Basin up to the Indus-Tsangpo suture zone and some of it may have underthrusted the Eurasian plate. The elevated topography of the Aravalli structure caused an uplift of the local crustal material and created a zone of northwest-southeast tension transverse to the Aravalli strike. The epicenters of 19-1-75 and 19-7-75 earthquakes occurred near the Kaurik-Chango Fault. The tectonic map of the Himalaya and surrounding areas by Gansser (1964, plate 1B) shows a general change in the trend of strike lines across this fault from ESE on the west side to ENE on the east. From the north-south trend, which is different from the Aravalli trend, of the nodal planes in the focal mechanism solutions of 19-1-75 and 19-7-75 earthquakes (some alternate solutions, Table 1, show northwest strike), and from the general change in the trend of the strike lines in the Himalaya across Kaurik-Chango Fault, it is inferred that the postulated northeast extension of the Aravalli Range is terminated on the western margin by the Kaurik-Chango Fault.

The focal mechanism solutions of 26-9-64, 16-12-66, 27-6-66, 12-1-65 and 27-3-64 earth-

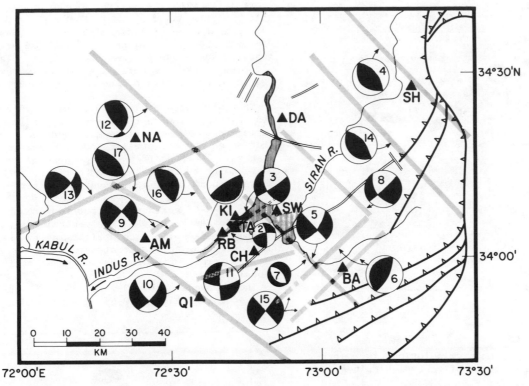

Fig. 14. Composite fault plane solutions, in northwestern Himalaya from microearthquakes, on an equal area projection of the upper focal hemisphere and associated seismic patterns, shown by shaded bars (after Armbruster et al., 1978).

quakes show thrust faulting with nodal planes paralleling the local structural trend. These solutions indicate crustal shortening and under-thrusting of the peninsular shield mass along the Himalayan Arc.

The focal mechanism solution of 15-8-66 earth-quake, in the Ganga Basin, provides further evidence in support of underthrusting along the Himalayan Arc. This earthquake has a normal fault plane solution with east-west striking nodal planes. The tensional character in the epicentral region is a consequence of bending of the litho-spheric plate as it dips in a subduction zone. Because of the lower density of the continental crust, the subduction along the Indus-Tsangpo suture zone has ceased since the closing of the Tethys Sea and has been replaced by a broad zone of crustal shortening and vertical crustal uplift.

Shillong Plateau

The Shillong Plateau lies between the Himalaya in the north and the Bengal Basin in the south. The east-west trending Dauki Fault separates the Shillong Plateau from the Bengal Basin. At the eastern margin, the Dauki Fault is joined by the northeast trending Haflong Fracture Zone. The Dhubri Fault is located at the western margin of the plateau.

The focal mechanism solutions of earthquakes near the Shillong Plateau are shown in Figure 16. The earthquakes of 19-6-63 and 21-6-63 occurred near the Dauki Fault at the southern edge of the Shillong Plateau. Focal mechanism solutions show thrusting along east-west striking nodal planes. The axis of pressure, P, trends in the north-south direction and suggests that the two events are caused by the stress system related to the collision of Indian and Eurasian continents. It appears that the activity along the Dauki Fault was rejuvenated by the continental collision. As a consequence, the Shillong Plateau was uplifted as a horst to its present elevation which varies from about 600 to 1,800 m. The near vertical east-west striking nodal planes in the solutions for events 21-6-63 and 19-6-63 are interpreted as the fault planes.

Event 18-6-65 has a thrust solution with a small component of strike slip faulting. The north-south trend of the axis of pressure indicates that this earthquake was also caused by the stress system related to the collision of northward moving Indian continent with Eurasia. The earth-quake occurred on the northeast trending Haflong Fracture Zone. Considering the northeast striking nodal plane to be the fault plane, the mechanism solution shows an upward movement of the northwest block relative to the southeast block.

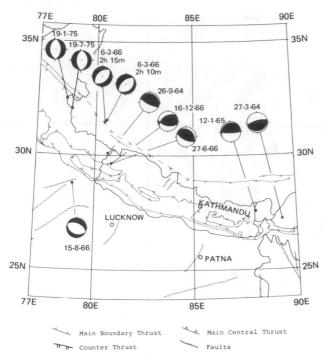

Fig. 15. Focal mechanism solutions for earth-quakes in the vicinity of central Himalaya.

Assam Syntaxis

At the northeastern extremity of the Himalaya, the convergence of the mountain ranges of the Himalayan Arc and the Burmese Arc forms what is described as the Assam syntaxis. The Burmese Arc consists of the Patkai, Naga, Manipur, Lushai, Chin, Arakan and other hill ranges. It continues toward the south through the Andaman and Nicobar Islands and curves eastward from Sumatra toward Java. Geologically the two flanks of the Assam syntaxis are not identical. Gansser (1964) observed that the western Burmese ranges have no affinities with the Lower or Higher Himalaya. He also noted that the Arakan Yoma Range is certainly no eastern equivalent of the Sulaiman Range.

On an epicenter map of the Himalaya, Figure 3, an aseismic northeast trending lineament about 80 km wide, separating the northeast Himalayan and Burmese seismic zones, can be identified in the Assam Valley. In marked contrast to the activity along the Himalayan and Burmese arcs, the seismicity within and near this lineament is very small.

The focal mechanism solutions of earthquakes in the vicinity of the Assam syntaxis are summarized in Figure 16. P-wave first motion data together with the P nodal planes corresponding to several of these solutions, previously unpublished, are presented on an equal area projection of the lower focal hemisphere in Figure 17.

The earthquakes of 18-2-64, 15-9-67, 1-9-64, 26-9-66, 21-10-64 and 14-3-67 are spatially

correlated with the MCT-MBT of the Himalayan flank of the Assam syntaxis and occur toward the north in the general dip direction of this thrust system. The fault plane solutions are of the thrust type. The strikes of the nodal planes dipping toward the north are generally parallel to the local strike of the MCT-MBT thrust system.

The earthquake of 19-2-70 occurred very close to the MCT-MBT thrust system. The fault plane solution is mainly strike slip and the sense of motion on the NNE striking nodal plane is left lateral. None of the nodal planes for an alternate solution for this event strikes in the direction of local fault system and structural trend.

The 15-6-65 earthquake occurred farther north, close to Namcha Barwa. The mechanism solution is mainly strike-slip with a right lateral sense of motion on the northeast striking nodal plane and left lateral motion on the southeast striking nodal plane.

The earthquakes of 29-7-70, 30-5-71, 29-12-71, 12-7-64, 13-7-64, 17-10-69, 22-1-64 and 27-2-64

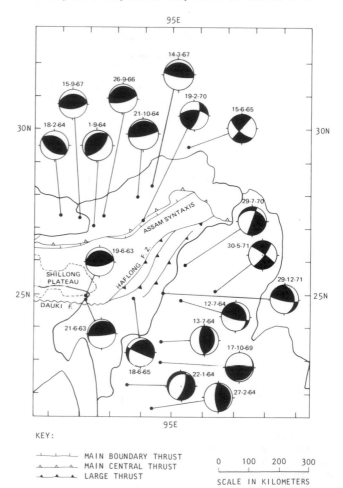

Fig. 16. Focal mechanism solutions for earth-quakes near Assam syntaxis.

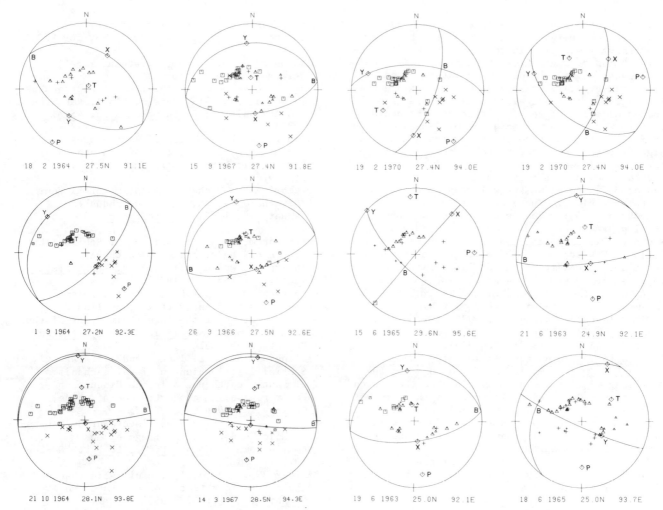

Fig. 17. P-wave first motion data, for earthquakes near Assam syntaxis, together with the nodal planes on an equal area projection of the lower focal hemisphere. See Figure 13 for explanation of letters and symbols.

occurred along the Burmese Arc. Events 29-7-70 and 30-5-71 occurred close to the syntaxial bend on the southeastern flank. The motion on near vertical NNE striking nodal plane is right lateral parallel to the trend of the Burmese Arc. Alternately, the northwest striking nodal plane may be interpreted as a sinistral arc-arc transform fault.

Mechanism solutions for events 29-12-71 and 12-7-74 show underthrusting of the lithospheric plate along the nodal plane dipping gently toward northeast.

Solutions for events 13-7-64 and 27-2-64 indicate thrust faulting. The axis of pressure is nearly horizontal and approximately perpendicular to the trend of the Burmese Arc in this region. Island arc-like structure indicated by the presence of volcanic activity, and the focal mechanism solutions suggest underthrusting of the lithospheric plate along the nodal plane dipping gently toward the east.

Further evidence in support of underthrusting

in this region comes from a study of the earthquake of 22-1-64, which has a normal fault plane solution with the axis of tension nearly horizontal and approximately perpendicular to the trend of the Burmese Arc. The earthquake resulted from tension caused by the bending of the lithospheric plate along the outer margin of the Burmese zone of subduction.

The focal mechanism solution for the event of 17-10-69 is somewhat difficult to interpret. The solution indicates thrust faulting, but unlike events 13-7-64 and 27-2-64, the axis of pressure, P, is parallel to the trend of the Burmese Arc. The earthquake may have resulted from a north-south compression caused by a local anomalous change in the direction of thrusting.

Conclusions

The continued convergence of the continents of Arabia and Eurasia, and India and Eurasia since the continental collision and suturing along the

Zagros suture zone and the Indus-Tsangpo suture zone, respectively, is taken up partly by crustal shortening (resulting in the building up of mountain chains of the Zagros, Kirthar, Sulaiman and Himalayan ranges) and partly by lateral mass movement (e.g., right lateral movement along the Main Recent Fault and along the Karakoram Fault). The parallelism of the direction of continental convergence and the orientation of P-axes derived from focal mechanism solutions suggests that the high stresses generated by continental collision caused a rejuvenation of activity along old orogenic belts of Elburz and Kopet Dagh ranges and along zones of weakness in the northern part of the Lut block. Therefore, it seems unnecessary to invoke the concept of a number of small plates (see, e.g., McKenzie, 1972 and Nowroozi, 1972) to explain the tectonics of the region. The occurrence of earthquakes in peninsular India, uplift of the Shillong Plateau, large scale tectonic deformation and earthquakes in Tibet, China and Mongolia may be explained as products of continental collision (Chandra, 1977, 1978; Molnar and Tapponnier, 1975, 1977, 1978; Tapponnier and Molnar, 1976, 1977; York et al., 1976; Ni and York, 1978; Ni, 1978).

The deformation pattern in southeast Zagros Folded Belt may be explained by postulating the colliding edge of the Arabian plate in the region north of the Oman peninsula to be in the form of a projected promontory. The focal mechanism solutions suggest a possible subsurface continuation of the Aravalli ranges up to the Indus-Tsangpo suture zone.

The Makran coast of Iran and Pakistan is identified as an active trench-arc system of lithospheric plate consumption.

Acknowledgments. I thank S. Khoury for critically reading the manuscript and making helpful suggestions. K. Jacob, J. Armbruster, L. Seeber, R. Quittmeyer, R. White and D. Ross provided preprints of their papers prior to publication. P. Molnar and R. Quittmeyer provided orientations of the mechanism axes for several fault plane solutions. Data for the determination of fault plane solutions were collected at the Lamont Doherty Geological Observatory of the Columbia University.

References

Akascheh, B., Seismizitat und Tektonik von Iran, J. Earth and Space Phys., 1, 8-24, 1972.

Akasheh, B., Mechanism of the earthquake of April 10, 1972 in Qir (Iran), Z. Geophys., 39, 1055-1061, 1973.

Ambraseys, N.N., The Buyin-Zara (IRAN) earthquake of September, 1962, Bull. Seism. Soc. Am., 53, 705-740, 1963.

Argand, E., La tectonique de l'Asie, C.R. Congr. Geol. Intern., 13e 1922 (Liege), 169-371, 1924.

Armbruster, J., L. Seeber and K.H. Jacob, The northwestern termination of the Himalayan mountain front: Active tectonics from micro-earthquakes, J. Geophys. Res., 83, 269-282, 1978.

Armbruster, J.G., L. Seeber and A. Farah, Personal Communication, 1979.

Banghar, A.R., Focal mechanisms of earthquakes in China, Mongolia, Russia, Nepal, Pakistan, and Afghanistan, Earthquake Notes, 45, 1-11, 1974a.

Banghar, A.R., Mechanism of earthquakes in Albania, China, Mongolia, Afghanistan and Burma-India border, Earthquake Notes, 45, 13-25, 1974b.

Banghar, A.R., Mechanism solution of Kinnaur (Himachal Pradesh, India) earthquake of January 19, 1975, Tectonophysics, 31, T5-T11, 1976.

Ben-Menahem, A., E. Aboodi and R. Schild, The source of the great Assam earthquake - an interplate wedge motion, Phys. Earth Planet. Inter., 9, 265-289, 1974.

Berberian, M., Contribution to the seismotectonics of Iran (Part II), published in commemoration of the 50th anniversary of the Pahlavi dynasty, Geological Survey of Iran, Report No. 39, 516 p, 1976.

Billington, S., B.L. Isacks and M. Barazangi, Spatial distribution and focal mechanisms of mantle earthquakes in the Hindu Kush - Pamir region: A contorted Benioff zone, Geology, 5, 699-704, 1977.

Braud, J. and L.E. Ricou, L'accident du Zagros ou Main Thrust, un charriage et un coulissement: Compt. Rend., t. 272, 203-206, 1971.

Canitez, N., The focal mechanisms in Iran and their relations to tectonics, Pure and Appl. Geophys., 75, 76-87, 1969.

Canitez, N. and S.B. Ucer, Computer determinations for the fault-plane solutions in the near Anatolia, Tectonophysics, 4, 235-244, 1967.

Chander, R. and J.N. Brune, Radiation pattern of mantle Rayleigh waves and the source mechanism of the Hindu Kush earthquake of July 6, 1962, Bull. Seism. Soc. Am., 55, 805-819, 1965.

Chandra, U., Stationary phase approximation in focal mechanism determination, Bull. Seism. Soc. Am., 60, 1221-1229, 1970.

Chandra, U., Combination of P and S data for the determination of earthquake focal mechanism, Bull. Seism. Soc. Am., 61, 1655-1673, 1971.

Chandra, U., Fault plane solution and tectonic implications of the Pattan, Pakistan earthquake of December 28, 1974, Tectonophysics, 28, T19-T24, 1975a.

Chandra, U., Seismicity, earthquake mechanisms and tectonics of Burma, 20°N - 28°N, Geophys. J.R. astr. Soc., 40, 367-381, 1975b.

Chandra, U., Earthquakes of peninsular India - A seismotectonic study, Bull. Seism. Soc. Am., 67, 1387-1413, 1977.

Chandra, U., Seismicity, earthquake mechanisms

and tectonics along the Himalayan mountain range and vicinity, Phys. Earth and Planet. Inter., 16, 109-131, 1978.

Chandra, U., Large scale Cenozoic tectonics of central and south-central Asia: products of continental collision, Phys. Earth Planet. Inter., 20, 33-41, 1979a.

Chandra, U., Earthquake mechanisms and tectonics of Iran (under preparation), 1979b.

Chatelain, J.L., S.W. Roecker, D. Hatzfeld and P. Molnar, Microearthquake seismicity and fault plane solutions in the Hindu Kush region and their tectonic implications, J. Geophys. Res., 85, 1365-1387, 1980.

Chaudhury, H.M., H.N. Srivastava and J. Subba Rao, Seismotectonic investigations of the Himalayas, Himalayan Geology, IV, Delhi-7, 481-491, 1974.

Chen, W. and P. Molnar, Seismic moments of major earthquakes and the average rate of slip in central Asia, J. Geophys. Res., 82, 2945-2969, 1977.

Chouhan, R.K.S. and V.K. Srivastava, Focal mechanisms in northeastern India and their tectonic implications, Pure and Appl. Geophys., 113, 467-482, 1975.

Das, S. and J.R. Filson, On the tectonics of Asia, Earth Planet. Sci. Letters, 28, 241-253, 1975.

Desio, A., Karakoram Mountains, in: Mesozoic-Cenozoic Orogenic Belts, A.M. Spencer (ed.), Geol. Soc. London Spec. Publ. 4, 255-266, 1973.

Farhoudi, G. and D.E. Karig, Makran of Iran and Pakistan, Geology, 5, 664-668, 1977.

Fitch, T.J., Earthquake mechanisms in the Himalayan, Burmese and Andaman regions and continental tectonics in Central Asia, J. Geophys. Res., 75, 2699,2709, 1970.

Gansser, A., Geology of the Himalayas, Interscience, London, 1964.

Gansser, A., The Indian Ocean and the Himalayas: a geological interpretation, Eclog. Geol., Helv., 59, 831-848, 1966.

Hedayati, A. and T. Hirasawa, Mechanism of the Hindu Kush earthquake of Jan. 28, 1964, derived from S wave data: The use of p^P phase for the focal mechanism determination, Bull. Earthquake Res. Inst., Tokyo Univ., 44, 1419-1434, 1966.

Ichikawa, M., Srivastava, H.N. and J. Drakopoulos, Focal mechanisms of earthquakes occurring in and around the Himalayan and Burmese mountain belts, Pap. Meteorol. Geophys. Tokyo, 23, 149-162, 1972.

Inst. of Geophysics, Progress report on the seismicity of north central Iran, Pub. No. 34, Institute of Geophysics, Tehran University, 1966.

Isacks, B. and P. Molnar, Distribution of stresses in the descending lithosphere from a global survey of focal-mechanism solutions of mantle earthquakes, Rev. Geophys., 9, 103-174, 1971.

Jacob, K.H. and R.C. Quittmeyer, The Makran region of Pakistan and Iran: Trench-Arc system with active plate subduction, in: Geodynamics of Pakistan, A. Farah and K.A. DeJong (eds.), Geological Survey of Pakistan, Quetta, 305-315, 1979.

Khattri, K.N., V.K. Gaur, J.C. Bhattacharji, A.R. Ansari and H. Sinvhal, Fault plane solution for Kinnaur earthquake of 19 January, 1975, Bull. Indian Soc. Earthquake Tech., 12, 155-160, 1975.

Khattri, K., K. Rai, A.K. Jain, H. Sinvhal, V.K. Gaur and R.S. Mithal, The Kinnaur earthquake, Himachal Pradesh, India, of 19 January, 1975, Tectonophysics, 49, 1-21, 1978.

McKenzie, D., Active tectonics of the Mediterranean region, Geophys. J.R. astr. Soc., 30, 109-185, 1972.

Molnar, P., T.J. Fitch and F.T. Wu, Fault plane solutions of shallow earthquakes and contemporary tectonics in Asia, Earth Planet. Sci. Letters, 19, 101-112, 1973

Molnar, P. and P. Tapponnier, Cenozoic tectonics of Asia: effects of a continental collision, Science, 189, 419-426, 1975.

Molnar, P. and P. Tapponnier, Relation of the tectonics of eastern China to the India-Eurasia collision: Application of slip-line field theory to large-scale continental tectonics, Geology, 5, 212-216, 1977.

Molnar, P., W.P. Chen, T.J. Fitch, P. Tapponnier, W.E.K. Warsi and F.T. Wu, Structure and tectonics of the Himalaya: A brief summary of relevant geophysical observations, in: Himalaya, Sciences de la Terre, Centre National de la Recherche Scientifique, Paris, 269-294, 1977.

Molnar, P. and P. Tapponnier, Active tectonics of Tibet, J. Geophys. Res., 83, 5361-5375, 1978.

Ni, J., Contemporary tectonics in the Tien Shan region, Earth and Planetary Science Letters, 41, 347-354, 1978.

Niazi, M., Fault rupture in the Iranian (Dasht-e-Bayaz) earthquake of August 1968, Nature, 220, 569-570, 1968.

Niazi, M., Source dynamics of the Dasht-e Bayaz earthquake of August 31, 1968, Bull. Seism. Soc. Am., 59, 1843-1861, 1969.

Nowroozi, A.A., Seismo-tectonics of the Persian Plateau, eastern Turkey, Caucasus, and Hindu Kush regions, Bull. Seism. Soc. Am., 61, 317-341, 1971.

Nowroozi, A.A., Focal mechanism of earthquakes in Persia, Turkey, West Pakistan, and Afghanistan and plate tectonics of the Middle East, Bull. Seism. Soc. Am., 62, 823-850, 1972.

Nowroozi, A.A., Seismotectonic provinces of Iran, Bull. Seism. Soc. Am., 66, 1249-1276, 1976.

Peive, A.V., V.S. Burtman, S.V. Ruzhentzev and A.I. Suvorov, Tectonics of the Pamir-Himalayan Sector of Asia, 22nd Int. Geol. Congr., Section 11, 441-464, 1964.

Pennington, W.D., A summary of field and seismic

observations of the Pattan earthquake –
28 December 1974, in: Geodynamics of Pakistan,
A. Farah and K.A. DeJong (eds.), Geological
Survey of Pakistan, Quetta, 143-147, 1979.

Prevot, R., D. Hatzfeld, S.W. Roecker and
P. Molnar, Shallow earthquakes and active
tectonics in eastern Afghanistan, J. Geophys.
Res., 85, 1347-1357, 1980.

Quittmeyer, R.C., A. Farah and K.H. Jacob, The
seismicity of Pakistan and its relation to
surface faults, in: Geodynamics of Pakistan,
A. Farah and K.A. DeJong (eds.), Geological
Survey of Pakistan, Quetta, 271-284, 1979.

Rastogi, B.K., Earthquake mechanisms and plate
tectonics in the Himalayan region, Tectonophysics, 21, 47-56, 1974.

Rastogi, B.K., Singh, J. and R.K. Verma, Earthquake mechanisms and tectonics in the Assam-
Burma region, Tectonophysics, 18, 355-366,
1973.

Ritsema, A.R., The fault plane solutions of
earthquakes of the Hindu Kush centre,
Tectonophysics, 3, 147-163, 1966.

Roecker, S.W., O.V. Soboleva, I.L. Nersesov,
A.A. Lukk, D. Hatzfeld, J.L. Chatelain and
P. Molnar, Seismicity and fault plane solutions
of intermediate depth earthquakes in the
Pamir-Hindu Kush region, J. Geophys. Res., 85,
1358-1364, 1980.

Schaffner, H.J., Die Grundlagen und
Auswerteverfahren zur Bestimmung von
Erdbebenmechanismen. Freiberger Forsch. 63
(Geophy.), 1959.

Schreiber, A., D. Weippert, H.P. Wittekindt and
R. Wolfart, Geology and petroleum potentials of
central and south Afghanistan, Am. Assoc.
Petr. Geol. Bull., 56, 1494-1519, 1972.

Seeber, L. and J. Armbruster, Seismicity of the
Hazara Arc in Northern Pakistan: Decollement
vs. Basement Faulting, in: Geodynamics of
Pakistan, A. Farah and K.A. DeJong (eds.),
Geological Survey of Pakistan, Quetta, 131-
142, 1979.

Shirokova, E.I., General features in the
orientation of principal stresses in earthquake foci in the Mediterranean – Asian
seismic belt, Izv. Earth Physics, No. 1, 22-36
(Engl. transl.), 1967.

Singh, D.D. and H.K. Gupta, Source dynamics of
two great earthquakes of the Indian subcontinent: The Bihar-Nepal earthquake of
January 15, 1934 and the Quetta earthquake of
May 30, 1935, Bull. Seism. Soc. Am., 70, 757-
773, 1980.

Sobouti, M., Sur le mechanisme au foyer dans
l'arc sismique entre l'Hindou-Koush et la
Méditerranée, Annales de Geophysique, 19, 1,
1963.

Srivastava, H.N., The crustal seismicity and the
nature of faulting near India Nepal-Tibet
trijunction, Himalayan Geology, 3, 321-393,
1973.

Stauder, W., Mechanism of the Rat Island earthquake sequence of February 4, 1965, with

relation to island arcs and sea-floor spreading, J. Geophys. Res., 73, 3847-3858, 1968.

Stevens, A.E., S-wave focal mechanism studies of
the Hindu Kush earthquake of July 6, 1962,
Can. J. Earth Sci., 3, 367-384, 1966.

Stocklin, J., Structural history and tectonics
of Iran - a review, Am. Assoc. Petr. Geol.
Bull., 52, 1229-1258, 1968.

Stocklin, J., Possible ancient continental
margins in Iran, in: The Geology of
Continental Margins, C.A. Burk and C.L. Drake
(eds.), Springer - Verlag New York Inc., 873-
887, 1974.

Stoneley, R., Evolution of the continental
margins bounding a former southern Tethys, in:
The Geology of Continental Margins, C.A. Burk
and C.L. Drake (eds.), Springer - Verlag New
York Inc., 889-903, 1974.

Takin, M., Iranian geology and continental drift
in the Middle East, Nature, 235, 147-150, 1972.

Tandon, A.N., Direction of faulting in the great
Assam earthquake of August 15, 1950, Ind. J.
Met. Geophys., 6, 61-64, 1955.

Tandon, A.N., Anantnag earthquakes (February to
April 1967), Ind. J. Met. Geophys., 23, 491-
502, 1972.

Tandon, A.N., Some typical earthquakes of north
and west Uttar Pradesh, Bull. Indian Soc. of
Earthquake Technology, 12, 74-88, 1975.

Tandon, A.N. and H.N. Srivastava, Fault plane
solutions as related to known geological faults
in and near India, Ann. di Geofis. 28, 13-27,
1975a.

Tandon, A.N. and H.N. Srivastava, Focal mechanisms of some recent Himalayan earthquakes
and regional plate tectonics, Bull. Seism.
Soc. Am., 65, 963-969, 1975b.

Tapponnier, P. and P. Molnar, Slip-line field
theory and large-scale continental tectonics,
Nature, 264, 319-324, 1976.

Tapponnier, P. and P. Molnar, Active faulting and
tectonics in China, J. Geophys. Res., 82, 2905-
2930, 1977.

Tchalenko, J.S. and J. Braud, Seismicity and
structure of the Zagros (Iran): The Main
Recent Fault between 33° and 35^ON, Phil. Trans.
Roy. Soc. London, 277, 1-25, 1974.

Valdiya, K.S., Tectonic framework of India: A
review and interpretation of recent structural
and tectonic studies, Geophys. Res. Bull., 11,
79-114, 1973.

Verma, R.K., M. Mukhopadhyay and M.H.
Ahluwalia, Earthquake mechanisms and tectonic
features of northern Burma, Tectonophysics,
32, 387-399, 1976.

Verma, R.K., M. Mukhopadhyay and B.N. Roy,
Seismotectonics of the Himalaya, and the continental plate convergence, Tectonophysics,
42, 319-335, 1977.

Wadia, D.N., The syntaxis of the northwest
Himalaya; its rocks, tectonics and orogeny,
Rec. Geol. Surv. Ind., 65, 189-220, 1931.

Wadia, D.N., Geology of India, 3rd ed. revised,
MacMillan and Co. Ltd., London, 1966.

White, R.S. and D.A. Ross, Tectonics of the western Gulf of Oman, J. Geophys. Res., 84, 3479-3489, 1979.

Wickens, A.J. and J.H. Hodgson, Computer re-evaluation of mechanism solutions, Publ. Dom. Obs. Ottawa, 33, 1-560, 1967.

STRUCTURE AND SEISMOTECTONICS OF THE HIMALAYA-PAMIR HINDUKUSH REGION AND THE INDIAN PLATE BOUNDARY

K. L. Kaila

National Geophysical Research Institute, Hyderabad 500007, India

Abstract. Deep Seismic Soundings (DSS) across the Great Himalaya-Pamir Hindukush region have revealed the deep crustal and upper mantle structure of the area. Along the Wular Lake-Nanga Parbat profile, the Moho which is at a depth of about 45 km near Sopur (34°17'N; 74°28'E) in the Kashmir valley, India, changes to 54 km depth in the region of Wular Lake, and it becomes still deeper towards NNE attaining some 64 km near Kanzalwan (34°39'N; 74°43'E). A deep fault then displaces the Moho downwards to a depth of 76 km, which again rises to about 62 km under the Nanga Parbat shot point (35°20'N; 74°48'E). A similar picture is revealed along two other profiles in this region. The updip of the Moho boundary in the region between Kanzalwan and Nanga Parbat is an indication of the upwarp of the Moho which is considered responsible for the uplift of the Nanga Parbat massif. The crustal block between Sopur and Kanzalwan in the Kashmir Himalaya is bounded by two large angle deep faults which extend down to the Moho boundary.

Based on DSS studies and seismotectonic data in the region, it is concluded that the northern boundary of the Indian plate does not lie along the Main Central Thrust in the Himalaya, nor the Indus Suture line, but falls very much north of the combined Indo-Tibetan block. A new Indian plate boundary is postulated which coincides with the southern margin of Tien Shan-Nan Shan mobile fold belt, passing south of the Ordos and Shanshi blocks, finally turning northeastwards, towards the southeast of Peking. The western boundary of the above proposed plate is formed by the southward continuation of Pamir-Alay fracture zone passing along the western boundary of the Badakshan mountains, joining further south with the fault zones of the Sulaiman and Kirthar ranges which in turn connect to the Owen Fracture zone in the Arabian sea. The eastern boundary of this plate divides southeastern China from Tibet. It passes along the Lung Men Shan thrust and the Arakan Yoma fold belt, continuing further southeastward to join with the Indonesian or Sunda trench system. The above postulated Indian plate boundary not only envisages a northward movement of the combined Indo-Tibetan block but also a simultaneous eastward movement.

Deep tectonic features in the Pamir-Hindukush region are also studied from earthquake hypocentral data, along two profiles, aligned transversely to the main geological trend of the area, (Profile I) Hindukush region and (Profile II) Pamir region. Along both these profiles, two systems of large angle deep faults are delineated, one system dipping northwest and the other in the southeast direction. Due to upward block movements along these two oppositely dipping fault systems, a zone of tension develops at great depth which gives rise to almost vertical deep fractures. The intermediate depth earthquakes in the Pamir-Hindukush region are considered to originate in relative movements along these deep fractures.

Introduction

The origin of the Himalaya, the highest mountain chain in the world, has been a most challenging problem to geologists. Although it is well known that the Himalaya consist of sedimentary rocks, deposited over millions of years in shallow seas yet it is not clear in what sea they were deposited and how they were sandwiched between the subcontinent of India and the Asian landmass to the north. To understand the evolution of the Himalaya, one has to understand first the relationship between the Indian and Tibetan landmasses, the history of Gondwanaland and the development of Asia. Holmes (1965) suggested that the Tibetan plateau was a median mass forming part of a geosynclinal block between the continents of Asia and India. Molnar and Tapponnier (1975) point out that Argand (1924), Holmes (1965) and others have suggested that the uniform and very high altitude in Tibet is due to underthrusting of India beneath the whole of Tibet causing almost double thickness of the crust. Such an idea according to Molnar and Tapponnier calls for a some 1000 km long, very shallow, fault plane dipping 0° to 5° separating the supposed underthrusting Indian plate from the overlying Tibetan crust. They further state that neither seismicity nor fault plane solutions in the Himalaya provide evidence for such a fault zone, and nowhere on the earth has been detected an inclined seismic zone that dips at such a shallow angle for such a long distance.

The Indus Suture line along the Indo-Tibetan border is a concept due to Gansser (1959, 1964, 1966) who describes it as a sudden root-like downbuckling, with orogenic effects on Middle Cretaceous marine sediments of Tethyan facies. Upper Cretaceous flysch contains ophiolites and huge exotic blocks of sediments with a pelagic facies unknown in the Himalaya. Gansser regarded these sediments as having been deposited in deep basins situated between the present Himalaya and the Tibetan plateau. These original basins have completely disappeared and are now evidenced by the Indus Suture line.

Some authors have sought the origin of Himalaya in the collision of two continents, India and Eurasia, with the Indus Suture line as the plate boundary. Santo (1972) concludes that the last stage 'D' in the schematic presentation of the collision of two continents by Dewey and Bird (1970) would be the situation prevailing at present in the Himalaya. Although, in principle, it seems reasonable to describe the tectonics of Eurasia as the interaction of blocks of lithosphere, it is not yet clear how successful this idea will be in practice because of the large number of blocks involved (Isacks et al., 1968). The interaction of blocks of lithosphere appears to be much more complex when all the blocks involved are continental than when at least one is an oceanic block. According to these authors the seismicity pattern in the Tibet-Himalayan region, quite diffused in areal extend, can be explained in two ways: either the lithosphere, being continental, is more heterogeneous in this region, or the old zones of weakness have been reactivated when the lithospheric blocks of the Indian and the Asian plate interacted. Because of its relatively low density, it may not be possible to underthrust a block of continental lithosphere into the mantle to depths of several hundred kilometers, which might explain the absence of intermediate and deep earthquakes in this region.

According to Meyerhoff and Meyerhoff (1972), India has been a part of Asia since Proterozoic or earlier. Near the base of the section in the Lower Himalaya, the Simla slates are an integral part of the Proterozoic - early Paleozoic Himalayan geosyncline (Gansser, 1964; Fuchs and Frank, 1970). The sedimentary section of the Lower Himalaya contains very few fossils, but equivalent and younger strata of the Tibetan zone, north of the Great Himalaya, are extremely fossiliferous. Because of the marked differences between the Lower Himalaya and Tibetan zones, Fuchs (1967, 1970) postulated the existence of a basement ridge between the two zones. The postulated ridge parallels the northeastern margin of the shield and extends northwestward to Kashmir. In northern Pakistan, in the Hazara region near the Salt Range, a well known sequence, transitional between that the Lower Himalaya and Tibetan zones, is exposed on the ridge (Fuchs, 1970). The

section of the Lower Himalaya and possibly the Tibetan zone contain beds of Gondwana facies found in the Indian shield (Gansser, 1964). The Gondwana beds occupy part of the section in Kashmir, where they are interbedded with Tethyan facies and in the Salt Range, where the Gondwana Nilawan group underlies Tethyan Zaluch group. Recently Sharma et al.(1979) have reported the occurrence of upper Gondwana floral elements from a dark arenaceous shale and sandstone sequence in Ladakh, north of the Indus Suture zone. These are the key stratigraphic and structural evidences which link India inseparably to the Tethys, the tropical zone, and to Central Asia (Meyerhoff and Meyerhoff, 1972).

The concept of the Indus Suture line as a plate boundary and the height of the Tibetan plateau are incompatible (Crawford, 1974). According to him the Indus Suture line is a relic of a fracture reaching down to the mantle, but for a period only represented by the faunas of the exotic blocks of the Tibetan Himalaya i.e., Permian-late Jurassic. In his opinion Tibet originally was part of a plate that included India, but submerged while India remained continental.

Molnar and Tapponnier (1975) consider most of the large-scale tectonics of Asia to be a result of the India-Eurasia continental collision, which apparently not only created the Himalaya but also rejuvenated an old orogenic belt (Tien Shan) 1000 km north of the Indus Suture line. This collision according to them caused important strike-slip faulting oblique to the Suture zone and as much as 1000 km from it, and perhaps ripped open two rift systems more than 2000 km away, such as the Baikal rift zone and the Shansi graben. When calculating convergence since the collision, they estimate that shortening and underthrusting of India beneath the Himalaya and Tibet probably accounts for at least 300 and 700 km. Probably another 200 and 300 km can be accounted for by thrusting and crustal thickening in the Pamir, Tien Shan, Altai, Nan Shan and other mountain belts. They further concluded that a major fraction of the convergence occurs along large-scale east-west trending strike slip faults in China and Mongolia. Movement of these latter may allow material lying between the stable portions of the Indian and Eurasian plates to move laterally out of the way of these two plates. They inferred that probably a total of 500 km and conceivably 1000 km of east-west motion could have occurred and could account for a comparable amount of shortening. Hence, the recognition of large strike slip motion, in their opinion, may obviate the need for postulating the underthrusting of India beneath the whole of Tibet. (op. cit. p. 419).

Deep Seismic Soundings in the Pamir-Himalayan Region

During September 1974, Soviet scientists fired experimental shots at two locations in the Pamir region. Two 5 ton explosions in the Quarrakol Lake (200 metres deep) and one 10 ton explosion in very shallow shot holes near Zorkol Lake were set off and were successfully recorded near Srinagar in the Kashmir Himalaya by Indian scientists (DSS group) from the National Geophysical Research Institute (NGRI), India. Recording of these experimental shots was carried out using two Russian made POISK 48-channel magnetic recorders along two selected spread locations northwest of Srinagar (shown by small dots in Figure 1), one spread being north of, and the other south of the Mansbal Lake with a geophone spacing of 200 metres. Figure 1 also shows the locations of the shot points in the Quarrakol Lake (Q-74) and near Zorkol Lake (Z-74). In August 1975, these two shots points were reactivated in USSR, using again two 5 ton shots in the Quarrakol Lake and two 10 ton shots in the vicinity of Zorkol Lake. Besides these, two more shot points were operated in Pakistan. Two shots were fired in the Nanga Parbat area in Lake Sangosar near Astor (Nanga Parbat shot point) and two shots fired in bore holes drilled near Lawrencepur (Attok shot point). These shots in Pakistan were arranged by the Italian scientists in collaboration with Pakistan scientists. All the eight shots were recorded by the DSS group of NGRI at two spread locations (I & II) in the Srinagar area; spread I (about 11 km long) near Tral, southeast of Srinagar and spread II (about 6 km long) near Sopur town, northwest of Srinagar, along the bank of the Wular Lake (Figure 1). Groups of geophones were bunched together to improve the signal to noise ratio. These recordings have provided wide angle reflection data at various shot-detector distance ranges of about 113-119, 153-163, 192-196, 244-248, 382-388, 430-440, 529-539 and 585-596 km.

The analysis of data along three DSS profiles across the Pamir-Himalaya yielded the velocity function shown in Figure 2 (Kaila et al.,(1978). In this figure is also shown for comparison the velocity function for the Pamir region, USSR, as given by Talvirsky (personal communication). The depth section along one of the profiles 1 across the Kashmir valley and the Great Himalayan ranges is given in Figure 3. This figure shows the crustal section as obtained from DSS and the geological section in Kumaon Himalaya (after Gansser, 1964) for comparison. The resemblance of the dip trends in both sections is remarkable, indicating that the Kashmir basin is structurally very similar to the Tethys geosyncline – may even be an extension of the same. Equally remarkable is the similarity in the order of the dips. Down to a depth of about 25 km, for which the information is available in Gansser's section, the dip angles seem to be of the same order, thus confirming the existence of large dips in the geosynclinal region. In both the sections, these dips become reversed on the northern side, and the Raksas High in the Kumaon Himalaya resembles structurally the Nanga Parbat high in the Kashmir region and both are located immediately south or west of the Indus Suture line. Hence, both these massifs, forming the border between the Himalayan geosyncline and the suture zone, also seem to have similar origins. It thus appears that the Tethys geosyncline extended considerably more northwestwards and included the Kashmir basin as well.

From the crustal cross-section along the Wular Lake-Nanga Parbat profile based on deep seismic soundings (Figure 3), it can be seen that there is a definite upwarp in the Moho boundary which just after the deep fault F_{21} rises from about 76 km (about 16 km NNE of Kanzalwan) to about 62 km under the Nanga Parbat shot point, giving an uplift of about 14 km. This upwarp in the Moho boundary must be responsible for the uplift of the Nanga Parbat massif. It may be mentioned here that Wadia (1966) also concluded from geological evidence that the rise of the Himalayan orogen from the Tethyan waters in the three-phased Tertiary uplift must have involved crustal upheaval and fold waves of about 13 km in vertical amplitude which is of the same order as the upwarp in the Moho boundary as observed from DSS data in the region of the Nanga Parbat massif. Due to a large upwarp of the Moho, the two flanks of the Nanga Parbat massif might have been faulted with deep faults extending upto the Moho boundary. These faults might have acted as channels for the flow of basic and ultrabasic material (ophiolites) to the surface. Is it not possible then that the Nanga Parbat uplift forms the earliest phase of the Himalayan orogeny and that this uplift divided the Kashmir Himalayan geosyncline into two parts, the western part extending upto Hazara and the eastern part comprising the region of Pir Panjal range and the Srinagar valley, and it caused this remarkable syntaxial bend rather than the presumed narrow tongue of the Peninsular shield?

On the basis of the DSS data, Kaila et al.(1978) arrived at the following conclusions regarding the structure of the Himalaya.
1) The crustal velocities in the Himalayan region are lower (by about 5 to 6%) than those in the Indian shield, although the velocity functions are quite close in the upper mantle in the two regions. Further, an appreciable thickening of the crust vis-a-vis the Indian shield is also indicated in this region.
2) The Moho and other shallower boundaries as inferred from the depth section along profile 1, dip gradually at 15 to 20 degrees towards NNE in the Kashmir Himalayan region from Sopur to Kanzalwan (Figure 3). The Moho, which is at a depth of about 45 km near Sopur descends rapidly to 54 km in the region of Wular Lake from where it continues to deepen further towards the NNE attaining a depth of about 64 km near Kanzalwan. Beyond the deep fault F_{21}, the Moho is at a depth of 76 km and then rises to about 62 km under the Nanga Parbat massif. This upwarp of the Moho could be responsible for the uplift of the Nanga Parbat massif. Along profile 2, the Moho is at a depth of about 53 km in the region about 10 km

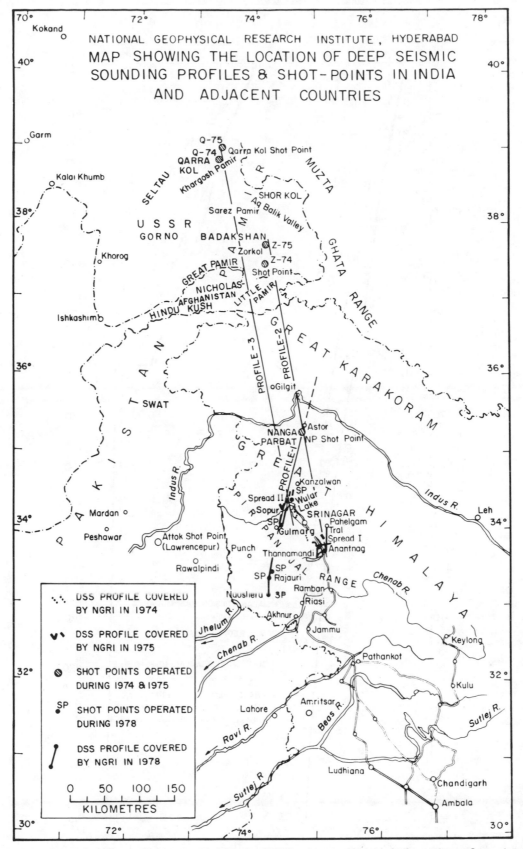

Fig. 1. Map showing DSS profiles in the Pamir-Himalayan region with location of various shot points in USSR, Pakistan and India which were operated during 1974, 1975 and 1978.

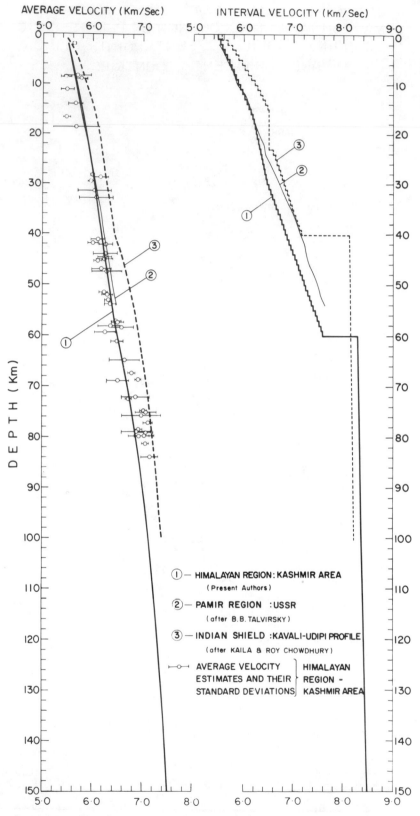

Fig. 2. Average and interval velocity functions for the Pamir–Himalayan region.

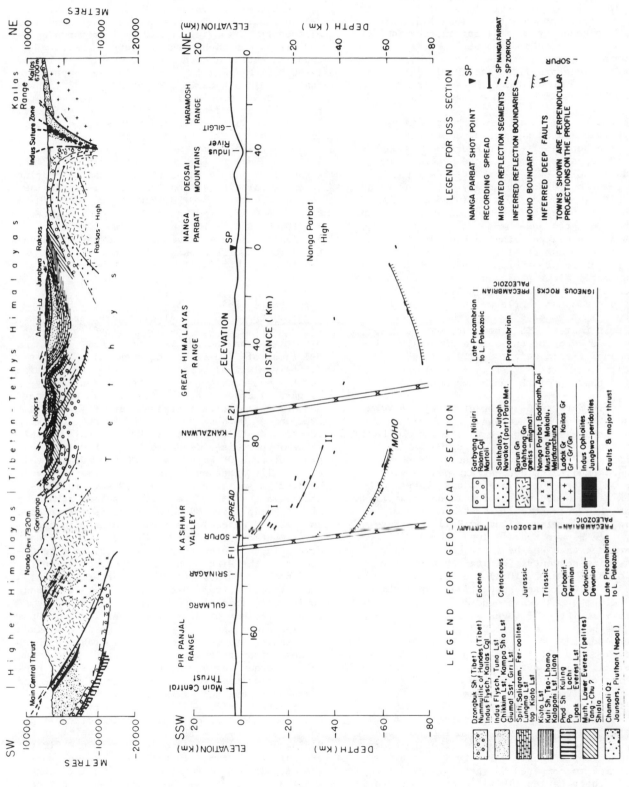

Fig. 3. Crustal depth section along Nanga Parbat–Wular Lake profile as obtained from Deep Seismic Soundings. For comparison, the geological section in Kumaun Himalaya (after Gansser, 1964) is reproduced at the top. The horizontal scales for the two sections are not same, although the dips shown are true dips, there being no vertical exaggeration.

NNW of Tral deepening NNW to a depth of about 64 km in the region some 17 km NE of Srinagar. However, along this profile which is not in the dip direction, there appears to be some flattening of the Moho boundary near Kanzalwan, the Moho lying there at a depth of about 70 km.

3)   From a comparison of the Moho depths at various places along profiles 1 and 2, it can be inferred that the Moho, under the Kashmir Himalaya, also dips to the SE with an average of about 10 degrees.

4)   The crustal block in the Kashmir Himalayan region between Sopur and Kanzalwan (along profile 1) and between Tral and Kanzalwan (along profile 2) appears to be bounded by two steep angle deep faults ($F_{11}-F_{12}$ and $F_{21}-F_{22}$) which extend down to the Moho boundary. The inferred deep fault $F_{11}-F_{12}$ may be associated with the fissure in this region through which Panjal trap might have erupted and the second deep fault $F_{21}-F_{22}$ may be associated with the boundary between the Precambrian and the Triassic.

5)   On the basis of several deep reflecting segments recorded along profiles 2 and 3, a rather deep reflector is inferred at a depth of about 140 km in the upper mantle which seems to extend from right beneath Nanga Parbat to the great Pamir ranges. This boundary may be associated with the top of the second order low velocity channel (Kaila et al., 1974a & b) where plasticity just sets in, or in other words, it indicates the top of the asthenospheric layer.

Upper Mantle Structure in the Himalaya. A very detailed study of P- and S-wave travel times from shallow earthquakes upto $\Delta = 55^{\circ}$ in the northerly azimuth from India was carried out by Kaila et al. (1968a & b). P-wave travel times of 39 shallow earthquakes and three nuclear explosions with epicentres in the north in Himalaya, Tibet, China and USSR as recorded in the Indian observatories were analysed statistically by the method of weighting observations given by Kaila and Narain (1970). The P-wave travel times from $\Delta = 2^{\circ}$ to 50° distance range can be represented by four straight line segments yielding apparent velocities of 8.35 ± 0.02, 10.11 ± 0.06, 11.22 ± 0.07 and 13.59 ± 0.08 km/sec respectively indicating abrupt velocity changes around 19°, 22° and 33°. The true velocity of P-waves at the top of the mantle has been found to be 8.31 ± 0.02 km/sec. A corresponding study of S-waves reveals that the travel times upto 55° can be represented by three straight line segments giving apparent velocities of 4.60 ± 0.01, 6.10 ± 0.04 and 7.63 ± 0.06 km/sec respectively with changes in slope of travel time curves occurring at 21° and 31°.

The P-residuals (J-B) in the north upto $\Delta = 19^{\circ}$ are mostly negative, varying from 1 to 10 seconds with a dependence on Δ values, indicating thereby a higher upper mantle velocity in the Himalayan region as compared to that used by Jeffreys-Bullen in their tables (1940). Between 19° to 33° there is a reasonably good agreement between the

J-B curve and the observation points. From $\Delta = 33^{\circ}$ to 55°, P-residuals are mostly positive with an average excess value of about 4 seconds. The J-B residuals for S-waves on the other hand are all negative, varying from 0 to 25 seconds upto 20° with a dependence on epicentral distance, thus indicating a higher upper mantle S-velocity (4.60 ± 0.01 km/sec) in this region as compared to that used in J-B tables. From 20° to 28°, the S-residuals are evenly distributed about the zero residual line and from 28° to 55° they are mostly positive with an average excess value of about 12 seconds.

P-wave travel time curves were used by Kaila et al. (1968a) to find the upper mantle velocity structure which in the north is found to consist of three velocity discontinuities at depths (below the crust) of 380 ± 20, 580 ± 50 and 1000 ± 120 km with true velocities below the discontinuities as 9.47 ± 0.06, 10.15 ± 0.07 and 11.40 ± 0.08 km/sec respectively. S-wave travel times curves are also found to be quite consistent with this upper mantle structure, determined from P-waves.

Upper Mantle Velocity Structure in the Hindukush Region. Upper mantle velocity structure in the Hindukush region has been determined to a depth of 310 km from the analysis of P- and S-wave travel times of 51 deep earthquakes using Kaila's (1969) analytical method. The velocity function for P-waves determined from the present study (Figure 4) reveals a velocity of 8.24 km/sec at a 55 km depth which increases linearly, with a gradient of 0.18 km/sec per 100 km, to 8.54 km/sec at a depth of about 220 km. At this depth of 220 km, there is a decrease in the P-velocity gradient to 0.08 km/sec per 100 km, but the velocity still increases linearly and attains a value of 8.62 km/sec at a depth of 310 km. This decrease in the P-wave velocity gradient at a depth of 220 km in the Hindukush region is interpreted as a second order low velocity channel in the upper mantle. For S-waves, however, such a decrease in the velocity gradient could not be detected in the Hindukish region, which may be within the accuracy of velocity determination. The S-velocity from 70 to 280 km depth is found to increase linearly from 4.61 to 4.77 km/sec with a gradient of 0.08 km/sec per 100 km. Upper mantle velocities for P- and S-waves in the Hindukush region are found to be considerably higher in comparison to those for the world average and various other regions of the earth. Higher upper mantle velocities were also obtained by Lukk and Nersesov (1965) and Kaila et al. (1968a, b & c) for the same area. Julian and Anderson (1968) had also made similar remarks while discussing Lukk and Nersesov's (1965) velocity model of the Hindukush region. According to them the first arrival travel times in this area are not consistent with those for the other models, suggesting that the earth may be significantly different in Asia than in Europe and North America.

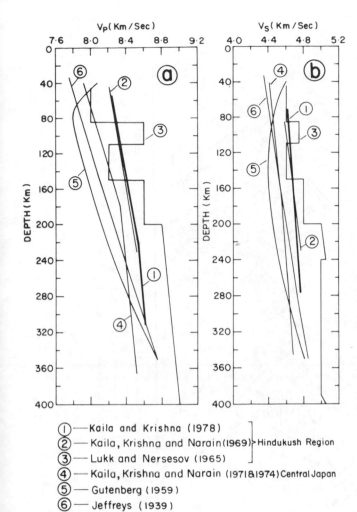

① —— Kaila and Krishna (1978)
② —— Kaila, Krishna and Narain (1969) ⎤ Hindukush Region
③ —— Lukk and Nersesov (1965)
④ —— Kaila, Krishna and Narain (1971&1974) Central Japan
⑤ —— Gutenberg (1959)
⑥ —— Jeffreys (1939)

Fig. 4. P- and S-wave velocity functions for the Hindukush region as determined from travel time data of deep earthquakes using Kaila's (1969) method. Velocity functions for other regions of the earth are also included for comparison.

Seismotectonics Of The Himalayan Belt. The seismicity of the Himalayan region in relation to its regional tectonics has been studied in detail on the basis of A values in the earthquake regression curve (log N = A - bM) using 0.5° x 0.5° grid averages (Kaila and Madhava Rao, 1977, 1979). This seismicity map (Figure 5) brings out a number of high seismic activity zones and a brief description of the same is given below.

The Srinagar high seismic activity zone extending northwest-southeast in the region of Punjab Himalaya, consists of two localized highs, one in the Anantnag-Kishtwar region (A = 5) southeast of Srinagar and the other in Muzaffarabad region (A = 5), northwest of Srinagar. The Anantnag-Kishtwar localized high falls within the Tethys Himalayan sub-province in the Anantnag region

and in the central crystalline subprovince in the Kishtwar region. The Jammu-Peshawar seismic high (A = 4) lies in a northwest-southeast direction almost parallel to the Srinagar high and extends south towards the Lahore-Sargodha ridge. In the Kumaon and western Nepal Himalayan region, the Kedarnath-Askot high seismic activity zone is depicted by two localized highs, one attaining a maximum A value of 5 east of Kedarnath and the other showing the highest A value of 6 east of Askot. The Kedarnath-Askot high runs in a northwest-southeast direction with its NNW extension towards the Spiti synclinorium parallel to the Zanskar thrust and its SW extension towards the Delhi-Hardwar ridge and the Moradabad fault area.

In the central Nepal Himalaya, the NW-SE trending Pokhara high seismicity zone is depicted by a maximum A value of 4 which falls within the crystalline sub-province of the Great Himalaya. The E-W Khatmandu-Everest seismic high in the Nepal Himalayan region shows the highest seismic activity depicted by an A value of 5 in the Khatmandu region over the crystalline sub-province. The NE-SW trending Taplejung-Kangchenjunga high at the border of eastern Nepal and the Sikkim Himalaya is characterized by an A value of 5 in the region of Taplejung. In the Bhutan-Assam region, the NW-SE trending Timphu-Dhubri high consists of three localized highs, one in the region of Timphu (A = 4), the second in the region of Dhubri (A - 4) and the third in the western part of Shillong plateau (A = 5) in the region of the Surma valley. The Tawang-Kangdu seismic high which lies in the crystalline sub-provinces of Bhutan and NEFA Himalaya indicates the highest A value of 5. A very active seismic zone which is associated with the Abor and Mishimi hills runs approximately in an east-west direction indicating the highest A value of 5.

Although high seismicity zones northwest of Everest are aligned parallel to the Himalaya, the seismic highs to the east are aligned almost transversely to the Himalayan structural trend, e.g., the Kangchenjunga-Taplejung high, Timphu-Dhubri and Tawang-Kangdu high. It may be interesting to investigate here whether these transversely aligned high seismicity zones have any relationship with the transverse structures in the Himalaya or adjoining Ganga basin. Valdiya (1973) states that transverse folds and faults, some of great dimensions, form a remarkable feature of the Himalaya and are seen throughout the length of the great mountain ranges, especially concentrated in eastern Nepal and adjoining Sikkim. The Arun anticline of eastern Nepal, and the Kangchenjunga-Dharanbazar syncline are examples of such structures which respectively fall in the eastern part of the NE-SW Khatmandu-Everest high and in the Kangchenjunga-Taplejung high which again has a NE-SW trend. It is quite possible that these two seismic highs and transverse structural folds in this region of the Himalaya are related to each other. Further, the Patna fault in the Gangetic plain may be related to the NE-SW trending east-

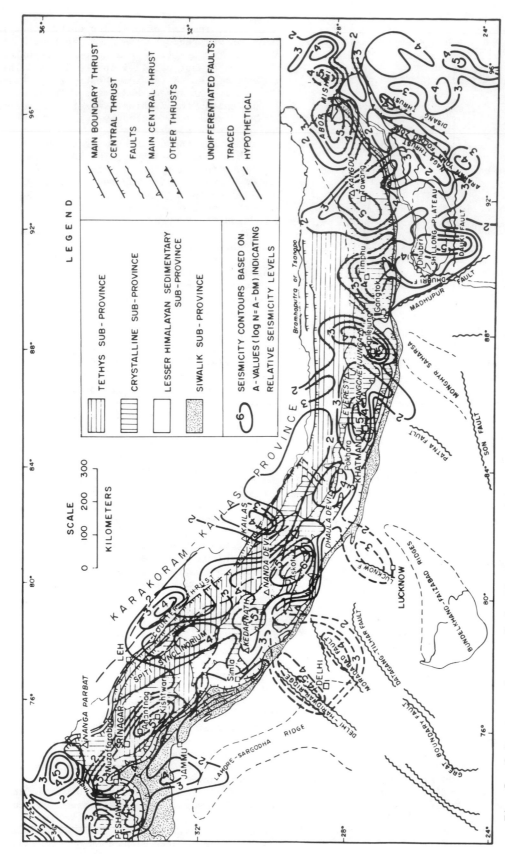

Fig. 5. Seismotectonic map of the Himalayan region based on A-value (log N = A – bM) from shallow earthquake data for the period from 1954 through 1967. Contours represent normalized A corresponding to 0.5 x0.5° grid averages as at the equator and for 14 years earthquake observation period. The tectonic features in the Himalaya and the Ganga basin are after Valdiya (1973). The undifferentiated faults shown are from the tectonic map of Eurasia (Yanshin, 1966).

ern part of the Khatmandu-Everest high. Similarly, the NW-SE trending Timphu-Dhubri seismic high may be related to the Madhupur fault which may be an active fault at the present time. According to Morgan and McIntyre (1959) the Madhupur fault is a tear fault with the eastern segment moving southwards. Abdel Gawad (1971) has inferred the existence of a 200 km long fault trending in a NW-SE direction between Kangdu and Takpashiri in Kemeng and adjoining Bhutan which falls in the region of the NW-SE Tawang-Kangdu seismic high zone and the two may be related to each other. He has also reported a large tear fault in the region east of Everest.

Further, the Kedarnath-Askot high (Figure 5) takes a southwestern diversion towards planes enclosing Delhi-Hardwar ridge and the Moradabad fault area which are transversely aligned to the Himalayan trend and both are seismically active at present. The southwestern extension of the Pokhara seismic high towards the SW-NE trending Lucknow fault is also an indication of its present day activity.

Seismotectonics of the Pamir-Hindukush Region.
The seismotectonic map of the Pamir-Hindukush region (Figure 6) by Kaila and Madhava Rao (1979) is based on A values determined for half degree grid averages using shallow earthquake data from 1954 through 1975 and the method of Kaila and Narain (1971). The brief description of the high seismicity zones in this region are given below.

In the Garm-Kulyab zone, the highest seismic activity is depicted by an A value of 9 (b = 1.8) right over the South Tienshan (Vakhsh) fault in the Garm and Surkhab valley area. The highest seismic activity of the central Pamir high associated with the Pamir mountain ranges is characterized by the highest A value of 7 (b = 1.4), northeast of Khorog. Towards the west of the Tarim basin, the Sinkiang region of China is characterized by a high level of A equal to 8.5 (b = 1.7) north of Yarkand. In western China, the northwest southeast trending Kungur Muztagh Ata high, associated with Mustagh Ata ranges, depicts a high A value of 6 (b = 1.3). Towards the east of the Karakorum anticlinorium, the east-west trending Aghil-Lokzung seismic high, associated with Aghil mountain ranges, indicates an A value of 5 (b = 1.1). In the Badakshan region of Afghanistan, the Faizabad-Zebak high is characterized by a high seismicity level of A equal to 7 (b = 1.4) southeast of Faizabad. The Baghlan-Khanabad high in the region of Kataghan is shown by a high A value of 7 (b = 1.4). The NE-SW trending Kabul-Jalalabad high is shown by two localized highs of A equal to 6 (b = 1.3), east of Kabul. One branch of the Kabul-Jalalabad high extends south where it attains a value of 5 in Waziristan. Another branch extends east and southeast showing there three localized highs one of which attains a value of A = 5 (b = 1.2) north of the Peshawar depression, the second shows a value of A = 6 (b = 1.3) west of Rawalpindi and

the third an A value of 4 north of Lahore. There are three localized high seismicity zones in the Srinagar high; one in the Muzaffarabad region (A = 6, b = 1.3), one west of Gilgit (A = 6, b = 1.3), one southeast of Srinagar (A = 5.5. b = 1.2).

Deep Tectonic Features of the Pamir Hindukush Region. Deep tectonic features in the Pamir-Hindukush region are delineated (Kaila and Madhava Rao, 1978, 1979) based on earthquake hypocentral data for the period from 1961 through 1975 along two profiles I and II of average width 200 km and 300 km respectively aligned transversely to the main geologic trend of the area (Figure 7). The earthquake hypocentral distribution versus depth has been studied, distances being measured from a north-east south-west trending reference fault in the Badakshan region, along the profile I. A similar study was made for profile II, taking the reference fault in this case as Wanch Akbayatal fault which is located between the north and the central Pamir zones. Assuming an average error of ±10 km in the location of epicentres and ±20 km in the determination of focal depth, the whole vertical section along the two profiles, is divided into 20 km wide x 40 km deep blocks. The cumulative number of earthquakes in each 20 x 40 km block is plotted at the centre of the block with reference to the horizontal base lines for the two profiles. For better averaging, 10 km overlap (50% overlap) was used in the horizontal direction. The frequency curves are thus drawn for every 40 km depth interval, showing the number of earthquakes occurring along the profile. The peaks of the earthquake frequency curves are shown by large dots along the base lines for every 40 km depth interval. Average curves without crossing any troughs are then drawn through these dots which are connected in the direction of the depth indicating the plausible location of deep faults along profile I and II (Figure 8 and 9). These faults have been numbered by Arabic numerals. The location, dip and the distribution of these faults as determined by the above technique from the earthquake hypocentral data show very good correspondence with the surface geological faults and their dip section as given by Desio (1964).

Along profile I that runs transverse to the Hindukush mountain trend, two systems of steep deep faults are delineated, one system dipping towards northwest and the other in the southeast direction. Between these two systems of faults with opposing dips, there is a zone lying below the eastern Hindukush which is characterized by two approximately vertical fractures, one extending from a depth of 100 to 300 km and the other extending from a depth of 100 to 260 km. Movements along these faults are considered responsible for intermediate depth earthquakes in the Hindukush region. Similarly, profile II (Figure 9) that runs transverse to the Karakorum and Pamir mountain trend also reveals two systems of

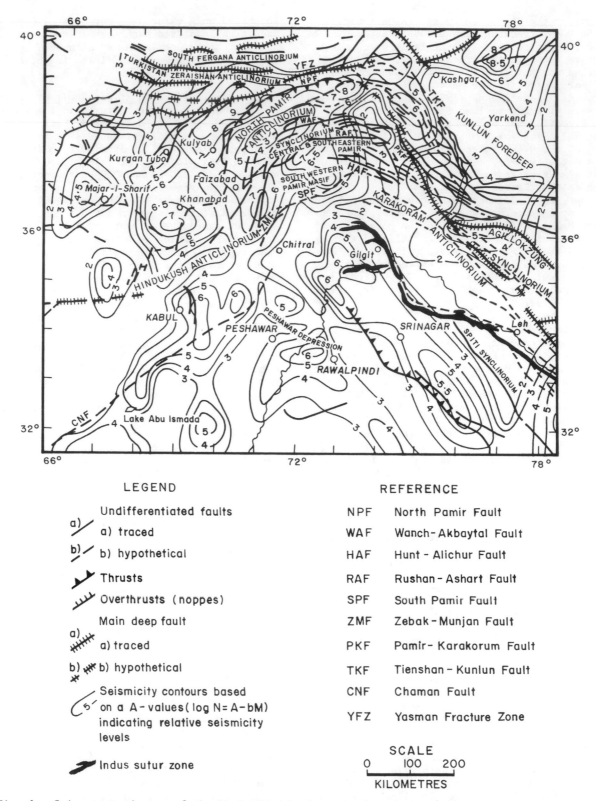

Fig. 6. Seismotectonic map of the Pamir-Hindukush region based on A-values as in Figure 5 but the earthquake data used is for a period of 22 years from 1954-75. The tectonic features are from the tectonic map of Eurasia (Yanshin, 1966).

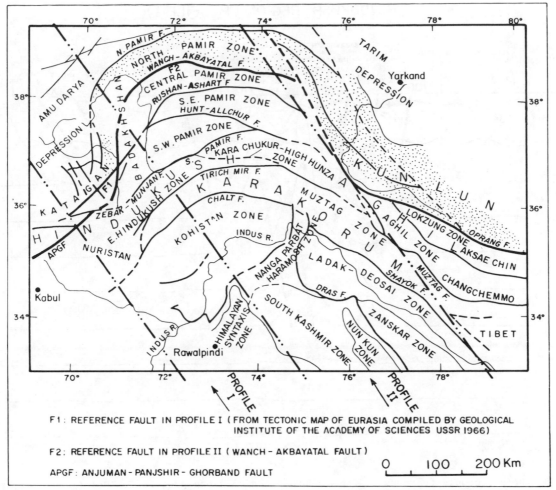

Fig. 7. Structural zones of Central Asia (after Desio, 1964) showing two profiles I and II with average widths 200 and 300 km respectively along with deep tectonic features have been delineated using earthquake hypocentral data.

steep deep faults dipping NNW and SSE respective-ly. Due to upward block movements along these two systems of oppositely dipping faults, a zone of tension develops at great depth where frac-tures appear and intermediate depth earthquakes originate through relative movement along these fractures. These deep fractures extend from an average depth of about 80 km to 280 km. It is concluded that intermediate depth earthquakes in the Hindukush and south Pamir regions are caused by relative movement along tension fractures de-veloped at great depth. As deep fractures are formed by development of tension in the region due to upward movement of earth blocks along the two systems of oppositely dipping faults, there should be a good correlation between the occur-rence of shallow and intermediate depth earth-quakes in the Hindukush and Pamir regions as re-vealed by the cross-correlation analysis carried out by Kaila and Madhava Rao (1978) who concluded

that the Pamir-Hindukush intermediate depth earth-quake zone is under tension.

Evolution of the Himalaya. From the seismicity map of Asia (Figure 10) and the detailed seismo-tectonic map of the Himalaya (Figure 5), it is concluded that there are two types of movements taking place at present in the Himalaya, one parallel to the Himalaya along the Main Central Thrust in the Kumaun and Punjab Himalaya and the second transverse to the Himalaya along trans-versely aligned faults and fracture zones which extend far north into the Tibet region. These two types of movements along two perpendicular directions in the Himalaya lead me to believe that at present the Himalaya do not form the northern plate boundary of the Indian plate and as such the Indian block is not subducting under the Himalaya in the region of Indus Suture zone. Further, if one examines the seismic plane in the

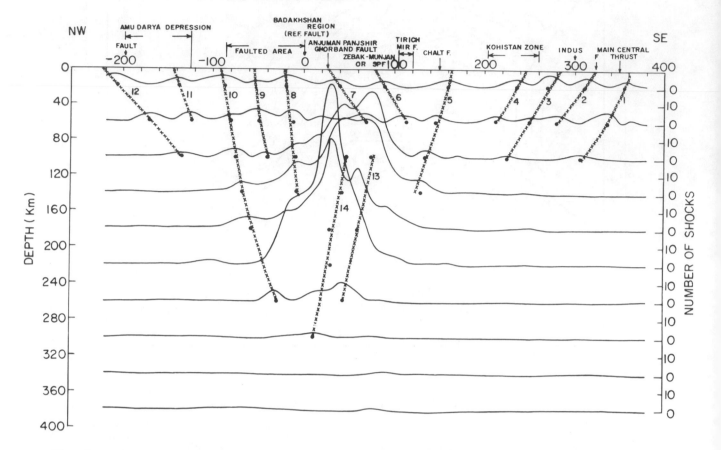

HINDUKUSH REGION (PROFILE I)

Fig. 8. Deep tectonic features of the Hindukush region along profile I. Continuous thin lines show the number of earthquakes, the positions of maxima on it are shown by thick dots along the respective base lines. The aligned crosses indicate the location of faults and fractures which are numbered by Arabic numerals.

Himalaya as delineated by plotting of earthquake epicentres with depth (Kaila and Narain, 1976), this plane is found to be dipping towards north between 30° to 60° (Figures 11 and 12). This may mean that the Main Central Thrust in the Himalaya, which I think is mainly responsible for the seismic activity in the Kumaon and Punjab Himalaya, has a dip between 30° to 60°.

Now sufficient field evidence has also been collected by Mehdi et al. (1972), who believe that the Main Central Thrust and Dar Martoli fault in the eastern Kumaun Himalaya are large angle faults which dip north at angles of 45° to 70°. These large angle faults cannot explain 300 km underthrusting of the Indian landmass under Tibet as proposed by Gansser and underthrusting of the Indian shield beneath Tibet as far as Altyn Tag as proposed by Holmes (1965) is just out of the question. Holmes had proposed such a hypothesis just to explain the almost double thickness of the crust under the Himalayan-Tibet region. If

we examine the deep crustal section (Figure 3) from deep seismic sounding studies in the Himalaya, it is quite obvious that most of reflectors including Moho boundary, dip north at about 17 degrees. Therefore, even if one starts with a small crustal thickness in the sub-Himalaya, this dip of the various reflectors under the lower and the Higher Himalaya will take Moho to a considerable depth, reaching almost 60 km under the Higher Himalaya and its depth may increase still further in the Tibet region. The above-mentioned arguments lead me to believe that the Himalaya are formed not as a result of continent to continent collision by underthrusting of the Indian landmass beneath Tibet as proposed by many workers (Holmes, 1965; Santo, 1972; Molnar and Tapponnier, 1975), but that the Himalaya were formed by almost vertical block movements along zones of weakness which developed parallel to the Himalaya during Neogene-Quaternary time due to large crustal thickness and sedimentary overloading in the

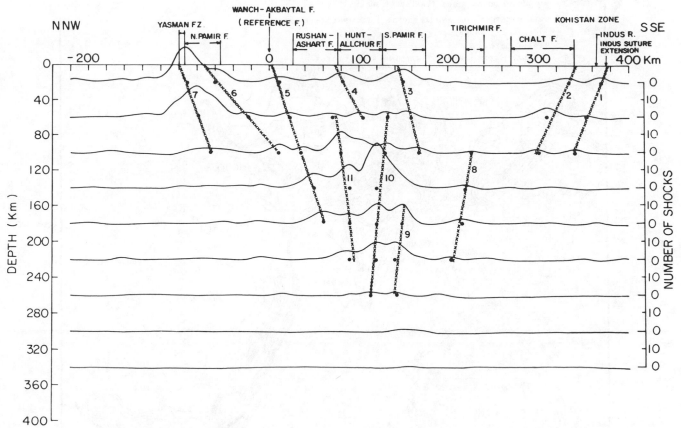

Fig. 9. Deep tectonic features of the Pamir region along profile II. Continuous thin lines show the number of earthquakes, the positions of maxima on it are shown by thick dots along the respective base lines. The aligned crosses indicate the location of faults and fractures which are numbered by Arabic numerals.

Trans-Himalayan Tibet block. Therefore, the large crustal thickness in the Tibet region is not a result of underthrusting of the Indian shield beneath Tibet but is the cause of the development of fracture zones parallel to the Himalaya along which the Himalayan block moved upwards sliding along steep faults. This idea is substantiated very well by recognition from DSS studies of $15°$ to $20°$ dip of the Moho boundary and other shallow reflectors under the middle and Higher Himalaya as if the accumulation of sediments in the Trans-Himalayan Tibet region caused a down-buckling of the Moho in that region. Gansser (1964) stated that the main elevation of the Himalaya was an event witnessed by the earliest man. He related the final rise of the Tibet plateau with the youngest, present-day, morphogenic rise of the Himalaya and its counterpart, the sinking of the Indian foreland. The Tibetan part of the Indo-Tibet plate is known to have started to rise before the main Himalayan orogeny, for the marine Eocene of Tibet passes up into continental deposits.

Petrushevsky (1971) also holds the view that prior to the Neogene, most of the Himalaya was a part of the epi-Proterozoic Indian platform which was reworked only during the epoch of Neogene-Quaternary activation of tectonic movements which gave rise to the formation of the Himalaya. Similarly Crawford (1974) opines that originally Tibet was a part of a plate including India, but submerged while India remained continental. Associated with the Indus Suture line, shallow sub-parallel fractures or rifts developed within which the Gondwana sediments of the Himalaya were preserved. Late in Phanerozoic time, according to Crawford, vigorous sea floor spreading in the NW Indian ocean led to the conditions in which the whole Indo-Tibetan plate closed up on the rest of Asia. Since the plate could not move further north, overthrusting developed along zones of weakness and thus produced the Himalaya.

The Newly Postulated Indian Plate Boundary. Kaila and Narain (1971) prepared a quantitative seismicity map of the Alpine-Himalayan belt in-

STRUCTURE AND SEISMOTECTONICS   285

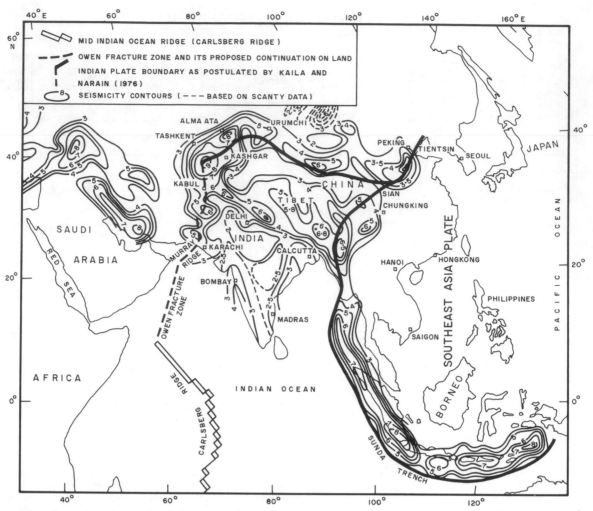

Fig. 10. Quantitative seismicity map of Asia and Far East based on A-values normalized to $2^\circ \times 2^\circ$ grid area as at equator and 14 years earthquake data set (after Kaila and Narain, 1971). The newly postulated Indian plate boundary is also shown.

cluding the Sunda arc region. If one scrutinizes the seismicity pattern in the Asia region (Figure 10), it is quite obvious that the highest seismicity lies along the southern side of the Tien Shan fold belt. Then it swings southeastward along the southern margin of the Tsaidam block and the Nan Shan fold belt. From Sian it veers northeastward into Tientsin southeast of Peking. This is the highest seismicity zone in Asia and is well demarcated. At its western end it swings southward along the western margin of the Badakshan mountains and joins the high seismicity zone in the region of the Sulaiman and Kirthar ranges. On the eastern side, the Tien Shan-Nan Shan high seismicity zone swings southwestward from Sian to join the Arakan Yoma high seismicity zone, finally passing southwards and joining with the Sunda island arc high seismicity zone.

Besides the above zone, the highest seismicity zone in Asia, there is another well demarcated high seismicity zone oriented NW-SE almost parall-

el to the Main Central Thrust in the Kumaun-Punjab Himalaya. However, in the Nepal-Bhutan-NEFA Himalaya the zones of high seismicity are aligned transversely to the Main Central Thrust. This transversely aligned seismicity in the Himalaya is also quite clear in the detailed seismotectonic map of Himalaya (Figure 5). The seismicity in the eastern Himalaya is most probably controlled by the transverse faults and fracture zones which may extend far into the Tibet region. Thus one can conclude that the seismicity in the Himalayan belt is not only smaller in magnitude as compared to Tien Shan-Nan Shan seismicity belt but it is also totally associated with the Main Central Thrust and other thrusts of the Himalayan mountain system On the other hand, seismicity in the eastern Himalaya depends to a great extent on the transversely aligned faults and fracture zones which may extend far into the Tibetan region.

According to Crawford (1974), along the northern edge of the Tarim Basin, a major tectonic

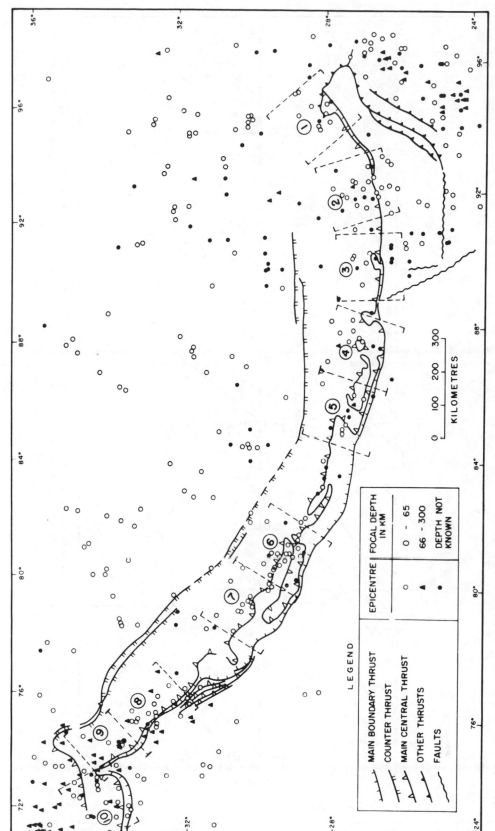

Fig. 11. The earthquake epicentral distribution map of the Himalayan region for the period from January 1954 through June 1972. The Arabic numbers in the closed circles represent profile numbers, taken transverse to the Main Central Himalayan Thrust, with width of about 200 km each demarcated by dashed lines.

LEGEND

MAIN BOUNDARY THRUST

COUNTER THRUST

MAIN CENTRAL THRUST

OTHER THRUSTS

FAULTS

EPICENTRE	FOCAL DEPTH IN KM
○	0 - 65
▲	66 - 300
●	DEPTH NOT KNOWN

KILOMETRES

0   100   200   300

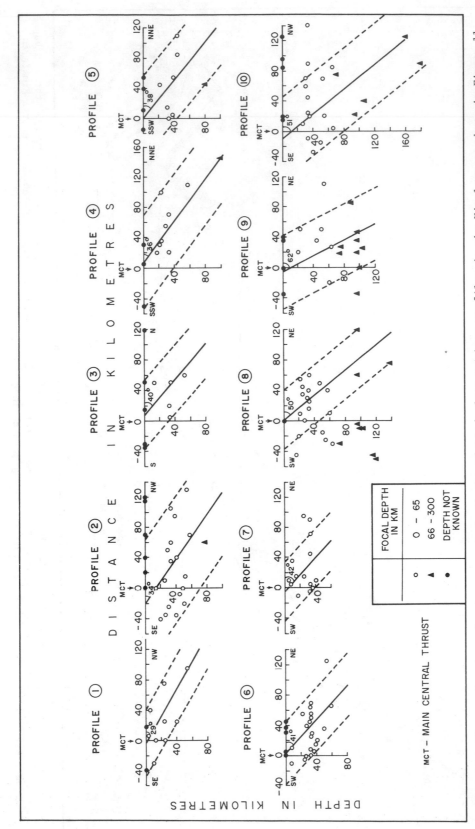

Fig. 12. The earthquake epicenters versus depth are plotted along various profiles in the Himalaya as shown in Figure 11, the horizontal distance is measured from the Main Himalayan Central Thrust. The outer dashed lines demarcate the seismic zone and the median continuous lines drawn are supposed to represent the seismic plane.

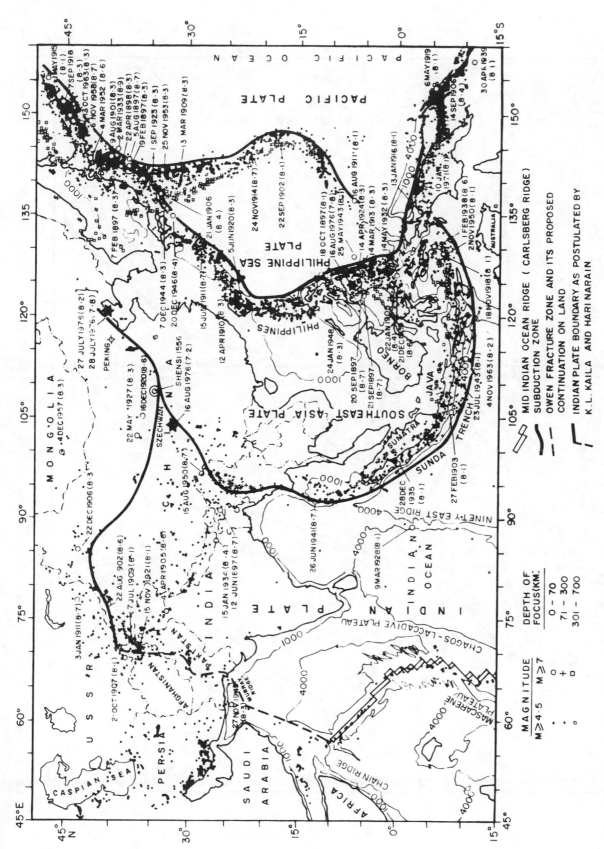

Fig. 13. Map showing the relationship between postulated new Indian plate boundary and the zones of most intense seismicity. The epicenters shown are for the period from July 1, 1963 upto December 31, 1972 taken from the World Seismicity Map prepared by United States Geological Survey (1974). Large open circles with data and year indicate major earthquakes with their magnitudes in brackets.

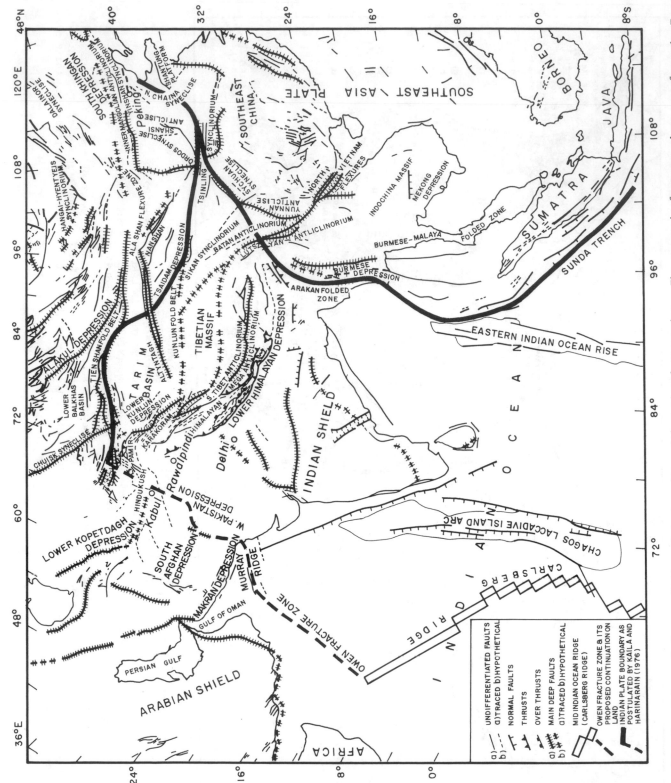

Fig. 14. Comparison of the new Indian plate boundary with regional tectonics. The tectonic features shown are from the Tectonic Map of Eurasia (Yanshin, 1966).

boundary separates the block from the southern-most ranges of the Tien Shan, which are thrust over it. Over 3000 km long, the Tien Shan include parallel mountain systems separated by long narrow depressions. In the east, these even extend below sea level in the Turfan-Hami basins of Sinkiang, yet immediately north of these troughs, the easternmost Tien Shan rise to over 5400 m. In the Soviet Tien Shan one depression contains the extraordinary lake known as Issyk Kul, 702 m deep, to the north and south of which the Tien Shan rise to 5000 m. Crawford thinks that all this indicates great and continual mobility along a relatively narrow belt now marked by the Tien Shan mountains and intra-montane depressions, a mobility accompanied by repeated volcanism in much of the Paleozoic. From its extreme seismicity, as is evident from Figure 10, the Tien Shan clearly constitutes a mobile belt which has been active since Paleozoic time and is therefore considered to form the northern boundary of the Indian plate and Issyk Kul and the Turfan Hami depressions may represent the remnants of the associated trench system.

The Tien Shan-Nan Shan seismic belt, a very well demarcated zone, shows the highest seismic activity in Asia with A value reaching 9 to 10, comparable to the seismicity in the Indonesian island arcs. This zone has been marked by a thick line on the map (Figure 13) showing the distribution of epicentres of all depth ranges from 1963 through 1972; this zone is postulated, as explained above, as the northern boundary of the Indian plate. In this figure, the Indian plate is considered to be subducting in the region of Tien Shan-Nan Shan high seismicity belt instead of in the Himalayan zone and the whole of India, Tibet, Kun Lun, and the Tarim Basin are considered as one continental block drifting northeastwards from the Carlsberg ridge (mid-Indian Ocean ridge). Two types of simultaneous movements of the above Indo-Tibetan landmass are postulated, one in a northeastern direction towards the Tien Shan-Nan Shan zone of subduction, the movement being along the Owen Fracture zone, Murray ridge, Kirthar and Sulaiman fault zones and the fracture zone on the western margin of Badakshan mountains. The other movement of the Indian plate including the Indo-Tibetan landmass is towards the east, subducting under the southeast China and Burma on land, and likewise along the Sunda trench in the sea. It is quite possible that the Tien Shan-Nan Shan seismicity zone not only acts as a plate boundary where the above modified Indian plate is subducting but there may also be eastwards movement along it. If such movement is possible then the Indian plate, as delineated above, need not penetrate too deep under either the Tien-Shan-Nan Shan subduction zone or under the southeastern China block and this may explain the absence of intermediate and deep earthquakes along the above postulated Indian plate boundary over Asia.

Large scale horizontal eastward movement of the

Tibetan block is also indicated by Molnar and Tapponnier (1975) along the Kun Lun strike-slip fault. They have analyzed in detail continental deformation in Asia and recognized several major left-lateral strike-slip faults, trending roughly east-west. The long linear valleys and adjacent ridges characteristic of active strike slip faulting were clearly well defined features as observed by them on ERTS photographs of this region. Fault plane solutions of earthquakes, surface faulting and associated en-echelon compressive features imply that the sense of all these faults is left lateral. The three most prominent of these faults recognized by them are Kang Ting fault (west of the Lung Men Shan thrusts), the Kun Lun fault (south of the Tsaidam basin, within the eastern Kun Lun mountains) and the Altyn Tagh fault separating Tibet from the Tarim basin and passing south of the Altyn Tagh (also called Astin Tagh or A-Erh-Chin).

The Indian plate boundary as postulated by Kaila and Narain (1976) is also shown along with the main structural lines in Asia (Figure 14) where it is seen to follow the main deep fracture in the Tien Shan mountain region. Then it cuts across the Altyn Tagh fault and passing south of the Tsaidam basin joins the Kun Lun fault zone and the Tsinling fold belt from where it takes a turn towards northeast along the trend of the faults reaching Tientsin southeast of Peking. From Sian it takes a southwestern bend along the Lung Men Shan thrust zone, cuts across the Kang Ting fault and joins the fracture zone in the Arakan Yoma fold belt, finally joining with the Indonesian trench system along the Sunda island arc. On the western side, the Tien Shan fracture zone (northern peripheral fracture) dividing the Pamir-Alay tectonic zone from the northern Pamir tectonic zone takes a southward bend separating the upper Amu Darya depression (Tuayev, 1961) from the mountainous area of Badakshan. The same fracture separates the Cenozoic formations from the older ones. In central Badakshan, the line separating the southern Amu Darya depression from the Tertiary deposits and from the mountainous region composed mostly of older rocks, crosses the Kokcha River west of Kakan (Desio, 1964).

Therefore, central Badakshan is situated to the east of the extension of the 'Northern Peripheral Fracture' of Pamir. On the other hand, the fracture to the south of the southwestern Pamir tectonic zone meets the Amu Darya near Ishkashim, a few kilometers from Zebak, where it joins the Zebak-Munjan fault. The latter continues westward with a western inflexion trending parallel to the Hindukush ridge for many kilometers. This tectonic line forms the southeastern boundary of Badakshan, as well as of the Pamir. It is obvious that the Pamir as well as Badakshan are situated between the above mentioned peripheral fractures. According to Desio (1964), the N-S striking metamorphic formations of the Jarm valley, may cross the Hindukush obliquely between the passes of Munjan and Anjuman continuing to Nuristan, where

the tectonic directions are aligned southeast or
southwest. Alternatively they may be truncated
between the two passes mentioned, by reasons of
the westward extension of the Zebak-Munjan fault.
In this case, the Nuristan zone would be independent of the Badakshan zone. In the absence of any
detailed information, Desio could not decide which
is the correct alternative. However, following
the first alternative, the peripheral fracture is
considered to extend southward across the Hindukush
joining with the fault system in the Sulaiman and
Kirthar ranges. These in turn continue towards
and join with the Owen Fracture zone following
the Murray Ridge. Thus the new Indian plate
boundary as postulated above is quite consistent
with the tectonic trends in Asia.

References

Abdel Gawad, M., Wrench movements in the Baluchistan arc and relation to Himalayan-Indian
ocean tectonics, Bull. Geol. Soc. Am., 82,
1235-1250, 1971.

Argand, E., La tectonique de l'Asie, 13th Int.
Geol. Congr., Abstract, 1924.

Crawford, A. R., The Indus Suture line, the
Himalaya, Tibet and Gondwanaland, Geol. Mag.,
3, 369-480, 1974.

Desio, A., Tectonic relationship between the
Karakorum, Pamir and Hindukush (Central Asia),
Report of the Twenty Second Session of IGC,
India (Himalayan Alpine Orogeny), XI, 192-213,
1964.

Dewey, J. F., and J. M. Bird, Mountain belts and
the new global tectonics, J. Geophys. Res., 75,
2625-2647, 1970.

Fuchs, G. R., Zum Bau des Himalaya, Osterr. Akad.
Wiss. Math. Nat. Kl., Denk., 113, 1-211, 1967.

Fuchs, G. R., The significance of Hazara to Himalayan Geology, Vienna, Geol. Bundesanstalt
Jahrb., Sonderbd., 15, 21-23, 1970.

Fuchs, G. R. and W. Frank, The geology of west
Nepal between the rivers Kali Gandaki and Thulo
Bheri, Vienna, Geol. Bundesanstalt Jahrb. Son-
derbd., 18, 1-103, 1970.

Gansser, A., Ausseralpine Ophiolit Probleme, Ecl.
Geol. Helv., 52, 659-680, 1959.

Gansser, A., The geology of the Himalayas, Interscience Publishers, New York, 1964.

Gansser, A., The Indian Ocean and the Himalayas -
a geological interpretation, Ecl. Geol. Helv.,
59, 831-848, 1966.

Gutenberg, B., The asthenosphere low velocity layer, Ann. Geofis., Rome, 12, 439-460, 1959.

Holmes A., Principles of physical geology, 2nd
revised ed., Nelson, London, 1965.

Isacks, B., J. Oliver and L. R. Sykes, Seismology
and the new global tectonics, J. Geophys. Res.,
73, 5855-5900, 1968.

Jeffreys, H., The times of P, S, and SKS and the
velocities of P and S, Monthly Notices, Roy.
Astr. Soc., Geophys. Suppl., 4, 498-533, 1939.

Jeffreys, H., and K. E. Bullen, Seismological

Tables, British Assn., Gray Milne Trust,
London, 1940.

Julian, B. R., and D. L. Anderson, Travel times,
apparent velocities and amplitudes of body
waves, Bull. Seism. Soc. Am., 58, 339-366, 1968.

Kaila, K. L., A new analytical method for finding
the upper mantle velocity structure from P and
S-wave travel times of deep earthquakes, Bull.
Seism. Soc. Am., 59, 755-769, 1969.

Kaila, K. L., P. R. Reddy and Hari Narain, P-wave
travel times from shallow earthquakes recorded
in India and inferred upper mantle structure,
Bull. Seism. Soc. Am., 58, 1879-1897, 1968a.

Kaila, K. L., P. R. Reddy and Hari Narain, Crustal structure in the Himalayan foothills area of
north India from P-wave data of shallow earthquakes, Bull. Seism. Soc. Am., 58, 597-612,
1968b.

Kaila, K. L., P. R. Reddy and Hari Narain, S-wave
travel time curves in the northernly azimuth
from India in the distance range 55° determined
from shallow earthquake data, Bull. NGRI, 6,
167-188, 1968c.

Kaila, K. L., V. G. Krishna and Hari Narain, Upper
mantle velocity structure in the Hindukush region from travel time studies of deep earthquakes
using a new analytical method, Bull. Seism. Soc.
Am., 59, 1949-1967, 1969.

Kaila, K. L., and Hari Narain, Interpretation of
seismic refraction data and the solution of the
Hidden Layer Problem, Geophysics, 35, 613-623,
1970.

Kaila, K. L., and Hari Narain, A new approach for
preparation of quantitative seismicity maps as
applied to Alpide Belt-Sunda Arc and adjoining
area, Bull. Seism. Soc. Am., 61, 1275-1291, 1971.

Kaila, K. L., V. K. Gaur and Hari Narain, Quantitative seismicity maps of India, Bull. Seism.
Soc. Am., 62, 1119-1132, 1972.

Kaila, K. L., V. G. Krishna and Hari Narain, Upper
mantle shear wave velocity structure in the Japan region, Bull. Seism. Soc. Am., 64, 355-374,
1974a.

Kaila, K. L., N. Madhava Rao and Hari Narain,
Seismotectonic maps of southwest Asia region
comprising eastern Turkey, Caucasus, Persian
Plateau, Afghanistan and Hindukush, Bull. Seism.
Soc. Am., 64, 657-669, 1974b.

Kaila, K. L., and Hari Narain, Evolution of the
Himalaya based on seismotectonics and deep
seismic soundings, Himalayan Geology Seminar,
(Section II: Structure, Tectonics, Sesimicity
and Evolution), New Delhi, India, 1-30, Sept.
13-17, 1976.

Kaila, K. L., and N. Madhava Rao, Detailed quantitative seismicity maps of the Himalayan Belt and
adjoining areas, Sixth World Conference on Earthquake Engineering, Section: Ground Motion, Seismicity, Seismic Risk and Zoning, 2, 514, 1977.

Kaila, K. L., and N. Madhava Rao, Deep tectonic
features of the Pamir-Hindukush region, Paper
presented in the Symposium on 'The lithosphere-
asthenosphere interaction, its role in tectonic
processes', organized by Inter-Union Commission

on Geodynamics (IUCG), Leningrad, USSR, Oct. 2-11, 1978.

Kaila, K. L., N. Madhava Rao and Hari Narain, Seismotectonics of the Pamir-Hindukush region, Paper presented in the Symposium on 'The lithosphere-asthenosphere interaction, its role in tectonic processes', organized by Inter-Union Commission on Geodynamics (IUCG), Leningrad, USSR, Oct. 2-11, 1978.

Kaila, K. L., and N. Madhava Rao, Seismotectonics of Himalayan Belt and the deep tectonic features of the Pamir Hindukush region, Geophys. Res. Bull., 17, 319-327, 1979.

Kaila, K. L., and V. G. Krishna, Upper mantle velocity structure in the Hindukush region, Monograph on 'The International Pamir-Himalayan Project', 1980.

Lukk, A. A., and L. Nersesov, Structure of the upper mantle as shown by observations of earthquakes of intermediate focal depth, Doklady Akad. Nauk SSSR, 162, 14-16, 1965.

Mehdi, S. H., G. Kumar and G. Prakash, Tectonic evolution of eastern Kumaun Himalaya: A new approach, Himalayan Geology, 2, 481-501, 1972.

Meyerhoff, A. A., and H. A. Meyerhoff, The new global tectonics: Major inconsistencies, Am. Assoc. of Pet. Geol. Bull., 56, 269-336, 1972.

Molnar, P., and P. Tapponnier, Cenozoic tectonics of Asia: Effects of a continental collision, Science, 189, 419-426, 1975.

Morgan, J. P., and W. G. McIntyre, Quaternary geology of the Bengal basin, Geol. Soc. Am. Bull., 70, 319-342, 1959.

Petrushevsky, B. A., On the problem of the horizontal heterogeneity of the earth's crust and upper most mantle in southern Eurasia, Tectonophysics, 11, 29-60, 1971.

Santo, T., On shallow earthquakes widely scattered in the continents, Pure & Appl. Geophys., 96, 94-105, 1972.

Sharma, K. K., K. R. Gupta and S. C. D. Sah, First record of upper Gondwana flora, north of Indus Suture Zone, Ladakh and its tectonic significance, paper presented at the Himalayan Geology Seminar, Dehradun, India, Nov. 6-8, 1979.

Tuayev, N. P., Main boundaries and geologic structure of the upper Amu Darya depression (English translation of A. G. U., Washington), Izvestiya Akad. Nauk SSSR., Ser. Geol., 5, 66-75, 1961.

USGS, World Seismicity Map, 1974.

Valdiya, K. S., Tectonic framework of India - A review and interpretation of recent structural and tectonic studies, Geophys. Res. Bull., 11, 79-114, 1973.

Wadia, D. N., The syntaxis of the northwest Himalaya: its rocks, tectonics and orogeny, Rec. Geol. Surv. India, 65, 189-220, 1931.

Wadia, D. N., The Himalayan geosyncline, Proc. Nat. Inst. Sci. India, 32A, 523-531, 1966.

Yanshin, A. L., Chief Editor, Tectonic Map of Eurasia, Geol. Institute, USSR, Academy of Sciences, Moscow, 1966.

A REVIEW OF THE LONG PERIOD SURFACE WAVES STUDIES IN THE HIMALAYA AND NEARBY REGIONS

Harsh K. Gupta and S.C. Bhatia

National Geophysical Research Institute, Hyderabad - 500007, India

Abstract. Gupta and Narain (1967) were the
first to estimate crustal thickness in the
Himalaya and Tibet Plateau region using surface
wave dispersion. Their finding of an enormously
thick crust (65 to 70 km) characterised by low
shear wave velocities has been confirmed by
recent investigations using sophisticated frequency time analysis (FTAN) technique. Rayleigh
wave attenuation studies suggest that the lower-
mostpart of the crust beneath the Tibet Plateau
is partially molten and uplift has been caused by
horizontal compression. Lack of mechanical
strength at shallow depths is probably responsible for the uniformity of Tibetan elevations
over wide areas in the center of an orogenic
zone. Shield-like upper mantle velocity structure has been inferred to exist below the Indogangetic Plains from a study of mantle Rayleigh
and Love wave phase and group velocities. Dispersion for interstation paths between Helwan
(HLW), Egypt; Shiraz (SHI), Iran; and New Delhi
(NDI), India is also characterised by shield
like behaviour and the experimental dispersion
curves have better concordance with theoretical
models featuring a shear wave low velocity zone
in the upper mantle. The upper mantle beneath
the Tibet Plateau also appears to have a normal
shield like structure. Installation of Seismic
Research Observatories (SRO), with instruments
having much longer dynamic range, has now made
it possible to investigate long period surface
waves from large earthquakes at short distances.
A typical example of SRO recording and its
analysis is presented.

Introduction

The origin of the Himalaya and the tectonics of
this region and the neighbouring south and central
Asia are believed to be due to continent to
continent collision of the Eurasian and the
Indian Plates (for example Dewey and Bird, 1970).
There exist certain difference of opinion about
the nature of the collision and the processes
involved. There are numerous papers relating the
observed regional surface tectonics with continental collision. Molnar and Tapponnier (1975)
in line with McKenzie (1972), believe that the
transmission of stress could be due to buoyancy
of continental crust, especially if it is hot, as
beneath Tibet. Molnar and Tapponnier (1977) have
explained the deformation in north-east China and
stability of south-east China as a consequence of
the India-Eurasia collision. Chandra (1979),
while considering the large scale Cenozoic
tectonics of central and south central Asia to be
a product of continental collision, has introduced
the concept of a rigid boundary zone between the
Eurasian and Indian continent. This rigid boundary zone has lbeen identified with the aseismic
lineaments in the Himalayan region. Chandra
(1979) observed that the high stresses generated
by the continental collision may be very extensive spatially and the entire Indian Peninsula
may be under a state of left lateral shear along
NNE – vertical planes. Bird (1978) has studied
the intracontinental sub-duction in Himalaya
from the angle of topographic stress, crustal
strength and the Himalayan metamorphism. It is
obvious that the deeper continental crustal
layers and upper mantle have been involved
actively in the origin of the entire Himalayan
belt. The information about the presently
existing crustal and mantle structure in the
region is therefore very relevant and important
in gaining a better understanding of the
tectonics of the Himalayan and neighbouring
regions.

Almost all the methods of exploration geophysics viz. gravity, magnetic, deep resistivity
sounding (to some extent), magnetotellurics and
seismic can be applied in investigating the
structure and physical properties of the crustal
and deeper portions of the earth with some limitations or constraints. The limitations are
either operational or interpretational which
prohibit deeper investigations. Seismic surface
waves provide a practical method of investigating
average physical properties of the crust and upper
mantle for a given geological province. There
are certain advantages of surface wave data acquisition and analysis compared to more conventional
seismic refraction and reflection surveys. Unlike
the refraction and reflection surveys the surface
wave investigation experiment is essentially
passive. The seismic reflection and refraction
methods generally do not provide information
regarding shear wave velocity distribution within
the earth, while the surface wave dispersion is
particularly sensitive to shear wave velocity

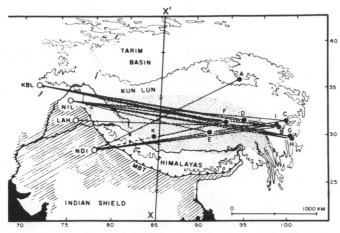

Fig. 1. Map showing the paths investigated by Chun and Yoshii (1977) in Tibetan Plateau region. Filled circles indicate epicenters and open circles indicate WWSSN recording stations. XX' line refers to the crustal section shown in Figure 4.

distribution. For remote, sparsely populated areas with extremely difficult or impossible logistics, such as the high Himalaya and the Tibet Plateau region, surface wave dispersion analysis provides one of the few practical methods to estimate average crustal and upper mantle velocity distribution.

Surface wave dispersion investigations received an impetus through the pioneering work of Ewing and Press (1954 a,b) and application of the Fourier transform to the analysis of earthquake generated surface waves by Sato (1958). In later years, numerous discoveries and improvements in surfacd wave analysis techniques have been made (for example Landisman 1977 a,b; Nyman et al, 1977). These newly developed surface wave analysis techniques have been successfully applied to long period surface waves to investigate crust and upper mantle across many paths in continents, e.g.: Bloch et al. (1969) for Southern Africa; Thomas (1969) for Australia; Sherburne (1975) for South America. In contrast to these investigations involving surface waves extending to long periods, till very recently, surface wave investigations in Eurasia were mostly confined to 60-70 sec periods(Nagmune, 1956; Shechkov, 1961; Tandon and Chaudhury, 1964; Santo, 1965; Gabriel and Kuo, 1966; Gupta and Narain, 1967; Gupta and Sato, 1968 and others). However, recently some studies extending to long periods have been carried out in Eurasia (Knopoff and Fouda, 1975; Gupta et al 1977 a,b and Chun and Yoshii, 1977).

Difficulties Encountered in Investigating Long Period Surface Waves

Gupta and Hamada (1975) have pointed out the difficulties encountered in investigating long period ($\geqslant$ 100 sec) surface waves at short

distance ($\simeq$ 1000 km) using conventional recording devices such as those used in the World Wide Standard Seismograph Network(WWSSN). It is observed that earthquakes with magnitude less than 6 do not generate adequate spectral excitation at periods greater than one minute, whereas, shallow large earthquakes produce very large amplitude crustal surface waves (period range 20-30 sec) with the oscillation maxima often exceeding the width of the seismogram. Severe overwriting and recrossing of the traces renders identification and separation of traces very difficult. It has been also observed that at times the delicate suspension of long period galvanometers, which supports the attached reflecting mirror, becomes jammed due to the large energetic impulse of the surface waves. As a consequence, there is a general paucity of long period surface wave dispersion information for short path lengths. The previous studies have mostly made use of world-circling surface wave passages and average dispersion has been calculated for huge portions of the globe. Attempts are often made to infer regional dispersion from dispersion along long intersecting pathspassing through different tectonic settings. Investigations reported by Kanamori (1970), Dziewonski (1971), Gupta and Santo (1973) are such examples. The interstation method (e.g. Landisman et al 1969; Gupta et al 1977 a) is quite suitable for regional dispersion investigation when the two stations are situated on a single geological province. However, lack of properly located seismicstations, occurrence of earthquakes in narrow belts and nonavailability of seismograms from network of stations in some countries make application of inter- station method for regions like the Himalaya and the Tibet Plateau very difficult.

Installation of High Gain Long Period (HGLP) and Seismic Research Observatory (SRO) instru-

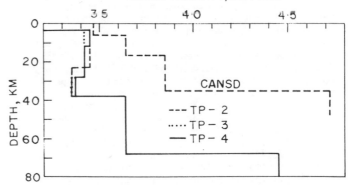

Fig. 2. Different shear wave velocity models (TP-2, TP-3, and TP-4) for the Tibetan Plateau as developed by Chun and Yoshii (1977). Models with a low velocity layer in the crust fit in with the observed surface wave dispersion pattern. Canadian Shield model (CANSD) of Brune and Dorman (1963) is also shown for reference.

ments (Peterson and Orsini, 1976) at a number of sites in the recent years has given hope that it shall now be possible to investigate long period surface wave at short distance. This is possible since the SRO and the HGLP have much larger dynamic range compared to the WWSSN system.

Himalaya and Tibet Plateau

As discussed earlier, the current tectonic activity in Asia, in accordance with the plate tectonics hypothesis, is considered to be a consequence of progressive continental collision between the Indian and Eurasian plates (for example Dewey and Bird, 1978). Since Cretaceous, India and Eurasia have been converging at a rate of 10 to 15 cm/year. However, since collision in the Eocene, this rate decreased by one half (Molnar and Tapponnier (1975). A good knowledge of the crustal and upper mantle velocity structure beneath Himalaya and Tibet Plateau region is essential for a better understanding of the tectonics at this unique site of continent-continent collision. In the absence of the other geophysical data, surface wave dispersion investigations have provided some useful information regarding crustal and upper mantle structure beneath Himalaya and Tibet Plateau.

Crustal structure beneath Himalaya and Tibet Plateau was estimated by Gupta and Narain (1977) from an analysis of fundamental mode Rayleigh and Love wave dispersion. They analysed seismograms recorded at WWSSN stations at Quetta (QUE), New Delhi (NDI), Shillong (SHL), Seoul (SEO) and Hong Kong (HKC) for an earthquake located east of Severnaya Zemlya in the northernmost USSR. The paths were entirely continental. Waves recorded at QUE, NDI and SHL, located just south of Himalaya and Tibet Plateau, pass through the high mountain ranges of Himalaya and Tibet Plateau. Whereas, the waves recorded at SEO and HKC do not pass through these high mountains and plateau. The distances between the epicenter and SEO and to the northern edge of Tibet Plateau are similar. Gupta and Narain (1967) assumed identical dispersion everywhere along the paths except Tibet Plateau and Himalaya, and by numerically subtracting the travel times of the surface waves of different periods recorded at SEO from those recorded at QUE, NDI and SHL, they obtained group velocities in the period range of 40 to 52.5 sec for the Himalaya and Tibet Plateau region. From an inversion of these data, they concluded that the crustal thickness in the Himalaya and Tibet Plateau region is 65 to 70 km. Recent results of Tung (1975), Knopoff (1976), Bird and Toksoz (1977), and Chun and Yoshii (1977), all based on surface wave dispersion investigations, confirm Gupta and Narain's (1967) finding of an extremely thick crust.

In another interesting study Chun and Yoshii (1977) have investigated group velocities of fundamental mode Rayleigh and Love waves for seventeen individual paths in the Tibet Plateau region shown in Figure 1. They used the moving window analysis technique of Landisman et al.

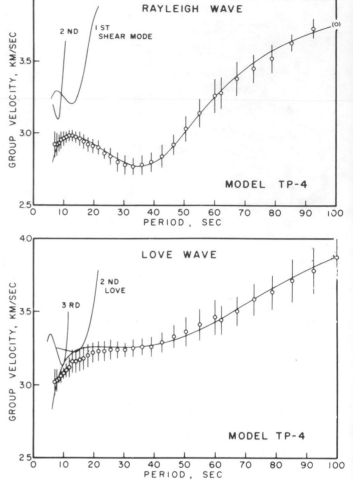

Fig. 3. Theoretical dispersion curves for model TP-4 shown in Figure 2. (adopted from Chun and Yoshii, 1977).

(1969). The periods investigated extended to 100 sec. Chun and Yoshii (1977) developed a number of theoretical models for the Tibet Plateau and computed dispersion curves. Persistent discrepancies were observed between the theoretical dispersion curves for models with monotonically increasing velocity with depth and the observed dispersion and it was found necessary to introduce a crustal low velocity layer. Some of the acceptable shear velocity models for Tibet Plateau are shown in Figure 2. The short period branch of the experimental Rayleigh wave dispersion curve fits TP-3 model better than the TP-4 model. Chun and Yoshii (1977) argue that the differences are very small and cannot be resolved within the accuracy limits of the observed data. The theoretical dispersion curves for model TP-4 and the experimental data points are shown in Figure 3. A vertical cross-section along 85°E (XX' in Figure 1) and the various models developed by different investigators for the Tibet Plateau are

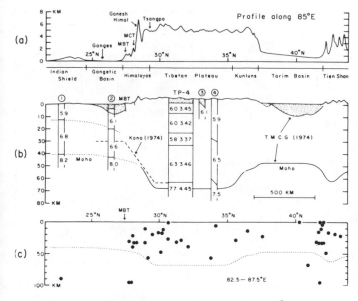

(a)

(b)

(c)

Fig. 4. Vertical cross section along 85°E
(profile XX' in Fig.1) covering (from south to
north) northern edge of the Indian Shield,
Gangetic Plains, Himalayas, Tibetan Plateau,
Kun Luns, Tarim Basin.
a. General topography with vertical exaggeration.
b. Crustal structure based on the estimates given
   by
   i) Bhattacharya, 1971
   ii) Chun and Yoshii, 1977
   iii) Chen and Molnar, 1975
   iv) Gupta and Narain, 1967
   The depth to Moho as shown by solid line is
   based on Tectonic Compiling Group, 1975.
c. Distribution of earthquake hypocenters with
   depth for the period 1966-1973 based on the
   hypocentral locations by ISC.
   (Diagram adopted from Chun and Yoshii, 1977)

shown in Figure 4 (Chun and Yoshii, 1977). The
hypocenters, plotted on a vertical plane, for the
period 1966-1973 are also included in Figure 4.
Chun and Yoshii conclude that a low velocity zone
located at an intermediate depth within the crust
is absolutely necessary to explain the observed
dispersion data. They support Gupta and Narain's
(1967) finding of an enormously thick crust
characterised with very low velocities. The low
velocities are probably caused by high temperatures. The reported widespread recent volcanism
as described by Norin (1977) and Kidd (1977) in
the Erik Norin Penrose Conference on Tibet
(Molnar and Burke, 1977) suggests that elevated
crustal temperatures indeed exist in the Tibet
Plateau.
    Gupta (1977, 1980), in a continued study of long
period surface wave dispersion, has suggested that
the upper mantle beneath the Tibet Plateau may
have normal shield structure. In Figure 7
Rayleigh wave group velocity dispersion curves
obtained for several trans-Eurasian paths are
displayed. Computer plots of digitized long-

period records for some of these paths are shown
in Figure 5. The high quality of data and capa-
bility of frequency time analysis makes it
possible to extract long period dispersion information to 200 sec period. Figure 6 shows typical
FTAN displays at Jerusalam and New Delhi.
Using the technique described by Gupta and Narain
(1967), a Rayleigh wave group velocity dispersion
curve, extending to 200 sec (TP(HOW) in Figure 7)
has been extracted for the Himalaya and Tibet
Plateau, region from this set of data. TP(HOW)
is in excellent agreement with the theoretical
dispersion curve TP-4 (Chun and Yoshii, 1977)
developed for the Tibet Plateau. Another fact,
evident from Figure 7, is a steep rise in the
experimental Rayleigh wave group velocities for the
Tibet Plateau (TP(HOW)). For periods in excess
of 100 sec, the group velocities are intermediate
between the 5.08 shield and 5.08 tectonic models
of Hamada (1972). It may also be noted that in
the period range of 100 to 200 sec TP(HOW) velocities have a trend similar to shield velocities,
and they are consistently lower by an amount
varying from 0.1 km/sec to 0.2 km/sec from the
remaining paths. This appears to be the contribution of about 70 km thick crust with relatively
low shear velocities. This leads to the conclusion that beneath an enormously thick crust of
Tibet Plateau which is characterized by low shear
velocities, lies an upper mantle which is quite
similar to other shield regions.

Attenuation in Tibet

    Bird and Toksoz (1977) have investigated
Rayleigh wave attenuation across Tibet. The Tibet
Plateau with an average elevation of 5.0 km above
MSL and a spread over an area of 7×10^5 km^2 is
covered with Cretaceous limestones. Its uplift
is Tertiary and is probably related to the Himalayan orogeny. There are various views regarding
the present structure and mode of deformation of
the Tibet Plateau. The uplift of Tibet could be
attributed to underthrusting of a second crustal
layer (Argand, 1924; Holmes, 1965; Powell and
Conaghan, 1975 and others), due to horizontal
compression and thickening (Dewey and Bird,
1970; Dewey and Burke, 1973) or due to low density
material in the mantle (Molnar and Tapponnier,
1975). Bird and Toksoz (1977) have studied
attenuation of Rayleigh waves along a number of
ray paths in Tibet and have compared it with
attenuation along ray paths that do not cross
Tibet. From an analysis of the data from 6
earthquakes along 14 ray paths across Tibet,
Bird and Toksoz (1977) report a strong reduction
in the amplitude of the long period Rayleigh
waves. The observed strong attenuation has been
attributed to a single attenuating layer centered
at a depth of 70 km. This layer should have a
very low value of Q, being of the order that is
found in case of partially molten rocks at such
high pressure. Bird and Toksoz (1977) argue that
the geological significance of the inferred
partially molten layer depends on whether it is

located in the crust or in the mantle. They are
of the opinion that Tibet has a crust of about
70 km thickness and its lowermost part is
partially molten. The crustal underthrusting
model is inconsistent with the inferred high
temperature of the Moho and the uplift has been
caused by horizontal compression. They suggest
that at present Tibet overlies an asthenosphere
which extends to lower crust and the consequential
weakness at shallow depths is responsible for the
uniformity of Tibetan elevations over wide areas
in the center of an orogenic zone.

Singh and Gupta (1979) have determined surface
wave attenuation values using long period records
of WWSSN stations for Tibet earthquakes of
July 14, 1973 and found Q values to vary from
21 to 1162 and 22 to 1110 for Rayleigh and Love
waves respectively. The period range of investigation is from 10 to 100 sec. The observed wide
variations in the attenuation data may be due to
the presence of heterogenity in the crust and
upper mantle. Yacoub and Mitchell (1977) have
obtained Rayleigh wave attenuation for the paths
crossing Eurasia at period varying from 4 to 50
sec and these values are of the same order as
obtained by Singh and Gupta (1979). However,
Yacoub and Mitchell (1977) and Singh and Gupta
(1979) have considered paths that include wide
continental portions besides the Tibet Plateau.

Indogangetic Plains

The Indogangetic Basin, which lies between the
Himalayan mountain ranges in the north and the
Peninsular Shield of India in the south, has a
width varying from 200 km in the east to 400 km
in the west. Extension of the Indian shield under
the Indogangetic Plains and an intimate relationship between the shield and the Himalayan rocks
has been inferred from geological studies (for
example Gansser, 1974; Valdia, 1976). Therefore,
understanding of the nature of the crust and
upper mantle beneath the Indogangetic Plains
assumes special importance in studying the tectonics of the Himalayan region. Gupta et al.
(1977b) have summarized results of the various
investigations   estimate physical properties and
velocity structure of the basin, the crust and
the upper mantle beneath the Indogangetic Plains.

The Indogangetic Basin has a thick pile of
sediments with an average thickness of about
3 km. A gradual increase in the thickness of the
sedimentary layers from the shield to the foot
hills of Himalaya was inferred from geodetic data
(Oldham, 1917). Variations in the sedimentary
thickness, uneven basement topography in the
form of basement upwarps and buried ridges have
been inferred from ground geophysical surveys and
borehole data (Aithal et al., 1964; Dutta et al.,
1964; Moolchand et al., 1964; Ray et al., 1964)
as well as from aeromagnetic surveys (Mathur and
Kohli, 1963). Qureshy (1969) and Choudhury
(1975) have also inferred a general increase in
the thickness of sedimentary pile, from the
Peninsular shield to the Himalayan foot hills,
on the basis of regional gravity anomalies.

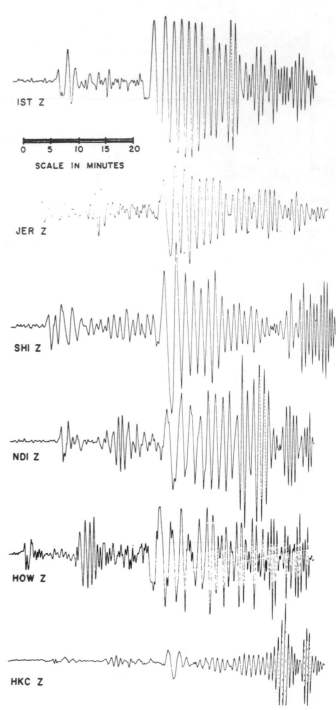

Fig. 5. Computer plot of digitized long period
records at a few WWSSN stations for an event in
Novaya Zemlya. The clarity with which the
long-period waves are recorded is remarkable.

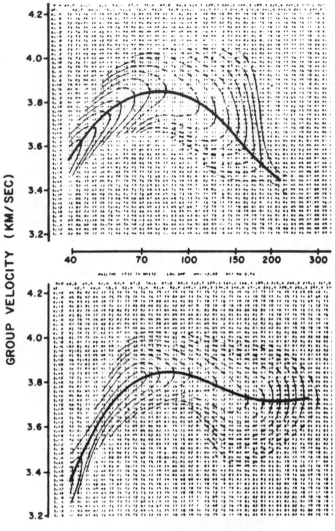

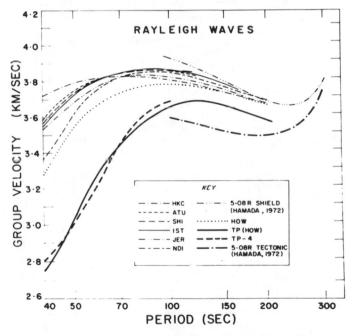

Fig. 6. Power as a function of period and
velocity generated by frequency-time analysis
(FTAN) of the records at Jerusalem (JER) and
New Delhi (NDI) shown in Figure 5. Power,
normalized to a maximum of 99 db, is contoured
at 1 db intervals. Minus signs indicate
relative maxima as a function of velocity.

At places the sedimentary pile is estimated to be
7.5 km thick or more.

Relatively much less is known about the crust
beneath the sedimentary layers of the Indogangetic Plains. Estimates of crustal properties
have been made from the study of earthquake
generated body waves (Chakravorty and Ghosh,
1960; Kaila et al., 1968; Tandon, 1954), surface
waves (Gabriel and Kuo, 1966; Chaudhury, 1966,
1969; Tandon and Chaudhury, 1964, 1967; Chatterjee,1971) and gravity surveys (Chaudhury, 1975;
Avasthi and Satyanarayana, 1970). The crustal

thickness is estimated to vary from 28 km (Kaila
et al, 1968) to 45 km (Tandon, 1954). These
reported variations are due to a combination of
actual variations, uncertainties of data and
occassional erroneous assumptions made in interpretation. The observational data is broadly
in agreement with an average crustal thickness
of 40 km for the Indogangetic Plains including at
3 km thick top sedimentary layer. Till recently,
practically no velocity information was available
for the upper mantle below the Indogangetic
Plains.

Gupta et aL, (1977b) have carried out seismic
surface wave dispersion studies extending to
200 sec period along a path covering almost
entire region of Indogangetic Plains (Figure 9).
For this study, they chose the New Hebrides Island
earthquake of July 9, 1964 (origin time 16:39:49;
epicenter 15.5°N, 167.6°E; depth 121 km; magnitude 6.6 (m_b)), which produced very well dispersed
Love and Rayleigh waves at Chiengmai (CHG),
Thailand and New Delhi (NDI), India. The epicen-
traldistances to CHG and NDI are 8412 and 10854 km
respectively. Gupta et al. (1977 b) generated digital time series of the selected portions of the
seismogram in accordance with the scheme outlined
by Gupta and Hamada (1975). Epicenter-to-station
dispersion was estimated through display-equalized
frequency time analysis (FTAN) of Nyman and
Landisman (1977). For estimating interstation
dispersion, a new technique "impulse response",

Fig. 7. Rayleigh wave group velocities inferred
from FTAN for a number of trans-Eurasian paths.
Seismograms at some of the stations are shown in
Figure 5. TP-4 is the theoretical model developed
by Chun and Yoshii (1967) for the Tibet Plateau.
Model 5.08 is after Hamada (1972). TP (HOW) is
the experimented dispersion curve for Tibet. For
details see the text.

developed by Gupta et al. (1977a) was used. The impulse response X_1 is obtained by taking inverse Fourier transform of the signal at the farther station X_2 divided by the Fourier transform of the signal at the nearest station X_1:

$$X_{i(t)} = \frac{1}{2\pi} \int_{-\infty}^{\infty} (F_2(\omega)/F_1(\omega)) \exp(i\omega t)d\omega$$

where

$$F_1(\omega) = \int_{-\infty}^{\infty} X_1(t)\exp(-i\omega t)dt$$

$$F_2(\omega) = \int_{-\infty}^{\infty} X_2(t)\exp(-i\omega t)dt$$

Being independent of the source, the path to the nearer station and the instrumental response, the impulse response method is superior to the cross-correlogram method of Landisman et al. (1969). Figure 10 shows the vertical component seismograms at CHG and NDI as well as the interstation impulse response obtained by Gupta et al.(1977b). Figure 11 shows FTAN display for the vertical component CHG–NDI impulse response.

With the exception of small sections in the

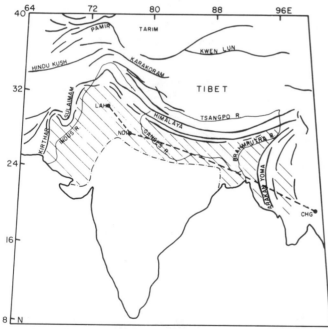

Fig. 8. A simplified map of the Himalaya and adjoining regions. The hatched portion shows the vast plains. The portion south of the chain line is the Peninsular shield of India. The dashed line shows the great circle paths between the WWSSN stations at Chiengmai (CHG) and New Delhi (NDI) and between New Delhi and Lahore (LAH).
(adopted from Gupta et al., 1977b)

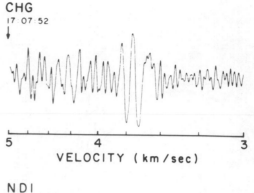

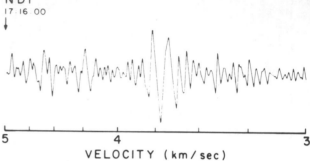

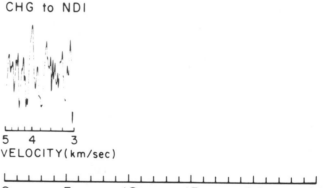

Fig. 9. The long period vertical component seismograms for WWSSN stations at Chiengemai (CHG) and New Delhi (NDI), and interstation (CHG to NDI) impulse response plotted on linear time scale. The GMT for the beginning of the record and the corresponding group velocities are also indicated.
(adopted from Gupta et al., 1977b)

Burmese mountains and northern Thailand, the great circle path between CHG and NDI is confined to the Indo-Gangetic Plains. The group and phase velocities for both Rayleigh and Love waves for this path (Figures 12 and 13) are, in general good agreement with theoretical dispersion curves computed for shield type upper mantle models (Kanamori, 1970; Hamada, 1972; Press, 1970; Block

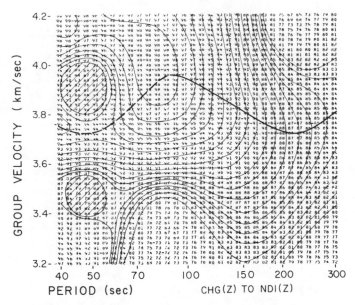

Fig. 10. Frequency-time analysis (FTAN) of the vertical component interstation "impulse response" between CHG and NDI. Contours represent power in db, normalised to a maximum of 99, at 1 - db contour intervals. Heavy line shows inferred fundamental mode Rayleigh wave group velocity dispersion. Minus signs indicate relative maxima as a function of velocity; relative minima are shaded.
(adopted from Gupta et al., 1977b)

et al., 1969; Brune and Dorman, 1963). The agreement is particularly good with model 5.08 Shield developed by Hamada (1972). Gupta et al. (1977b) have pointed out that the agreement of the observed Rayleigh wave phase velocities with model 5.08 shield is not as good as that for the Rayleigh wave group velocities or Love wave phase and group velocities. This may have been caused by relatively larger effect of lateral refraction of phases observed for Rayleigh waves. Model 5.08 Shield, shown in Figure 14, was modified from models developed by Kanamori (1970) and Press (1970) and is in good agreement with the experimental surface wave dispersion data. The Rayleigh and Love wave phase velocities shown in Figure 12 and 13 are consistently slower than the Candian Shield Model (CANSD) of Brune and Dorman (1963). Gupta et al. (1977b) point out that similar observations have been made for the shield areas of Africa, South America and elsewhere. Model 5.08 Shield has a crust similar to the crustal model developed by Chaudhury (1969) to explain crustal surface wave dispersion in the Indogangetic Plains, which is responsible for the slow Love wave phase and group velocities for periods less than 70 sec (Figure 13).
Rayleigh wave phase velocities in the period range of 10 to 45 sec for the path NDI-LAH (Figure 9) were determined by Gabriel and Kuo

(1966) and were reported to be "unusually high". Gupta et al.(1977b) point out that there is a basement upwarp and sedimentary layer thickness reduces to only 400 meters along the section between NDI and LAH. This sedimentary thinning is responsible for the high phase velocities reported by Gabriel and Kuo (1966). The geological deduction that the Peninsular hield of India extends below the Indogangetic Plains (Gansser, 1974; Valdia, 1976) earlier lacked supporting geophysical evidence which has been provided by Gupta et al. (1977b).

Middle East Region Neighbouring to the Indo-gangetic Plains.

Rayleigh wave phase velocities along three profiles in the Arabian Peninsula have been reported by Knopoff and Fouda (1975). The periods investigated extend to 170 sec and the phase velocities reported by them are slightly lower than those for the Candian Shield (Brune and Dorman, 1963). Existence of a low velocity layer for shear waves in the upper mantle with its top at depths between 100 and 140 km has been inferred by Knopoff and Fouda (1975) from inversion of their phase velocity data. Gupta et al. (1977a) have reported interstation Rayleigh and Love wave phase and group velocities for paths between Helwan (HLW), Egypt ; Shiraz (SHI), Iran; and New Delhi (NDI), India. For some of these sections, using FTAN, dispersion could be studied to 200 to

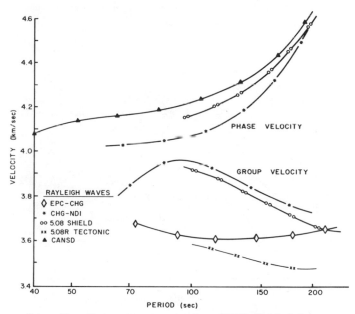

Fig. 11. Epicenter to station (CHG) Rayleigh wave group velocities and interstation (CHG to NDI) phase and group velocities as determined by Gupta et al (1977b). Phase and/or group velocities corresponding to models CANSD (Brune and Dorman, 1963), 5.08 Shield and 5.08 R Tectonic (Hamada, 1972) are also shown for comparison. (adopted from Gupta et al., 1977b).

300 sec periods. The interstation phase and group velocities for Rayleigh and Love waves are best explained by the model 5.08 shield developed by Hamada (1975). Gupta et al. (1977a) point out that there is an excellent agreement between the experimental data points and the model 5.08 shield, which features a shear wave low velocity zone (LVZ) in the upper mantle. Whereas, models without a LVZ do not fit in the experimental data. This is concordant with the conclusion of Knopoff and Fouda (1975) that it is essential to include a LVZ in the upper mantle for the SHI-HLW path. A shear wave low velocity zone thus can be considered to be an essential feature for the upper mantle in these regions.

Seismic Research Observatories

Many of the earlier mentioned problems encountered in long period surface wave dispersion investigations at short distances from the focus are solved while using the recordings made at the Seismic Research Observatories (SRO). The dynamic recording range of most conventional seismograph systems, like the WWSSN, is quite limited being about 44 dB. This corresponds to about two orders of earthquake magnitude, which is not a very large slice in the earthquake magnitude scale. The digital recording system used at SRO provides 66 dB of resolution plus 60 dB of automatic gain control. This total of 126 dB of recording range corresponds to about 6 order of magnitude (Peterson and Orsini, 1976). SRO instrumentation has been

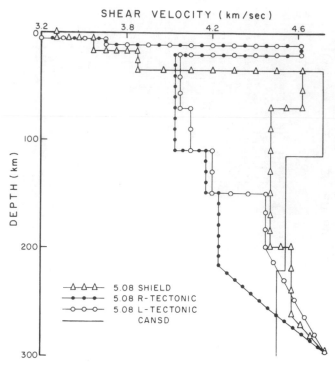

Fig. 13. Various shear wave velocity models. The 5.08 shear velocity models are after Hamada (1972) and CANSD shear velocity model is after Brune and Dorman (1963).
(adopted from Gupta et al 1977b)

installed at a number of sites. Installations at Mashad, Iran; Shillong, India; and Chiang Mai, Thailand are of particular interest for investigating the Himalayan and nearby territory. Figure 14 shows computer plots of three component SRO recordings of May 29, 1976 China-Burma border earthquake (Origin time 14:00:18.5; Epicenter, 24.5°N, 98.7°E; Depth 10 km; m_b 6.0; M_s 7.0). The epicentral distance to Mashad (36.3°N, 59.6°E) is 3970 km. In Figure 14, the clarity with which the long period surface waves are recorded is remarkable. For comparison, a section of the long period vertical component seismogram recorded at Hyderabad is shown in Figure 15. Hyderabad is a quiet site and epicentral distance is similar to Mashad. Figure 16 shows results of FTAN for the vertical component at Mashad shown in Figure 15. Rayleigh wave group velocities for periods exceeding 200 sec can be unambiguously inferred. More than one half of the total path length from this epicenter to Mashad corresponds to Tibet Plateau region. The inferred Rayleigh wave group velocities are plotted in Figure 17 along with the experimental curve TP(HOW) and theoretical curves TP-4 and 5.08. As expected, China/Burma border to Mashad experimental group velocities are in between the group velocities for the Tibet Plateau and the average trans-Eurasian group velocities. This further confirms the conclusions drawn in earlier sections.

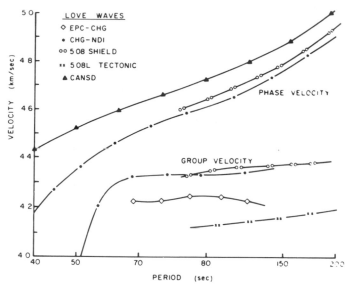

Fig. 12. Epicenter to station (CHG) Love wave group velocities and interstation (CHG to NDI) phase and group velocities as obtained by Gupta et al (1977b). Phase and/or group velocities corresponding to models CANSD (Brune and Dorman, 1963), 5.08 Shield and 5.08L Tectonic (Hamada, 1972) are also shown for comparison.
(adopted from Gupta et al., 1977b)

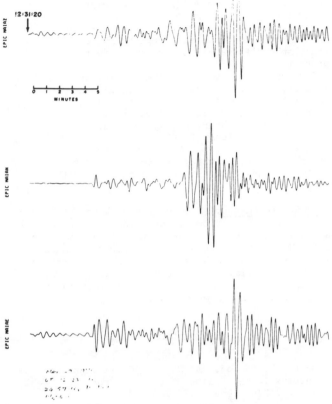

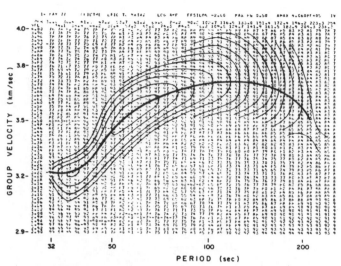

Fig. 16. Frequency-time analysis of the vertical component seismogram at Mashad (top, Figure 14) for the Burma/China border earthquake of May 29, 1976. Power, normalised to a maximum of 99 db is contoured at 1 db intervals. Minus signs indicate relative maximas. The thick line shows inferred group velocities. Rayleigh wave group velocities for periods exceeding 200 sec are unambiguously inferred from the SRO recordings of a medium size earthquake at a rather short distance.

Fig. 14. Computer plots of the three component long period SRO recordings at Mashad for the May 29, 1976 Burma/China border earthquake. Major part of the path length corresponds to the Himalaya and Tibet Plateau region.

Conclusions

Recent long period surface wave studies have been extremely useful in estimating the physical properties of the crust and upper mantle below Himalaya, Tibet Plateau and the neighbouring regions. Continued developments in surface wave analysis techniques and use of carefully selected data are primarily responsible for the success of these investigations. Main conclusions drawn could be summarized as follows :

1. Gupta and Narain's (1967) finding of an enormously thick crust (65-70 km), characterised by low shear wave velocity, below the Himalaya and Tibet Plateau region has been confirmed by later investigations. Chun and Yoshii (1977) have developed a number of models of the crust under the Tibet Plateau. Acceptable models require a thick crust (70 km) and a low velocity zone located at an intermediate depth.

2. Rayleigh wave attenuation studies (Bird and

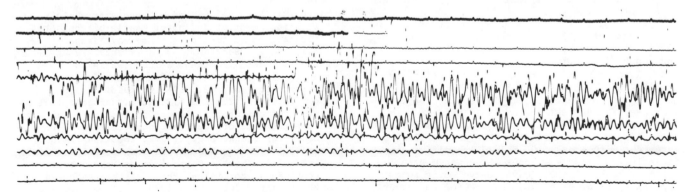

Fig. 15. Section of a seismogram for the Burma/China border earthquake of May 29, 1976, as recorded by NGRI observatory. The mixing up of the traces makes the analysis difficult.

LONG PERIOD SURFACE WAVES STUDIES   303

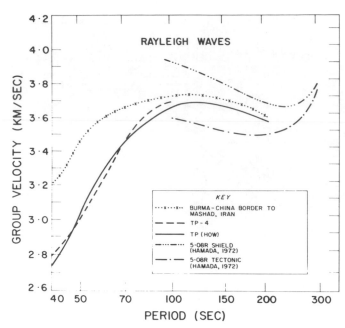

RAYLEIGH WAVES

KEY
······|······· BURMA-CHINA BORDER TO MASHAD, IRAN
—— —— —— TP-4
————————— TP (HOW)
······|······ 5·08R SHIELD (HAMADA, 1972)
—·—·—·— 5·08R TECTONIC (HAMADA, 1972)

Fig. 17. Rayleigh wave dispersion curve obtained for the path from Burma/China to Mashad, Iran from SRO recording using FTAN. Remaining symbols are same as in Figure 7. For details see the text.

Toksoz, 1977) in Tibet Plateau indicate that the lowermost part of the crust is partially molten and the uplift has been caused by horizontal compression. Experimental data do not favour the hypothesis that uplift of the Tibet Plateau has been caused by underthrusting. The inferred mechanical weakness at shallow depths seems to be responsible for uniformity of Tibet Plateau elevations over wide areas.

3. Detailed inter-station Rayleigh and Love wave phase and group velocity measurements extending beyond 200 sec period reveal that shield like velocity structure exists beneath the Indogangetic Plains (Gupta et al.,1977b). The experimental data fits in best with the theoretical model 5.08 Shield developed by Hamada (1972). This model features a low velocity zone in the upper mantle. This confirms the earlier geological deduction that the Peninsular Shield of India extends below the Indogangetic Plains (Gansser, 1974; Valdia, 1976). This deduction previously lacked supporting geophysical evidence.

4. Results of the detailed investigations by Knopoff and Fouda (1975) for the inter-station path HLW-SHI, and by Gupta et al (1977a) for the inter-station paths HLW-SHI and SHI-NDI indicate shield like upper mantle velocity structure in middle-east Asia. These dispersion data have an excellent agreement with model 5.08 Shield and differ significantly from models without an upper mantle low velocity zone. A shear wave low velocity zone appears to be an essential feature for the upper mantle below these regions.

5. The upper mantle beneath Tibet also appears

to be normal (Gupta, 1977). It is tempting to infer that the upper mantle does not vary significantly from one region to another for the entire south-central Asian region, whereas remarkable differences in crustal structure and properties exist.

6. Earlier, it was difficult to investigate long period surface waves at short distances, basically due to limitation of the dynamic range of the conventional long period seismographs. This problem has been largely overcome due to recent installations of SRO equipment with much larger dynamic range. It is hoped that much more would be learnt in the years to follow from long period surface wave investigations using HGLP and SRO recordings.

Acknowledgements. Much of the work reported here was carried out during visits of one of the authors (HKG) to the University of Texas at Dallas, USA. Mr. K. Suryaprakasam, Mr. G.B. Navinchander and Mr. K. Ramana Rao assisted in the preparation of the manuscript. Authors are thankful to the Director, NGRI for his continued interest in the long period surface wave studies.

References

Aithal, V.S., G. Gopalkrishnan, P. Nath, Y.K. Murty, B. Nandan, and D. Das, Gravity and magnetic surveys in the Plains of Punjab, Proc. Int. Geol Congr., 22nd Sess., 179-188, 1964.

Argand, E., La tectonique de l'Asie, Thirteenth International Geological Congress, 1, Part 5, 171-372, 1924.

Avasthi, D.N., and M. Satyanarayana, Inferences regarding the upper mantle from the gravity data in Punjab and Ganga plains, Proc. Second Symp. U.M.P. Dec 1970, GRB and NGRI Pub. No. 11, 25-39, 1970.

Bhattacharya, S.N., Seismic surface-wave dispersion and crust-mantle structure of Indian Peninsula, Indian J. Met. Geophys. 22, 179-186, 1971.

Bird, P., Initiation of Himalayan intracontinental subduction, J. Geophys. Res., 83, 4975-4987, 1978.

Bird, P., and M.N. Toksoz, Strong attenuation of Rayleigh waves in Tibet, Nature, 266, 161-163 1977.

Bloch, S., A.L. Hales, and M. Landisman, Velocities in the crust and upper mantle of southern Africa from multi-mode surface wave dispersion, Bull. Seism. Soc. Am., 59, 1599-1629, 1969.

Brune, J., and J. Dorman, Seismic waves and earth's structure in the Canadian Shield, Bull. Seism. Soc. Am., 43, 167-210, 1963.

Chakravorthy, K.C., and D.P. Gosh, Seismological study of the crustal layers in Indian region from the data of near earthquakes, Proc. World Conf. Earthquake Eng., Tokyo. 1633-1642, 1960.

Chandra, U., Large - scale Cenozoic tectonics of Central and South- Central Asia : products of continental collision, Phys. of the Earth and Plant. Int., 20, 33-41, 1979

Chatterjee, S.N., On the dispersion of Love waves and crust-mantle structure in the Gangetic Basin, Geophys. J.R. Astron. Soc., 23, 129-138, 1971.

Chaudhury, H.M., Seismic surface wave dispersion and the crust across the Gangetic Basin, Indian J. Meterol. Geophys., 17, 385-393, 1966.

Chaudhury, H.M., Dispersion of short period surface waves and the surficial layer in the region west of Delhi, Indian J. Meterol. Geophys., 20, 227-234, 1969

Chen, W.P., and P. Molnar, Short period Rayleigh wave dispersion across the Tibetan Plateau, Bull. Seism. Soc. Am., 65, 1051-1057, 1975.

Chen, W.P., Presentation in Erik Norin Penrose Conference on Tibet, Report eidted by P. Molnar and K. Burke, Geology, 5, 461-463, 1977

Choudhury, S.K., Gravity and crustal thickness in the Indogangetic Plains and Himalayan region, India, Geophys. J.R. Astron. Soc., 40, 441-452 1975.

Chun, K., and T. Yoshii, Crustal structure of the Tibet Plateau: a surface wave study by a moving window analysis, Bull. Seism. Soc. Am., 67, 735-750, 1977.

Data, A.N., T.S. Balakrishnan, A.P. Ghosh, V.C. Mohan, and R. Nath, Seismic surveys in the Punjab Plain, Proc. Int. Geol. Congr., 22nd Sess., II, 248-259, 1964.

Dewey, J.F., and J.M. Bird, Mountain belts and the new global tectonics, J. Geophys., Res., 75, 2625-2647, 1970.

Dewey, J.F., and K.C.A. Burke, Tibetan, Variscan, and Precambrian basement reactivation: Products of the continental collision, J. Geol., 81, 683-692, 1973.

Dziewonski, A., On regional differences in dispersion of mantle Rayleigh waves, Geophys. J., 22, 289-325, 1971.

Dziewonski, A., M. Landisman, S. Bloch, Y. Sato, and S. Asano, Progress report on recent improvements in the analysis of surface waves observations, J. Phys. Earth, 16, 1-26, 1968.

Dzienwonski, A., S. Bloch, and M. Landisman, A technique for the analysis of transient seismic signals, Bull. Seism. Soc. Am., 59, 427-444, 1969.

Ewing, M., and F. Press, An investigation of mantle Rayleigh waves, Bull. Seism. Soc. Am., 44, 159-184, 1954a.

Ewing, M., and F. Press, Mantle Rayleigh waves from the Kamchatka earthquake of November, 4, 1952, Bull. Seism. Soc. Am., 44, 471-474. 1954b.

Gabriel, V.G., and J.T. Kuo, High Rayleigh wave phase velocities for the New Delhi, India-Lahore, Pakistan profile, Bull. Seism. Soc. Am., 56, 1137-1145, 1966.

Gansser, A., Himalaya, Mesozoic, Cenozoic orogenic belts, Geol. Soc. London, Special publication no. 4, 207-218, 1974.

Gupta, H.K., and H. Narain, Crustal structure of the Himalayan and the Tibet Plateau regions from surface wave dispersion, Bull. Seism. Soc. Am., 57, 235-248, 1967.

Gupta, H.K., and Y. Sato, Regional characteristics of Love wave group velocity dispersion in Eurasia, Bull. Earthquake Res. Inst., Tokyo Univ., 57, 235-248, 1968.

Gupta, H.K., and T. Santo, World wide investigation of the mantle Rayleigh wave group velocities Part I: Dispersion data for new 31 great circle paths, Bull. Seism. Soc. Am., 63, 271-281, 1973.

Gupta, H.K., and K. Hamada, Rayleigh and Love wave dispersion upto 140 sec period range in the Indonesia-Philippine region, Bull. Seism. Soc. Am., 65, 507-521, 1975.

Gupta, H.K., Presentation in Erik Norin Penrose Conference on Tibet Report eidted by P. Molnar and K. Burke, Geology, 5, 461-463, 1977.

Gupta, H.K., Investigation of long period surface waves using High Gain Long Period (HGLP), Seismic Research Observatory (SRO) and World Wide Standard Seismic Net-work (WWSSN) data, manuscript under preparation, 1980.

Gupta, H.K., D.C. Nyman, and M. Landisman, Shield like upper mantle structure inferred from long period Rayleigh and Love wave dispersion in the Middle East and South East Asia, Bull. Seism. Soc. Am., 67, 103-119, 1977a.

Gupta, H.K., D.C. Nyman, and M. Landisman, Shield like upper mantle velocity structure below the Indogangetic Plains: Inferences drawn from long period surface wave dispersion studies, Earth Planet. Sc. Letters, 34, 51-55, 1977b-

Hamada, K., Regionalized shear velocity models for the upper mantle inferred from surface wave dispersion data, J. Phys. Earth, 20, 301-326, 1972.

Holmes, A., Principles of physical geology, Nelsons London, 1288p, 1965.

Kaila, K.L., P.R. Reddy, and H. Narain, Crustal structure in the Himalayan foothills area of north India, from P wave data of shallow earthquakes, Bull. Seism. Soc. Am., 58, 597-612, 1968.

Kanamori, H., Velocity and Q of mantle waves, Phys. Earth Planet. Int., 2, 259-275, 1970.

Kidd, W.S.F., Presentation in Erik Norin Penrose Conference on Tibet, Report edited by P. Molnar and K. Burke, Geology, 5, 461-463, 1977.

Knopoff, L., Regionalization of the Arctic region, Siberia and Eurasian continental area, Final Report, IGPP, Univ. of California, 33p, 1976.

Knopoff, L., and A.A. Fouda, Upper mantle structure under the Arabian peninsula, Tectonophysics, 26, 121-134, 1975.

Kono. M., Gravity anomalies in east Nepal and their implications to the crustal structure of the Himalayas, Geophys. J. Roy. Astr. Soc., 39, 283-299, 1974.

Landisman, M., A. Dziewonski, and Y. Sato, Preliminary report on recent improvements in the analysis of surface waves, Fourth International Symposium on Geophysical Theory and Computers, Nuovo Cimento, 6, 126-131, 1968.

Mathur, L.P. and G. Kohli, Exploration and development for oil in India, World Pet. Congr., 6, 633-658, 1963.

McKenzie, D.P., Active tectonics of the Mediterranean region, Geophy. J.Roy. Astr. Soc., 30, 109-185, 1972.

Molnar, P. and K. Burke, Erik Norin Penrose Conference on Tibet (edited), Geology, 5, 461-463, 1977.

Molnar, P. and P. Tapponnier, Cenozoic tectonics

of Asia : Effects of a continental collision, Science, 189, 419-426, 1975.

Mool Chand, A.N., Datta, R.S. Chellam, A.P. Ghosh, A.M. Awasthi, D.N. Aswasthi, and V.C. Garg, Seismic surveys in western Uttar Pradesh, Proc. Int. Geol. Congr., 22nd Sess., II, 260-276, 1964.

Nagmune, T., On the travel time and dispersion of the surface waves, Geophys, Mag., 27, 93-104, 1956.

Norin, E., Presentation in Erik Norin Penrose Conference on Tibet, Report edited by P. Molnar and K. Burke, Geology, 5, 461-463, 1977.

Nyman, D.C., Dispersion analysis and frequency-time filters, Geophys. J. Roy. Astr. Soc.. (in press).

Nyman, D.C., and M. Landisman, The display equalized filter for frequency-time analysis, Bull. Seism. Soc. Am., 67, 393-404, 1977.

Nyman, D.C., H.K. Gupta, and M. Landisman, The relationship between group velocity and phase velocity forfinite discrete observations,, Bull. Seism. Soc. Am., 67, 1249-1258, 1977.

Oldham, R.D., The structure of the Himalayas and of the Gangetic Plains as elucidated by geodetic observations in India, Geol. Surv. India, Mem. 42 (2), 1-153, 1917.

Peterson, J., and N.A. Orsini, Seismic Research Observatories : Upgrading the world-wide seismic network, EOS Trans. AGU, 57, 548-557, 1976.

Powell, C. Mc A., and P.J. Conaghan, Tectonic models for the Tibetan Plateau, Geology, 3, 727-731, 1975.

Press, F., Earth models consistent with geophysical data, Earth Planet. Inter., 3, 3-22, 1970.

Qureshy, M.N., Thickening of a basalt layer as a possible cause for the uplift of the Himalayas - A suggestion based on gravity data, Tectonophysics, 7, 137-157, 1969.

Ray, S., S. Lyngdoh, K.K. Gheevarghese, and S.N. Sengupta, Seismic prospecting for oil in the southwestern part of the Brahmaputra valley of Upper Assam, Proc. Int. Geol. Congr., 22nd Sess., II, 277-291, 1964.

Santo, T., Lateral variation of Rayleigh wave dispersion character, Part II: Eurasia, Pure App. Geophys., 62, 67-75, 1965.

Sato, Y., Attenuation, dispersion, and the wave guide of the G waves, Bull. Seism. Soc. Am., 48, 231-251, 1958.

Shechkov, B.N., Structure of the earth's crust in Eurasia from the dispersion of the surface waves, Izvestiya Acad. Sci. USSR Geophys. Ser., (English Translation), 450-453, 1961.

Sherburne, R.W., Crust-mantle structure in continental South America and its relation to sea floor spreading, Ph.D. thesis, The Pensylvania State University, 145p, 1975.

Singh, D.D., and H.K. Gupta, Source mechanism and surface wave attenuation studies for Tibet earthquake of July 14, 1973, Bull. Seism. Soc. Am., 69, 730-750, 1979.

Tandon, A.N., Study of the great Assam earthquake of August 1950 and its aftershocks, Ind. J. Meteorol. Geophys., 5, 95-137, 1954.

Tandon, A.N., and H.M. Chaudhury, Thickness of earth's crust between Delhi and Shillong from Surface wave dispersion, Ind. J. Met. Geophys. 15, 467-479, 1964.

Tandon, A.N., and H.M. Chaudhury, Phase velocity of Rayleigh waves over the Punjab, Ind. J. Meteorol. Geophys., 19, 431-434, 1968.

Tectonic Map Compiling Group, Institute of Geology, Academia Sinica, A preliminary note on the basic tectonic features and their developments in China, Scientia Geologica Sinica, 1, 1-17, 1974.

Thomas, L., Rayleigh wave dispersion in Australia, Bull. Seism. Soc. Am., 59, 167-181, 1969.

STATE OF STRESS IN THE NORTHERN PART OF THE INDIAN PLATE

R. N. Singh

Theoretical Geophysics Group, National Geophysical Research Institute, Hyderabad 500 077, India

Abstract. Variations of shear stress with depth in the northern part of the Indian plate has been obtained using the flow law of olivine, a representative rock-forming mineral of the lithosphere, and the prevailing thermodynamic conditions. Estimates of thickness of the elastic brittle zone and the basal shear stress at the lithosphere-asthenosphere boundary have been made. In the western part of the shield, the thickness of the brittle zone is approximately 23-30 km whereas in the eastern part, it is 29 - 36 km for strain rates of 10^{-16} to 10^{-11} s^{-1} and these estimates are approximately the same for both wet and dry olivine models of lithosphere. However, the estimates of the basal shear stress are significantly different in the above two models. In the wet olivine model the basal shear stresses in the western part, for the above creep rates, fall into the range .13 - 3.56 bars and in the eastern part, this range is .26 - 9.66 bars. For the dry olivine model the above estimates, respectively, are .41 - 10.0 bars and 1.03 - 25.29 bars. Applications of some of these results to the Himalayan orogeny have been discussed.

Introduction

Knowledge of the state of stress is needed for variety of reasons. However, its important uses lie in evaluating seismic hazards and in deciphering the mechanism of plate motion. In the northern part of the Indian plate, such knowledge is also desirable, as the existing stresses would be related to the thermomechanical processes associated with the Himalayan orogeny. The Himalayan orogeny is understood, clearly now, to be a typical case of continent-continent collision (Argand, 1924; Wegener, 1929; Gansser, 1966; Dewey and Bird, 1970). It has involved the subduction of some part of the Indian shield (Valdiya, 1976) leading to the observed crustal thickening (Chauhan and Singh, 1965; Gupta and Narain, 1967; Kono, 1974) and transmission of crustal compression to the Tibetan and adjoining regions and its further distribution in the vast territory of China (Molnar and Tapponnier, 1976). Sykes (1978) in his classic survey of intra-

plate seismotectonics has stated that "stresses in peninsular India, much like those north of the Alps, may be a plate-boundary-related phenomenon". Thus it is necessary to know the nature of stress in the northern part of the Indian shield region, forming northern part of Indian plate.

Stress determinations are made by various means. Some of these techniques are hydraulic fracturing, geological mapping, grain size palaeopiezometry, seismicity, earthquake focal mechanism analysis, plate kinematics and flow laws of rock-forming minerals. In the present article, we shall be concerned with the application of the last-named method. In this method the physico-mathematical model of flow laws of various materials, constituting the lithosphere, are used. These laws relate the rate of strain with the applied stress, temperature, pressure and other properties describing the composition of the rocks. We shall use here the flow law of olivine, taking it as the most representative rock-forming mineral of the lithosphere (Kirby, 1977). The behaviour of olivine depends, in addition to those mentioned above, upon the presence of water in small amounts. Thus it would be necessary to consider both the 'wet' and 'dry' olivine flow-law. For fuller discussion on creep laws of the various other materials, we shall make reference to Kirby and Rayleigh (1973), Stocker and Ashby (1974), Weertman (1970, 1976), Nicolas and Poirier (1976), Carter (1976), and Kohlstedt, Goetz and Durham (1976).

Table 1. Creep Parameters

Constant	Dry Olivine	Wet Olivine
B (s^{-1})	$1.44 \cdot 10^{10}$	$1.67 \cdot 10^{10}$
α $(cm^2/dyne)$	$6.93 \cdot 10^{-10}$	$6.93 \cdot 10^{-10}$
E K Cal/mol	95	91
n	3.6	3.2
V (cm^3/mol)	11.0	11.0

Table 2. Heat generation constants and surface heat flux

Locality	A_o (10^{-13} cal/cm^{3}s)	D(km)	Surface heat flux (10^{-6} cal/cm^{3}s)
Khetri	5.24	14.8	1.76
Singhbhum	2.75	14.8	1.30

Olivine Rheology - Steady State Creep

Creep takes place after the applied stress has crossed the yield strength of the materials. In the viscous fluid model, the creep rate, $\dot{\varepsilon}_\nu$, is proportional to the applied stress, τ, i.e.,

$$\dot{\varepsilon}_\nu = \nu^{-1} \tau \qquad (1)$$

where ν is called viscosity and is related to the transport of the linear momentum. In microphysical considerations, it means that the transport of momentum is taking place through diffusion of atoms and vacancies via the Herring-Nabbaro and Cobble mechanism. More explicitly in the Herring-Nabbaro creep mechanism, we have

$$\nu = \propto (\frac{kT}{\Omega}) \; (L^2/D) \qquad (2)$$

where k, t, Ω, L and D are the Boltzman constant, temperature, atomic volume, grain size and diffusion coefficient respectively and in the Herring-Nabbaro-Cobble creep mechanism, we have

$$\nu = \propto (\frac{kT}{\Omega}) \; (L^2/D) \; \{1+(\eta\delta/L)(D_B/D)\}^{-1} \qquad (3)$$

where δ and D_B are the effective thickness of the grain boundary and the grain boundary diffusion constant respectively. For the stress, temperature and grain size conditions prevailing in the lithosphere, these two mechanisms would not be operative.

Experimental rheological data reported by Kohlstedt and Goetz (1974) and Post (1973) have

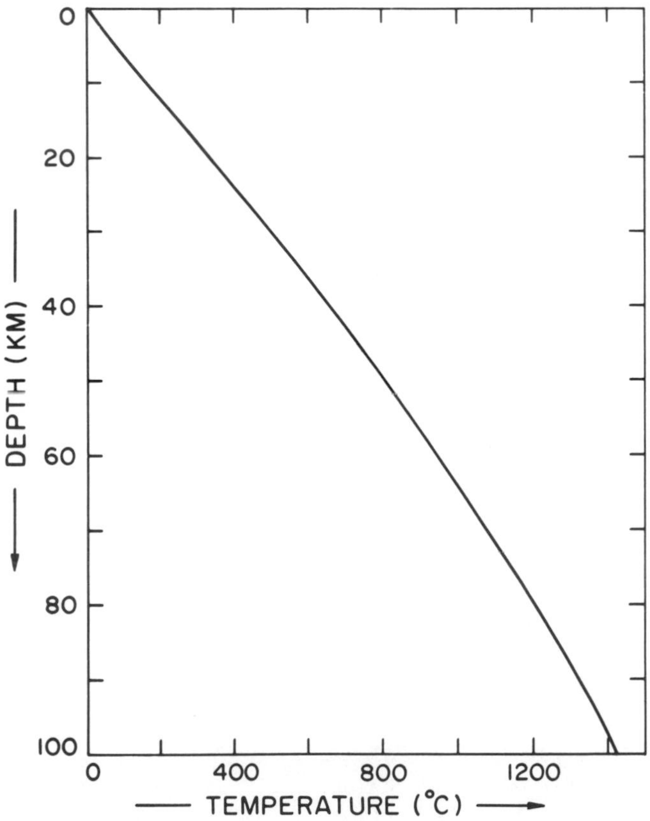

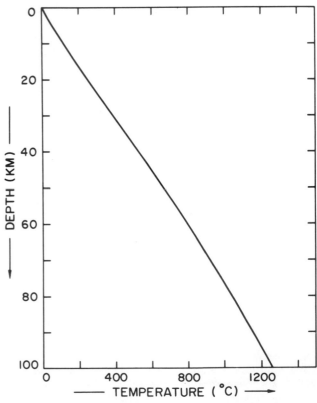

Fig. 1a. Temperature distribution with depth below Khetri.

Fig. 1b. Temperature distribution with depth below Singhbhum.

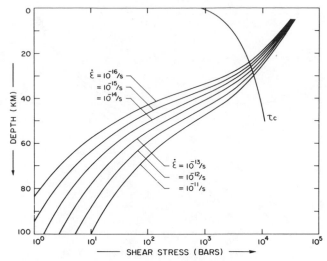

Fig. 2a. Shear stress distribution with depth below Khetri - Dry olivine model.

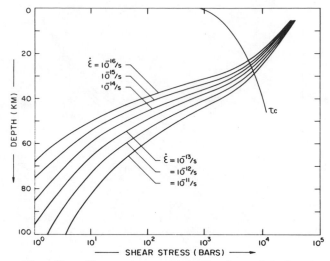

Fig. 2b. Shear stress distribution with depth below Khetri - Wet olivine model.

clearly shown that for the conditions prevailing in the lithosphere the strain rate is non-linearly related with stress. Thus the viscosity becomes function of stress also. Kirby (1977) has recently proposed the following functional relationship between stress, τ, and strain rate $\dot{\varepsilon}$,

$$\dot{\varepsilon} = B\{Sinh\ (\alpha\tau)\}^n\ exp\{-(E+PV)/RT\} \qquad (4)$$

where E, P, V and R are activation energy, pressure, activation volume and universal gas constant, respectively. B, N and α are constants. B, n, E and V differ for various earth materials. For small values of $\alpha\tau$, we can take in (4) sinh $\alpha\tau \approx \alpha\tau$ and for larger values of $\alpha\tau$ we can take, instead, sinh $\alpha\tau \approx exp\ (\alpha\tau)$. In table 1, we have given the values of these constants for both 'wet' and 'dry' model of olivine (Kirby, 1977).

Pressure is determined by using the following formula:

$$P = \rho gZ$$

Here Z, ρ, g are the depth, density and gravity respectively. We take the value of density as 3.45 gm/cm^3.

Temperature is obtained by solving the following steady state heat diffusion equation and using the heat flow data for the region under study.

$$\frac{d}{dZ}(K\ \frac{dT}{dZ}) = -A(z) \qquad (5)$$

where thermal conductivity K, is given by following Schatz and Simmons (1972):

$$K = (30.6 + 0.21\ T)^{-1} \quad for\ T \leqslant 500^\circ\ K$$
$$= (30.6 + 0.21\ T)^{-1} = 5.5 \times 10^{-6}\ (T-500)$$
$$for \quad T \geqslant 500^\circ\ K \qquad (6)$$

and heat source distribution A(z) is taken as (Lachenbruch, 1970):

$$A(z) = A_O\ exp\ (-z/D) \qquad (7)$$

Estimates of A_O and D are available for three locations in the northern part of the Indian shield. Fortunately one is in the western part and the other two are in the east. Two localities in the eastern part are nearby and we shall, for our purpose here, select the most easterly one. The localities are Khetri (~22° N, 85.5° E) and Singhbhum (~27.5° N, 76° E) respectively. Table 2 gives the estimates of A_O, D and surface heat flow, C_O for these localities (Verma and Gupta, 1975; Rao et al., 1976).

Lastly, to determine the shear stress profiles at these locations we would also need the estimates of shear strain rate. Kirby (1977) has provided such a range. Creep rates ranging $(10^{-14} - 10^{-16})$ s^{-1} are associated with the asthenospheric convection whereas those within $(10^{-11} - 10^{-14})$ s^{-1} are associated with vertical uplift.

All these estimates would be used in calculating the temperature and stress profiles.

Thermal and Shear Stress Structure

Figures (1a, 1b) give the temperatures below Khetri and Singhbhum regions respectively. The crustal temperatures in the Indian shield are generally higher than the average shield region, as already noted by Rao et al., (1976), Negi and Pandey (1976) and Singh and Negi (1979). We have given here the temperature profile for the whole lithospheric region. We find that at the base of the lithosphere i.e., at the depth of 100 km, the temperature in Khetri and Singhbhum are around 1425° C and 1260° C respectively,

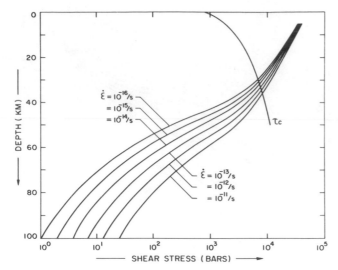

Fig. 3a. Shear stress distribution with depth below Singhbhum - Dry olivine model.

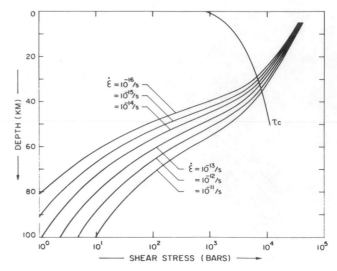

Fig. 3b. Shear stress distribution with depth below Singhbhum - Wet olivine model.

which are considerably higher than that of average shield as given in Green and Ringwood (1967). Further if the lithosphere/asthenosphere boundary is taken as the 1300° C - isotherm (McKenzie, 1978) then the thickness of the lithosphere would be less than 100 km in the western part and more than 100 km in the eastern part of the shield.

Figures (2a, 2b) show the shear stress profile for both 'dry' and 'wet' olivine models of the lithosphere respectively, in the Khetri and for Singhbhum region, the corresponding results are shown in Figures (3a, 3b). We should point out that these profiles give the optimal estimate of the possible tectonic stress regime. In these diagrams we have also shown the depth distribution of the critical shear stress. Above this curve the lithosphere behaves elastically and below it responds in ductile way. This curve is estimated by using the following formula (Kirby, 1977):

$$(kb) = 0.8 + 0.6 \ p \ (kb) \qquad (8)$$

Conclusion

i. The thickness of the elastic zone: Table 3 shows the thickness of the leastic (brittle) zone in both areas for both models of olivine. In the Khetri region for the creep rates under consideration, the thickness falls within the 24 - 30 km range for 'dry' olivine model and 23 - 29 km for the 'wet' olivine model. For the Singhbhum region these ranges are 29.5 - 36 and 29 - 35.5 km respectively. We do not find much change in the estimates of 'dry' and 'wet' olivine models. Earthquakes associated with predominantly brittle fracture would occur within this zone. Some recent studies of the earthquakes in these regions mention that the earthquake focal depths fall in this elastic zone (Reddy, 1971; Chandra, 1977; Singh and Gupta, 1979).

These results can also be used to infer the crustal thickness below the Himalaya. As mentioned in the introduction it is believed that thrusting during the Himalayan orogeny, shortened part of the crust and led to crustal doubling.

Table 3. Thickness of Elastic Zone (km)

Strain rate (s^{-1})	Khetri		Singhbhum	
	Dry olivine model	Wet olivine model	Dry olivine model	Wet olivine model
10^{-11}	30	29	36	35.5
10^{-12}	29	27.5	34.5	34.0
10^{-13}	27	26	33.0	32.5
10^{-14}	26	25	31.5	31.0
10^{-15}	25	24	30.5	30.0
10^{-16}	24	23	29.5	29.0

Table 4. Basal Shear Stress(bars)

Strain rate (s^{-1})	Khetri		Singhbhum	
	Dry olivine model	Wet olivine model	Dry olivine model	Wet olivine model
10^{-11}	10.0	3.56	25.29	9.66
10^{-12}	5.3	1.73	13.34	4.70
10^{-13}	2.80	.84	7.04	2.29
10^{-14}	1.48	.41	3.71	1.12
10^{-15}	.78	.20	1.96	.54
10^{-16}	.41	.13	1.03	.26

That part of the lithosphere which shows brittle fracture could easily be that layer which participated in this crustal doubling. Thus the thickness of the crust, participating in this process along the main central thrust would be 26 km in the western part and 31 km in the central part, for a strain rate $\sim 10^{-14}$ s^{-1}. as the temperature would have been higher at the time of initial thrusting than at the present, the calculated estimate gives the higher limit. Bird (1978) has taken this thickness of 35 km all along the length of the Himalaya. We thus infer from our calculations that, if crustal doubling has taken place to give the abnormal thickness of the Himalayan crust, then in the western part, one would have a thickness of 52 km and in the central part, a thickness of 62 km. Bird (1978) has obtained an average estimate of 60-70 km. We should stress, however, that these estimates have been arrived at using a large number of assumptions and these conclusions should be viewed in that light.

ii. Basal shear stress: Table 4 shows the estimates of the basal shear stress at the base of the lithosphere at a depth of 100 km. In the Khetri region for the 'dry' olivine model, the basal shear stress falls within the range .41-10 bars for creep rates under consideration and for 'wet' olivine model within .13 - 3.56 bars. In the Singhbhum region similar ranges are 1.03 - 25.29 bars and .25 - 9.66 bars, respectively. In these estimates there is a considerable difference between the estimates of 'dry' and 'wet' olivine models.

Thus for the strain rates associated with large asthenospheric cells having velocities 5 cm/yr, the basal shear stress in the western part is .41 bar and in the eastern part it is 1.03 bars. But if the secondary small convective cells with higher velocities are induced in the asthenosphere below the region under consideration, the strain rate would be higher and the basal shear stress could increase upto 10 bars in the western part and 25.29 bars in the eastern part. These shear stresses would be reflected in the shear stress regime in the Himalayan region, and would have played a crucial role in

reaching the balance of force of the lithospheric plate. Bird (1978) has proposed, based on the rock mechanical and topographic arguments, that a shear stress of the order of 200 bars is active on Main Boundary fault. Thus in the northern part of the shield, the shear stress would have highly inhomogeneous distribution in the lower part of the lithosphere.

In conclusion we would remark that, as stresses in the northern part of the Indian shield region, Himalayan ranges and the Tibetan region are coupled, the observational and theoretical studies related to their common aspects, should be undertaken to resolve various problems associated with Himalayan orogeny.

Acknowledgements. Author would like to thank Prof. Hari Narain, Vice-Chancellor, Banaras Hindu University and Dr. H. K. Gupta of National Geophysical Research Institute for suggesting the study reported in the paper. Author is thankful to the Director, National Geophysical Research Institute, Hyderabad for according permission to publish this work.

References

Argand, E., La tectonique de l'Asie, C. R. Congr. Geol. Intern., 13e, 1922 (Liège), 169-371, 1924.
Bird, P., Initiation of intra-continental subduction of Himalaya, J. Geophys. Res., 83, 4975-4987, 1978.
Carter, N. L., Steady state flow of rocks, Rev. Geophys. Sp. Phys., 14, 301-360, 1976.
Chandra, U., Earthquakes of peninsular India, A seismotectonic study, Bull. Seism. Soc. Am., 67, 1387-1413, 1977.
Chauhan, R. K. S. and R. N. Singh, Crustal studies in Himalayan region, J. Ind. Geophys. Un., 2, 51-57, 1965.
Dewey, J. F. and J. M. Bird, Mountain belts and the new global tectonics, J. Geophys. Res., 75, 2625-2647, 1970.
Gansser, A., The Indian ocean and the Himalayas, A geological interpretation, Ecl. Geol. Helv., 59, 831-848, 1966.

Green, D. H. and A. E. Ringwood, The stability fields of aluminous pyroxene peridotite and garnet peridotite and their relevance in upper mantle structure, Earth Planet. Sc. Lett., 3, 151-160, 1967.

Gupta, H. K. and H. Narain, Crustal structure in the Himalayan and Tibetan Plateau region from surface wave dispersion, Bull. Seism. Soc. Am., 57, 235-248, 1967.

Kirby, S. H., State of stress in the lithosphere, Inference from the flow laws of olivine, Pageoph., 115, 245-258, 1977.

Kirby, S. H. and V. B. Rayleigh, Mechanisms of high temperature solid-state flow in a minerals and Ceramics and their bearing on the creep behaviour of the mantle, Tectonophysics, 19, 165-195, 1973.

Kohlstedt, K. L., and V. Goetze, Low stress high temperature creep in olivine single crystal, J. Geophys. Res., 2045-2051, 1974.

Kohlstedt, K. L., C. Goetze and W. B. Durham, Experimental deformation of single crystal olivine with the application to flow in the mantle, The Physics and Chemistry of Minerals and Rocks (Ed. R.G.J. Sterns), Wiley, London, 35-45, 1976.

Kono, M., Gravity anomalies in East Nepal and their implications to the crustal structure of the Himalayas, Geophys. Jour. R. Astr. Soc., 39, 283-299, 1974.

Lachenbruch, A. H., Crustal temperature and heat production implications of the linear heat flow relation, J. Geophys. Res., 75, 3291-3300, 1970.

McKenzie, D. P., Active tectonics of the Alpine-Himalayan belt: The Acgean Sea and surrounding regions, Geophys. J. R. Astr. Soc., 55, 217-254, 1978.

Molnar, P. and P. Tapponnier, Cenozoic tectonics of Asia: Effects of a continental collision, Science, 189, 419-426, 1975.

Negi, J. G., and O. P. Pandey, Correlation of heat flow and crustal topography in the Indian region, Geophy. J. R. Astr. Soc., 45, 201-217, 1976.

Nicholas, A. and J. P. Poirier, Cystalline Plasticity and solid state flow in metamorphic rocks, Wiley, London, 444, 1976.

Post, R., The flow laws of Mt. Burnett dunite, Thesis, University of California, Los Angeles, California, 1974.

Rao, R. U. M., G. V. Rao and H. Narain, Radioactive heat generation and heat flow in Indian region, Earth Planet. Sci. Lett., 30, 57-64, 1976.

Reddy, P. R., Crustal and upper mantle structure in India from body wave travel time studies of shallow earthquakes, Ph. D. Thesis, Andhra Univ., 1971.

Schatz, J. F. and G. Simmons, Thermal conductivity of earth materials, J. Geophys. Res., 77, 6966-6983, 1972.

Singh, D. D. and H. K. Gupta, Source dynamics of two great earthquakes of the Indian subcontinent: The Bihar-Nepal earthquake of Jan.15, 1934 and the Quetta earthquake, May 30, 1935. (In press), 1979.

Singh, R. N. and J. G. Negi, High moho temperatures in the Indian shield region (In press), 1979.

Stocker, R. L. and M. F. Ashby, On the rheology of the upper mantle, Rev. Geophys. Sp. Phys., 11, 391-426, 1973.

Sykes, L. R., Intraplate seismicity, reactivation of pre-existing zones of weakness, Alkaline Magmatism and other Tectonism Post dating continental fragmentation, Rev. Geophys. Sp. Phys., 16, 621-688, 1978.

Valdiya, K. S., Himalayan transverse faults and folds and their parallelism with subsurface structures of north Indian plains, Tectonophysics, 32, 353-386, 1976.

Verma, R. K. and M. L. Gupta, Present status of heat flow studies in India, Geophys. Res. Bull., 13, 247-255, 1975.

Wegener, A., The origin of continents and oceans (4th ed.) New York, Dover Press, 246 (translated by B. J. Wegener), 1929.

Weertman, J., The creep strength of the Earth's mantle, Rev. Geophys. Sp. Phys., 11, 391-426, 1970.

Weertman, J., Creep laws for the mantle of the Earth, Phil. Trans. Roy. Soc., London, A288, 9-26, 1978.

GREATER INDIA'S NORTHERN EXTENT AND ITS UNDERTHRUST OF THE TIBETAN PLATEAU: PALAEOMAGNETIC CONSTRAINTS AND IMPLICATIONS

Chris T. Klootwijk*

Research School of Earth Sciences, Australian National University
Canberra A.C.T. 2600, Australia

Abstract. Palaeomagnetic constraints are
summarized for three categories of models for the
northern extent of the Indian fragment of Gondwanaland.
1. Models which postulate the presence of exposed Gondwana blocks north of the Indus-Tsangpo
suture zone (ITS). The only palaeomagnetic constraint available to date is a preliminary result
for Upper Devonian sedimentary ironstones from
Chitral (Eastern Hindu Kush). When interpreted in
terms of a primary magnetization, this result
suggests an Eurasian rather than a Gondwana
affinity for the Eastern Hindu Kush.
2. Models which advocate an enlarged Indo-Pakistan continental block, "Greater India", underthrust along the Main Central Thrust (MCT) beneath
its leading edge and the Tibetan Plateau. Comparison of palaeomagnetic results from the Tibetan
Sedimentary Series (north central Nepal) of Greater
India's former leading edge, with results from the
Indo-Pakistan subcontinent suggests a counterclockwise rotational underthrusting of Greater
India beneath its former leading edge over 10 to
15 degrees. Generally accepted values for crustal
shortening within the Himalayan belt, and the
above magnitude of intracontinental underthrusting
constrain a minimum estimate for the northern
extent of Greater India. Within the updated
Gondwana reconstruction of Smith and Hallam, this
estimate satisfies constraints imposed from facies
analyses of the juxtaposed western Australian
continental margin.
3. Models which envisage a slightly enlarged
Indo-Pakistan continental block, reduced to its
present outline along the ITS through crustal
shortening within the Himalaya and minor subduction along the ITS. Models in this category
differ in their interpretation of the post-
collisional convergence of India with Asia. This
convergence is thought to be taken up, either

through extensive crustal shortening within the
Tibetan Plateau and adjacent mountain belts, or
through lateral displacement within southern Asia,
of lithospheric blocks out of the way of the
indenting Indo-Pakistan continental block.
Available palaeomagnetic data have no direct
bearing on the former process, but can be interpreted indirectly in support of the latter.

Introduction

It has been recognised over the last decade
that the northern outline of late Palaeozoic Gondwana extended considerably beyond the classical
configurations of Du Toit (1937) and Smith and
Hallam (1970). Stratigraphic and structural
correlation studies indicate that fragments of
Gondwana's northern rim drifted away during late
Palaeozoic-early Mesozoic times (Dewey et al.
1973; Stöcklin, 1974, 1977; Hsü, 1977; Hsü et al.,
1977; Boulin and Bouyx, 1977; Şengör, 1979) to
amalgamate with Eurasia during the late Mesozoic.
With closure of the Palaeotethys, the Neotethys
thus opened up in the wake of these fragments.

Palaeomagnetic results from central and
northern Iran (Becker et al., 1973; Soffel et al.
1975; Wensink et al., 1978) and from central
Afghanistan (Krumsiek, 1976) support this concept.
Such palaeomagnetic conclusions have not been
reached as yet for the more western and eastern
regions of the Alpine-Himalayan belt. Widely
different opinions exist about the former northern
extent of the Indian fragment of Gondwanaland.

Models

Currently prevalent models (see summaries by
Powell and Conaghan (1975), Klootwijk and Conaghan
(1979) and Klootwijk (1979a) can broadly be
classified in three categories (Fig. 1).

Category 1. Models that advocate exposed Gondwana
elements to extend far north of the ITS. Crawford
(1974a,b; Fig.1a) interprets such elements as a
coherent extension of a much larger Indo-Pakistan
continental block. Others (Stöcklin, 1977; Boulin
and Bouyx, 1977 (Fig.1b); Şengör, 1979) envisage

*Now at: Département des Sciences de la Terre,
        Université Paris 7,
        2 Place Jussieu, 75320,
        PARIS CEDEX 05
        FRANCE

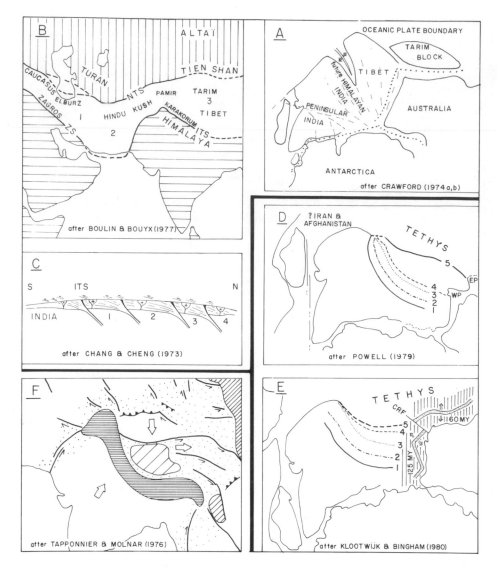

Fig. 1. Various models for the extent of the Indian fragment of Gondwanaland. See text for discussion.
Category 1 models:
a) After Crawford (1974a,b) showing part of East Gondwanaland during Permian to late Jurassic times. The Indus suture is interpreted as a former intracontinental opening to the mantle with development of oceanic crust, now closed and subducted. Associated rifting within Peninsular India allowed preservation of Gondwana sediments. Stippled: zones of subsequent separation of Gondwanic fragments.
b) After Boulin and Bouyx (1977). Vertical hatching = Angara elements. Horizontal hatching = Gondwana elements. White = zone of supposed Gondwana elements, which broke off Gondwana and collided with Angaraland prior to being caught by the Arabian and Greater Indian fragments of Gondwana during closure of the Neotethys. Available palaeomagnetic data support a Gondwana origin for central and northern Iran (1: Becker et al., 1973; Soffel et al., 1975; Wensink et al., 1978) and for central Afghanistan. (2: Krumsiek, 1976), but are not available as yet for the Tarim block and the Tibetan Plateau (3). Preliminary palaeomagnetic results from the Eastern Hindu Kush (Klootwijk and Conaghan, 1979) suggest an Eurasian rather than a Gondwana affinity. NTS = generalized location of the Northern (Palaeo-) Tethys. ZG = Zagros suture zone. ITS = Indus-Tsangpo suture zone.
c) After Chang and Cheng (1973), modified according to Powell and Conaghan (1975). Successive accretion since late Palaeozoic of small continental blocks of southern origin, onto Angaraland. 1 = Southern Tibetan plate, 2 = Northern Tibetan plate, 3 = Chang Tang plate, 4 = Tsaidam plate, ITS = Indus-Tsangpo suture zone.
Category 2 models:
d) After Powell (1979). 1 = Main Boundary Thrust (MBT). 2 = Indus-Tsangpo suture zone (ITS). 3 =

Reflection of the MBT about the ITS, and is separated from 1 by twice the present distance between 1 and 2. 4 = Connection between the inferred minimum northward extent of Greater India off western Australia, according to Veevers (1971), and NW. Pakistan in an arc parallel to 3; preferred model of Powell (1979). 5 = the present Kun Lun - Nan Shan mountain front. According to Powell and Conaghan (1973, 1975) and Veevers et al., 1975, Greater India's underthrust of the Tibetan Plateau has proceeded up to this zone. Line 5 represents a minimum estimate for Greater India's northern extent, not corrected for crustal shortening within the Himalyan belt, nor for the former leading edge of Greater India now situated between the MCT and the ITS. W.P. - Wallaby Plateau, E.P. = Exmouth Plateau.
e) After Klootwijk and Bingham (1980). 1 to 5 are five lines relevant to determining the northern extent of Greater India with the Smith and Hallam reconstruction of Gondwanaland, adapted after Griffiths (1974), and Norton and Molnar (1977). 1 = the present Himalayan front (MBT), 2 = the Indus Tsangpo suture zone (ITS), 3 = a reflection of the ITS which is separated from 2 by a 300 km wide belt representing: a) 200 km crustal shortening within the Himalayan region excepting underthrusting along the MCT, b) 50 km subduction along the ITS, and c) 50 km northward translational underthrusting along the MCT. The zone between lines 3 and 4 and between lines 3 and 5 respectively, represent that part of Greater India underthrust along the MCT that corresponds with rotational underthrusting values of 10 and 15 degrees respectively. Rift zones (stippled) and assumed continental lithosphere (vertical shading) according to Veevers and Cotterill (1978). CRF = Cape Range fracture zone.
Category 3 models:
f) After Molnar and Tapponnier (1975) and Tapponnier and Molnar (1976). Indentation of a rigid Indo-Pakistan subcontinent into a rigid-plastic southern Asia induces lateral displacement of small lithospheric fragments within southern Asia. Horizontal hatching = region of major crustal thickening. Dotted = region where strike slip faulting occurs. Oblique hatching = region where normal faulting and crustal thinning occurs.

---

the existence in southern Asia of a separate zone of amalgamated fragments which broke off Gondwanaland in about Triassic times and coalesced with southern Angaraland by the early Jurassic, prior to their Late Cretaceous to Palaeogene capture by northwards drifting Arabia and Indo-Pakistan. Chang and Cheng (1973, Fig. 1c) envisage this Mesozoic-Cenozoic accretion as only the later cycles of a Phanerozoic process of successive collisions with the leading edge of Asia of small continental blocks derived from the south.

Other models, which all recognise the ITS as the farthest northern exposed extent of the Indo-Pakistan continental fragment within Asia, group into two categories.

Category 2. Models which advocate an enlarged Indo-Pakistan continental block, i.e. Greater India, underthrust along the MCT beneath its former leading edge and beneath the Tibetan Plateau (Powell and Conaghan, 1973, 1975; Veevers et al., 1975; Powell, 1979). Powell and Conaghan originally postulated Greater India's underthrust to have proceeded up to the Kun Lun - Nan Shan mountain belt, beneath the full width of a rather undisturbed Tibetan Plateau (Fig. 1d, Line 5). Recently Powell (1979) reduced this estimate considerably (Fig. 1d, Line 4).

Category 3. Models which envisage an Indo-Pakistan continental block reduced to its present outline along the ITS through crustal shortening over several hundred kilometers within the Himalayan belt and with only minor subduction along the ITS. Dewey and Burke (1973) interpret the twice normal crustal thickness of the Tibetan Plateau as the result of basement reactivation and large scale contraction. Şengör and Kidd (1979) interpret the Tibetan Plateau as a huge accretionary sedimentary prism with minor occurrence of subduction related plutonic and volcanic rocks. Molnar and Tapponnier (1975, 1978) interpret the Tibetan Plateau as formed by successive accretion of continental and/or island arc fragments between late Palaeozoic and middle Mesozoic (Chang and Cheng, 1973), and acting as a hydrostatic head transferring the India-Asia convergence into strike-slip faulting mainly north and east of the Tibetan Plateau (Fig. 1f).

Palaeomagnetic Tests

Models of the various categories can be tested by palaeomagnetic methods, but such tests are limited at present by a restricted data set.

A. Models of Category 1

Models in this category are constrained only by preliminary results from Upper Devonian sedimentary ironstones from Chitral in the Eastern Hindu Kush (Klootwijk and Conaghan, 1979). These data indicate a palaeolatitude of 3°S. When interpreted in terms of a primary magnetization this result suggests an Eurasian rather than a Gondwana affinity. Pending forthcoming results of a more extensive study, this preliminary interpretation does not support the models of this category. Other palaeomagnetic data from Cretaceous (Burtman and Gurariy, 1973) and Palaeogene (Bazhenov et al. 1978) sediments of the outer Pamir and Transalai and from Upper Cretaceous to Lower Tertiary red beds from south central Tibet (Zhu Xiangyuan et al., 1977; Molnar and Wang-ping Chen, 1978) are not useful for a palaeomagnetic test of this category of models. These results either represent regions outside the belt of supposed Gondwana fragments, and/or are stratigraphically too young.

B.   Models of Category 2

This category of models can be tested through comparison of palaeomagnetic data from the Indo-Pakistan subcontinent and Eurasia (see summaries by respectively Klootwijk (1979b) and Irving (1977)) with recently obtained palaeomagnetic data from the central Asian segment of the Alpine-Himalayan belt (Klootwijk, 1979b; Klootwijk and Conaghan, 1979; Klootwijk et al., 1979ab; Bingham and Klootwijk, 1980; Klootwijk and Bingham, 1980). Determination of the magnitude of Greater India's underthrust beneath the Tibetan Plateau is facilitated particularly by two favourable conditions:
1. Northward directed underthrusting along the MCT (Fig.2) proceeded at first beneath Greater India's former leading edge which is preserved at the surface as the Tibetan Sedimentary Series and the underlying Tibetan Slab.
2. Seafloor spreading data (McKenzie and Sclater, 1971; Sclater and Fischer, 1974; Molnar and Tapponnier, 1975; Johnson et al., 1976; Powell, 1979) show that post-collisional convergence of India with Asia was characterized by a counterclockwise rotational movement, which may have resulted into more pronounced underthrusting along the MCT to the east (Colchen and Le Fort, 1977; Molnar et al., 1977; Powell, 1979).

The counterclockwise rotational component of Greater India's underthrusting has been restrained through comparison of palaeomagnetic data from the Tibetan Sedimentary Series (north central Nepal) with results from the Indo-Pakistan subcontinent. The magnitude of a northwards translational component (if any) was beyond resolution of such a palaeomagnetic comparison.

Magnitude of Intracontinental Underthrusting

Upper Palaeozoic to Upper Mesozoic carbonates and sandstones from the Tibetan Sedimentary Series of the Thakkhola region in north central Nepal (Fig. 2), reveal two magnetic components relevant to this determination:
1. Primary magnetic components ranging in age from Middle? Permian to Early Cretaceous.
2. A secondary magnetic component acquired between 50 m.y. to 60 m.y. ago, upon initial collision of Greater India with southern Asia. Directions and pole positions (Fig. 3) for the primary and the "collision" components are systematically displaced with respect to comparable data from the Indo-Pakistan subcontinent. The collision component shows an overall 10 degrees and the primary components an about 15 degrees greater declination. These differences have been interpreted as lower and upper limits for Greater India's counterclockwise rotational underthrust of its former leading edge along the MCT (Bingham and Klootwijk, 1980; Klootwijk and Bingham, 1980).

There is consensus from geological data (Le Fort,

1975), deep seismic soundings (Kaila and Narain, 1976), gravity data (Kono, 1974; Warsi and Molnar, 1977) and fault plane solutions (Fitch, 1971; Molnar et al., 1973, 1977) for a northward dip of the MCT of about 15 degrees. As a first approximation the pivot point for the counterclockwise rotational underthrusting has been assumed to lie at the western end of the Indus-Kohistan seismic zone (Armbruster et al., 1978; Seeber and Armbruster, 1979). This zone represents the subsurface continuation of the Main Central Thrust-Main Boundary Thrust complex (Fig.2) beyond the Western Himalayan Syntaxis. Under these assumptions the surface projection of the frontal part of underthrust Greater India is shown in Figure 2 for rotational underthrusting magnitudes of 10 and 15 degrees respectively, reaching depths at the northern "15°" line ranging from 60 km north of Mount Kailas to 120 km north of Lhasa. A relation between this underthrusted slab and the widespread calcalkaline volcanism in southern Tibet (Fig. 2., Gansser, 1977; see also Dewey and Burke, 1973; Kidd, 1975; Sengör and Kidd, 1979) has been pointed out before (Bingham and Klootwijk, 1980), assuming a Neogene age for the latter (Gansser, 1977). New radiometric data, however, suggest an early Tertiary or older age for these volcanics (Gansser pers.comm., 1980; Zhang Yuquan pers.comm., 1980). If so, their geographical coincidence with the surface projection of the subducted continental slab may be purely fortuitou

Constraints on Greater India's Northern Extent

The above constraints on the magnitude of Greater India's underthrust, in combination with generally accepted magnitudes of crustal shortening within the Himalayan belt, allow for a minimum estimate of Greater India's northern extent (Fig.1e).
This estimate is based on:
1. 200 km of crustal shortening within the Himalayan belt excluding underthrusting along the MCT.
2. 50 km continental subduction along the ITS, and
3. A similar estimate for northwards translational underthrusting along the MCT.
4. Rotational underthrusting along the MCT over 10 and 15 degrees respectively.

The combined contributions of 1, 2 and 3 account for the position of line 3 in Figure 1e. This line reflects the ITS at an uniform distance of 300 km to the north. Line 4 and line 5 represent additional contributions corresponding with rotational underthrusting over 10 and 15 degrees respectively. The latter estimate in particular satisfies remarkably well, within the updated (Griffiths, 1974) Gondwana reconstruction of Smith and Hallam (1970), with constraints from facies analyses of western Australia's continental margin (Veevers and Cotterill, 1978). These facies data suggest the former existence of an extensive continental lithospheric block off the western

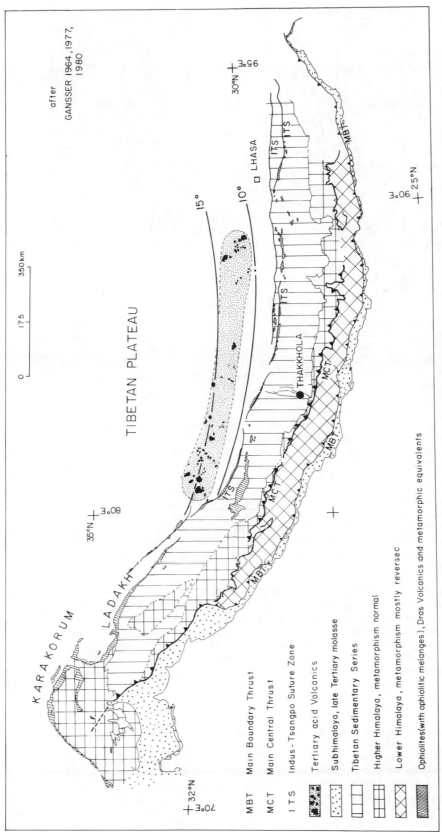

Fig. 2. Main structural features of the Himalaya after Gansser (1964, 1977, 1980). The dashed line indicates the Indus – Kohistan seismic zone (IKSZ), i.e. the westwards continuation in the subsurface of the Main Boundary Thrust – Main Central Thrust zone according to seismic observations (Armbruster et al., 1978; Seeber and Armbruster, 1979). The two full lines represent the surface projection of the frontal part of the underthrusted slab of Greater India according to counterclockwise rotational underthrusting over 10 and 15 degrees respectively around an assumed pivot point at the western end of the IKSZ. Northwards dip of the underthrust slab is taken at a uniform 15 degrees, as discussed in the text.

PALEOMAGNETIC CONSTRAINTS AND IMPLICATIONS   317

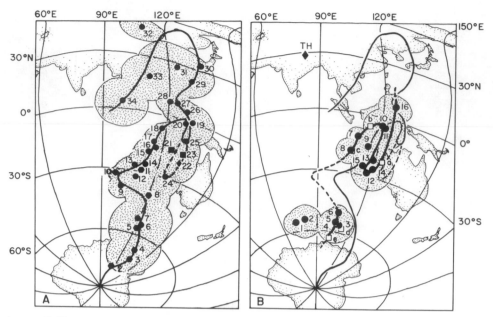

Fig. 3. Comparison of the apparent polar wander path for the Indo-Pakistan subcontinent (3A, see Klootwijk (1979a) and Klootwijk and Radhakrishnamurty (1980)), with pole positions according to the secondary collision component and the primary magnetic components observed in the Tibetan Sedimentary Series of the Thakkhola region (3B, see Klootwijk and Bingham (1980)). The eastward displacement of the latter with respect to the former reflects rotational underthrusting of Greater India beneath its former leading edge.

Legend 3A:

2 = 5-22.5 My (DSDP), 3 = 25-35 My (DSDP), 4 = 35-45 My (DSDP), 5 = 45-55 My (DSDP), 6 = Sanjawi Lst. (Palaeocene-Eocene), 7 = Upper Brewery Lst. (base Upper Palaeocene), 8 = 55-65 My (DSDP), 9 = Upper normal Deccan Traps (60-65 MY), 10 = Lower reversed Deccan Traps (60-65 My), 11,12 = Tirupati beds (Lower Cretaceous), 13 = 65-75 My (DSDP), 14 = Satyavedu beds (Lower Cretaceous), 15 = Sylhet Traps (≈Rajmahal Traps), 16, 17 = Goru Fm.-Parh Lst. (Aptian-Albian to Santonian-Campanian), 18 = Rajmahal Traps (100-105 My), 19 = Loralai Lst. (Middle-Upper Jurassic), 20 = Chiltan Lst. (Upper Jurassic), 21 = Mean Middle Jurassic pole for Australia transferred to the India plate, 22 = as above, mean Lower Jurassic pole, 23 = as above the mean Middle Jurassic pole of Antarctica, 24 = Parsora beds (Upper Triassic), 25 = Pachmarhi beds (Upper Triassic), 26 = Kamthi beds, Wardha Valley (Upper Permian-Lower Triassic), 28 = Panchet beds (as above), 29 = Kamthi beds, Wardha Valley (Upper Permian-Lower Triassic), 30 = Talchir beds (Permo-Carboniferous). 31 = Krol-A limestone (Permo-Triassic), 32 = Blaini Limestone (Permo-Carboniferous), 33 = Alozai Formation (Permo-Carboniferous), 34 = Lower Blaini Diamictite (Permo-Carboniferous).

Legend 3B:

Secondary "collision" component (1-6; 50-60 My) and primary component (7-16); Lower Aptian to Middle? Permian). Results a-e are of low accuracy (α95 in excess of 10°) and have not been considered in the determination of the magnitude of the differential rotation (collision component : 10°, primary component: 15°). Collision component results 1 and 2, which are slightly off the main grouping, were also discarded. Collision component: 1, 2 = Kagbeni Sst., 2, 3, 6 = Thinigaon Lst., d = Thini Chu Fm., e = Lumachelle Fm. Primary components: 7 = Dzong Sst. (Lower Aptian), 8, 9 = Kagbeni Sst. (Wealden), 10,a,b = Lumachelle Fm. (Dogger), 11, 12, c = Jomosom Lst. (Lias), 13 = Jomosom Qzt. (Rhaetian), 14 = Thinigaon Lst. (Norian), 15 = Thinigaon Lst. (Ladinian - Carnian), 16 = Thini Chu Fm. (Middle?Permian).

Australian margin as far north as the Cape Range fracture zone, with possibly a more limited strip of continental lithosphere off northwestern Australia. Seafloor spreading data indicate that the latter rifted away from about 155 to 160 m.y. ago (Larson, 1975; Larson et al., 1979), and that the former drifted away from about 120 to 125 m.y. ago onwards (Markl, 1974; Larson, 1977;

Veevers and Cotterill, 1978; Larson et al., 1979; Johnson et al., 1980).

Line 5 in Figure 1e resembles but falls somewhat north of Powell's (1979) model for Greater India's northern extent (Fig. 1d, line 4), but clearly falls short of Powell and Conaghan's original model (Fig. 1d, line 5), which is based on a postulated underthrusting of Greater India beneath the full width of the Tibetan Plateau (Powell and Conaghan, 1973, 1975; Veevers et al., 1975).

It has to be emphasized that the model advanced here (Fig. 1e, line 5) represents only a minimum estimate for the northern extent of Greater India (post-Neothethys formation, category 2 and 3) and has no bearing whatsoever on a possibly more extensive Gondwana India (pre-Neothethys formation, category 1).

C. Implications for Models of Category 3

Greater India's palaeolatitudinal position at the time of initial contact of India with Asia is shown in Figure 4 according to palaeomagnetic results for the secondary collision component observed in the Tibetan Sedimentary Series (Fig. 3, poles 3-6). The palaeoposition of the juxtaposed southern boundary of central Asia is not well determined. This boundary is constrained only by palaeomagnetic data from the Ladakh Intrusives (Fig. 4, heavy line 1), which represent the batholithic foundations of an island arc off southern Asia (Frank et al., 1977), and to some extent by Upper Cretaceous to Lower Tertiary red bed data from south central Tibet (Zhu Xiangyuan et al., 1977; Molnar and Wang-ping Chen, 1978). The former data are in good agreement with the palaeoposition of Greater India's northern boundary at the time of collision. The Tibetan red bed data show a palaeoposition of the sampled locality at 8°N (Fig. 4, dashed line 2), but poor age control reduces their usefulness for further delineation of central Asia's southern boundary at the time of collision.

Figure 4 evidently shows a mismatch of more than 2500 km between the observed equatorial palaeolatitude of collision, represented in Figure 4 by the above determined northern boundary of Greater India (Fig. 1e, line 5), and the expected palaeolatitude of this zone at about 30°N according to Eurasian palaeomagnetic data (Irving, 1977) under the assumption of no post-collisional deformation within central Asia. The preliminary palaeomagnetic results from Upper Devonian ironstones from the Eastern Hindu Kush (Klootwijk and Conaghan, 1979) can be interpreted in further support of this mismatch (Fig. 5). Palaeomagnetic results from Palaeogene sediments from the outer Pamir and the Transalai (Bazhenov et al., 1978; Klootwijk, 1979b) clearly indicate crustal shortening within Asia north of this zone, which to date has proceeded over more than 1000 km.

The more than 2500 km mismatch therefore may be accounted for by:
a. Closure of small ocean-floored basins.
b. Crustal shortening within the Pamir, the Tibetan Plateau and the Tien Shan and Nan Shan mountain belts to the north.
c. Lateral displacement of lithospheric blocks out of the way of the advancing Indo-Pakistan continental block.

Magnitudes for each of these effects have not been established as yet from direct observation.

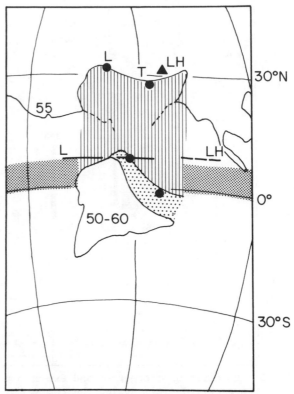

Fig. 4. Comparison of palaeolatitudinal positions of Greater India and Asia at the time of initial collision (about 55 My), which occurred at low northern palaeolatitudes (finely stippled zone). Greater India is positioned according to the secondary collision component (Fig.3A:3-6) from the Tibetan Sedimentary Series of the Thakhola region (location indicated by dot T). Asia is positioned according to Irving's (1977) mean pole position for Eurasia (55 My). Its southern boundary is shown under assumption of no post-collisional deformation within central Asia. The Palaeoposition of Greater India upon collision thus obtained agrees well with the contemporaneous palaeoposition of the Ladakh Intrusives (heavy line L through dot, Klootwijk et al., 1979a), which are situated directly north of the ITS. The dotted line LH represents the palaeoposition obtained from Upper Cretaceous - Lower Tertiary red beds in south central Tibet, east of Lhasa (Zhu Xiangyuan et al., 1977, Molnar and Wang-ping Chen, 1978). The coarsely stippled and the vertically striped areas indicate the magnitude of post-collisional crustal shortening within Greater India and southern Asia respectively.

There are, however, palaeomagnetic constraints on the first two effects which allow for an indirect estimation of the third effect.

Former existence of an island arc has been

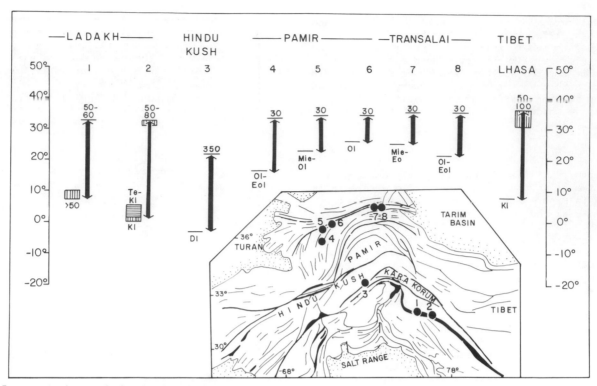

Fig. 5. Comparison of observed and expected palaeolatitudes in south central Asia, north of the ITS, for various periods (My) during the Phanerozoic. Localities are shown in a tectonic sketch after Gansser (1964). For each locality, the left column represents observed palaeomagnetical palaeolatitudes and the right column palaeolatitudes expected according to palaeomagnetic data for Eurasia (Irving, 1977), assuming that no subsequent displacement has occurred relative to Eurasia. The length of the double-headed arrow indicates the magnitude of convergence between the sampled locality and Eurasia since the indicated period (stratigraphic or radiometric ages). 1 = Ladakh intrusives at Kargil (Klootwijk et al., 1979a). The vertically hatched palaeolatitude range represents magnetic components acquired upon collision. The other two palaeolatitudes represent components of undetermined younger ages. 2 = Upper Cretaceous Dras Flyschoids at Khalsi-Lamayuru (vertical hatching, Klootwijk, 1979b) and Upper Cretaceous to Lower Tertiary Indus Molasse at Khalsi (horizontal hatching, Klootwijk, 1979b). 3 = Upper Devonian sedimentary ironstones of Chitral, Eastern Hindu Kush (Klootwijk and Conaghan, 1979). 4-8: Palaeomagnetic results from Palaeogene sediments of the outer Pamir and the Transalai, according to Bazhenov et al. (1978), as summarized by Klootwijk (1979b). 4 = Oligocene, southern Darvaz (outer Pamir). 5 = Upper Oligocene - Lower Miocene, Chil'Dara (Peter 1 Range). 6 = Oligocene, Khipsiun (Peter 1 Range). 7 = Eocene-Lower Miocene, Kyzylart (Transalai). 8 = Upper Eocene - Oligocene, Khatyn-Kanysh (Transalai). 9 = Upper Cretaceous to Lower Tertiary red beds from south central Tibet, east of Lhasa (not indicated on sketch map, Zhu Xiangyuan et al. (1977), Molnar and Wang-ping Chen (1978)).

suggested for the western part of the convergence zone (Frank et al., 1977; Tahirkheli et al., 1979), but it is not clear whether subduction beneath southern Tibet was of continental arc or of island arc type (Molnar and Burke, 1977; Tahirkheli et al., 1979). As for the western part of the collision zone, the extreme case that the total India-Asia convergence during the period between initial contact (about 55 m.y. ago) and the distinct slow-down in convergence at about 40 m.y. ago (Molnar and Tapponnier, 1975; Peirce, 1978; Klootwijk et al., 1979a) was consumed by closure of marginal basins, would lead to a maximal magnitude for effect a. of about 1500 km. Crustal shortening within the Pamir block has been estim-

ated at 200 to 300 km (Ulomov (1974) in Molnar and Tapponnier (1975)).

As for the eastern part of the collision zone, contribution of effect a. was minor in case of Andean type subduction, but may have amounted up t 1800 km in case of former presence of offshore island arcs. Crustal shortening within the Tien Shan and Nan Shan has been estimated at about 200 to 300 km (Molnar and Tapponnier, 1975; Tapponnier and Molnar, 1976). The magnitude of crustal shortening within the Tibetan Plateau is disputed. Dewey and Burke (1973) argue that the twice averag crustal thickness has resulted from contraction through basement reactivation. This may account for about 800 km of crustal shortening at maximum.

Their argument may find support in Kidd's estimate (in Şengör and Kidd, 1979) from interpretation of Landsat photographs, that crustal shortening was probably not less than 30%. On the other hand, Powell and Conaghan (1975) argue that geological sections over the Tibetan Plateau (Terman, 1974) show no evidence of extreme crustal shortening. Absence of pronounced body wave dispersion was interpreted similarly by Molnar and Tapponnier (1975). Normal faulting on north-south trending faults is, however, well pronounced in the Tibetan Plateau (Molnar and Tapponnier, 1978; Ni and York, 1978) and crustal shortening over at least a few hundred kilometers cannot be excluded (Molnar and Tapponnier, 1975).

As a direct consequence of these as yet vague estimates, at least for the western part of the convergence zone over 1000 km (Klootwijk, 1979a; Klootwijk and Bingham, 1980) of the more than 2500 km mismatch may have to be attributed to lateral displacement of small lithospheric blocks within Asia, out of the way of the advancing Indo-Pakistan continental block. This conclusion and also the above discussed palaeomagnetic evidence for about 1000 km crustal shortening since the Palaeogene within Asia north of the outer Pamir-Transalai zone (Fig.5) support Molnar and Tapponnier's (1975) continental deformation model.

Conclusions

Comparison of recently obtained palaeomagnetic data from the Himalayan belt with results from the Indo-Pakistan subcontinent and Eurasia shows:
1) Intracontinental underthrusting along the Main Central Thrust may have proceeded over several hundred kilometers.
2) A palinspastic reconstruction of Greater India's northern boundary, based on the magnitude of 1) and on generally accepted estimates for crustal shortening within the Himalayan belt, satisfies within the updated Gondwana reconstruction of Smith and Hallam, facies constraints from western Australia's continental margin.
3) Palaeomagnetic data from the Himalayan belt and central Asia can be interpreted in support of Molnar and Tapponnier's continental collision model.

References

Armbruster, J., L. Seeber and K.H. Jacob, The northwestern termination of the Himalayan mountain front: Active tectonics from micro-earthquakes, Jour. Geophys.Res., 83, 269-282, 1978.

Bazhenov, M.L., V.S. Burtman, and G.Z. Gurariy, Palaeomagnetic results for the Palaeogene of the Outer Pamir and Transalai, Dokl.Akad.Nauk.USSR., 242, 1137-1139, 1978 (in Russian).

Becker, H., H. Förster, and H. Soffel, Central Iran a former part of Gondwanaland? Palaeomagnetic evidence from Infra Cambrian rocks and iron ores of the Bafq area, Central Iran, Zeitschr.f. Geophys., 39, 953-963, 1973.

Bingham, D.K., and C.T. Klootwijk, Palaeomagnetic constraints on Greater India's underthrust of the Tibetan Plateau, Nature, 284, 336-338, 1980.

Boulin, J., and E. Bouyx, Introduction à la géologie de l'Hindou Kouch occidental, Mém.h. Sér.Soc.Géol.Fr., 8, 87-105, 1977.

Burtman, V.S., and G.Z. Gurariy, Character of folded areas in the Pamirs and Tien Shan from the geophysical data, Geotectonics, 1, 90-92, 1973.

Chang, Chen-fa and Cheng Hsi-lan, Some tectonic features of the Mt. Jolmo Lungma area, southern Tibet, China, Sci.Sinica, 16, 257-265, 1973.

Colchen, M., and P. Le Fort, Some remarks and questions concerning the geology of the Himalaya, In: Colloq. Int. C.N.R.S., Ecologie et Geologie de l'Himalaya 2, 131-137, 1977.

Crawford, A.R., The Indus Suture line, the Himalaya, Tibet and Gondwanaland, Geol.Mag., 111, 369-380, 1974a.

Crawford, A.R., A greater Gondwanaland, Science, 184, 1179-1181, 1974b.

Dewey, J.F., W.C. Pitman, III, W.B.F. Ryan, and J. Bonnin, Plate tectonics and the evolution of the Alpine system, Geol.Soc.America Bull., 84, 3137-3180, 1973.

Dewey, J.F., and K.C.A. Burke, Tibetan, Variscan and Precambrian basement reactivation: Products of continental collision, Jour.Geol., 81, 683-692, 1973.

Du Toit, A.L., Our Wandering Continents, Oliver and Boyd, Edinburgh, 366 p, 1937.

Fitch, T.J., Earthquake mechanisms in the Himalayan, Burmese and Andaman regions and continental tectonics in central Asia, Jour.Geophys.Res., 75, 2699-2709, 1971.

Frank, W., A. Gansser, and V. Trommsdorff, Geological observations in the Ladakh area (Himalayas), A preliminary report, Schweiz. Mineral.Petrogr.Mitt., 57, 89-113, 1977.

Gansser, A., The Geology of the Himalaya, Interscience Publishers, London, New York, Sydney, 289 p., 1964.

Gansser, A., The great suture zone between the Himalaya and Tibet: A preliminary account, Colloq.Int. C.N.R.S., Ecologie et Géologie de l'Himalaya 2, 181-191, 1977.

Gansser, A., The significance of the Himalayan suture zone, Tectonophysics, 62, 181-191, 1980.

Griffiths, J.T., Revised continental fit of Australia and Antarctica, Nature, 249, 336-338, 1974.

Hsü, K.J., Tectonic evolution of the Mediterranean basins, In: The Ocean Basins and Margins 4A: The Eastern Mediterranean, (eds.) A.E.M. Nairn, W.H. Kanes and F.G. Stehli, 29-75, 1977.

Hsü, K.J., I.K. Nachev, and V.T. Vuchev, Geological evolution of Bulgaria in light of plate tectonics, Tectonophysics, 40, 245-256, 1977.

Irving, E., Drift of the major continental blocks since the Devonian, Nature, 270, 304-309, 1977.

Johnson, B.D., C.McA. Powell, and J.J. Veevers, Spreading history of the eastern India Ocean and Greater India's northward flight from

Antarctica and Australia, Geol.Soc.America Bull., 87, 1560-1566, 1976.

Johnson, B.D., C.McA. Powell, and J.J. Veevers, Early spreading history of the Indian Ocean between India and Australia, Earth Planet.Sci. Lett., 47, 131-143, 1980.

Kaila, K.L., and H. Narain, Evolution of the Himalaya based on seismotectonics and deep seismic soundings, In: Himalayan Geology Seminar, New Delhi, preprint 30 p., 1976.

Kidd, W.S.F., Widespread late Neogene and Quaternary calcalkaline vulcanism in the Tibetan Plateau, EOS, Trans.American Geophys.Un., 56, 453, 1975.

Klootwijk, C.T., A summary of palaeomagnetic data from Extrapeninsular Indo-Pakistan and south central Asia: Implications for collision tectonics, In: Structural Geology of the Himalaya, (ed.) P.S. Saklani, Today and Tomorrow Publishers, Delhi, 307-360, 1979a.

Klootwijk, C.T., A review of palaeomagnetic data from the Indian fragment of Gondwanaland, In: Geodynamics of Pakistan, (eds.) A. Farah and K.A. DeJong, Geol.Surv.Pakistan, Quetta, 41-80, 1970b.

Klootwijk, C.T. and P.J. Conaghan, The extent of Greater India 1: Preliminary palaeomagnetic data from Upper Devonian of the Eastern Hindu Kush, Chitral (Pakistan), Earth Planet.Sci.Lett., 42, 167-182, 1979.

Klootwijk, C.T. and D.K. Bingham, The northern extent of India 3: Palaeomagnetic data from the Tibetan Sedimentary Series, Thakkhola region, Nepal Himalaya, submitted to Earth Planet.Sci. Lett., 1980.

Klootwijk, C.T., and C. Radhakrishnamurty, Phanerozoic palaeomagnetism of the Indian plate and the India-Asia collision, In: Int.Geodynamics Program, Final Report Working Group 10, (eds.) M.W. McElhinny and D. Valencio, in press, 1980.

Klootwijk, C.T., M.L. Sharma, J. Gergan, B. Tirkey, S.K. Shah, and V.K. Agarwal, The extent of Greater India 2: Palaeomagnetic data from the Ladakh Intrusives at Kargil, northwestern Himalaya, Earth Planet.Sci.Lett., 44, 47-64, 1979a.

Klootwijk, C.T., R.N. Ullah, K.A. De Jong, and H. Ahmed, A palaeomagnetic reconnaissance of northeastern Baluchistan, Pakistan, Jour.Geophys. Res., in press, 1980.

Kono, M., Gravity anomalies in east Nepal and their implications to the crustal structure of the Himalaya, Royal Astron.Soc.Geophys.Jour., 39, 283-300, 1974.

Krumsiek, K., Zur Bewegung der Iranisch-Afghanischen Platte, Geol.Rundschau, 65, 909-929, 1976.

Larson, R.L., Late Jurassic seafloor spreading in the eastern Indian Ocean, Geology, 3, 69-71, 1975.

Larson, R.L., Early Cretaceous break-up of Gondwanaland off Western Australia, Geology, 5, 57-60, 1977.

Larson, R.L., J.C. Mutter, J.B. Diebold, G.B. Carpenter, and P. Symonds, Cuvier Basin: A product of ocean crust formation by Early Cretaceous rifting off Western Australia, Earth Planet.Sci.Lett., 45, 105-114, 1979.

LeFort, P., Himalayas: the collided range. Present knowledge of the continental arc, American Jour. Sci., 275-A, 1-44, 1975.

Markl, R.G., Evidence for the breakup of Eastern Gondwanaland by the Early Cretaceous, Nature, 251, 196-200, 1974.

McKenzie, D., and J.G. Sclater, The evolution of the Indian Ocean since the Late Cretaceous, Royal Astron.Soc.Geophys.Jour., 25, 437-528, 1971.

Molnar, P., J. Fitch, and F.T. Wu, Fault plane solutions of shallow earthquakes and contemporary tectonics in Asia, Earth Planet.Sci.Lett., 19, 101-112, 1973.

Molnar, P., and P. Tapponnier, Cenozoic tectonics of Asia: effects of a continental collision, Science, 189, 419-426, 1975.

Molnar, P., and K. Burke, Erik Norin Penrose Conference on Tibet, Geology, 5, 461-463, 1977.

Molnar, P., Wang-ping Chen, Fitch, T.J., Tapponnier, P., W.E.K. Warsi, and Wu, R.T., Structure and tectonics of the Himalaya:A brief summary of geophysical observations, In: Colloq.Int.C.N.R.S. Ecologie et Geologie de L'Himalaya 2, 269-294, 1977.

Molnar, P., and P. Tapponnier, Active tectonics of Tibet, Jour.Geophys.Res., 83, 5361-5375, 1978.

Molnar, P., and Wang-ping Chen, Evidence of large Cenozoic crustal shortening of Asia, Nature, 273, 218-220, 1978.

Ni, J., and J.E. York, Late Cenozoic tectonics of the Tibetan Plateau, Jour.Geophys.Res., 83, 5377-5384, 1978.

Norton, I. and P. Molnar, Implications of a revised fit between Australia and Antarctica for the evolution of the eastern Indian Ocean, Nature, 267, 338-340, 1977.

Peirce, J.W., The northward motion of India since the Late Cretaceous, Royal Astron.Soc.Geophys. Jour., 52, 277-311, 1978.

Powell, C.McA. and P.J. Conaghan, Plate tectonics and the Himalayas, Earth Planet.Sci.Lett., 20, 1-12, 1973.

Powell, C.McA. and P.J. Conaghan, Tectonic models of the Tibetan Plateau, Geology, 4, 727-731, 1975.

Powell, C.McA., A speculative tectonic history of Pakistan and surroundings: some constraints from the Indian Ocean, In: Geodynamics of Pakistan, (eds.) A. Farah and K.A. DeJong, Geol.Surv. Pakistan, Quetta, 5-24, 1979.

Sclater, J.G., and R.L. Fischer, Evolution of the east central Indian Ocean with emphasis on the tectonic setting of the Ninetyeast Ridge, Geol.Soc.America Bull., 85, 683-702, 1974.

Seeber, L., and J. Armbruster, Seismicity of the Hazara Arc in northern Pakistan: Décollement versus basement folding, In: Geodynamics of Pakistan, (eds.) A. Farah, and K.A. DeJong, Geol.Surv.Pakistan, Quetta, 131-142, 1979.

Sengör, A.M.C., Mid-Mesozoic closure of Permo-

Triassic Tethys and its implications, Nature, 279, 590-593, 1979.

Şengör, A.M.C., and W.S.F. Kidd, Post-collisional tectonics of the Turkish-Iranian Plateau and a comparison with Tibet, Tectonophysics, 55, 361-376, 1979.

Smith, A.G., and A. Hallam, The fit of the southern continents, Nature, 225, 139-144, 1970.

Soffel, H., H. Forster, and H. Becker, Preliminary polar wander path of central Iran, Jour.Geophys. 41, 541-543, 1975.

Stöcklin, J., Possible ancient continental margins in Iran, In: The Geology of Continental Margins, (eds.) C.A. Burk and C.L. Drake, Springer Verlag, New York, 873-887, 1974.

Stöcklin, J., Structural correlation of the Alpine ranges between Iran and central Asia, Mém.h.Sér. Soc.Géol.Fr., 8, 333-353, 1977.

Tahirkheli, R.A.K., M. Mattauer, F. Proust, and P. Tapponnier, The India Eurasia suture zone in northern Pakistan: Synthesis and interpretation of recent data at plate scale, In: Geodynamics of Pakistan, (eds.) A. Farah, and K.A. DeJong, Geol.Surv.Pakistan, Quetta, 125-130, 1979.

Tapponnier, P., and P. Molnar, Slip-line field theory and large-scale continental tectonics, Nature, 264, 319-324, 1976.

Terman, M.J., Tectonic map of China and Mongolia, Geol.Soc.America, Boulder Colarado, 2 sheets, scale 1:5,000,000, 1974.

Veevers, J.J., Phanerozoic history of Western Australia related to continental drift, Geol. Soc.Australia, Jour., 18, 87-96, 1971.

Veevers, J.J., C.McA. Powell, and B.D. Johnson, Greater India's place in Gondwanaland and in Asia, Earth Planet.Sci.Lett., 27, 383-387, 1975.

Veevers, J.J., and D. Cotterill, Western margin of Australia: evolution of a rifted arch system. Geol.Soc.America Bull., 89, 337-355, 1978.

Warsi, W.E.K. and P. Molnar, Gravity anomalies and plate tectonics in the Himalaya, In: Colloq.Int. C.N.R.S., Ecologie et Géologie de l'Himalaya, 2, 463-478, 1977.

Wensink, H., J.D.A. Zijderveld, and J.C. Varekamp, Palaeomagnetic and ore mineralogy of some basalts of the Geirud Formation of Late Devonian-Early Carboniferous age from the southern Alborz, Iran, Earth Planet.Sci.Lett., 41, 441-450, 1978.

Zhu Xiangyan, Liu Chun, Ye Sujuan, Lin Jinlu, Remanence of red beds from Linzhou Xizang, and the northward movement of the Indian Plate, Sci. Geol.Sinica, 1, 44-51, 1977 (in Chinese with an English abstract).